Aboveground Storage Tanks

Philip E. Myers

McGraw-Hill

New York   San Francisco   Washington, D.C.   Auckland   Bogotá
Caracas   Lisbon   London   Madrid   Mexico City   Milan
Montreal   New Delhi   San Juan   Singapore
Sydney   Tokyo   Toronto

Library of Congress Cataloging-in-Publication Data

Myers, Philip E.
    Aboveground storage tanks : Chevron research and Technology /
Philip E. Myers.
        p.   cm.
    Includes index
    ISBN 0-07-044272-X
    1. Oil storage tanks. I. Title.
TP692.5.M94 1997
665.5′42—dc21
                                     96-51117
                                       CIP

McGraw-Hill

A Division of The McGraw-Hill Companies

13 14 IBT/IBT 1 9 8 7 6 5 4 3 2 1

ISBN 0-07-044272-X

*The sponsoring editors for this book were Zoe Foundotos and Robert
Esposito, the editing supervisor was David E. Fogarty, and the
production supervisor was Suzanne W. B. Rapcavage. It was set in
Century Schoolbook by Estelita Green of McGraw-Hill's Professional
Book Group composition unit.*

Printed and bound by IBT Global.

 This book is printed on recycled, acid-free paper containing a minimum of 50% recycled, de-inked fiber.

McGraw-Hill books are available at special quantity discounts to use
as premiums and sales promotions, or for use in corporate training programs. For more information, please write to the Director of Special
Sales, McGraw-Hill, 11 West 19th Street, New York, NY 10011. Or contact your local bookstore.

Contents

ABOUT THE AUTHOR

Philip E. Myers (Orinda, California) worked at Chevron Research and Technology Company in Richmond, California as a pressurized equipment technology specialist. His responsibilities included the maintenance, application, and development of technology in the areas of ASTs and pressure vessels. He is a Registered Professional Chemical Engineer and graduated from the University of California.

Preface

Aboveground storage tanks (ASTs) have been around since the inception of industrial processing, but surprisingly, very little practical engineering or general information is readily available to the tank inspector, engineer, or operator. Why this is can only be speculated on. Perhaps, the concept of a tank is so simple that it fosters a belief that there is little complexity to them and they do not warrant expenditures of resources. Perhaps, the tank owner believes that it is appropriate to relegate all tank issues to the care of the manufacturer. Perhaps, it is because they are generally reliable pieces of equipment or are considered infrastructure.

Whatever the case, for those who have had to deal with ASTs, understanding the complex issues and problems and implementing good design, inspection, operational or environmental solutions to AST problems have been all but simple. Well-intentioned individuals and companies in need of sound engineering information frequently make major blunders in areas of design, inspection, or safety. This often results in high costs, shortened equipment life, ineffective inspection programs, environmental damage, or accidents and injuries as well as the threats of more national and state legislation.

In recent years there has been an increasing polarization between industry, environmental groups, regulators, and the public. Each facility which operates with tanks carries much more risk than just damaging its equipment. Regardless of cause, injuries, fatalities, and incidents all create a kind of press that can be used against the entire industry with no real benefit. So rather than proper application of industry standards to maintain facility integrity on a site-specific basis, we are seeing a trend where the design, inspection, and operation of facilities is being politically controlled or regulated. This is the worst possible way of running these facilities because it does not address the fundamental causes of the problems and it creates inefficiencies of mammoth proportion. It also directs resources away from where they are more needed for the public good such as higher risk operations or in other places in the facility. The political approach to controlling tank facilities penalizes the companies willing to do things right while not really fixing the fundamental problems. However, this is not to say that there should be no responsibility to operate these facilities carefully, safely, and in accordance with recognized and generally accepted good practices. In large part the situation we are in now is a result of the industrial reticence to speak up on issues, to promote information such as contained in this book to reduce the incidents which form the basis of regulations, and to be more proactive in the regulatory process than simply writing industry-recommended practices or standards.

The purpose of the book then is to help break the cycle described above by introducing appropriate information that will make any tank facilities safer, more reliable, and not in need of more stringent regulations. Specifically, this book can help any individual, company, or industry using ASTs to improve their performance in the areas of safety (both employee and the public), environmentally responsible operations, and implementation of good practices. Fortunately, this can be done with relatively small expenditures of time and effort when armed with knowledge and experience.

This book covers fundamental principles of aboveground storage tanks as well as more advanced principles such as seismic engineering needed for work in susceptible areas. It will be of interest to engineers, inspectors, designers, regulators, and owners as well as to any other person involved in any of the many specialized topics related to tanks. Each topic is treated from a perspective that the reader knows nothing and works up to a fairly advanced level so that the reading may be selective as appropriate and as needed. Where the topic becomes extremely advanced or where only unproved theories exist, then this is noted and further references are made available. One of the best sources of information about tanks, petroleum related issues, and all kinds of problems associated with the petroleum business is the American Petroleum Institute (API). This organization has produced numerous high-quality standards, recommended practices, and publications from which the reader may have access to the state of the art in these topics.

Philip E. Myers

1

Fundamentals

1.1 Overview

Although the word *tank* identifies only a single type or piece of equipment in an industrial facility, tanks have been used in innumerable ways both to store every conceivable liquid, vapor, or even solid and in a number of interesting processing applications. For example, they perform various unit operations in processing such as settling, mixing, crystallization, phase separation, heat exchange, and as reactors. Here we address the tank primarily as a liquid storage vessel with occasional discussion regarding specialized applications. However, the principles outlined here will, in many ways, apply generally to tanks in other applications as well as to other equipment.

1.1.1 Above- and belowground tanks

The most fundamental classification of storage tanks is based upon whether they are above- or belowground. Aboveground tanks have most of their structure aboveground. The bottom of the tank is usually placed directly on an earthen or concrete foundation. Sometimes these tanks are placed on grillage, structural members, or heavy screen so that the bottom of the tank can be inspected on the underside or leaks can be easily detected. The aboveground tank is usually easier to construct, costs less, and can be built in far larger capacities than underground storage tanks.

Another class of aboveground tanks is called *elevated tanks*. These tanks are elevated by structural supports. However, they are almost exclusively relegated to the domain of the municipal water supply companies. Although the same effect can be accomplished by placing tanks on hills, where this is not possible, they are elevated by placing

1

them on structural supports. The familiar municipal water supply tanks can be seen almost everywhere throughout the country. Because the municipal water supply is considered a vital public resource, elevation of these tanks is used instead of pumps to provide a 100 percent reliable pressurized source at sufficient flow.

Another important class of tanks is the underground tank. These are usually limited to between 5000 and 20,000 gal with most being under 12,000 gal. They have been used to store fuels as well as a variety of chemicals. They require special consideration for the earth loads to which they are subjected. Buoyancy must also be considered, and they are often anchored into the ground so that they do not pop out during periods when groundwater surrounds them. In addition, because they are underground, they are subject to severe corrosion. Placement of special backfill, cathodic protection, and coatings and liners are some of the corrosion prevention measures that ensure good installations. In addition, because of the regulatory requirements, most underground tanks must have a means of being monitored for leakage, which often takes the form of a double-wall tank with the interstitial space being monitored.

At retail refueling stations such as service stations, marinas, and convenience stores, the fire protection codes prohibit the use of aboveground tanks. This naturally resulted from the high traffic levels and potential for accidents. Another reason for underground tanks in chemical and processing plants has been that land values were at a premium. Underground storage tanks provided a good answer to space constraints. An additional factor lending itself to underground storage of fuels and hazardous chemicals is the relatively constant temperature. Evaporation losses are reduced because of the insulating effect of the underground installation. These factors led to widespread use of underground tanks for retail fuel-dispensing facilities.

However, in the 1970s and 1980s, the discovery of groundwater contamination in many areas of the United States resulted in the 1984 Subtitle 1 to the Resource Conservation and Recovery Act (RCRA) which required the Environmental Protection Agency (EPA) to develop regulations for all underground tanks. These rules apply to any tank with 10 percent or more of its volume underground or covered. Tanks in basements or cellars are not subject to these rules, however.

1.1:2 Number and capacity of tanks

Because there is no uniform regulation requiring registration of tanks, the number of tanks in existence is not known. However, the American Petroleum Institute (API) conducted a survey indicating there are about 700,000 petroleum storage tanks. Although the count may not be precise, the EPA has estimated that there are approximately 1.3 million regulated underground storage tanks with an unknown number of exempt underground tanks used for home heating oil and farm fuel storage tanks.

1.2 Basic Concepts

1.2.1 Units

Although most of the world has adopted the International System of units (called SI units), the U.S. system of units is still strongly entrenched in the United States. However, most of the standards maintenance organizations are attempting to convert their standards to the SI system so that they may be useful throughout the world and ready for the inevitable shift to the SI system.

Domestic regulations are now generally using metric values to specify tank sizes and vapor pressures. The U.S. industry, however, still uses the gallon or barrel for tank capacities. One barrel (1 bbl) is 42 U.S. gal. For low pressures, inches of water column (in WC) or ounces per square inch (osi) are common. For higher pressures, the units of pounds per square inch absolute or gauge (psia or psig) are used. Additionally, in the petroleum and chemical industries, specialized units such as Reid vapor pressure are often used.

Although this book uses primarily the U.S. system of units, App. 1A provides conversion factors between the systems for the most common units needed for tank work.

1.2.2 Physical properties of stored liquid

Since most tanks store liquids rather than gases, this discussion applies to liquid properties only. Although there are many characteristics unique to the stored liquid, we address only the primary physical properties here. Characteristics and properties such as corrosiveness, internal pressures of multicomponent solutions, tendency to scale or sublime, and formation of deposits and sludges are vital for the designer and the operator of the tank and will be treated in appropriate places throughout this book. Also excluded from the following discussion are unique properties and hazards of aerosols, unstable liquids, and emulsions.

A good source of information for liquid properties for a wide range of compounds is Perry's *Chemical Engineers' Handbook*.[22]

Density and specific gravity. The density of a liquid is its mass per unit volume. Water has a density of 1.000 g/cm^3 (62.4 lb/ft^3) at 4°C, whereas mercury has a density of about 13.5 g/cm^3 at 4°C. Obviously, the

TABLE 1.1   Specific Gravity of Liquids

Liquid	Sp. gr.	Liquid	Sp. gr.	Liquid	Sp. gr.
Acetic acid	1.06	Fluoric acid	1.50	Petroleum oil	0.82
Alcohol, commercial	0.83	Gasoline	0.70	Phosphoric acid	1.78
Alcohol, pure	0.79	Kerosene	0.80	Rape oil	0.92
Ammonia	0.89	Linseed oil	0.94	Sulfuric acid	1.84
Benzene	0.69	Mineral oil	0.92	Tar	1.00
Bromine	2.97	Muriatic acid	1.20	Turpentine oil	0.87
Carbolic acid	0.96	Naphtha	0.76	Vinegar	1.08
Carbon disulfide	1.26	Nitric acid	1.50	Water	1.00
Cottonseed oil	0.93	Olive oil	0.92	Water, sea	1.03
Ether, sulfuric	0.72	Palm oil	0.97	Whale oil	0.92

density of a liquid plays an important role in the design of a tank. All things being equal, greater densities mean thicker required tank shell thicknesses. Some selected specific gravities are shown in Table 1.1.

Specific gravity is a measure of the relative weight of one liquid compared to a universally familiar liquid, water. More specifically, it is the ratio of the density of the liquid divided by the density of water at 60°F:

$$\text{Specific gravity} = \frac{\text{weight of a unit volume of liquid at } 60°F}{\text{weight of a unit volume of water at } 60°F} \quad (1.1)$$

In the petroleum industry, a common indicator of specific gravities is known as the *API gravity*. It is usually applied to the specific gravity of crude oils. The formula for the API gravity is

$$\text{Degrees API} = \frac{141.5}{\text{specific gravity}} - 131.5 \quad (1.2)$$

On inspection of the above equation, it can be seen that the higher the specific gravity, the lower the API gravity. Incidentally, water having a specific gravity of 1 has an API gravity of 10.

Another common indicator of specific gravities that is used in the chemical industry is *degrees Baumé.*
For liquids heavier than water,

$$\text{Degrees Bé} = \frac{140}{\text{specific gravity}} - 145 \quad (1.3)$$

For liquids lighter than water,

$$\text{Degrees Bé} = \frac{140}{\text{specific gravity}} - 130 \quad (1.4)$$

When tank operators change the stored liquid, care must be exercised if there is a significant increase in the specific gravity of the new liquid because the effective hydrostatic pressure acting on the tank walls will be greater if the design liquid level is not reduced.

Temperature. Temperature is measured on two types of scales, absolute and relative. The two most common relative scales are the Celsius or (Centigrade) and Fahrenheit scales. Each scale is divided into uniform increments of 1 degree each. However, the 1°C is 1.8 times larger than 1°F. The Celsius scale defines 0°C as the freezing point (triple point) of water and 100°C as the boiling point. The Fahrenheit scale is arbitrarily defined by assigning it a temperature of 32°F at the freezing point of water and 212°F at the boiling point of water.

When one is working with various properties or using various handbooks, it is necessary to convert to and from the absolute temperature scales. Absolute zero is the lowest temperature possible and is set at zero for the absolute temperature scales. As with the Celsius and Fahrenheit scales, there are two corresponding absolute temperature scales.

The absolute temperature scale that corresponds to the Celsius scale is the *Kelvin scale,* and for the Fahrenheit scale it is called the *Rankine scale.* The Celsius scale reads 0°C when the Kelvin scale reads 273 K, and the Fahrenheit scale reads 0°F when the Rankine scale reads 460°R. These relationships are shown in Fig. 1.1.

Tanks are used to store liquids over a wide temperature range. Cryogenic liquids such as liquefied hydrocarbon gases can be as low as −330°F. Some hot liquids such as asphalt tanks can have a normal storage temperature as high as 500 to 600°F. However, most storage temperatures are at ambient or a little above or below ambient temperatures.

At low temperatures, materials selection becomes an important design problem to ensure that the material has sufficient toughness and to preclude transition of the tank material to a brittle state. At high temperatures, corrosion is accelerated, and thermal expansion of the material must be accounted for.

Vapor pressure and boiling point. Vapor pressure is important in liquid storage tank considerations because it affects the design and selection of the tank and its roof, evaporation losses, and resulting pollution; and for flammable liquids, vapor pressure becomes important in classifying liquids for the purpose of characterizing fire hazardousness. The boiling point is also important to know since liquids should usually be stored at temperatures well below the boiling point. Flam-

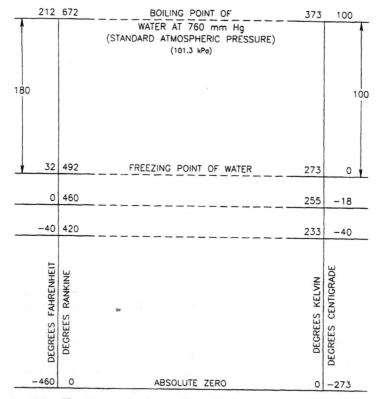

Figure 1.1 The temperature scales. (*From Kirkbride, ref. 13.*)

mable and combustible liquids are expressly prohibited by the fire codes to be stored at temperatures above their boiling points. It is also used to classify the degree of fire hazardousness of the liquid. The lower the boiling point, the lower the vapor pressure will be for liquids stored at ambient temperatures. The boiling point of a few selected substances is shown in Table 1.2.

The vapor pressure of a pure liquid is the pressure of the vapor space above the liquid in a closed container. It is a specific function of temperature and always increases with increasing temperature (see Fig. 1.2).

If the temperature of a liquid in an open container is increased until its vapor pressure reaches atmospheric pressure, boiling occurs. In fact, this is the definition of the boiling point. The boiling point of pure water is 100°C. Note that the temperature of a pure liquid will not increase beyond its boiling point as heat is supplied; rather, all the liquid will evaporate at the boiling point. Since standard atmospheric pressure is 14.7 psia, this is also the vapor pressure of the liquid when

TABLE 1.2   Boiling Points of Various Substances at Atmospheric Pressure

Substance	Boiling point, °F	Substance	Boiling point, °F	Substance	Boiling point, °F
Aniline	363	Ether	100	Sulfur	833
Alcohol	173	Linseed oil	597	Sulfuric acid	590
Ammonia	−28	Mercury	676	Water, pure	212
Benzene	176	Naphthalene	428	Water, sea	213.2
Bromine	145	Nitric acid	248	Wood alcohol	150
Carbon bisulfide	118	Oil of turpentine	315		
Chloroform	140	Saturated brine	226		

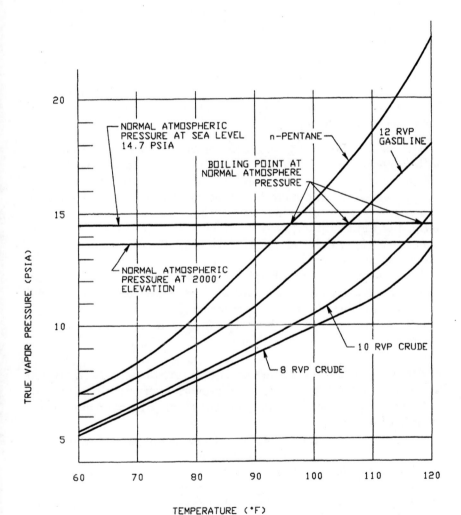

Figure 1.2 Vapor pressure and temperature. (*From W. B. Young, Floating Roofs—Their Design and Application, no. 73-PET-44, American Society of Mechanical Engineers, New York, 1973.*)

it boils. Remember, however, that atmospheric pressure varies with altitude. Water, e.g., will boil at 100°C at sea level and at approximately 98°C at 2000-ft altitude. Barometric pressure changes will slightly alter the boiling point of liquids in tanks as well.

In the fire codes, the boiling point is an important physical property that is used to classify the degree of hazardousness of the liquid. If a mixture of petroleum liquids is heated, it will start to boil at some temperature but will continue to rise in temperature over a boiling temperature range. Since the mixture does not have a definite boiling point, the NFPA fire codes define a comparable value of boiling point for the purposes of classifying liquids based on the 10 percent point of a distillation performed in accordance with ASTM D 86, *Standard Method of Test for Distillation of Petroleum Products.*

Vapor pressure has also become a means of regulating storage tank design by the Environmental Protection Agency. Since increasing vapor pressure tends to result in an increase of emissions, the EPA has specific maximum values of vapor pressure for which various tank designs may be used.

Flash point. A liquid does not really burn—it is the vapor that mixes with oxygen in the atmosphere above the liquid that burns. As a liquid is heated, its vapor pressure and consequently its evaporation rate increase. The minimum temperature at which there is sufficient vapor generated to allow ignition of the air-vapor mixture near the surface of the liquid is called the *flash point.* Even though evaporation occurs below the flash point, there is insufficient vapor generated to form a flammable mixture.

For flammable and combustible liquids, the flash point is the primary basis for classifying the degree of fire hazardousness of a liquid. The National Fire Protection Association (NFPA) classifications 1, 2, and 3 are the most to the least fire hazard liquids, respectively. In essence, low-flash-point liquids are high-fire-hazard liquids.

Pressure. To begin the discussion of tank pressures, it is instructive to consider some very fundamental properties of pressures. *Pressure* is defined as force per unit area, and the standard system of measurement is shown in Fig. 1.3.

Pressure can be expressed as an absolute or relative value. Although atmospheric pressure constantly fluctuates, a standard value of 14.7 psia has been adopted as the nominal value at sea level. The a in *psia* stands for "absolute," meaning that atmospheric pressure is 14.7 psi above a complete vacuum. Most ordinary pressure-measuring instruments do not actually measure the true pressure, but measure a pressure that is relative to barometric or atmospheric pressure. This relative pressure is called the *gauge pressure*; and the atmospheric

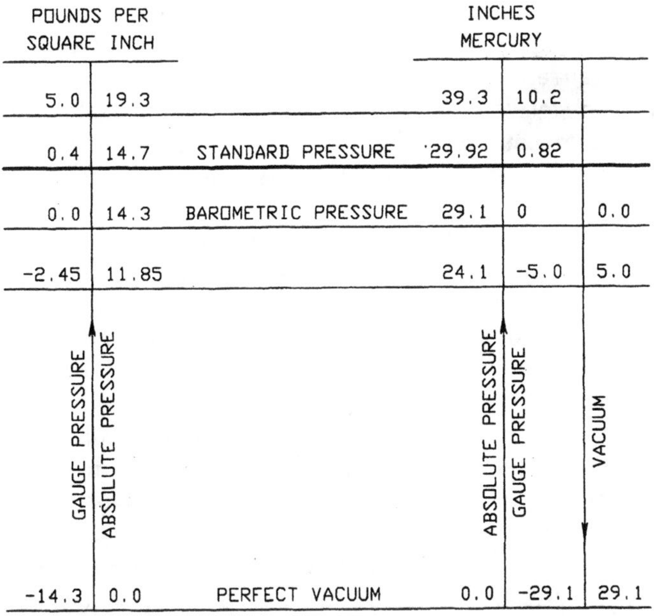

POUNDS PER SQUARE INCH			INCHES MERCURY		
5.0	19.3		39.3	10.2	
0.4	14.7	STANDARD PRESSURE	·29.92	0.82	
0.0	14.3	BAROMETRIC PRESSURE	29.1	0	0.0
-2.45	11.85		24.1	-5.0	5.0
GAUGE PRESSURE	ABSOLUTE PRESSURE		ABSOLUTE PRESSURE	GAUGE PRESSURE	VACUUM
-14.3	0.0	PERFECT VACUUM	0.0	-29.1	29.1

Figure 1.3   Standard system of pressure measurement. (*From Kirkbride, ref. 13.*)

pressure is defined to be 0 psig, where the g indicates that it is relative to atmospheric pressure. Vacuum is the pressure below atmospheric pressure and is therefore a relative pressure measurement as well. For tank work, inches of water column or ounces per square inch (osi) are commonly used to express the value of pressure or vacuum in the vapor space of a tank, because the pressures are usually very low relative to atmospheric pressure. The table below compares these common measures of pressure.

Pressure, psi	in WC	osi
1	27.68068	16
0.03612628	1	0.578205
0.0625	1.730042	1

Although both cylindrical shapes and spherical shells have simple theories to determine the strength and thus thickness of tanks, the region of the tank that is most complex to design is the roof-to-shell junction. This is due to the fact that when there is internal pressure that exceeds the weight of the roof plates and framing of the roof, this junction wants to separate from the shell. When tanks are subjected to pressures sufficient to damage them, the roof-to-shell junction is usually the first area to show damage.

Internal and external pressure. The difference in pressure between the inside of a tank, or its vapor space, and the local barometric pressure, or atmospheric pressure, is called the *internal pressure*. When the internal pressure is negative, it is simply called a *vacuum*. To be precise, the pressure is measured at the top of the liquid in the tank because the liquid itself exerts hydrostatic pressure which increases to a maximum value at the base of the tank.

Since tanks can be large structures, even small internal pressures can exert large forces which must be considered in the design and operation of tanks. For example, a 100-ft-diameter tank with only 1-in WC internal pressure exerts a force of almost 41,000 lb on the roof of the tank.

When the vapor space of a tank is open to the atmosphere or is freely vented, then the internal pressure is always zero or atmospheric because no pressure buildup can occur. This, of course, does not apply to dynamic conditions that occur in explosions or deflagrations. However, most tanks are not open to the atmosphere but are provided with some form of venting device, usually called a *pressure-vacuum valve,* or *PV valve.* A primary purpose of these valves is to reduce the free flow of air and vapors into and out of tanks, thereby reducing fire hazards and/or pollution. However, these valves are designed to open when the internal pressure builds up to some level in excess of atmospheric pressure and to keep the internal pressure from rising high enough to damage the tank. Typical flat-bottom tank pressures range from 1 in WC to several pounds per square inch. Conversely, the vacuum portion of the valve prevents the vacuum inside the tank from exceeding certain limits. Typical internal vacuum is 1 or 2 in WC.

Internal pressure is caused by several potential sources. One source is the *vapor pressure* of the liquid itself. All liquids exert a characteristic vapor pressure which varies with temperature. As the temperature increases, the vapor pressure increases. Liquids with a vapor pressure equal to atmospheric pressure boil. Another source of internal pressure is the presence of an inert gas blanketing system. The internal pressure is regulated by PV valves or regulators. Inert gas blankets are used to pressurize the vapor space of a tank to perform specialized functions such as to keep oxygen out of reactive liquids.

The most fundamental limitation on pressure is at 15 psig. If containers are built to pressures exceeding this value, they are termed *pressure vessels* and are covered by the American Society of Mechanical Engineers' (ASME) Boiler and Pressure Vessel Code. For all practical purposes, tanks are defined to have internal pressures below this value.

External pressure implies that the pressure on the outside of the tank or vessel is greater than that on its interior. For atmospheric tanks, the development of a vacuum on the interior will result in external pressure. External pressure can be extremely damaging to tanks because the surface area of tanks is usually so large and this generates high forces. The result of excessive external pressure is a buckling of the shell walls or total collapse. There have been cases where high wind velocities during hurricanes have developed sufficient external pressure to knock down and collapse tanks.

Miscellaneous properties.   Other properties such as viscosities, freezing or solidification temperature, pour point, and cubic rate of thermal expansion are all important for the tank designer or operator to consider and understand.

1.3   Tank Classification

There are many ways to classify a tank, but there is no universal method. However, a classification commonly employed by codes, standards, and regulations is based on the internal pressure of a tank. This method is useful in that it depends on a fundamental physical property to which all tanks are subjected—internal or external pressure.

1.3.1   Atmospheric tanks

By far the most common type of tank is the atmospheric tank. Although called atmospheric, these tanks are usually operated at internal pressures slightly above atmospheric pressure, perhaps up to ½ psig and usually only a few inches above. The fire codes define an atmospheric tank as operating from atmospheric up to ½ psi above atmospheric pressure.

1.3.2   Low-pressure tanks

Ironically, low pressure in the context of tanks means tanks designed for a higher pressure than atmospheric tanks. In other words, these are relatively high-pressure tanks. These tanks are designed to operate from atmospheric pressure up to 15 psig.

1.3.3   Pressure vessels
(high-pressure tanks)

Since high-pressure tanks (vessels operating above 15 psig) are really pressure vessels, the term *high-pressure tank* is not used by those working with tanks. Because pressure vessels are a specialized form of container and are treated separately from tanks by all codes, standards, and regulations, they are not covered in detail in this book.

However, a few words are in order to clarify the relationship between pressure vessels and tanks.

When the internal design pressure of a container exceeds 15 psig, it is called a *pressure vessel*. The ASME Boiler and Pressure Vessel Code is one of the primary standards that has been used throughout the world to ensure safe storage vessels.

Various substances such as ammonia and hydrocarbons are frequently stored in spherically shaped vessels which are often referred to as *tanks*. Most often the design pressure is 15 psig or above, and they are really spherical pressure vessels and their design and construction fall under the rules of the ASME Boiler and Pressure Vessel Code.

1.4   Major Tank Components

There is no clear way of classifying tanks based upon a single criterion such as shape or roof type; however, the vapor pressure of the substance stored or internal design pressure is the broadest and most widely used method adopted by codes, standards, and regulations, as explained above. To a large extent, the vapor pressure determines the shape and, consequently, type of tank used. Some of the key components that determine tank type are described below. (See Figs. 1.4 to 1.10.)

1.4.1   Fixed-roof tanks

The roof shape of a tank may be used to classify the type of tank and is instantly self-explanatory to tank fabricators and erectors. To understand why, it is helpful to have a brief understanding of the effect of internal pressure on plate structures including tanks and pressure vessels. If a flat plate is subjected to a pressure on one side, it must be made quite thick to resist visible bending or deformation. A shallow cone roof deck on a tank approximates a flat surface and is typically built of $^3/_{16}$-in-thick steel. It is, therefore, unable to withstand more than a few inches of water column. The larger the tank, the more severe the effect of pressure on the structure. As pressure increases, the practicality of fabrication practice and costs force the tank builder to use shapes which are more suitable for internal pressure. The cylinder is an economical, easily fabricated shape for pressure containment. Indeed, almost all tanks are cylindrical on the shell portion. The problem with cylinders is that the ends must be closed. As discussed, the relatively flat roofs and bottoms or closures of tanks do not lend themselves to much internal pressure. As internal pressure increases, the tank builders use domes or spheres. The sphere is the most economical shape for internal pressure storage in

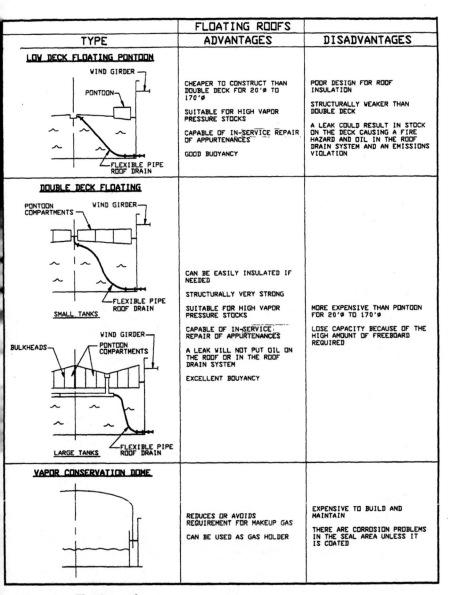

| TYPE | FLOATING ROOFS | |
	ADVANTAGES	DISADVANTAGES
LOW DECK FLOATING PONTOON	CHEAPER TO CONSTRUCT THAN DOUBLE DECK FOR 20'ø TO 170'ø SUITABLE FOR HIGH VAPOR PRESSURE STOCKS CAPABLE OF IN-SERVICE REPAIR OF APPURTENANCES GOOD BOUYANCY	POOR DESIGN FOR ROOF INSULATION STRUCTURALLY WEAKER THAN DOUBLE DECK A LEAK COULD RESULT IN STOCK ON THE DECK CAUSING A FIRE HAZARD AND OIL IN THE ROOF DRAIN SYSTEM AND AN EMISSIONS VIOLATION
DOUBLE DECK FLOATING SMALL TANKS LARGE TANKS	CAN BE EASILY INSULATED IF NEEDED STRUCTURALLY VERY STRONG SUITABLE FOR HIGH VAPOR PRESSURE STOCKS CAPABLE OF IN-SERVICE REPAIR OF APPURTENANCES A LEAK WILL NOT PUT OIL ON THE ROOF OR IN THE ROOF DRAIN SYSTEM EXCELLENT BOUYANCY	MORE EXPENSIVE THAN PONTOON FOR 20'ø TO 170'ø LOSE CAPACITY BECAUSE OF THE HIGH AMOUNT OF FREEBOARD REQUIRED
VAPOR CONSERVATION DOME	REDUCES OR AVOIDS REQUIREMENT FOR MAKEUP GAS CAN BE USED AS GAS HOLDER	EXPENSIVE TO BUILD AND MAINTAIN THERE ARE CORROSION PROBLEMS IN THE SEAL AREA UNLESS IT IS COATED

Figure 1.4 Floating roofs.

terms of required thickness, but it is more difficult to fabricate generally than dome- or umbrella-roof tanks.

Cone-roof tanks.   Cone-roof tanks are cylindrical shells with a vertical axis of symmetry. The bottom is usually flat, and the top is made in

FIXED ROOFS		
TYPE	**ADVANTAGES**	**DISADVANTAGES**
SELF SUPPORTED CONE ROOF 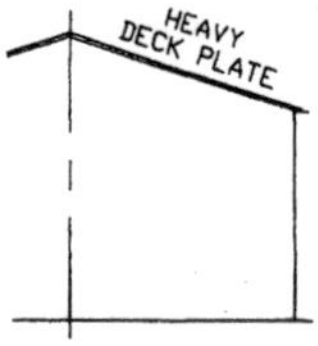	MINIMUM INTERNAL OBSTRUCTIONS RELATIVELY INEXPENSIVE SUITABLE FOR INTERNAL PROTECTIVE COATING MAKES COST-EFFICIENT CONVERSION TO INTERNAL FLOATING ROOF	MAY REQUIRE HEAVIER ROOF DECK PLATE ONLY SUITABLE FOR SMALL TANKS
SUPPORTED CONE ROOFS **CENTER SUPPORTED CONE ROOF** 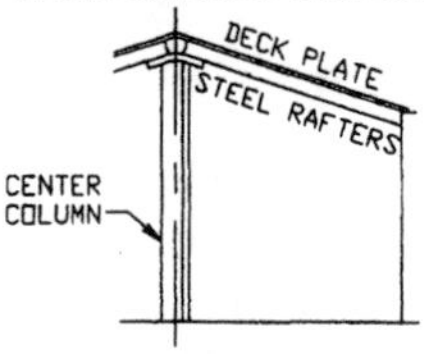	SIMPLE STRUCTURAL DESIGN MINIMAL INTERNAL OBSTRUCTIONS RELATIVELY INEXPENSIVE MAKE COST-EFFICIENT CONVERSION TO INTERNAL FLOATING ROOF	LESS IDEAL FOR INTERNAL PROTECTIVE COATING TANK DIAMETER LIMITED BY SPAN OF RAFTERS
SUPPORTED CONE ROOF 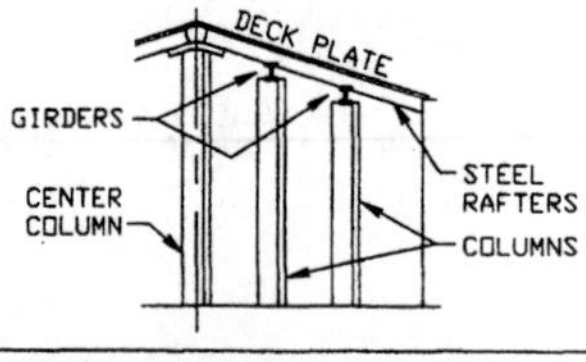	SIMPLE STRUCTURAL DESIGN RELATIVELY INEXPENSIVE SUITABLE FOR ANY DIAMETER TANK CAN BE FRANGIBLE FOR EMERGENCY VENTING	POOR FOR INTERNAL PROTECTIVE COATING MANY INTERNAL OBSTRUCTIONS DIFFICULT TO INSPECT MAKES COSTLY CONVERSION TO INTERNAL FLOATING ROOF
EXTERNALLY SUPPORTED CONE ROOFS ROOF SUPPORT COLUMNS CAN BE DESIGNED THE SAME AS INTERNALLY SUPPORTED CONE ROOFS 	SAME ADVANTAGES AS CENTER SUPPORTED AND SUPPORTED CONE ROOFS LESS INTERNAL OBSTRUCTIONS GOOD FOR INTERNAL COATINGS	SAME DISADVANTAGES AS CENTER SUPPORTED AND SUPPORTED CONE ROOFS MORE EXPENSIVE THAN INTERNALLY SUPPORTED ROOFS IS NOT FRANGIBLE
DOME OR UMBRELLA ROOF 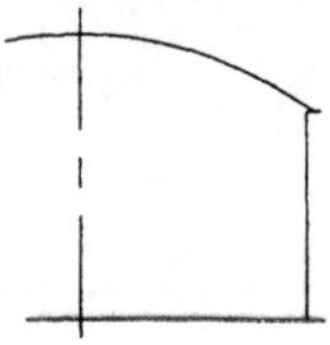	EXCELLENT DESIGN FOR INTERNAL COATING EXCELLENT DESIGN FOR HIGH CORROSION SERVICES SUCH AS SULFUR	MORE EXPENSIVE THAN CONE ROOF SUITABLE FOR ONLY SMALL AND MEDIUM TANKS ROOF DECK PLATE ONLY STRUCTURAL SUPPORT, EXCEPT FOR LARGER DIAMETER TANKS NOT SUITABLE FOR HIGH VAPOR PRESSURE STOCKS UNLESS VAPOR RECOVERY IS USED IS NOT FRANGIBLE

Figure 1.5 Fixed roofs.

the form of a shallow cone. These are the most widely used tanks for storage of relatively large quantities of fluid because they are economical to build and the economy supports a number of contractors capable of building them. They can be shop-fabricated in small sizes but are most often field-erected. Cone-roof tanks typically have roof rafters and support columns except in very small-diameter tanks.

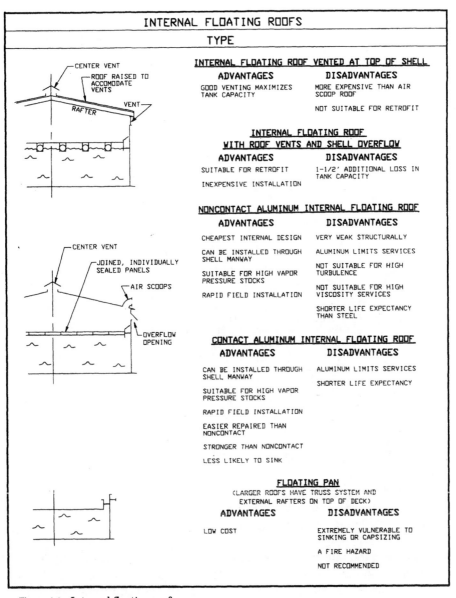

Figure 1.6 Internal floating roofs.

Umbrella-roof tanks. These are similar to cone-roof tanks, but the roof looks like an umbrella, thus its name. They are usually constructed to diameters not much larger than 60 ft. These tank roofs can be a self-supporting structure, meaning that there are no column supports that must be run to the bottom of the tank.

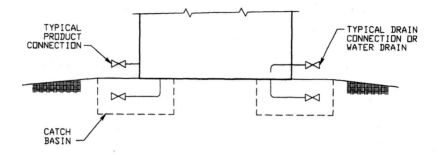

CHARACTERISTICS

USES:

MOSTLY SMALL TANKS, 20 FT. DIAMETER OR LESS. SUITABLE FOR FIELD RUN TANKS, GAUGE TANKS, TREATING TANKS, ETC. WIDELY USED BY THE CHEMICAL INDUSTRY.

ADVANTAGES:

SIMPLE AND ECONOMICAL TO FABRICATE AND INSTALL IN SMALL SIZES.

BOTTOM CONNECTIONS ARE ACCESSIBLE FOR INSPECTION AND MAINTENANCE, JUST AS THE CONE UP AND SINGLE SLOPE DESIGNS ARE.

DISADVANTAGES:

DIFFICULT TO THOROUGHLY DRAIN DUE TO LOW SPOTS (BIRD BATHS) CAUSED BY SETTLING OF FOUNDATION AND WARPING OF BOTTOM PLATES.

A SIPHON DRAIN WILL NOT DRAIN COMPLETELY SINCE IT MUST CLEAR THE BOTTOM.

Figure 1.7   Tank bottom designs, flat-bottom tanks.

Dome-roof tanks.   These are similar to umbrella-roof tanks except that the dome more nearly approximates a spherical surface than the segmented sections of an umbrella roof.

Aluminum geodesic dome-roof tanks.   These fixed-roof tanks are becoming popular. As they are often the economical choice, they offer superior corrosion resistance for a wide range of conditions and are clear-span structures not requiring internal supports. They can also be built to essentially any required diameter.

1.4.2   Floating-roof tanks

All floating-roof tanks have vertical, cylindrical shells just as a fixed-cone-roof tank does. These common tanks have a cover that floats on the surface of the liquid. The floating cover or roof is a disk structure that

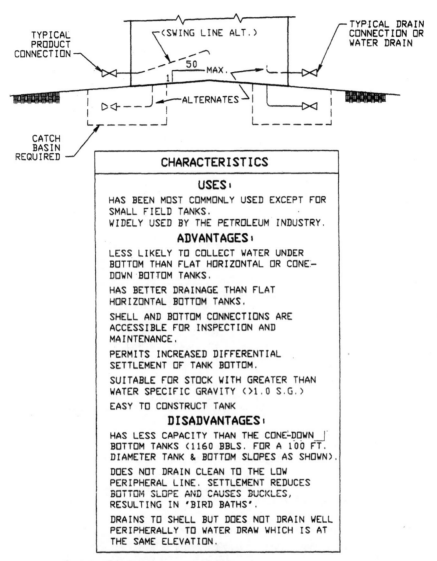

Figure 1.8 Tank bottom designs, cone-up bottoms.

has sufficient buoyancy to ensure that the roof will float under all expected conditions, even if leaks develop in the roof. It is built with approximately an 8- to 12-in gap between the roof and the shell, so it does not bind as the roof moves up and down with the liquid level. The clearance between the floating roof and the shell is sealed by a device called a *rim seal*. The floating roof may be of any number of designs, as described in the section on roofs. The shell and bottom are similar to those of an ordinary vertical cylindrical fixed-roof tank. The two cate-

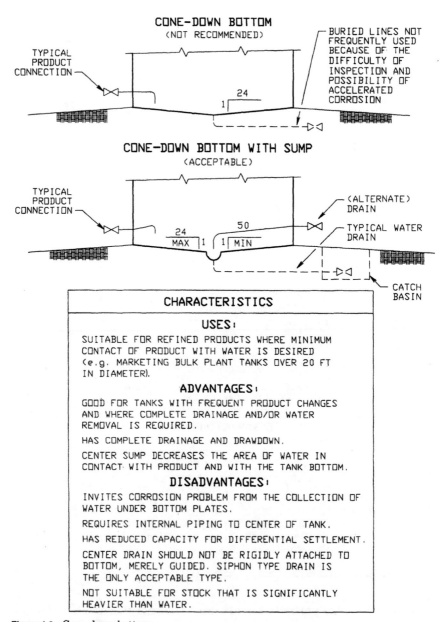

Figure 1.9 Cone-down bottom.

gories of floating-roof tanks are *external floating roof* (EFR) and *internal floating roof* (IFR). If the tank is open on top, it is called an EFR tank. If the floating roof is covered by a fixed roof on top of the tank, it is called an internal floating-roof (IFR) tank. The function of the cover is to reduce evaporation losses and air pollution by reducing the surface area

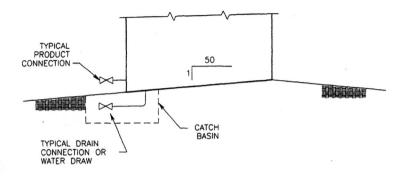

TYPICAL
PRODUCT
CONNECTION —

50
1

TYPICAL DRAIN
CONNECTION OR
WATER DRAW

CATCH
BASIN

CHARACTERISTICS

USES:

SUITABLE FOR TANKS LESS THAN 100 FT. DIAMETER.

GOOD FOR TANKS WITH FREQUENT PRODUCT CHANGE AND WHEN COMPLETE DRAINAGE AND/ OR WATER REMOVAL IS REQUIRED.

ADVANTAGES:

HAS IMPROVED DRAINAGE OVER CONE-UP AND FLAT HORIZONTAL BOTTOM TANKS.

BOTTOM CONNECTIONS ARE ACCESSIBLE FOR INSPECTION AND MAINTENANCE.

DISADVANTAGES:

INSTALLED COST IS MORE THAN CONE-UP OR CONE-DOWN BOTTOM TANKS DUE TO DESIGN AND COST OF FOUNDATION AND ERECTION OF SHELL.

SHALLOW SLOPE MAKES PROBLEMS FOR SEDIMENT CONTAINING TANKS. THE SEDIMENT CAN FORM POCKETS OF WATER THAT DO NOT DRAIN.

Figure 1.10   Tank bottom designs, plan A slope bottom.

of liquid that is exposed to the atmosphere. Note that fixed-roof tanks can easily be converted to internal floating-roof tanks by simply installing a floating roof inside the fixed-roof tank. Conversely, external floating-roof tanks can be easily converted to internal floating-roof tanks simply by covering the tank with a fixed roof or a geodesic dome.

EFR tanks have no vapor space pressure associated with them and operate strictly at atmospheric pressure. IFR tanks, like fixed-roof tanks, can operate at or above atmospheric pressure in the space between the floating roof and the fixed roof.

The fundamental requirement for floating roofs is dependent upon whether the roof is for an internal or external application. The design conditions of the external floating roof are more severe since they must handle rainfall, wind, dead-load, and live-load conditions, comparable to and at least as severe as those for building roofs.

The following paragraphs describe the roof type for both internal and external floating-roof tanks.

External floating roof

Pontoon roofs. These roofs are common for floating roofs from approximately 30- to 100-ft diameter. The roof is simply a steel deck with an annular compartment that provides buoyancy.

Double-deck roofs. These roofs are built for very small floating roofs up to about 30-ft diameter. They are also used on diameters that exceed about 100 ft. They are very strong, durable roofs, because of the double deck, and are suitable for large-diameter tanks.

Internal floating roofs

Pan roof. These are simple sheet-steel disks with the edge turned up for buoyancy. These roofs are prone to capsizing and sinking because a small leak can cause them to sink.

Bulkhead pan roof. This roof has open annular compartments at the periphery to prevent the roof from sinking, should a leak develop.

Skin and pontoon roof. These are usually constructed of an aluminum skin supported on a series of tubular aluminum pontoons. These tanks have a vapor space between the deck and the liquid surface.

Honeycomb roof. The honeycomb roof is made from a hexagonal cell pattern similar to a beehive in appearance. The honeycomb is glued to a top and bottom aluminum skin that seals it. This roof rests directly on the liquid.

Plastic sandwich. This roof is made from rigid polyurethane foam panels sandwiched inside a plastic coating.

1.4.3 Tank bottoms

As mentioned earlier, the shapes of cylindrical tank closures (e.g., top and bottom) are a strong function of the internal pressure. Because of the varying conditions to which a tank bottom may be subjected, several types of tank bottoms have evolved. Tank bottom classifications may be broadly classified by shape as

- Flat-bottom
- Conical
- Domed or spheroid

Because flat-bottom tanks only appear flat but usually have a small designed slope and shape, they are subclassified according to the following categories:

- Flat
- Cone up
- Cone down
- Single-slope

Tank bottom design is important because sediment, water, or heavy phases settle at the bottom. Corrosion is usually the most severe at the bottom, and the design of the bottom can have a significant effect on the life of the tank. In addition, if the stock is changed, it is usually desirable to remove as much of the previous stock as possible. Therefore, designs that allow for the removal of water or stock and the ease of tank cleaning have evolved. In addition, specialized tank bottoms have resulted from the need to monitor and detect leaks since tank bottoms in contact with the soil or foundation are one of the primary sources of leaks from aboveground tanks.

Flat.   For tanks less than 20 to 30 ft in diameter, the flat-bottom tank is used. The inclusion of a small slope as described above does not provide any substantial benefit, so they are fabricated as close to flat as practical.

Cone up.   These bottoms are built with a high point in the center of the tank. This is accomplished by crowning the foundation and constructing the tank on the crown. The slope is limited to about 1 to 2 in per 10-ft run. Therefore, the bottom may appear flat, but heavy stock or water will tend to drain to the edge, where it can be removed almost completely from the tank.

Cone down.   The cone-down design slopes toward the center of the tank. Usually, there is a collection sump at the center. Piping under the tank is then drained to a well or sump at the periphery of the tank. Although very effective for water removal from tanks, this design is inherently more complex because it requires a sump, underground piping, and an external sump outside the tank. It is also particularly prone to corrosion problems unless very meticulous attention is paid to design and construction details such as corrosion allowance and coatings or cathodic protection.

Single-slope.   This design uses a planar bottom but it is tilted slightly to one side. This allows for drainage to be directed to the low point on the perimeter, where it may be effectively collected. Since there is a constant rise across the diameter of the tank, the difference in elevation from one side to the other can be quite large. Therefore, this design is usually limited to about 100 ft.

Conical bottoms.   Tanks often use a conical bottom which provides for complete drainage or even solids removal. Since these types of tanks are more costly, they are limited to the smaller sizes and are often found in the chemical industry or in processing plants.

1.5   Small Tanks

Numerous types of small tanks have developed as a result of increasingly stringent regulations regarding leaks, spills, and containment. The numerous categories of design type can be broadly grouped as follows.

1.5.1   Single-wall tanks

These tanks are usually cylindrical with either a vertical or horizontal orientation. Horizontal tanks are generally supported by two saddle supports. They use more plot space than the vertical tanks but have the advantage that leaks can be seen as they occur. Also, water can easily be drained from a drain valve located on the bottom.

1.5.2   Double-wall tanks

These have become common for both above- and underground applications, since the outer tank can contain a leak from the inner tank and also serves as a means of detecting leaks. They are usually cylindrical tanks with either vertical or horizontal orientation.

1.5.3   Diked or unitized secondary containment tanks

Small tanks can be built with a secondary containment dam built integrally with the tank. They may be either vertically or horizontally oriented in both cylindrical and rectangular shapes. The secondary containment dikes may be open or closed. Closing the dikes makes access to the primary containment tank more difficult but keeps out rainwater.

1.5.4   Vaulted tanks

This term refers to tanks which are installed inside a concrete vault. The vault, which is itself a liquid-tight compartment, reduces the fire protection requirements as the NFPA and the International Fire Code Institute (IFCI) recognize these tanks as fire-resistant, aboveground storage tanks. The vault provides a 2-h fire wall—thermal protection that minimizes tank breathing losses and pollution, secondary containment, and ballistic protection.

1.6 Engineering Considerations

1.6.1 Materials of construction

Tanks are constructed from a number of different materials based upon the cost of the material, ease of fabrication, resistance to corrosion, compatibility with stored fluid, and availability of material. Sometimes specialized composites and techniques are used in tank construction, and these are the exception. The more common materials are described below.

Carbon steel, or mild steel, is by far the most common material for tank construction. It is readily available; and because of the ease with which it is fabricated, machined, formed, and welded, it results in low overall costs.

Stainless steel is another important material used for storage of corrosive-liquid tanks. Although the material cost, usually for the austenitic group of stainless steels, is significantly more than that of steel, it has the same ease of fabricability and availability as carbon steel.

Fiberglass reinforced plastic (FRP) tanks are noted for their resistance to chemicals where stainless-steel or aluminum tanks are not acceptable. The fabrication and construction techniques are somewhat more specialized than those for metals fabrication. Because of the lack of fire resistance, FRP tanks are not normally used to store flammable or combustible liquids. FRP tanks have been used to store water, water-treating chemicals, fire-fighting foam, wastes, lubricants, and nonflammable chemicals and corrosives.

Aluminum tanks are suitable for a limited number of materials. Historically, they were used for cryogenic applications because aluminum remains ductile at temperatures much lower than those of carbon steel. However, nickel steels and stainless steels have largely supplanted the market for aluminum tanks. Aluminum is still used for some acids, fertilizers, and demineralized water applications. However, in general, the use of aluminum for storage tanks has been low.

Concrete tanks have been used in the water and sewage treatment business for a long time. However, because of the relatively high cost, they are not in common use today.

1.6.2 Tank selection criteria

The selection of tanks is a complex process of optimizing an array of information to yield a particular design, as indicated in Fig. 1.11. As shown, once the specific liquid or liquids to be stored are established, the physical properties then determine the range of possible tank

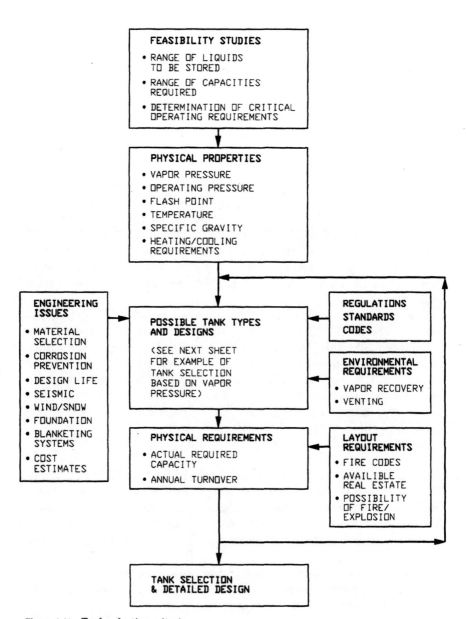

Figure 1.11 Tank selection criteria.

types. Although vapor pressure is a major component in tank selection, other properties such as flash point, potential for explosion, temperature, and specific gravity all factor in the selection and design of tanks. A simplified example of tank selection by fluid stored is shown in Table 1.3. In addition to fundamental physical properties influencing tank selection, size, regulations, current best practices, and external loads (such as wind, snow, and seismic loads), as well as numerous additional engineering issues, play a role. Material selection,

TABLE 1.3 Storage Tank Type for Liquid Chemicals, 25°C (77°F)

Chemical	Tank type	Chemical	Tank type
Acetaldehyde	H	Ethylene diamine	A
Acetamide	A	Ethylene dichloride	L
Acetic acid	A	Ethylene glycol	A
Acetone	L	Elene glycol monoethyl	A
Acetonitrile	L	Ether	L
Acetophenone	A	Formic acid	A
Acrolein	L	Freons	H
Acrylonitrile	L	Furfural	A
Allyl alcohol	L	Gasoline	A
Ammonia	H	Glycerine	A
Benzene	L	Hydrocyanic acid	L
Benzoic acid	A	Isoprene	L
Butane	L	Methyl acrylate	A
Carbon disulfide	L	Methyl amine	A
Carbon tetrachloride	A	Methylchloride	A
Chlorobenzene	L	Methyl ethyl ketone	A
Chloroethanol	A	Methyl formate	L
Chloroform	L	Naphtha	A
Chloropicrin	L	Nitrobenzene	A
Dichlorosulfonic acid	A	Nitrophenol	A
Cumene	A	Nitrotoluene	A
Cyclohexanone	A	Pentane	L
Cyclohexane	L	Petroleum oil	A
Dichloromethane	L	Propane	H
Diesel oil	A	Pyridine	A
Diethyl ether	L	Styrene	A
Dimethylformanide	A	Sulfuric acid	A
Dimethyl phthalate	A	Sulfur trioxide	L
Dioxane	L	Tetrachloroethane	A
Epichlorohydrin	A	Tetrahydrofuran	L
Ethanol	L	Toluene	A
Ethyl acetate	L	Trichloroethylene	L
Ethyl benzene	A	Xylene	A

Key:
A = atmospheric, less than 0.5 psig
L = low pressure, less than 15 psig but greater than 0.3 psig
H = high pressure, greater than 15 psig

SOURCE: *Ecology and Environment*, 1982.

corrosion prevention systems, and environmental requirements and considerations may also influence the selection. Layout of the tank within existing or new facilities is always a major consideration, as the limited plot space for the tank is a factor in the type of tank selected. The requirements of the fire authority having jurisdiction will almost always have property line and public way setbacks, as well as distance requirements between tanks and equipment or other tanks. The ultimate selection criteria are keenly dependent on the actual site-specific conditions, local regulations, cost considerations, required operating life, space available, potential for fires and explosion, and many other factors. These should all be evaluated and considered by the responsible engineering designer and documented for the protection of the plant owner/operator.

1.6.3   Special engineering considerations

Since tanks are used in so many different ways, some specialized applications have developed that have become fairly commonplace. Some of these applications are described below.

Cryogenic tanks.   Low-temperature tanks are used for liquefied hydrocarbon gases (LHGs); liquefied natural gas (LNG); various liquefied gases such as air, nitrogen, or oxygen and ammonia; and other refrigerated liquids. As a general rule, larger quantities of liquids that are stored which have high vapor pressures favor low-temperature or cryogenic storage. Although the end use of these products may be in the gaseous state, a higher quantity can be stored in the liquid state, so it is often the preferred method for storing large quantities of gases. However, many of these compounds cannot be liquefied at ordinary temperatures and pressures. Not only must they be cooled to substantial temperatures below ambient conditions, but also they may need to be kept under pressure. Because of these requirements, these tanks often need accessory cooling systems. Some of these systems cool the vapor in the tank space and return it to the tank, while others cool the liquid. In addition, to reduce the size of the cooling equipment and conserve energy, these tanks must be insulated. For flammable tanks such as LNG or LHG, additional fire code requirements may call for a secondary containment tank to be provided which can contain the contents of the inner tank, should it fail.

Careful material selection is required to prevent brittle failure of tanks at low temperatures. In addition, for existing tanks whose service temperatures are reduced, it is essential that an engineering analysis be performed to ensure that they are not subject to brittle failure at these lower temperatures.

Heated tanks. Many compounds will freeze, solidify, or thicken to the point where they cannot be transferred through piping and equipment unless they are maintained at minimum temperature. Examples are heavy oils, asphalts, sulfur, highly concentrated salt or caustic soda solutions, or even molasses and food stuffs. The storage tanks for these fluids must be heated and maintained at a minimum temperature. There are several ways to heat tanks, as shown in Fig. 1.12. Heat

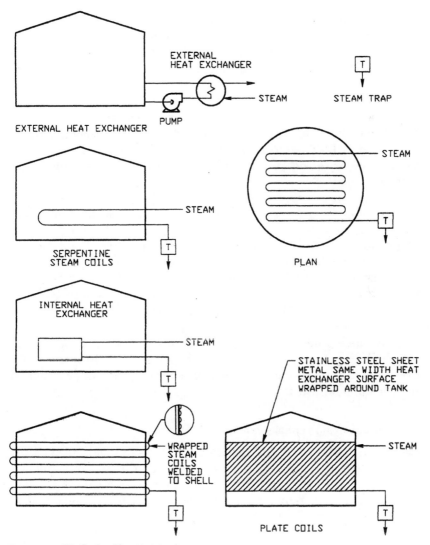

Figure 1.12 Methods of heating tanks.

transfer is a rather complex engineering optimization problem to minimize the heat transfer surfaces, since these tend to be costly. This is done by mixing and/or pumping the fluids, varying the insulation requirements of the tank, picking the proper heat transfer medium such as steam, or optimizing the type of heat transfer surface (extended or finned surface versus tubular). Of particular importance in establishing the heating requirements is the rate of bringing a cold tank up to temperature versus the rate of heat needed to maintain a minimum temperature under design conditions. If the heat-up rate is too high, a large heating system will be required, whereas if it is too small, the tank will take excessively long to heat up or may not heat up to its design minimum under adverse conditions. The temperature of the heating fluid must be kept low where the product is heat-sensitive and will not suffer degradation. Also, temperature levels of the heat transfer surface must be limited where stress corrosion cracking of carbon steel can occur, as in some carbonate or caustic solutions. Other design requirements involve the prevention of stratification of hot and cold layers and the ability to remelt or heat up the tank quickly and efficiently. These considerations usually involve use of mixers and eductors or consideration of thermal circulation occurring within the tank.

Since the exposed surface area of a tank is relatively large, heated tanks are almost always insulated. Another reason for insulating them is that the external corrosion rate of the steel due to atmospheric conditions increases with increasing temperature. Insulation, if properly installed, reduces external corrosion. Many types of insulation systems are available. The difficulty with most of them is that if rainwater gets into them, it becomes trapped inside and tends to accelerate corrosion. Many of the recent improvements have been directed at keeping water out of the insulation.

An extremely important safety consideration for both tanks is that lower-boiling-point liquids must not be introduced into the tank. These liquids can boil and cause a froth-over or a violent evolution of vapor, causing failure of the tank.

1.6.4 Various design considerations

Most of the design codes and standards for tanks provide good assurance that critical factors are not overlooked by the unwary designer. In particular, the tank standards issued by the American Petroleum Institute (API) have reduced the risks of the catastrophic results that can occur when one fails to consider materials selection, brittle fracture, insufficient welding, joining, or inspection methods, fabrications methods, and the like. In fact, these standards are recognized world-wide to be of the highest caliber. As a result, they have been used in industries such as chemicals, pulp and paper, food, and a host of others. Even with API standards at the designer's disposal, she or he is faced with many site-specific considerations that can have a substantial impact on the design life of the tank as well as on its safe operation. All these considerations should be documented in records maintained by the owner or operator of the tank. Some of the following design elements are briefly covered by this list:

- Determination of which standard and code to use in building the tank
- Compliance with fire codes
- Materials selection
- Linings and coatings
- Cathodic protection
- Wall thickness selection
- Appurtenance design (ladders, internal piping, instrumentation)
- Establishing and designing for external anticipated loads (e.g., seismic or snow loads)
- Venting and emergency venting
- Foundation and settling criteria
- Pollution prevention (secondary containment, leak detection, seal selection)
- Fabrication, erection, inspection, and testing
- Electrical area classification

1.7 Regulations

Regulations and laws are mandatory requirements with which a tank owner or operator must comply, or else suffer the consequences of civil or even criminal liability. Most regulatory requirements are channeled through an agency whose general responsibility is the safety, well-being, and protection of the public or the environment. The authority having jurisdiction may be a federal, state, local, or regional agency; an individual such as a fire chief or marshal; a labor department; health department; building official or inspector; or any others having statutory authority.

The general rule of thumb is that most tank facilities are subject to multiple authorities having jurisdiction over them. When this is so and the rules have overlapping or even conflicting provisions, the facility must comply with all the requirements of the multiple authorities. In short, one authority's requirements do not preempt the requirements or even satisfy the other agency, even if they accomplish exactly the same thing. There are currently so many examples of this type of inefficiency in government regulations, particularly in the environmental arena, that the administrative cost burden to facilities alone subject to the multiple, overlapping compliance requirements is driving many businesses away or out of business.

For aboveground tanks, there is no comprehensive regulation or program as there is for underground tanks under the RCRA program. Instead, most tanks have been unregulated or controlled only if they contained flammable liquids under the jurisdiction of local fire authorities. This is likely to change as there have been calls for national legislation. In recent years, some piecemeal legislative regulation has started to directly or indirectly affect the installation and operation of tanks.

Unfortunately, the federal regulations tend not to be aimed at spill prevention but rather at spill response. This can be understood by considering the fact that they were enacted after some of the most calamitous oil spills in history (e.g., Ashland Oil, Exxon's *Valdez*). They therefore address issues such as containment of spills, financial liability and responsibility, discharge of contaminated storm water, and reporting and response requirements.

At present, the framework from which all regulations that affect the petroleum, chemical, and petrochemical industries are derived is essentially the result of nine statutes:

- Clean Air Act (CAA)
- Clean Water Act (CWA)
- Resource Conservation and Recovery Act (RCRA)
- Comprehensive Environmental Response, Compensation, and Liability Act (CERCLA)
- Superfund Amendment and Reauthorization Act, Title III (SARA, Title III)
- Occupational Safety and Health Act (OSHA)
- Toxic Substances Control Act (TSCA)
- Hazardous Materials Transportation Act (HMTA)
- Safe Drinking Water Act (SDWA)

They are also shown in greater detail in Fig. 1.13. These statutes address the issues of

Clean Air Act (CAA)	Clean Water Act (CWA)	Resource Conservation and Recovery Act (RCRA)
Regulates: • Emissions of pollutants to atmosphere • emissions by treatment technology unless air quality requires tighter limits	Regulates: • Discharges of waste waters to receiving waters and to POTWs • Discharges by treatment technology unless water quality requires tighter limits	Regulates • Generation, transportation, storage, treatment, and disposal of hazardous solid wastes • Storage of fuels in underground tanks - nonhazardous waste
Comprehensive Environmental Response, Compensation, and Liability Act (CERCLA)	Superfund Amendment and Reauthorization Act Title III (SARA Title III)	Occupational Safety and Health Act (OSHA)
Regulates: • Clean up of leaking landfills • Reporting spills of certain chemicals • Responsibility and liability for contaminated disposal cleanup	Regulates: • Emergency response plans • Right-to-know issues • Chemical release reporting	Regulates: • Employee right-to-know • Employment free of recognized hazards • Specific standards for job and industry safety
Toxic Substances Control Act (TSCA)	Hazardous Materials Transportation Act (HMTA)	Safe Drinking Water Act (SDWA)
Regulates: • Commercial use of most chemicals • Use and disposal of asbestos, PCBs, and CFCs • Reporting of all adverse health effects • Use, labeling, and documenting for chemicals that pose risk to health or the environment	Regulates: • Hazardous materials when transported in commerce • Activities associated with identifying and classifying the material, marking, labeling and placarding, packaging, and documentation the hazardous material • Loading, unloading and incidental storage • Reporting of unintentional releases and injury or death to a person or property damage exceeding $50,000	Regulates: • Enforceable quality standards to drinking water • Protection of groundwater sources

Figure 1.13 Summary of major environmental laws. (*From Chevron Environmental Health and Safety Regulatory Summary and Desk Reference.*)

- Limiting exposure of substances that may be harmful to human health or the environment to acceptably low levels

- Assigning financial as well as criminal responsibility for damaging human health or the environment

- Reporting of data, incidents, and information that may affect the regulatory agencies for enforcing the regulations associated with the statutes listed above

Chapter 2 deals in detail with the topic of codes and standards and provide a complete listing of relevant standards organizations, addresses, and documents available for all facets of tank work.

The relationship of codes and standards to the regulatory framework should be clearly understood. By themselves, codes and standards have no authority. However, a study of both regulations and codes show that the government jurisdictions that have authority over tanks usually rely on the codes and standards to form the technical basis of their requirements. In many cases, they simply refer to the code or standard by name, which then elevates it to a legal requirement. Frequently, the authority having jurisdiction will often add other requirements, such as registration fees, to its technical requirements to support the administrative costs of implementing the inspection and enforcement of its responsibilities.

Identifying and using codes and standards is one of the first steps to take when you are considering a new tank installation, regulating it, or simply trying to understand it.

1.8 Spills, Leaks, and Prevention

Leaks and spills from aboveground storage tanks (ASTs) facilities have had greater impact on changing the way tanks are regulated and will be regulated as well as upon the design and operation of tanks than any other single factor. Leaks and spills have had a substantial environmental impact, not only upon localized public health and social issues but also upon public awareness. Because leaks and spills are associated with groundwater which is then associated with public water supplies and irrigation, it is a natural consequence that the public has had little tolerance for environmental accidents, which continue to occur with regular frequency. An example is East Meadow, Long Island, where a service station leaked 50,000 gal of gasoline into the aquifer. Hydrocarbon fumes leaked into the basements of a number of surrounding homes and caused them to be abandoned, with almost total loss of property value. In spite of what I believe to be a reasonably good environmental performance record for storage facilities (which could be far worse), there is an accelerating trend to bullet-proof every aspect of tank facilities without regard to the costs, economic impact, or site-specific conditions. Let us review the fundamental aspects of leaks and spills.

1.8.1 Causes of spills and leaks

There are numerous causes of tank leaks and spills. Some examples, as well as causes of bulk storage and handling facilities and piping, are noted in Table 1.4. Although the causes of spills are numerous, they can be categorized as follows.

TABLE 1.4 Causes of Leaks and Spills

Leak or spill source	Characteristics	Root causes	Preventive measures
Corrosion	Most common in tank bottoms, underground piping Low rate Lack of warning—may continue for years undetected Large volumes released over long periods Common	Corrosion Materials selection Costs of corrosion prevention methods	Careful design and engineering Inspection per API 653 Tank management program
Operations ■ Overfills and transfers	Larger quantities released Quickly discovered Hazardous potential for fires Relatively common	Operator error Instrumentation and equipment failure Lack of training Failure to maintain overfill systems	Tank management program with written operating procedures Training and drills Periodic testing of instrumentation
■ Roof drains	Large volumes released Easily discovered Usually occurs in stormy weather Relatively rare	Equipment failure Failure to use secondary containment properly	
■ Leaks	Leaks relatively common in piping, valves and fittings, pump seals, or penetrations through secondary containment areas		
Tank breakage ■ Brittle fracture	Occurs in cold weather Catastrophic failure mode Entire tank contents can empty Extremely rare	Materials selection Poor fabrication details Failure to hydrotest	Careful design and actual solution Assessment of brittle fracture and seismic effects after each major charge of service
■ Seismic	Damage to piping Tearing of repads and appurtenances Loss of tank contents Relatively rare	Ground acceleration	Assessment of fabrication details per API 653 Documentation of all work and engineering performed on all tasks

TABLE 1.4 Causes of Leaks and Spills (*Continued*)

Leak or spill source	Characteristics	Root causes	Preventive measures
Maintenance	Corrosion leaks (as covered above) Instrumentation malfunction	Poor tank management programs	Establish tank management program Periodically test all instrumentation Establish API 653 program Document all work on all tanks
Vandalism	Damage from opening valves Damage from gunfire Ignition of tank contents Bombs or explosions	Poor security	Improve security systems
Piping	Major cause of leaks Failure to provide sufficient capacity to diked areas Leakage of product through second containment penetration Poor and improper tank materials selection, corrosion prevention, design Charge of services	Inadequate or no design engineering or periodic assessment	Comply with API Standard 570
Fire and explosion	Spills inside and outside secondary containment Fire likely to spread to all leaking tanks Rim seal fires relatively minor Spill fires can quickly become dangerous	Improper design or operation	Set up tank management program Ensure compliance with NFPA Conduct fire and safety audits Document results Establish emergency command system and resources Conduct process management safety review

Corrosion.   Corrosion is one of the most prevalent and insidious causes of leaks. Because the large surface area of either aboveground or underground tanks cannot be easily inspected, leaks that develop tend to go on for long periods with large underground contamination pools resulting. Corrosion can be mitigated by proper foundation design and material selection, use of lining and coatings for both topside and bottomside corrosion, cathodic protection, and chemical inhibition. A good tank inspection program such as API 653 which requires periodic internal inspections is one of the best ways to ensure that leaks do not go on undetected for long periods of lime.

Operations

- Overfill of tanks due to any number of reasons is a common occurrence. This results from inoperative or failed equipment such as level alarms, instrumentation, and valves, as well as operator error or lack of training. A comprehensive tank management program addresses these kinds of problems.

- Another occurrence is the leakage of product through the roof drains on external floating-roof tanks and out at the roof drain discharge nozzle. These spills are sometimes caused by equipment failures, but also result from operator error or lack of training. Because it is easier to leave secondary containment valves and roof drain valves open (so that they do not need to be opened in periods of rainfall), the effectiveness of this equipment is reduced, and this may be classified as operator error. A comprehensive tank management program addresses these kinds of problems.

- Another significant cause of contamination is draining the water bottoms from tanks. This water escapes through the secondary containment system or stagnates in pools on the ground, resulting in contamination. A change in operation, as well as inclusion of the procedures for disposal of tank water bottoms in an overall tank management program, eliminates this form of leak and spill.

Tank breakage.   Tank breakage, due to either brittle fracture or ductile tearing during earthquakes or some uncorrected settlement problems, results in sudden and total loss of the tank contents. Fortunately, these occurrences are relatively rare. Proper engineering and design regarding materials selection, application of the various codes and standards, and carefully detailed designs can prevent these incidents.

Maintenance.   The lack of maintenance and investment in inspection programs results in not only poor housekeeping but also unnoticed leaks due to corrosion, leaking flanges and valves, and inoperative instrumentation that could prevent spills.

Vandalism.   Vandalism is a surprisingly significant cause of spills. If a facility adopts a tank management program aimed at preventing leaks and spills, one element addresses facility security.

Piping.   Piping is a major cause of leaks and spills. It has many of the same characteristics as tank leaks and spills. Indeed, piping is always connected to tankage. Perhaps the most significant leaks have resulted from pressurized underground piping.

Design.   Design can create almost any combination of the problems described above. It can also result in catastrophic and major spills. The best prevention is to use established codes such as those provided by the API.

Fire and explosion.   Most fires are associated with spills from overfills or leaking pump seals that form pools and clouds of flammable material. Periodic fire compliance and safety reviews will prevent many problems. Compliance with API and NFPA codes is good insurance. If fires do develop, then a program of emergency response and preparedness, as well as the establishment of an incident command system, can mitigate unforeseeable disasters. Careful design details and operational procedures are more effective than pumping resources into costly fire-fighting equipment in general.

1.8.2 Spill anatomy and remediation

Contrary to past arguments that leaks or spills from aboveground tanks would stay near the surface, Fig. 1.14 shows that they go straight down and spread out. Various obstacles such as clay lenses, rock, or impermeable layers of soil simply divert the downward path. Slow leaks from tank bottoms tend to form a narrow plume, as shown, while larger spills cover much wider areas. When the contaminant reaches groundwater, it tends to be dispersed in the direction of the groundwater current and movement.

The properties of the material spilled, of course, play a key role in the dispersion of the spill. Not only are the various hazardous properties (toxicity, flammability, reactivity) important because they can damage the environment and pose health hazards, but these properties have a significant effect on dispersion, and consequently on the potential for cleanup. Most petroleum products are relatively insoluble in water and therefore tend to remain in distinct and separate phases when they contact groundwater. Since petroleum is lighter than water, it tends to collect in pools above the groundwater and can be withdrawn by drilling wells into the high point of these pockets.

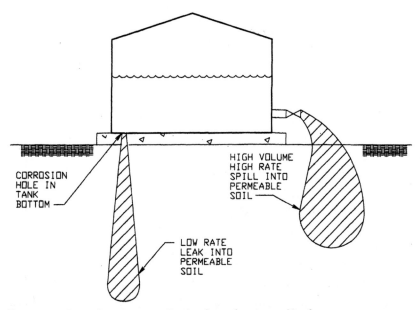

Figure 1.14 Groundwater contamination from aboveground tanks.

However, it also moves with the groundwater and often seeps into creeks, lakes, or rivers. On the other hand, miscible products such as alcohols, oxygenates used for motor fuels, and various chemicals can mix with the groundwater. Once mixed, the entire groundwater current that passes through the spill region must be treated in order to remove the contaminant. Products with specific gravities greater than 1.0, such as halogenated hydrocarbons, will tend to sink to the bottom of the aquifer. In spite of efforts to remove spilled chemicals by withdrawal from the ground through wells, a substantial amount remains embedded in the soil and groundwater in the form of coated particles or dissolved in the moisture of the soil and aquifers. Sometimes portions of the groundwater stream are pumped to the surface for treatment and reinjection of neutralizing chemicals or oxidizers, such as air injected into the ground to reduce the effects of the spill and leak. None of these methods is 100 percent effective.

Obviously, it is far better to prevent a leak or a spill than to clean one up.

1.8.3 Spill prevention and detection

The fundamental rule of leak and spill prevention is to reduce the possibility for contamination by directing resources as close to the source as possible. This is graphically illustrated in Fig. 1.15. In addi-

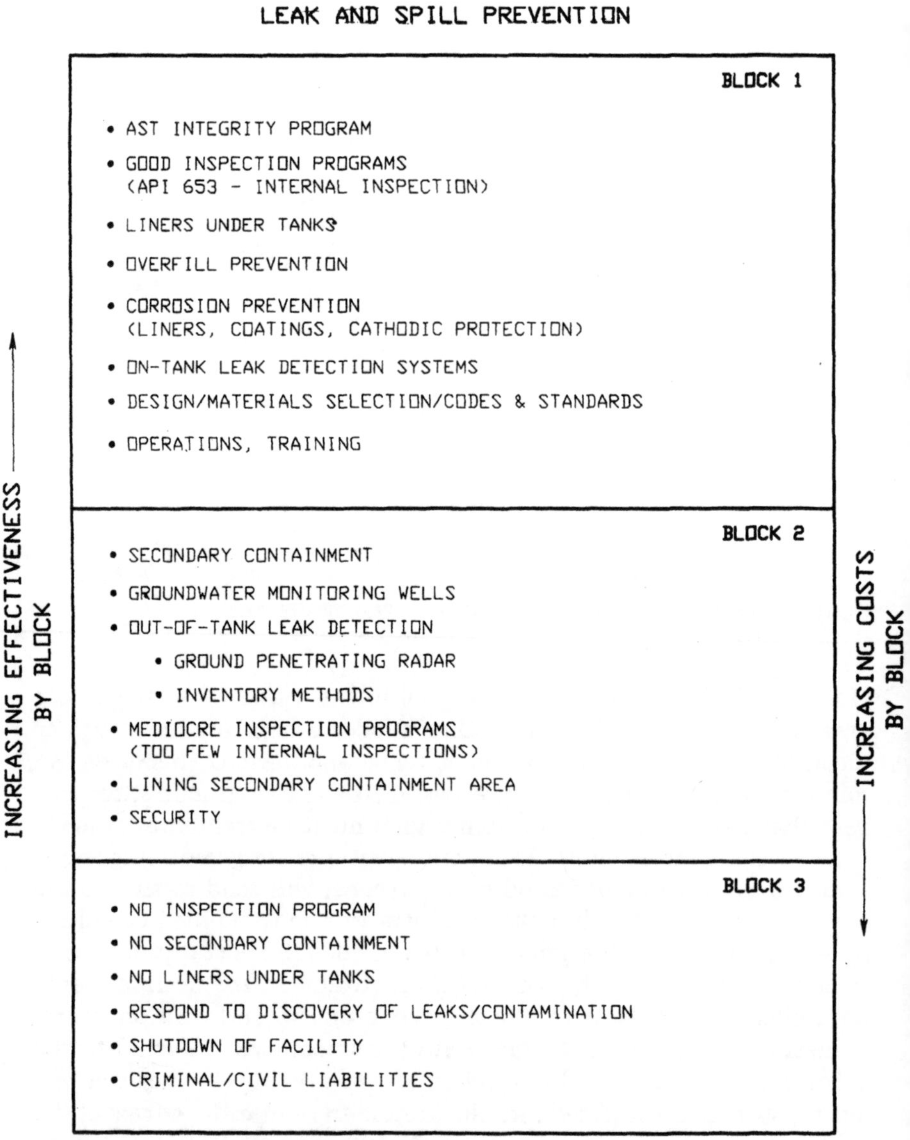

LEAK AND SPILL PREVENTION

INCREASING EFFECTIVENESS BY BLOCK

INCREASING COSTS BY BLOCK

BLOCK 1

- AST INTEGRITY PROGRAM
- GOOD INSPECTION PROGRAMS
  (API 653 - INTERNAL INSPECTION)
- LINERS UNDER TANKS
- OVERFILL PREVENTION
- CORROSION PREVENTION
  (LINERS, COATINGS, CATHODIC PROTECTION)
- ON-TANK LEAK DETECTION SYSTEMS
- DESIGN/MATERIALS SELECTION/CODES & STANDARDS
- OPERATIONS, TRAINING

BLOCK 2

- SECONDARY CONTAINMENT
- GROUNDWATER MONITORING WELLS
- OUT-OF-TANK LEAK DETECTION
  - GROUND PENETRATING RADAR
  - INVENTORY METHODS
- MEDIOCRE INSPECTION PROGRAMS
  (TOO FEW INTERNAL INSPECTIONS)
- LINING SECONDARY CONTAINMENT AREA
- SECURITY

BLOCK 3

- NO INSPECTION PROGRAM
- NO SECONDARY CONTAINMENT
- NO LINERS UNDER TANKS
- RESPOND TO DISCOVERY OF LEAKS/CONTAMINATION
- SHUTDOWN OF FACILITY
- CRIMINAL/CIVIL LIABILITIES

Figure 1.15 Leak and spill prevention.

tion to increasing the effectiveness of a spill and leak prevention program, the costs are lower if the focus is on preventing the occurrence in the first place. Unfortunately, the regulatory trend is to require methods which respond to leaks after they occur. In addition to being more costly, this type of requirement is a disincentive to prevent the leaks in the first place because of additional funds required.

Leak and spill prevention really comprises a system of management or a program embodying many facets which, when all working together, virtually eliminate the possibility of leaks and spills. In all cases, documentation and recordkeeping of all aspects of the tank management system are good practice and is tending toward being required. Some of the elements of this type of program follow.

Engineering controls.   The design, engineering, and maintenance of storage tanks strongly affect the potential for leaks and spills. For example, corrosion resistance can be increased substantially by the use of coatings or cathodic protection and proper foundation design. Instrumentation can be designed that reduces the likelihood of overfills, fires, or other accidents. Good engineering practice can virtually eliminate brittle fracture. Control of fabrication and inspection of new or repaired tanks can reduce the chances of all kinds of failures. Other parameters subject to engineering controls are foundation and settlement, seismic capability, fire resistance, and resistance to vandalism.

Operational controls.   Standard written operating instructions go a long way toward ensuring that operators not only know what to do but also have sufficient understanding to act effectively in the event of a leak or spill. These instructions should include information on the materials stored and their properties, notification of the appropriate authorities in the event of spills, emergency shutdown procedures, and availability and use of emergency and protective equipment. Operational controls should have provisions for inspection. One of the most effective leak prevention methods available is a daily walkthrough in the plant by an experienced operator. As a result, operational controls should include provisions for regular training, addressing contract personnel as well. Spill response and planning should be an integral part of the operational controls.

Secondary containment.   One of the most effective methods for mitigating large and catastrophic spills is the use of secondary containment. In this concept, the entire tank field is surrounded with a dike wall or impoundment area within which the volume of the largest tank can be contained. Thus, even if the tank were to fail from a sudden and total release, the contents would be captured in the secondary containment area for immediate removal and disposal. In fact, any facility near a navigable waterway must use secondary containment, according to the Spill Prevention Control and Countermeasures (SPCC) rules enacted by the EPA. However, if the drain valves that allow rainwater to escape are left open, the secondary containment area will fail to operate as intended. Only operator procedure, knowledge, training, and practice will ensure that these systems work as intended.

Leak detection. These methods may be subclassified as to whether they are on the tank or not on the tank. On-the-tank leak detection systems operate immediately upon leakage. Some of the more common methods as described below.

On-tank leak detection systems

Leak detection bottoms. Tanks with leak detection bottoms have a means of directing any leaks to the outside of the tank perimeter, where they can be visually observed. Before any significant contamination can occur, these leaks are discovered, and the tank is taken out of service to address the leak.

Precision mass and volumetric methods. These methods use very precise measurements of pressure and/or level in the tank to detect leaks. The tank must be isolated so that no liquid enters or leaves the tank. While in principle these systems seem appropriate, they become less reliable as the tank size increases.

Hydrocarbon sensors. Hydrocarbon sensors placed directly below the tank bottoms can be effective. However, old contamination or contamination from other tanks or piping can yield misleading results. In addition, the low permeability of some areas in the soil can prevent the migration of vapors to the sensing ports under the tank bottom.

Tracer methods. Tracer methods use chemical markers injected into the contents of the tank. Instrumentation capable of detecting the chemical marker is then used to determine the presence of a leak caused by seepage of the marker into the ground. This, like the hydrocarbon sensing method, is a method generically referred to as *soil vapor monitoring,* and it suffers from the same weaknesses that have to do with undertank soil permeabilities.

Sensing cables. These methods use wires stranded into a cable. When liquid contacts the wires, there is a flow of electric current between the wires indicating a leak. Variations of this method include changes of impedance or of other properties such as conductivity. These systems tend to be very costly and are prone to work only once or during the first contamination.

Acoustic emissions. By listening to the "sound" emitted from leaking tanks, it is possible to estimate both the existence and the location of holes in tank bottoms. Much work needs to be done in this area before it can be considered reliable.

Off-tank leak detection systems

Monitoring wells. Monitoring wells are drilled near the tank site. They depend on relatively large losses and resulting contamination before detection of a leak is possible.

Inventory reconciliation. This is a method of detecting discrepancies in receipts and disbursements of product through metered piping.

However, these methods are relatively inaccurate, and a substantial leak can escape detection by this method. The advantage of this method is that it requires little capital investment because most of the metering is usually already in place. Again, this method, when it works, is associated with a substantial amount of lost product and contamination.

Use of liners. The use of impermeable liners and membranes, often called *release prevention barriers* (RPBs), under tanks is extremely effective as a leak detection and prevention method and may be the most effective method. On new tanks, it is relatively easy to install these systems; and large numbers of tanks today are being built with this type of system. However, for existing tanks, it could be very costly, if not impractical, to install liners. For existing tanks, the combination of other methods, as well as an effective inspection program, can be effective as a substitute for an RPB.

There has been much debate about using liners to line the entire secondary containment area in addition to the area under the tank. Lining the entire secondary containment area is costly and probably ineffective, for the following reasons:

- Major spills and leaks into a secondary containment area are not left for long periods where they can permeate the ground but are cleaned up immediately. Most secondary containment areas are relatively impermeable for the short duration for which spills reside.

- Lining the secondary containment area is difficult to achieve completely. At walls, partitions, piping penetrations, and equipment foundations, there are joints and cracks which permit fluids to migrate under the liner anyway. Once under the liner, they cannot be cleaned up, and from this perspective, it is worse to use a liner.

- According to the API, it may be cheaper to remediate than to provide lining.

Inspection programs. One of the most effective ways to reduce leaks and spills resulting from mechanical failure or corrosion is to implement an inspection program. The API has issued API Standard 653, which provides a rational and reasonable approach to the problem of inspecting tanks. Because tanks are very costly to empty, remove from service, clean, and prepare for internal inspection, requiring the entry of inspectors into the interior of tanks has been avoided. As a result, there have been numerous long-term leaks resulting from corrosion that has gone undetected. API Standard 653 provides a basis for scheduling internal inspections based upon anticipated corrosion rates. For existing tanks which cannot easily be fitted with leak detection systems or liners, the use of this type of inspection program with appropriately spaced internal inspections is effective in reducing leaks.

Appendix 1A.   Conversion Factors

Temperature				
To convert from	To:	°C	°F	°R	K
°C		—	1.8(°C) + 32	1.8(°C) + 459.7	°C + 273.2
°F		(°F−32)/1.8	—	°F + 459.7	(°F + 459.7)/1.8
°R		(°R−491.7)/1.8	°R−459.7	—	°R/1.8
K		K−273.2	1.8(K)−459.7	1.8(K)	—

Length				
To convert from	To:	cm	m	in	ft
			Multiply by:		
cm		1.000	0.0100	0.3937	0.03281
m		100.0	1.000	39.37	3.281
in		2.540	0.0254	1.000	0.08333
ft		30.48	0.3048	12.00	1.000

Area				
To convert from	To:	cm^2	m^2	in^2	ft^2
			Multiply by:		
cm^2		1.000	1.000×10^{-4}	0.1550	1.076×10^{-3}
m^2		10,000	1.000	1550	10.76
in^2		6.451	6.451×10^{-4}	1.000	6.944×10^{-3}
ft^2		929.0	0.09290	144.0	1.000

Volume							
To convert from	To:	in^3	ft^3	U.S. gal	Imp gal	cm^3	L	bbl (42's)
					Multiply by:			
in^3		1.000	5.787×10^{-4}	4.329×10^{-3}	3.607×10^{-3}	16.39	0.01639	1.031×10^{-4}
ft^3		1728	1.000	7.481	6.232	2.832×10^4	28.32	0.1781
U.S. gal		231.0	0.1337	1.000	0.8326	3785	3.785	0.02381
Imp. gal		277.3	0.1605	1.200	1.000	4543	4.543	0.02857
cm^3		0.06102	3.531×10^{-5}	2.642×10^{-4}	2.201×10^{-4}	1.000	1.000×10^{-2}	6.290×10^{-6}
L		61.02	0.03531	0.2642	0.2201	1000	1.000	6.290×10^{-3}
bbl (42's)		9700	5.614	42.00	34.97	1.590×10^5	159.0	1.000

Force				
To convert from	To:	Poundals	lb	dyne	g
			Multiply by:		
Poundals		1.000	0.03108	13,830	14.10
lb		32.17	1.000	4.448×10^5	453.6
dyne		7.233×10^{-5}	2.248×10^{-6}	1.000	1.020×10^{-3}
g		0.07093	2.205×10^{-3}	980.7	1.000

Density			
To convert from	To:	Sp gr	lb/gal	lb/ft^3
			Multiply by:	
Sp gr		1.000	8.347	62.43

Density (*cont.*)

To convert from	To:	Sp gr	lb/gal	lb/ft³
			Multiply by:	
lb/gal		0.1198	1.000	7.481
lb/ft3		0.01602	0.1337	1.000

Pressure

To convert from	To:	si	psf	atm	kg/cm2	inHg	mmHg	ft of H$_2$O (60°F)
					Multiply by:			
psi		1.000	144.0	0.06804	0.07031	2.036	51.70	2.307
psf		6.944×10^{-3}	1.000	4.726×10^{-4}	4.882×10^{-4}	0.01414	0.3592	0.01602
atm		14.70	2,116	1.000	1.033	29.92	760.0	33.90
kg/cm2		14.22	2,048	0.9678	1.000	28.96	735.5	32.81
inHg		0.4912	70.73	0.03342	0.03453	1.000	25.40	1.133
mmHg		0.01934	2.785	1.316×10^{-3}	1.360×10^{-3}	0.03937	1.000	0.04461
ft of H$_2$O (60°F)		0.4335	62.43	0.02950	0.03048	0.8826	22.41	1.000

Rate of Flow

To convert from	To:	L/s	gal/min	gal/h	ft³/s	ft³/min	ft³/h	bbl/h	bbl/day
					Multiply by:				
L/s		1.000	15.85	951.2	0.03532	2.119	127.1	22.66	543.8
gal/min		0.06308	1.000	60.00	2.228×10^{-3}	0.1337	8.019	1.429	34.30
gal/h		1.052×10^{-3}	0.01667	1.000	3.713×10^{-5}	2.228×10^{-3}	0.1337	0.02382	0.5716
ft³/s		28.30	448.9	2.693×104	1.000	60.00	3600	641.1	1.538×104
ft³/min		0.4717	7.481	448.9	0.01667	1.000	60.00	10.69	256.5
ft³/h		7.862×10^{-3}	0.1246	7.481	2.778×10^{-4}	0.01667	1.000	0.1781	4.272
bbl/h		0.04415	0.6997	42.00	1.560×10^{-3}	0.09359	5.615	1.000	24.00
bbl/day		1.840×10^{-3}	0.02917	1.750	6.498×10^{-5}	3.899×10^{-3}	0.2340	0.04167	1.000

Energy, Heat, and Work

To convert from	To:	Btu	gcal	ft•lb	hp•h	kWh
				Multiply by:		
Btu		1.000	252.0	777.5	$3.928 \times 10-4$	$2.928 \times 10-4$
g•cal		$3.968 \times 10-3$	1.000	3.086	$1.558 \times 10-6$	$1.162 \times 10-6$
ft•lb		$1.286 \times 10-3$	0.3241	1.000	5.050×10^{-7}	3.767×10^{-7}
hp•h		2547	6.417×10^5	1.980×10^6	1.000	0.7457
kWh		3415	8.605×10^5	2.655×10^6	1.341	1.000

Power

To convert from	To:	Btu/h	ft•lb/ min	ft• lb/s	hp	kW	kg•cal/s	g•cal/s	tons of refrig.
					Multiply by:				
Btu/h		1.000	12.96	0.2160	3.928×10^{-4}	2.928×10^{-4}	6.999×10^{-5}	0.06999	8.333×10^{-5}
ft•lb/min		0.07715	1.000	0.01667	3.033×10^{-5}	2.260×10^{-5}	5.402×10^{-6}	5.402×10^{-3}	6.431×10^{-6}
ft•lb/s		4.630	60.00	1.000	1.820×10^{-3}	1.356×10^{-3}	3.241×10^{-4}	0.3241	3.858×10^{-4}
hp		2547	33,000	550.0	1.000	0.7457	0.1782	178.2	0.2122
kW		3415	44,250	737.6	1.341	1.000	0.2390	239.0	0.2845

				Power:				
To convert from	To:	Btu/h	ft•lb/ min	ft• lb/s	hp	kW	kg•cal/s	g•cal/s	tons of refrig.
					Multiply by:				
kg•cal/s		1.428×10^4	1.851×10^5	3086	5.610	4.183	1.000	1000	1.191
g•cal/s		14.28	185.1	3.086	5.610×10^{-3}	4.183×10^{-3}	0.0010	1.000	1.191×10^{-3}
tons of . refrig.		1.200×10^4	1.555×10^5	2592	4.712	3.514	0.8400	840.0	1.000

References

1. "A Survey of API Members' Aboveground Storage Tank Facilities," American Petroleum Institute, Washington, D.C., July 1994.
2. *Aboveground Storage Tank Guide,* Thompson Publishing Group, Washington, 1994.
3. *API Design, Construction, Operation, Maintenance and Inspection of Terminal and Tank Facilities,* API Standard 2610, American Petroleum Institute, Washington, D.C., 1994.
4. G. M. Barrow, *General Chemistry,* Wadsworth Publishing, Belmont, Calif., 1972.
5. R. P. Benedetti, *Flammable and Combustible Liquids Code Handbook,* 5th ed., Quincy, Mass., 1994.
6. N. H. Black, *An Introductory Course in College Physics,* rev. ed., Macmillan, New York, 19 .
7. R. A. Christensen and R. F. Eilbert, "Aboveground Storage Tank Survey," Final Report, Entropy Limited, April 1989, API.
8. R. S. Dougher and T. F. Hogarty, "The Net Social Costs of Mandating Out-of-Service Inspections of Aboveground Storage Tanks in the Petroleum Industry," API Research Study 048, American Petroleum Institute, December 1989.
9. "Estimated Costs of Benefits of Retrofitting Aboveground Petroleum Industry Storage Tanks with Release Prevention Barriers," API Research Study 065, American Petroleum Institute, September 1982.
10. *Guide for Inspection of Refinery Equipment,* 4th ed., American Petroleum Industry.
11. D. M. Himmelblau, *Basic Principles and Calculations in Chemical Engineering,* 2d ed., Prentice-Hall, Englewood Cliffs, N.J., 1967.
12. "Inland Oil Spills: Stronger Regulation and Enforcement Needed to Avoid Future Incidents," GAO, February 1989.
13. C. G. Kirkbride, *Chemical Engineering Fundamentals,* McGraw-Hill, New York, 1947.
14. William L. Leffler, *Petroleum Refining for the Non-Technical Person,* PennWell Books, Tulsa, Okla.
15. J. B. Maxwell, *Databook on Hydrocarbons,* D. Van Nostrand Co., New York, 1950.
16. NFPA 395, *Storage of Flammable and Combustible Liquids on Farms and Isolated Construction Projects,* National Fire Protection Association, Quincy, Mass.
17. NFPA 30, *Flammable and Combustible Liquids Code,* National Fire Protection Association, Quincy, Mass.
18. NFPA 30A, *Automotive and Marine Service Station Code,* National Fire Protection Association, Quincy, Mass.
19. NFPA 321, *Basic Classification of Flammable and Combustible Liquids,* 1991 ed., National Fire Protection Association, Quincy, Mass.
20. NFPA 325M, *Fire Hazard Properties of Flammable Liquids, Gases, and Volatile Solids,* 1991 ed., National Fire Protection Association, Quincy, Mass.

21. E. Oberg, F. D. Jones, and H. L. Horton, *Machinery's Handbook,* 20th ed., 4th printing, Industrial Press, New York, 1978.
22. R. H. Perry (ed.), *Chemical Engineers' Handbook,* 4th ed., McGraw-Hill, New York, 1963.
23. R. Norris Shreve, *Chemical Process Industries,* 4th ed., McGraw-Hill, New York, 1967.
24. "Technology for the Storage of Hazardous Liquids: A State-of-the-Art Review," New York State Department of Environmental Conservation, Albany, N.Y., rev. March 1985.
25. "Toxic Substance Storage Tank Containment," Ecology and Environment, Inc., Buffalo, N.Y.; and Whitman, Requardt & Associates, Baltimore, Md., Noyes Publications, 1985.

2

Codes, Standards, and Regulations

SECTION 2.1 CODES AND STANDARDS

2.1.1 Overview of Codes and Standards

Industry standards and codes have been developed primarily on a voluntary basis by national standards-setting bodies by industries affected by them. Because the standards-setting bodies in most cases represent the interests of all parties, they must be consensus standards. The standards have been relied upon heavily by both industry and authorities to ensure safe and effective equipment and designs. In fact, the purpose of these standards and codes has been to provide acceptable, practical, and useful standards that ensure quality, safety, and reliability in equipment, practices, operations, or designs.

Because of the widely varied historical beginnings of the industrial organization, numerous standards-writing bodies have been created. There has been some overlap of coverage between organizations, but on the whole, the current system in place in the United States has been extremely effective in promoting the excellent standards that have used as a model by other nations throughout the world. Table 2.1.1 is a listing of several organizations whose work is useful to the tank engineer, including cities and telephone numbers. Table 2.1.2 is a listing of some of the more important tank codes and standards produced by these organizations.

Most of these organizations continue to exist because much work is needed to maintain the standards as technology changes, as the regulations change, as environmental concerns and rules develop, and as public opinion toward the various industries changes. These organizations have periodic meetings approximately one or more times per

TABLE 2.1.1   List of Standards Organizations

Organization and location	Phone fax
American Chemical Society	(202) 872-4414
Washington, DC 20036	(202) 872-6337
American Institute of Steel Construction	(312) 670-2400
Chicago, IL 60601-2001	(312) 670-5403
American Conference of Governmental Industrial Hygienists	(513) 661-7881
Cincinnati, OH 45211-4438	(513) 661-7195
American Institute of Chemical Engineers	(212) 705-7338
New York, NY 10017	(212) 752-3294
American National Standards Institute	(212) 642-4900
New York, NY 10036	(212) 398-0023
American Petroleum Institute	(202) 682-8000
Washington, D.C. 20005	(202) 682-8031
American Society of Mechanical Engineers	(212) 705-7722
New York, NY 10017	(212) 705-7739
American Society for Testing and Materials	(215) 299-5400
Philadelphia, PA 19103-1187	(215) 977-9679
American Welding Society	(305) 443-9353
Miami, FL 33126	(305) 443-7559
Association of Iron and Steel Engineers	(412) 281-6323
Pittsburgh, PA 15222	(412) 281-4657
Building Officials and Code Administrators International	(708) 799-2300
Country Club Hills Rd, IL 60478-5795	(708) 799-4981
Chemical Manufacturers Association	(201) 887-1100
Washington, DC 20037	(202) 887-1237
International Conference of Building Officials	(310) 699-0541
Whittier, CA 90601	(310) 692-3853
National Fire Protection Association	(617) 770-3000
Quincy, MA 02259-9990	(617) 770-0700
Petroleum Equipment Institute	(918) 494-9696
Tulsa, OK 74101	
Society of the Plastics Industry	(202) 371-5200
Washington, DC 20005	(202) 371-1022
Steel Tank Institute	(708) 438-8265
Lake Zurich, IL 60047	(708) 438-8766
Underwriters Laboratories	(708) 272-8800
Northbrook, IL 60062	

year in which specialized committees and subcommittees address specific topics. Most are open to membership from all affected parties and from the membership develop consensus standards.

Most of the standards organizations produce different levels of standards which can be generalized into the following basic categories:

- *Standards.* These are considered to be mandatory practices that must be complied with so that the equipment manufactured may be considered in compliance or may be marked as complying with the standard. Standards are also often called *codes.*

TABLE 2.1.2   List of Tank-Related Standards and Codes

Organization and code no.			Title	Comment
ACI			*Guide for Protection of Concrete against Chemical Attack by Means of Coatings and Other Corrosion-Resistant Materials*	Construction
ACI			*Manual of Concrete Practices*	Construction
ACI		344	*Design and Construction of Circular Prestressed Concrete Structures*	Construction
ANSI		249.1	*Safety in Welding and Cutting*	Safety
ANSI		288.2	*American National Standard for Respiratory Protection*	Safety
ANSI		B31.3	*Chemical Plant and Petroleum Piping*	Construction
ANSI		B31.4	*Liquid Transportation Systems for Hydrocarbons, Liquid Petroleum Gas, Anhydrous Ammonia, and Alcohols*	Construction
API	Specification	12B	*Bolted Tanks for Storage of Production Liquids*	Construction
API	Specification	12D	*Field Welded Tanks for Storage of Production Liquids*	Construction
API	Specification	12F	*Shop Welded Tanks for Storage of Production Liquids*	Construction
API	Standard	620	*Design and Construction of Large Welded, Low Pressure Storage Tanks*	Construction
API	Standard	650	*Welded Steel Tanks for Oil Storage*	Construction
API	Publication	1604	*Abandonment or Removal of Used Underground Service Station Tanks*	Environmental
API	Bulletin	1615	*Installation of Underground Petroleum Storage Systems*	Construction
API	Publication	1621	*Bulk Liquid Stock Control at Retail Outlets*	Operation
API	Bulletin	1628	*Underground Spill Clean-up Manual*	Environmental
API	Standard	2000	*Venting Atmospheric and Low Pressure Storage Tanks*	Fire protection
API	Recommended Practice	2003	*Protection against Ignitions Arising out of Static, Lighting, and Stray Currents*	Fire protection
API	Publication	2009	*Safe Practices in Gas and Electric Cutting and Welding in Refineries, Gasoline Plants, Cycling Plants, and Petrochemical Plants*	Fire protection
API	Publication	2013	*Cleaning Mobile Tanks in Flammable or Combustible Liquid Service*	Operation
API	Publication	2015	*Cleaning Petroleum Storage Tanks*	Operation
API	Publication	2015A	*A Guide for Controlling the Lead Hazard Associated with Tank Cleaning and Entry*	Safety

TABLE 2.1.2   List of Tank-Related Standards and Codes (*Continued*)

Organization and code no.			Title	Comment
API	Publication	2021	*Fighting Fires in and around Flammable and Combustible Liquid Atmospheric Petroleum Storage Tanks*	Fire protection
API	Publication	2023	*Safe Storage and Handling of Petroleum-Derived Asphalt Products and Crude Oil Residues*	Safety
API		2026	*Safe Descent onto Floating Roofs of Tanks in Petroleum Service*	Safety
API		2027	*Ignition Hazards Involved in Abrasive Blasting of Atmospheric Storage Tanks in Hydrocarbon Service*	Safety
API	Bulletin	2202	*Dismantling and Disposing of Steel from Tanks which Have Contained Leaded Gasoline*	Safety
API	Publication	2207	*Preparing Tank Bottoms for Hot Work*	Fire protection, safety
API		2219	*Safe Operating Guidelines for Vacuum Trucks in Petroleum Service*	Safety
API		2220	*Improving Owner and Contractor Safety Performance*	Safety
API			*Guide for Inspection of Refinery Equipment:* *Chapter II: Conditions Causing Deterioration or Failures* *Chapter III: General Preliminary and Preparatory Work* *Chapter IV: Inspection Tools* *Chapter V: Preparation of Equipment for Safe Entry and Work* *Chapter VI: Pressure Vessels* *Chapter XI: Pipes, Valves, and Fittings* *Chapter XII: Foundations, Structures, and Buildings* *Chapter XII: Atmospheric and Low Pressure Storage Tanks* *Chapter XIV: Electrical Systems*	Inspection
API			*Chapter XV: Instruments and Control Equipment* *Chapter XVI: Pressure Relieving Devices* *Chapter XVII: Auxiliary and Miscellaneous Equipment* *Appendix: Inspection of Welding*	
API	Recommended Practice	2350	*Overfill Protection for Petroleum Storage Tanks*	Operation
API	Standard	2510	*Design and Construction of LPG Installations at Marine Terminals, Natural Gas Plants, Refineries, and Tank Farms*	

TABLE 2.1.2   List of Tank-Related Standards and Codes (*Continued*)

Organization and code no.		Title	Comment
API	2517	*Evaporation Loss for External Floating Roof Tanks*	Environmental
API	2518	*Evaporation Loss From Fixed Roof Tanks*	
API	2519	*Evaporation Loss from International Floating Roof Tanks*	Environmental
API		*Guide for Follow-up Inspection of Interior Tank Coatings*	Inspection
ASME		*Boiler and Pressure Vessel Code: Section II: Materials Specifications Section V: Nondestructive Examination Section VIII: Pressure Vessels Section X: FRP Pressure Vessels*	Construction
AWWA	D100 -84	*AWWA Standard for Welded Steel Tanks for Water Storage*	Construction
AWWA		*Standard for Painting and Repainting Steel Tanks, Standpipes, Reservoirs, and Elevated Tanks for Water Storage*	Construction
CGA	G-7.2	*Commodity Specification*	
NACE	RP-01-69	*Control of External Corrosion on Underground or Submerged Metallic Piping Systems*	Construction
NACE	No. 1	*Surface Preparation for Tank Linings*	Construction
NACE	No. 2	*Surface Preparation for Some Tank Linings and Heavy Maintenance*	Construction
NACE	No. 3	*Surface Preparation for Maintenance*	Maintenance
NACE	No. 4	*Surface Preparation for Very Light Maintenance*	Maintenance
NACE	RP-03-72	*Method for Lining Lease Production Tanks with Coal Tar Epoxy*	Construction
NACE	RP-0193-93	*External Cathodic Protection of On-Grade Metallic Storage Tank Bottoms*	Construction
NFPA	11	*Foam Extinguishing Systems*	Fire protection
NFPA	11A	*High Expansion Foam Systems*	Fire protection
NFPA	11B	*Synthetic Foam and Combined Agent Systems*	Fire protection
NFPA	12	*Carbon Dioxide Extinguishing Systems*	Fire protection
NFPA	12A	*Halogenated Fire Extinguishing Agent Systems*	Fire protection
NFPA	16	*Installation of Foam-Water Sprinkler Systems and Foam-Water Spray Systems*	Fire protection
NFPA	17	*Dry Chemicals Extinguishing Systems*	Fire protection

TABLE 2.1.2   List of Tank-Related Standards and Codes (*Continued*)

Organization and code no.			Title	Comment
NFPA		30	*Code for Flammable and Combustible Liquids*	Fire protection
NFPA		43A	*Liquid and Solid Oxidizing Materials*	Fire protection
NFPA		49	*Hazardous Chemical Data*	Fire protection
NFPA		58	*Storage and Handling of LPG*	Fire protection
NFPA		59	*Storage and Handling of LPG at Utility Gas Plants*	Fire protection
NFPA		68	*Explosion Venting*	Fire protection
NFPA		69	*Explosion Preventing Systems*	Fire protection
NFPA		70	*National Electrical Code*	Fire protection
NFPA		72A	*Installation, Maintenance, and Use of Local Proprietary Protective Signaling Systems*	Fire protection
NFPA		72C	*Installation, Maintenance, and Use of Auxiliary Protective Signaling Systems*	Fire protection
NFPA		72D	*Installation, Maintenance, and Use of Proprietary Protective Signaling Systems*	Fire protection
NFPA		72E	*Automatic Fire Detectors*	Fire protection
NFPA		77	*Recommended Practice on Static Electricity*	Fire protection
NFPA		78	*Lighting Protection Code*	Fire protection
NFPA		231	*General Indoor Storage*	Fire protection
NFPA		231A	*General Outdoor Storage*	Fire protection
NFPA		325M	*Fire Hazard Properties of Flammable Liquids*	Fire protection
NFPA		327	*Cleaning Small Tanks and Containers*	Fire protection
NFPA		329	*Underground Leakage of Flammable and Combustible Liquids*	Fire protection
NFPA		419M	*Code for Explosive Materials*	Fire protection
NFPA		495	*Identification of Fire Hazards of Materials*	Fire protection
NFPA		1221	*Installation, Maintenance, and Use of Public Fire Service Communications*	Fire protection
PEI	Recommended Practice	200	*Recommended Practices for Installation of Aboveground Petroleum Storage Systems for Motor Vehicle Fueling*	Construction
SSPC		5-63	*White Metal Blast*	Construction
SSPC		10-63	*Near-White Metal Blast*	Construction
SSPC		6-63	*Commercial Blast*	Construction
SSPC		7-63	*Brush Off Blast*	Construction
STI	Recommended Practice	892	*Recommended Practice for Corrosion Protection of Underground Piping Networks Associated with Liquid Storage and Dispensing Systems*	Construction

TABLE 2.1.2   List of Tank-Related Standards and Codes (*Continued*)

Organization and code no.			Title	Comment
STI	Recommended Practice	893	*Recommended Practice for External Corrosion Protection of Shop Fabricated Aboveground Storage Tank Floors*	Construction
STI		F911	*Standard for Diked Aboveground Steel Tanks*	Construction
STI		R912	*Installation Instructions for Factory Fabricated Aboveground Storage Tanks*	Construction
STI		F921	*Standard for Aboveground Tanks with Integral Secondary Containment*	Construction
STI		R931	*Double Wall Aboveground Storage Tank Installation and Testing Instructions*	Construction
UL	UL 142		*Steel Aboveground Tanks for Flammable and Combustible Liquids*	Construction
UL	UL 2085		*Insulated Aboveground Tanks for Flammable and Combustible Liquids*	Construction

Organizations:
ACI   American Concrete Institute
API   American Petroleum Institute
ASME   American Society of Mechanical Engineers
AWWA   American Water Works Association
NACE   National Association of Corrosion Engineers
NFPA   National Fire Protection Association
SSPC   Steel Structures Painting Council
STI   Steel Tank Institute
UL   Underwriter Laboratories

- *Recommended Practices (RPs).*   These are advisory documents that provide technological background and practices which may be useful for the specific application at hand.

- *Publications or bulletins.*   These are primarily for the purpose of informing the user of general aspects of the industry technology or practices.

- *Specifications.*   They are considered interchangeable with standards. Specifications may also be a component of standards or codes.

The use of the word *shall* or *should* has become extremely important for any organization involved with standards maintenance. When a standard is legally mandatory, the word *shall* has a legal basis for enforcing compliance whereas *should* is considered advisory. Most standards make use of the word *shall* where no exceptions are allowed. The *should* requirements tend to apply to situations where the applicability of the standard requirement is dependent on the specific circumstances. In these cases, good practices and engineering judgment are used to determine whether compliance is required.

2.1.2   The Organizations

Although many of the organizations listed in Table 2.1.1 have contributed in some way to storage tank technology, there is space here only to mention the most important ones.

2.1.2.1   The American Petroleum Institute

The American Petroleum Institute (API) was founded in 1919 to represent the domestic petroleum industry and to establish standardization in this industry. The API has a membership of over 250 countries that are involved with the exploration, producing, transportation, refining, and marketing of petroleum and its products.

The API has played a key role in the technology that affects storage tanks and has done much to improve the safety and environmental reliability as well. An indication of the contribution of API to storage tank technology can be gleaned by perusing Table 2.1.2.

2.1.2.2   The fire protection organizations and codes (application and jurisdiction of U.S. fire codes)

There are two established, recognized fire protection organizations that are currently used to produce the following codes in the United States:

1. Uniform Fire Code (UFC)

2. National Fire Protection Association (NFPA)

Uniform Fire Code.   The UFC is a product sponsored by the International Fire Code Institute (IFCI). This organization was created by the Western Fire Chiefs Association (WFCA) and the International Conference of Building Officials (ICBO) in 1991 and is made up only of fire chiefs. Its rules and standards are based solely upon the voting of the member chiefs. It is updated once every year. Since this code is controlled solely by the members from WFCA and

ICBO, it is not a consensus code and represents a point of view of only these organizations.

The NFPA. The NFPA is a consensus code development and maintenance organization made up of state and local fire marshals; volunteer experts from state, local, and federal agencies; educational institutions; insurers; manufacturers; fire protection engineers; industrial representatives; and a host of others. The approximately 280 NFPA codes are updated every 3 years and are written and maintained by over 200 technical committees.

Neither the UFC nor the NFPA fire code in and of itself has authority. However, by citation by the local regulatory agency one or the other of these two codes applies as regulation. The local agency that usually decides which code applies is the agency responsible for the building standards and codes. Some states such as California have opted to use the UFC. Others such as Louisiana are under the jurisdiction of NFPA. However, the jurisdictional limits are not necessarily drawn by state. Some states have both codes applicable within different geographic areas of the state. Although most jurisdictions adopt one or the other of these codes verbatim, many modify these codes with additional requirements or changes to provide for the unique circumstances of the jurisdiction.

Because of possible conflicts, there are no cases where both codes simultaneously apply to the same geographic region.

NFPA codes work hand in hand with American Petroleum Institute codes when combustible or flammable liquid storage is involved. API and NFPA codes tend to be complementary and provide an excellent technological foundation for the safe storage of liquids. Most local regulation of storage tanks is based upon these NFPA codes.

2.1.2.3 American Society of Mechanical Engineers

This technical society of mechanical engineers and students conducts research and develops boiler, pressure vessel, and power codes. It develops safety codes and standards for equipment and conducts short course programs and the Identifying Research Needs Program. The ASME maintains 19 research committees and 38 divisions.

SECTION 2.2 REGULATIONS AND LAWS

2.2.1 Introduction

Understanding the principles and structure of the federal laws has become necessary for industrial personnel such as managers, engineers, and operations people to effectively cope with and be effective in their jobs. Although the regulatory controls have largely bypassed aboveground storage tanks, they may soon permeate every aspect of tank design, installation, maintenance, inspection, and operation. The information in this section is intended to provide the framework of understanding needed to effectively deal with environmental issues.

2.2.2 Historical Background

The federal statutes have roots that go back as far as the 18th century; however, only in the last few decades have they begun to create a significant impact on the way industry operates. In the 1950s and 1960s, public awareness about the effects of pollution on health and the environment triggered scientific research into the effects of the limited resources of the world and the delicate ecosystems and life systems that depend on each other for survival. At the same time, various organizations advocating the control of industrial operations to reduce or eliminate pollution increased in number and visibility. There was initially significant resistance by industry to make any changes that could raise the costs of production. However, today, environmental compliance is a fact of doing business, and the business communities seek to achieve the goals of compliance by the most efficient means possible.

2.2.3 The Legal Effect of Industry Codes and Standards

The are numerous industry organizations which publish standards and codes that constitute good practice, such as the API and NFPA. Although many of these are voluntary, they may be substantially binding as they are often incorporated by reference into statutes, regulations, or permits. They also are used to benchmark whether there has been evidence of reasonable care or the absence of it in negligence actions involving criminal and civil liability.

When private codes and standards are named in regulations, statutes, or permits, they, too, have the same effect as if they were government standards. Therefore, violation of incorporated codes and standards can lead to the same liabilities as violation of government statutes, regulations, or requirements. Even though private codes are not enforceable, failure to comply with them may bring the same liabilities.

In civil cases involving negligence, the defendant is held to a standard of reasonable care which is usually defined as what a reasonable person in that industry would do. Whether a defendant has been negligent is determined by a court or jury. In making this determination, evidence of reasonable care is often determined by whether codes and standards normally used were observed or violated.

2.2.4 Uniform Regulatory Programs for Aboveground Storage Tanks

At present there is no uniform federal program which regulates aboveground storage tanks. Instead, there is a complex, confusing, and overlapping network of miscellaneous federal statutes and regulations that directly or indirectly govern tanks as well as local requirements imposed by state and local authorities. For the most part, the applicable rules are determined by tank content, size, and location.

Several bills have been unsuccessfully introduced into Congress. In 1993 the Daschle bill was introduced to regulate tanks to include release detection, secondary containment, and inspection. The Moran and Robb bills were also introduced that would impose permit fees, release detection, inspections, and corrosion prevention measures as well as reporting requirements. EPA has also proposed several regulations that affect storage tanks.

These bills were probably introduced because major incidents involving groundwater contamination by hydrocarbons occurred in areas represented by the bill's sponsors.

Incidents involving tank leaks and failures have contributed to the increased probability that tanks will eventually be regulated by a uniform federal program that would probably be similar in nature to the current underground storage tank program (Subtitle I of the Resource Conservation and Recovery Act, 42 USC 6991). This program will most likely have provisions for tank registration and fees, financial assurance, release detection, leak detection, commissioning and decommissioning procedures, etc.

2.2.5 Federal Rules Covering Air Pollution

The Clean Air Act is the source of nearly all federal regulations involving the regulation of air pollution from tanks. It has several distinct facets and provisions which apply to storage tanks. The rules as they apply to tanks are summarized in Tables 2.2.1, 2.2.2, and 2.2.3.

2.2.5.1 New source performance standards

New Source Performance Standards (NSPS) rules (EPA Rule 40 CFR 60) apply to new sources or sources that have undergone major modi-

TABLE 2.2.1   Summary of Exemption Levels under Various CAA Rules

Tank Capacity and Vapor Pressure Cutoffs (gal, psia)

Regulation	Source category	0< <10,000 gal	10,000≤ <20,000 gal	20,000≤ <40,000 gal	40,000≤	47,000≤
40 CFR 60 Kb (new tanks)	VOL storage	N/A	N/A	4.0 psia	0.75 psia	
40 CFR 61 Y	Benzene storage	N/A	All vapor pressures	All vapor pressures	All vapor pressures	
HONeshaps (existing tanks)	SOCMI	N/A	N/A	1.9 psig	0.75 psia	
HONeshaps (next tanks)	SOCMI	N/A	1.9 psia	1.9 psia	0.10 psia	
NESHAP (MACT floor)	Petroleum refineries	Group 1	Group 1	Group 1	Group 1	1.5 psia (Group 2)

Notes:

Group 1 tanks have no requirements other than documentation; Group 2 tanks are subject to refinery MACT requirements. Tanks under 10,000 gal are exempt from all requirements.

VOL = volatile organic liquids

SOCMI = Synthetic Organic Chemical Manufacturing Industry

MACT = Maximum achievable control technology

fications since July 23, 1984. The NSPS standards cover tanks which store volatile organic liquids. The rules have specific criteria for size and vapor pressure which limit their scope.

2.2.5.2   Reasonably available control technology (RACT)

RACT rules apply to existing tanks depending upon the attainment status (degree of air quality) for the location considered. They apply to tanks storing volatile organic compounds.

2.2.5.3   National Emission Standards for Hazardous Air Pollution (NESHAP)

NESHAP rules apply to both new and existing tanks storing any of 189 listed hazardous air pollutant compounds. Maximum achievable control technology (MACT) is applied to the fittings of these tanks to minimize emissions of the listed compounds. Whether a tank is regulated depends on how much of the hazardous air pollutant (HAP) it is emitting on an annual basis.

2.2.5.4   SARA, Title III and CERCLA

These are rules that address requirements for spill reporting and response for a list of hazardous substances based on whether a minimum threshold is located on the site.

TABLE 2.2.2 Summary of Rim Seal Requirements under Various CAA Rules

Regulation	Source category	Tank type	Seal system primary seal type.				Seal gap measurement	Visual from fixed-roof hatch	Internal out-of-service	Notes
			Vapor-mtd	Mech-shoe	Liq-mtd	Notes				
40 CFR 60 Kb (NSPS)	VOL storage	EFRT		ok w/sec	ok w/sec		Annual			
		IFRT	ok w/sec	ok alone	ok alone			Annual	10 years	†
40 CFR 61 Y	Benzene storage	EFRT		ok w/sec	ok w/sec		Annual			
		IFRT	ok w/sec	ok alone	ok alone	*		Annual	10 years	†
HONeshaps	SOCMI	EFRT		ok w/sec	ok w/sec		Annual			
		IFRT	ok w/sec	ok alone	ok alone			Annual	10 years	†
		EFRT w/dome	ok w/sec	ok alone	ok alone			Annual	10 years	†

*Tanks with IFRs and continuous seals installed prior to July 28, 1988 are not required to add secondary seals.
†IFRTs with double-seal systems are allowed the option of 5 years between visual inspections, but the internal inspection must also then be on a 5-year schedule.

TABLE 2.2.3 Summary of Required Fittings Control under Various CAA Rules

Fitting Controls—Variances from Kb

Regulation	Source category	Tank type	Sloted guide pole		Unslotted guide pole	Access hatch	Automatic guage (gauge float well)	Sample well (sample pipe/wall)
			Gasketed cover?	Internal float?				
40 CFR 60 Kb (NSPS)	VOL storage	EFRT	Req'd.	*	Gasketed cover	Gasketed	Gasketed	Gasketed cover
		IFRT	Req'd.	*	Gasketed cover	Bolted, gasketed	Bolted, gasketed	Slit-fabric well only
40 CFR 61 Y	Benzene storage	EFRT	Req'd.	*	Gasketed cover	Gasketed	Gasketed	Gasketed cover
		IFRT	Req'd.	*	Gasketed cover	Bolted, gasketed	Bolted, gasketed	Slit-fabric well only
HONeshaps	SOCMI	EFRT	Req'd.	Req'd.	Gasketed cover	Bolted, gasketed	Bolted, gasketed	Gasketed cover
		IFRT	Req'd.	*	Gasketed cover	Bolted, gasketed	Bolted, gasketed	Slit-fabric well only
		EFRT w/dome	Req'd.	Req'd.	Gasketed cover	Bolted, gasketed	Bolted, gasketed	Slit-fabric well only

*The regulation does not expressly require slotted guide poles to have floats, but it does require that all openings be "…equipped with a gasketed cover, seal or lid…(i.e., no visible gaps)." This language has been construed to require floats in slotted guide poles.

†The requirement for all openings to be equipped with a gasketed cover applies to all other openings except those for roof drains and leg sleeves.

‡Roof drains on EFRTs which open directly into the stored product are to have a slotted-membrane fabric cover.

2.2.5.5 Federal rules covering water pollution

The Clean Water Act (CWA) is the counterpart of the Clean Air Act for water pollution control technology. Again, there are several components that bear on the design or operation of storage tank facilities.

2.2.5.6 Spill Prevention, Control, and Countermeasures (SPCC) Program, 40 CFR 112

This applies to tank facilities which are near navigable waters. The definition applies to almost all facilities and essentially requires that tanks storing petroleum compounds be surrounded with a secondary containment area that can be used to prevent leaks or spills from escaping to the waters of the United States.

2.2.5.7 Oil Pollution Act of 1990 (OPA90)

This act requires facilities that store petroleum products or hazardous substances to have response plans for onshore facilities as well as to assume financial responsibility for worst-case spill scenarios.

2.2.5.8 National Pollution Discharge Elimination System (NPDES)

It applies to water quality of storm water runoff including the runoff from tank facilities. It is essentially a monitoring and permitting control system.

2.2.5.9 OSHA rules

The labor law provisions of OSHA specifically impact storage tanks.

OSHA Flammable and Combustible Liquid Standard 29 CFR 1910.106. This covers the storage of liquids with flash points below 200°F. It essentially duplicates the flammable and combustible liquid storage requirements in the national fire codes.

OSHA Standard for Occupational Exposure to Benzene, 29 CFR 1010.1028. It has specific provisions regarding the storage of benzene in tanks.

OSHA Chemical Process Safety Management (PSM) Standard 29 CFR 119. This requires in-depth "what if" analysis of the potential for accidents, explosions, or fires by a systematic review of process data and equipment flow diagrams. Tanks, if connected via piping, to the process are considered subject to these rules. These rules also contain requirements for good engineering practice such as performing periodic inspections of tanks and the documentation of the findings and results of the inspections.

OSHA confined-space entry rules. On January 14, 1993, OSHA adopted the final confined-space entry rules promulgated in 58 FR 4462 which required compliance on April 15, 1993 (29 CFR 1910.146). These rules require employees to set up ccomprehensive programs at all facilities to include identification, testing, permitting, training, emergency response, rescue, and other activities associated with entry into confined spaces. Figures 2.2.1 and 2.2.2 help specify and identify requirements for aboveground storage tanks.

2.2.5.10   Codes and standards

Codes and standards are developed and maintained by a handful of organizations. The organizations that have had probably the greatest impact on the development of standards directly related to storage vessels are

- National Fire Protection Association
- American Petroleum Institute
- American Society of Mechanical Engineers

2.2.5.11   Outlook

The common thread behind all the regulations governing aboveground storage tanks is that there are several control options:

1. External floating roof with primary and secondary seal
2. Internal floating roof with one or two seals
3. Fixed roof with vapor recovery

It does not appear to be the intent of the regulators to limit the application of these three technologies to particular processes or construction sites. Instead, the driving force behind these regulations is to eliminate the storage of volatile organic or hazardous materials in fixed-roof storage tanks without the above-listed control options. Therefore, it is incumbent upon the owner or operator to economize, negotiate, and implement the most strategic, cost-effective, and ver-

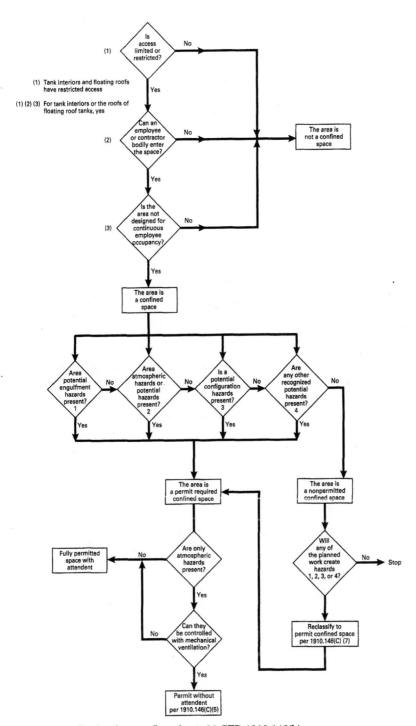

Figure 2.2.1   Confined-space flow chart, 29 CFR 1910.146(b).

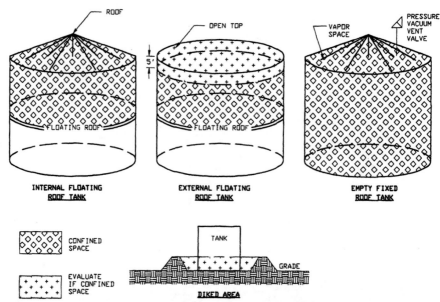

Figure 2.2.2 Confined spaces for tanks.

satile control option for each specific application. In other words, the owner or operator must show that the emission control option selected for a specific installation represents the best practices, given all factors taken into consideration including emissions, costs, existing processes, and facilities.

3

Materials Considerations

3.1 Materials Selection

This section is based upon API 650, but the principles are applicable to other codes such as API 620 or AWWA D-100.

Carbon steels used for storage tanks have specified minimum yield strengths of approximately 30 to 60 kilopounds per square inch (ksi). Tables 3.1 to 3.3 give the some acceptable steels generally used in tanks fabricated in accordance with API 650 as well as compositions and design properties.

Unless special circumstances require a specific material, the buyer of a tank should leave the carbon steel material selection to the tank supplier. The supplier will take into account factors affecting the material selection such as allowable stresses, plate cost, availability, as well as purchasing terms and mill delivery in order to provide the lowest-cost tank that meets the buyer's requirements.

3.2 Principles of Material Selection

3.2.1 Experience

Probably the most appropriate means of selecting a steel that is appropriate for the tank and its service conditions is to rely on past experience with similar conditions. Most tanks are built of plain carbon steels, although the chemical industry uses both stainless and aluminum storage tanks as well as nonmetallic tanks. In cryogenic applications, nickel-steels, aluminum, or stainless steel may be used.

3.2.2 Code requirements

The code that governs tank design often has very specific material selection requirements and limitations. Most modern codes have

TABLE 3.1 API 650 Steel Plate Requirements

Group	Plate specification	Grade	Requirements	Thickness limit, in	ASTM spec.
I	A 283	C		<1½	A 6
As rolled, semikilled, or killed	A 285	C		<1	A 20
	A 131	A		<½	A 6
	A 36	—		<½	A 6
II	A 131	B		<1	A 6
As rolled, semikilled, or killed	A 36	Modified	Mn 0.8–1.2%	<½	—
	A 442	55		<1½	A 20
	A 442	60		<1½	A 20
III	A 573	58		<1½*	A 20
As rolled, killed, fine-grain practice	A 516	55		<1½*	A 20
	A 516	60		<1½*	A 20
IIIA	A 131	CS		<1½†	A 6
Normalized, killed, fine-grain practice	A 573	58		<1½*	A 20
	A 516	55		<1½*	A 20
	A 516	60		<1½*	A 20
IV	A 573	65		<1½*	A 6
As rolled, killed, fine-grain practice	A 573	70		<1½*	A 6
	A 516	65		<1½*	A 20
	A 573	70		<1½*	A 20
	A 662	B		<1½*	A 20
IVA	A 662	C		<1½	A 20
As rolled, killed, fine-grain practice	A 573	70	$C_{max} = 0.2\%, Mn_{max} = 1.6\%$	<1½*	A 6
V	A 573	70		<1½	A 6
Normalized, killed, fine-grain practice	A 516	65		<1½*	A 20
	A 516	70		<1½*	A 20

Normalized or quenched and tempered, killed, fine-grain practice, reduced C	A 633	C		<1¾‡	A 6
	A 633	D½		<1¾‡	A 6
	A 537	I		<1¾‡	A 20
	A 537	II		<1¾‡	A 20
	A 678	A	Quenched and tempered	<1⅜§	A 6
	A 678	B	Quenched and tempered	<1⅜§	A 6
	A 737	B	Quenched and tempered	<1½	A 20

*Insert plates and flanges up to 4 in.
†Insert plates and flanges up to 2 in.
‡Insert plates up to 4 in.
§Insert plates up to 2½ in.

Generic ASTM Specifications:

A6   Rolled steel structural plate, shapes, sheet piling, and bars, generic
A20   Steel plate for pressure vessels, generic
A29   Carbon and alloy steel bars, hot rolled and cold finished, generic
A505 Alloy steel sheet and strip, hot rolled and cold rolled, generic
A510 Carbon steel wire rod and coarse round wire, generic
A568 Carbon and HSLA, hot rolled and cold rolled steel sheet and hot rolled strip, generic
A646 Premium-quality alloy steel blooms and billets for aircraft and aerospace forgings
A711 Carbon and alloy steel blooms, billets and slabs for forging

Note: All plates greater than 1½ in. thick shall be killed, made to fine-grain practice, normalized, and tempered or quenched and tempered, and shall be impact-tested.

TABLE 3.2 API 650 Steel Plate Compositions

Group	Specification	Grade	Composition										
			C_{max}	Mn_{max}	P_{max}	S_{max}	Si_{max}	Cu_{max}	Cr_{max}§	Ni_{max}§	Mo_{max}§	V_{max}§	Nb_{max}§
I As rolled, semikilled	A 283	C	0.24	0.90	0.04	0.05	0.05	0.20†					
	A 285	C	0.28	0.90	0.35	0.04							
	A 131	A	0.26		0.05	0.05	0.05						
	A 36		0.25		0.04	0.05	0.05	.20†					
II As rolled, semikilled or killed	A 131	B	0.21	0.8–1.1	0.04	0.04	0.35						
	A 36	Modified	0.25	0.8–1.2	0.04	0.05	—						
	A 442	55	0.24*	0.8–1.1*	0.05	0.04	0.15–0.40						
		60	0.27*	0.8–1.1*	0.05	0.04	0.15–0.40						
III As rolled, killed, fine-grain practice	A 573	58	0.23	0.6–0.9*	0.04	0.05	0.10–0.35						
	A 516	55	0.20	0.6–1.2*	0.04	0.04	0.15–0.40						
	A 516	60	0.23	0.6–1.2*	0.035	0.04	0.15–0.40						
IIIA Normalized, killed, fine-grain practice	A 131	CS	0.16	1.0–1.35*	0.04	0.04	0.10–0.35						
	A 573	58	0.23	0.06–0.09*	0.04	0.05	0.10–0.35						
	A 516	55	0.20*	0.06–0.98*	0.035	0.04	0.15–0.40						
	A 516	60	0.23*	0.06–0.98*	0.035	0.04	0.15–0.40						
IV As rolled, killed, fine-grain practice	A 573	65	0.24	0.85–1.2	0.04	0.05	0.15–0.40						
	A 573	70	0.27	0.85–1.20	0.04	0.05	0.15–0.40						
	A 516	65	0.26	0.85–1.20	0.035	0.04	0.15–0.40						
	A 516	70	0.28	0.85–1.5	0.035	0.04	0.15–0.40						
	A 662	B	0.19	0.85–1.5	0.035	0.04	0.15–0.40						
IVA As rolled, killed, fine-grain practice	A 662	C	0.20	1.0–1.6	0.035	0.04	0.15–0.50						
	A 573	70	0.28*	0.85–1.2	0.04	0.05	0.15–0.40						
V Normalized, killed, fine-grain practice	A 573	70	0.28	0.85–1.2	0.04	0.05	0.15–0.40						
	A 516	65	0.26	0.85–1.20	0.035	0.04	0.15–0.40						
	A 516	70	0.31	0.85–1.25	0.035	0.04	0.15–0.40						

	A 131	EH 36	0.18	0.9–1.6	0.04	0.04	0.04	.1–.5	.35	.25	.40	.08	.1	0.05
VI Normalized, quenched and tempered, killed, fine-grain practice, reduced C														
	A 633	C	0.20	1.15–1.5	—	0.04	0.05	—	—					0.05
	A 633	D	0.20	0.7–1.6	—	0.04	0.05	—	0.35	0.25	0.25	0.08	0.15	—
	A 537	I	0.24	0.7–1.6	0.035	0.04	0.15–0.5		0.35	0.25	0.25	0.08	0.5	—
	A 537	II	0.24	0.7–1.6	0.035	0.04	0.15–0.5		0.35	0.25	0.25	0.08	—	—
	A 678	A	0.16	0.9–10.5	0.04	0.05	0.15–0.5		0.20†					
	A 678	B	0.20	0.9–10.5	0.04	0.05	0.15–0.5		0.20†					
	A 737	B	0.20	1.15–1.5	0.035	1.30	0.15–0.5							

Maximum Permissible Alloy Content

Alloy	Heat analysis, %	Notes
Columbium	0.05	1, 2, 3
Vanadium	0.10	1, 2, 4
Columbium (£0.05%) plus vanadium	0.10	1, 2, 3
Nitrogen plus vanadium	0.015	1, 2, 4
Copper	0.35	1, 2
Nickel	0.50	1, 2
Chromium	0.25	1, 2
Molybdenum	0.08	1, 2

Notes:

1. When not included in the material specification, the use of these alloys, or combinations of them, shall be at the option of the plate producer, subject to the approval of the purchaser. These elements shall be reported when requested by the purchaser.

2. On product analysis, the material shall conform to these requirements, subject to the product analysis tolerances of the specification.

3. When columbium is added either singly or in combination with vanadium, it shall be restricted to plates of 0.50-in. max. thickness, unless combined with 0.15% minimum silicon.

4. When nitrogen (£0.015%) is added as a supplement to vanadium, it shall be reported, and the minimum ratio of vanadium to nitrogen shall be 4:1.

*Limiting values vary with plate thickness. Maxima listed here are based on maximum plate thickness shown in Table 3.1.

†Applicable only if copper bearing steel specified.

‡Upper limit of Mn may be exceeded provided C + Mn £ 0.4%

§These elements may not be reported on the mill sheet unless intentionally added.

TABLE 3.3   API 650 Steel Plate Design Stresses, ksi

Group	Specification	Grade	Min. yield strength	Min. tensile strength	Product design stress	Hydrostatic test stress
I As rolled, semikilled	A 283	C	30	55	20.0	22.5
	A 285	C	30	55	20.0	22.5
	A 131	A	34	58	22.7	24.9
	A 36		36	58	23.2	24.9
II As rolled, semikilled or killed	A 131		34	58	22.7	24.9
	A 36	Modified	36	58	23.2	24.9
	A 442	55	30	55	20.0	22.5
	A 442	60			21.3	24.0
III As rolled, killed, fine-grain practice	A 573	58	32	58	21.3	24.0
	A 516	55	30	55	20.0	22.5
	A 516	60	32	60	21.3	24.0
IIIA Normalized, killed, fine-grain practice	A 131	CS	34	58	22.7	24.9
	A 573	58	32	58	21.3	24.0
	A 516	55	30	55	20.0	22.5
	A 516	60	32	60	21.3	24.0
IV As rolled, killed, fine-grain practice	A 573	65	35	65	23.3	26.3
	A 573	70	42	70	28.0	30.0
IVA As rolled, killed, fine-grain practice	A 662	C	43	70*	28.0	30.0
	A 573	70	42	70*	28.0	30.0
V Normalized, killed, fine-grain practice	A 573	70	42	70*	28.0	30.0
	A 516	65	35	65	23.3	26.3
	A 516	70	38	70	25.3	28.5
VI Normalized, quenched and tempered, killed, fine-grain practice, reduced C	A 131	EH 36	51	71*	28.4	30.4
	A 633	C	50	70*	28.0	30.0
	A 633	D	50	70*	28.0	30.0
	A 537	I	50	70*	28.0	30.0
	A 537	II	60	80†	32.0	34.3
	A 678	A	50	70*	28.0	30.0
	A 678	B	60	80†	32.0	34.3
	A 737	B	50	70*	28.0	30.0

*By agreement between purchaser and manufacturer, the tensile strength may be increased to 75 ksi minimum and 90 ksi maximum. The design allowable stress is then two-thirds of yield strength or two-fifths of tensile strength, whichever is less. The hydrostatic test stress is three-fourths of yield strength or three-sevenths of tensile strength, whichever is less.

†By agreement between purchaser and manufacturer, the tensile strength may be increased to 85 ksi minimum and 100 ksi maximum. The design allowable stress is then two-thirds of yield strength or two-fifths of tensile strength, whichever is less. The hydrostatic test stress is three-fourths of yield strength or three-sevenths of tensile strength, whichever is less.

included provisions in the material selection criteria that ensure materials with sufficient toughness under the service conditions to prevent brittle fracture.

The construction of new tanks in the different temperature categories is covered by API Standards 650 and 620. API Standard 650 covers the construction of storage tanks for ambient temperature operation. Appendix M of API 650 covers tank design temperatures between 200°F and 500°F by applying a derating factor for the allowable stresses of the steels. For low-temperature tanks, API Standard 620 gives specific material requirements that allow for operation to as low as −270°F.

3.2.3 Brittle fracture

The susceptibility of the material to brittle fracture is one of the most important material selection considerations. Brittle fracture is the tensile failure of a material showing little deformation or yielding. Brittle fractures typically start at a flaw and can propagate at speeds of up to 7000 feet per second (fps), resulting in catastrophic failures such as the Ashland Oil Company incident in 1988. Almost all major tank design codes have been revised to reflect the current understanding of brittle fracture.

3.2.4 Corrosion

Corrosive effects in tanks may be divided into internal and external ones. External corrosion is usually minimized by the use of coatings for carbon steel tanks. Although more corrosion-resistant materials such as stainless steel, aluminum, and plastics often do not require external coatings to resist atmospheric corrosion, in some circumstances external coatings are required. Examples are the coating of stainless steel in warm seacoast environments to prevent pitting and stress corrosion cracking and the coating of plastics to prevent ultraviolet degradation.

3.2.5 Other factors

Factors such as material toughness and strength can also affect material selection. An example is the need to have resistance to brittle fracture in tanks used at cryogenic temperatures. Nickel-steel, stainless steel, or aluminum is usually selected for these applications. Also, factors such as material availability, fabricability, and cost can affect the choice of materials.

3.3   Carbon Steel Selection Procedure

Steel selection generally entails the following three steps.

3.3.1   Determine the design metal temperature

The selection of a design metal temperature is the first and most important step in ensuring that materials are selected which are tough enough to prevent brittle fractures under the service conditions. Brittle fracture is covered in detail below.

The *design metal temperature* (DMT) is the basis for establishing the required toughness for the selected steel. It may be determined by measurement or by experience. However, all major codes accept the use of the procedure for determining the DMT described below, in the absence of better data. The standards set the design metal temperature at the lowest one-day mean ambient temperature (LODMAT) for a given location plus 15°F. Figure 3.1 shows the LODMAT lines for the United States and parts of Canada.

To use this chart, the temperature is read or interpolated directly for the selected location. To this number 15°F is added. The addition of 15°F accounts for the fact that the liquid contents act as a large thermal mass and prevent the tank wall from actually dropping to the ambient air temperature. The resulting temperature is the design metal temperature. This temperature is then used to select the steel for the tank or to assess an existing tank's suitability for service. This chart should not be used for refrigerated tanks since the wall temperature will be controlled by the refrigerated liquid temperature.

3.3.2   Determine which materials are acceptable

Ensuring a material with adequate toughness.   There are two ways to ensure that a selected steel has adequate toughness for the design metal temperature of the tank. The first is to proof-test each plate by impact toughness testing (using the Charpy V-notch method) samples at or below the DMT.

The preferred design approach is to select a material whose transition temperature is below the design metal temperature, so that impact testing is not required. The transition temperature is the temperature above which the material is ductile, and failure can be assumed to be in the ductile mode (as opposed to brittle mode). Most companies prefer to use a steel with inherent toughness because the overall cost of impact testing is high. It adds a premium of about 10 percent to the cost of the plate. In addition, mills will not normally guarantee steel shipments that are not normalized, so that the risk of plate's not meeting the

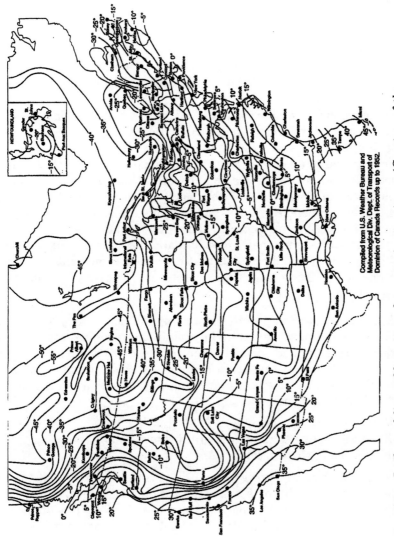

Figure 3.1 Isothermal lines of lowest one-day mean temperatures. (*Courtesy of the American Petroleum Institute, API 650, 9th ed., fig. 2-2.*)

Compiled from U.S. Weather Bureau and
Meteorological Div. Dept. of Transport of
Dominion of Canada Records up to 1952.

required toughness would be that of the tank manufacturer. The steel toughness depends on temperature, plate thickness, and quality. Most common tank steels have toughness that decreases with temperature.

Note that older codes do not have toughness requirements because they were unknown at the time the codes were written. Some current codes do not have toughness requirements. For example, the API 12 series codes do not have these requirements, nor do Underwriters Laboratories (UL) tank standards. It is recommended that when no provisions are given, a code such as API 650, 7th edition or later, be used to establish the material requirements regardless of whether the code to which the tank is constructed has any material toughness requirements.

Material groupings in API 650.   API 650 groups materials into six classifications. The groupings are shown in Tables 3.1 to 3.3. The purpose of this grouping is to allow selection of a steel that has adequate toughness based upon the design metal temperature and the required thickness. Groups I to III are plain carbon steels. Groups IV to VI are higher-strength materials. Within each of the two broad groupings, the toughness increases as the group number increases. For example, group III materials are tougher than group II materials, and group VI materials are tougher than group V materials. Group VI is composed of high-strength low-alloy steels that may be quenched and tempered.

Some of the terminology that is useful in understanding this table is as follows:

Killed or semikilled: In the context of API 650, a fully killed steel is an aluminum killed steel. The aluminum deoxidizes the steel and results in a finely dispersed aluminum oxide that promotes a fine-grain structure. A fine-grain structure yields a tougher material than one with a coarse grain. Semikilled steels are silicon deoxidized, and the grain structure is larger.

Normalizing: It consists of heat-treating the steel with subsequent air cooling to refine the grain structure and improves uniformity of the microstructure.

Quenching and tempering plate: Plate is heated to a specified temperature and held for a time, then cooled at a rate sufficient to increase the strength and hardness of the steel. The quenching process produces hard, brittle material. The tempering relieves the residual stress resulting from tempering and improves ductility.

API has taken the variables of inherent material toughness (as grouped by chemistry and mill practice, described above), design metal temperature, and material thickness and created an exemption chart, reproduced here as Fig. 3.2. This chart shows which materials,

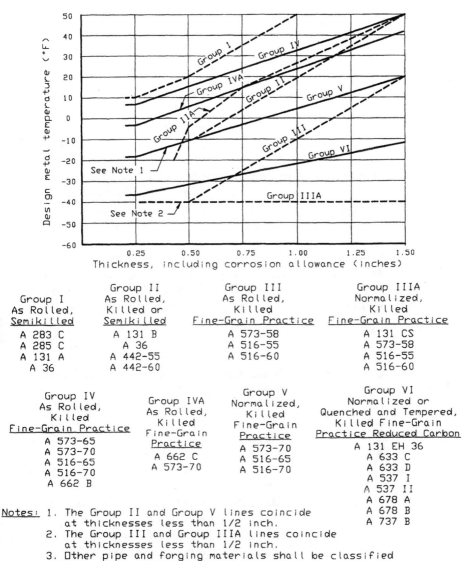

Thickness, including corrosion allowance (inches)

Group I As Rolled, Semikilled	Group II As Rolled, Killed or Semikilled	Group III As Rolled, Killed Fine-Grain Practice	Group IIIA Normalized, Killed Fine-Grain Practice
A 283 C	A 131 B	A 573-58	A 131 CS
A 285 C	A 36	A 516-55	A 573-58
A 131 A	A 442-55	A 516-60	A 516-55
A 36	A 442-60		A 516-60

Group IV As Rolled, Killed Fine-Grain Practice	Group IVA As Rolled, Killed Fine-Grain Practice	Group V Normalized, Killed Fine-Grain Practice	Group VI Normalized or Quenched and Tempered, Killed Fine-Grain Practice Reduced Carbon
A 573-65	A 662 C	A 573-70	A 131 EH 36
A 573-70	A 573-70	A 516-65	A 633 C
A 516-65		A 516-70	A 633 D
A 516-70			A 537 I
A 662 B			A 537 II
			A 678 A
			A 678 B
			A 737 B

Notes: 1. The Group II and Group V lines coincide
   at thicknesses less than 1/2 inch.
2. The Group III and Group IIIA lines coincide
   at thicknesses less than 1/2 inch.
3. Other pipe and forging materials shall be classified
   under the material groups shown:
   a. Group IIA – API Spec 5L, Grades A, B, and X42;
      ASTM A 106, Grades A and B; ASTM A 53, Grades A and B;
      ASTM A 181; and ASTM A 105.
   b. Group VI – ASTM A 524, Grades I and II.

Figure 3.2 API 650 material groups and properties. (*Courtesy of the American Petroleum Institute.*)

graded by thickness, have sufficient toughness without impact testing for a given design metal temperature. When a point is above the curve in the figure, the material is considered to have adequate toughness at that condition and impact testing is exempted. If a point falls below the curve, then impact testing is required to prove that the material has sufficient toughness to operate at the design conditions.

It is important to realize that the material toughness requirements apply to all shell plates, shell reinforcement plates, bottom sketch plates (bottom plates that have the shell welded to them), shell insert plates, and plates used for manhole and nozzle necks and plate flanges. There are no toughness requirements for roofs, floating roofs, or bottom plates (exclusive of sketch plates) either because a failure here would not result in a catastrophic loss of contents or because these areas are not stressed sufficiently to cause a potential brittle fracture.

API 650 Appendix A toughness requirements are less stringent due to the lower allowable stresses that are used. API 620 for tanks gives materials that are suitable to temperatures down to −35°F and its appendixes have provisions allowing much lower temperatures.

3.3.3 Optimize costs

Once the array of acceptable materials has been determined based on the overall code requirements, the practical issues come to the fore. These are cost, delivery, mill-supplier relationship, availability of material, any owner or operator special requirements or input, and location.

Since minimum steel mill orders are not usually placed for less than 40,000 lb, the size of the tank may require a search for existing warehoused plate. Costs for warehouse steel may be less than those for mill-ordered plate. For special materials or large quantities, mill orders are usually placed. Mill orders are almost always required for larger tanks that use the higher-strength materials.

For tanks under ½ in thick, A36 is widely used because of its low cost, weldability, and availability in plate and structural shapes. When the toughness of A36 is insufficient, then a modified A36 (specified in API 650) may be used. The modified A36 requires that the manganese be within 0.8 to 1.2 percent. The A283C is also widely used for water storage and petroleum plates, but A36 has generally become the steel of choice for tanks under ½ in thick. The A283C has a lower allowable stress and therefore may be less economical when required thicknesses are taken into consideration. The A131B is a common choice where improved toughness is required in thinner plate. The A285C and A131C are not common materials for tanks.

For tanks over ½ in, the A573-70 is a common choice because of economics. Although it costs 5 to 6 percent more, it has about a 20 percent higher allowable stress than A36 (28 versus 23.2 ksi). The A516, which has replaced its precursor, A212, is often used where good toughness is required.

For larger tanks the higher-strength steels are used because of the economies which result from the higher allowable stresses. Although a loss in toughness often goes with an increase in strength, the steels permitted in API 650 are made by practices that enhance both the strength and the notch toughness. Since there is a 1.75-in maximum limit on storage tank shells, this often limits the choice of steels in large-capacity tanks to the highest-strength steels listed in groups V and VI. These are the normalized and quenched and tempered steels made to fine-grain practice. A common steel for these applications is A537 class 2. The A633D and A633C are also used.

For some cases, the selection of a material may depend on the stress-relieving requirements for nozzles. Because postweld heat-treated insert plates are an additional cost for the tank supplier, he or she may choose to use a shell material that does not have to be stress-relieved to reduce overall costs. Where stress relieving is required for stress corrosion cracking resistance, this option should not be used.

A final, but nontrivial, matter to be understood is that different steels are frequently used in different areas of the tank. The bottom may be made of one material, and each of the shell courses may be made of other material groups on the same tank.

The tank supplier will often work up two or three designs based on different materials for a specific bid on a tank and will determine the most cost-effective alternative based upon the bids she or he receives from steel mills on each of the alternatives.

3.4 Brittle Fracture Considerations

3.4.1 Introduction

Brittle fracture of tanks can result in sudden, catastrophic failures. The most publicized case in recent history occurred in Floreffe, Pennsylvania, in January 1988. A reconstructed tank with 120-ft diameter failed in this mode during initial filling and spilled its contents into the Monongahela River. There have been numerous brittle fracture incidents. The 1988 failure is only the most recent, well-publicized one.

For brittle fracture to occur, there must be the simultaneous occurrence of three factors:

- A steel that is notch-brittle at some designated temperature
- A notch or geometric discontinuity that causes high local stresses
- A stress at the notch

The absence of any of these factors greatly reduces the probability of brittle fracture.

As material properties, *brittleness* is the opposite of *toughness*. Brittleness is the material property that indicates the propensity of a material to fracture without yielding or deformation. Although brittle fractures are infrequent, they tend to be catastrophic due to rapid fracture propagation and sudden loss of contents under pressure.

3.4.2 Inherent material toughness

In spite of the quality of material and the fabrication and inspection practices, there will be defects and flaws in the as-built tank. The notch toughness of a material is an inherent material property that indicates how resistant the tank will be to brittle fracture.

Notch toughness is the ability of a material with a flaw to absorb energy when loaded. This property is also dependent on temperature.

3.4.3 The Charpy V-notch test

In an attempt to quantify and qualify the material requirements necessary to prevent failure by brittle fracture, a number of tools have been developed from the simplest of mechanical tests to very sophisticated fracture mechanics mathematical models and analyses. For plain carbon steels the Charpy V-notch (CVN) test has proved to be the most widespread method because of its simplicity.

It is important to understand that there is no real correlation between material toughness and the CVN energy. The CVN impact energy is related to thickness, yield strength, and toughness from which the specimen came. Generally, higher-yield-strength materials will require higher CVN energies to be considered adequate for service. Similarly, specimens from thicker plates require higher test energies to ensure adequacy. The required CVN energies are based upon theoretical considerations of fracture mechanics modified by experience. The required impact test energies for API 650 tanks are shown in Fig. 3.3.

Since the energy absorbed varies with various materials, it is essential to select proper energy values based upon the material being considered. For A-283 steel at room temperature, 15 ft•lb is considered adequate.

Charpy energies are simply the most common indicator of notch toughness. Various tank codes rely on the Charpy V-notch impact test to determine the notch toughness of a plate material and its ability to resist brittle fracture at a given design metal temperature.

The CVN test procedure is outlined by the procedure in ASTM A-370. It uses a $\frac{3}{8}$-in square bar that is 2 in long with a notch machined

Plate Material and Thickness from Table 2-3 (Inches)[1]	Average of Three Specimens	
	Longitudinal (ft·lb)	Transverse (ft·lb)
Group I, II, III, IIIA		
To maximum thicknesses of	15	14
Par. 2.2.2, 2.2.3, 2.2.4, 2.2.5		
Group IV, IVA, V, VI		
(Except Quenched and Tempered)		
To 1.5 Inclusive	30	20
Over 1.5 to 1.75 Inclusive[2]	35	25
Over 1.75 to 2 Inclusive[2]	40	30
Over 2 to 4 Inclusive[2]	50	40
Group VI		
(Quenched and Tempered)		
To 1.5 Inclusive	35	25
Over 1.5 to 1.75 Inclusive[2]	40	30
Over 1.75 to 2 Inclusive[2]	45	35
Over 2 to 4 Inclusive[2]	50	40

[1]For plate ring flanges, the acceptance requirements for all thicknesses shall be those required to 1.5 inches inclusive.
[2]Interpolation permitted to nearest ft·lb.

Figure 3.3 Minimum requirements for acceptance. (*Courtesy of the American Petroleum Institute.*)

into it. See Fig. 3.4. The sample is taken from a specified location of the plate. A pendulum hammer strikes the sample, and the energy absorbed is recorded. This can be done at various temperatures. Tough materials deform more than brittle materials before breaking and therefore absorb more energy from the impact.

3.4.4 Variables affecting notch toughness

Temperature. Ferritic steels are brittle at low enough temperatures. However, they become ductile as the temperature is increased. If a plot is made measuring absorbed energy versus test temperature, a curve like the one shown in Fig. 3.5 results. This shows that at lower temperatures the notch toughness of a material is reduced, because less energy is absorbed up to the point of specimen fracture. It may, therefore, be subject to brittle fractures at lower service temperatures.

The *transition temperature* is defined as the temperature at which the fracture changes from predominantly ductile to brittle. Ductile failure has a dull appearance and occurs by shear failure. The appearance of the fracture face failing by brittle fracture is shiny. This failure mode is called *cleavage*. Another definition of transition temperature is the temperature where the distribution of shear and cleavage failure is

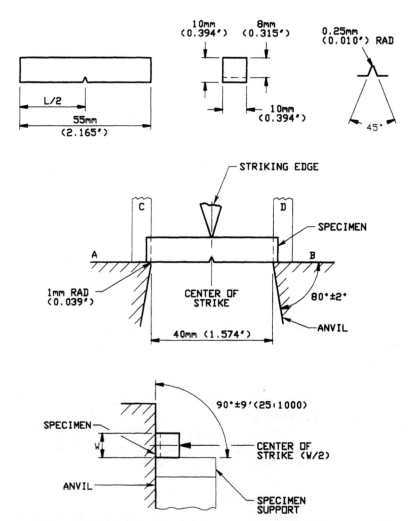

Figure 3.4 Standard Charpy V-notch impact test specimen. (*Courtesy of the American Society for Metals.*)

equal. For low-strength carbon steels, a value of approximately 15 ft•lb as measured by the CVN test has been a criterion to ensure that the metal will be used above its transition temperature. This point on the curve is not well defined but represents a temperature above which the toughness of the material improves markedly.

These are some variables that affect the transition temperature:

Composition. The carbon content of carbon steels has the greatest effect. Toughness is decreased with increasing carbon content. As manganese content is increased, the toughness increases.

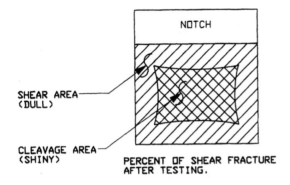

PERCENT OF SHEAR FRACTURE
AFTER TESTING.

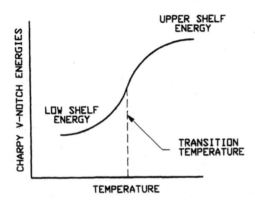

Figure 3.5 Energy transition curves.

Deoxidation. To reduce carbon levels in steel to a specific value, oxygen is mixed with the molten steel to convert the carbon to carbon monoxide, which bubbles to the surface. Steels are called *killed* if they are deoxidized because the steel lies quietly in the mold without evolution of gases. Rimmed or capped steels are not normally deoxidized by the addition of aluminum or silicon. To produce a killed steel, elements such as aluminum or silicon are added to the ladle before pouring which combine with the oxygen and end up in the slag. Fully killed steels are tougher than rimmed or capped steels. For tank plates, most codes require the use of killed steels.

Grain size. The finer the grain size, the tougher the steel. The addition of aluminum to a fully killed steel is made to "fine-grain practice." See the ASTM 20 definition.

Heat treatment. Normalized or quenched and tempered steels have lower transition temperatures than as-rolled plates. The reason is that the grain size is small and uniform.

Welding. Welding usually produces a higher transition temperature in the heat-affected zone compared to the base material. High carbon content in the base metal is detrimental because it causes brittle, hard, heat-affected zones.

Other. There are other phenomena such as temper embrittlement of low-alloy steels that are usually not a consideration for aboveground storage tanks.

Loading rate. If a Charpy impact test is run dynamically (rapid loading rates), the results will be considerably different from those run statically (low loading rates). An example of the difference in behavior of mild steel when slowly loaded or impact-loaded is shown in Fig. 3.6. This illustrates the difference between the energy absorption for a Charpy V-notch specimen that is loaded slowly and one that is loaded quickly. The CVN slow-bend test specimen is identical to the impact specimen of Fig. 3.4. Rather than being struck with a hammer, the specimen is loaded slowly, at a rate comparable to the standard tension test. The total energy to fracture is obtained by measuring the area under the load-displacement diagram. Figure 3.6 indicates a downward shift in the 15-ft•lb transition temperature of about 160°F between the CVN impact test and the slow-bend test, illustrating that

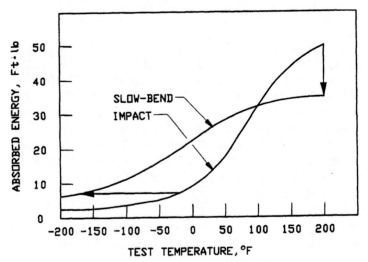

Figure 3.6 Slow-bend and impact CVN test results for A36 steel.

materials are much tougher under static than dynamic loading conditions. The vertical distance between the two CVN energy absorption curves may be considered the safety factor against brittle fracture for a statically loaded structure, such as a storage tank, when the toughness criterion is based on impact test data.

Many pre-API 650 tank steels have a high static Charpy transition temperature.* In fact it has been found that the transition temperature for some old vintage steels used in storage tanks is well above 120°F. This is not consistent with the API 653 provision that tanks with design metal temperatures at or above 60°F not have brittle fracture propensity. However, since the loads in a tank are static, it is reasonable to adjust the impact energy found for the high-transition-temperature steels by taking into consideration this phenomenon. This provision was included in API Standard 653 to allow for the many existing tanks that are in service with unknown high-transition-temperature steels but that have relatively high ambient temperatures.

For almost any other case, the dynamic CVN test is the applicable one, and the exception above was noted to explain why there might be an inconsistency in the codes where transition temperature is involved.

Tensile stresses. Research conducted by the API shows that a sustained primary membrane stress of at least 7 ksi is needed to drive brittle fracture. Since this is well below allowable stresses, most tanks operating full will be well above this value. The exception is small diameter tanks which typically operate under low membrane stresses.

However, even above 7 ksi, the possibility of brittle fracture is reduced by reducing the presence of discontinuities. This can be done by careful design and attention to details which minimize stress raisers. Piping loads, support brackets, and weld details are examples where stress increases might occur. Other examples of discontinuities, weld cracks, and flaws are

- Improperly welded temporary erection brackets which have been left in place.
- Deep undercutting and weld flaws in tank seams.
- Arc strikes which have not been ground out.
- Cracked welds caused by poor repair techniques. One example of this is butt patches. Because of contraction due to material cooling,

---

*J. M. Barson and S. T. Rolfe, ASTM STP 466, 1970, pp. 281–302.

high stresses are introduced which can crack the welds. By pre-heating the patch and controlling the cooling rate there is less likelihood of cracking.

Thickness. Adequate toughness can be assumed to exist for plates under ½ in thick even if the steel material is unknown. Figure 3.7 is an exemption curve that illustrates the inherent decrease of toughness with increasing plate thickness. When the plate exceeds this thickness in welded tanks, the notch toughness decreases rapidly. Preheating steels prior to welding can improve notch toughness, however.

For a given material the fracture toughness decreases with increasing thickness. There are two reasons for this. First, mill practice results in less rolling for thicker sections, and grain refinement is likely to be higher in thin sections. Second, the thick sections restrain the plasticity of the crack tip, reducing the ability of the material to plastically deform and arrest a propagating crack. Thicker sections therefore require better fracture toughness control in order to provide the needed resistance to brittle fracture. This is reflected in the exemption curve in Fig. 3.2.

Chemistry. Carbon increases the transition temperature and decreases the fracture energy. For maximum toughness, the carbon content should be kept low. Manganese reduces the transition temperature.

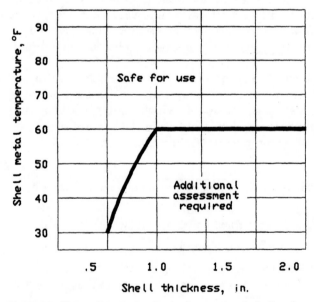

Figure 3.7 Exemption curve for tanks constructed of carbon steel of unknown toughness. (*Courtesy of the American Petroleum Institute.*)

In hardened or tempered steels, manganese can have the opposite effect. Up to 0.04 percent sulfur has little effect on the transition temperature for silicon killed (semikilled) steels. For silicon or aluminum killed steels, the presence of sulfur can substantially reduce the toughness. Phosphorus reduces the steel toughness significantly.

There are other residual and alloying elements that affect toughness, but the most important effects for carbon steel plates used for tanks are covered by the above list.

3.5 Fitness for Service for Existing Tanks

3.5.1 Background

It should be emphasized that brittle fractures are not common. However, the results of a brittle fracture failure can be catastrophic in terms of life, property damage, and liability. It is therefore useful to have tools available to assess the fitness for service where the integrity of an existing tank may be questionable. Two approaches are covered later in this section.

Prior to the late 1970s, little was known about the effects of flaws on the susceptibility of tanks to brittle fracture. Many welded storage tanks were built to codes such as API Standard 12C, *All Welded Storage Tanks*. This only required two-thirds penetration of horizontal butt-welded joints. This requirement was not changed until 1978 when API Standard 650 required full penetration for horizontal welds for 3 in on either side of vertical seams. Two-thirds penetration was accepted for the remainder of the horizontal seam. Vertical seams had always required full penetration and fusion, and in 1980, the 7th edition of API 650 specified that the horizontal seams should have similar requirements.

Therefore, for tanks built prior to the late 1970s there is a significant probability that very large discontinuities or defects will exist in the vicinity of vertical weld seams in steels that have low toughness. Discovery of defects or flaws during tank inspection also brings to issue what to do about the flaws. Many flaws were built into the original construction but are discovered because the current inspection techniques such as wet fluorescent magnetic-particle inspection are more advanced than those techniques applied when the tanks were built. Original flaws built into the tank but which have not grown may be acceptable. Presumably, the tank has survived the hydrostatic testing requirements and has operated satisfactorily during its service life. However, if welding modifications have been undertaken, the possibility of new flaws or high residual stress which introduced the likelihood of brittle fracture should be taken into consideration.

There are two basic approaches to addressing the suitability for service issues. The first method depends upon experience and is the approach taken by API 653. API 653 uses a flowchart or decision tree to base acceptability for service on past experiences and a history of storage tank failures. The basic variables used in the decision tree approach are changes in service temperatures, increasing material thickness, and detrimental effects of repair welds.

The second approach to assessing the risks of operating existing in service tanks with defects is to undertake a fracture mechanics assessment. Fracture mechanics relates material toughness, crack size, and stresses that govern the fracture process.

3.5.2 API 653 method

The current technology for the assessment of susceptibility to brittle fracture is based upon statistics that were collected by API. For this reason it is important to understand that the risks are based upon probabilities and experience. Therefore, there are no absolute assurances that brittle fracture will not occur in any given situation.

Several underlying trends emerged from these statistics that form the basis for assessment of brittle fracture of existing tanks.

1. Once a storage tank has been operated at some temperature with product or under hydrotest, it is unlikely to fail by brittle fracture if it continues to operate under similar or less severe conditions and crack growth has not occurred. Experience shows that brittle fracture does not occur in these tanks unless the service conditions become more severe. This does not apply to situations where fatigue and fatigue crack growth are significant.

2. Tanks built to the 7th edition of API 650 or later have adequate consideration for material toughness. No tank built to these standards has ever failed by brittle fracture.

3. Repairs, modifications, reconstruction, and changes to a tank can increase its susceptibility to brittle fracture. The provisions of API 653 should be complied with to prevent this possibility.

4. Brittle fracture has not occurred in tanks with shell plates less than $\frac{1}{2}$ in thick. Any tank that uses $\frac{1}{2}$-in-thick or less plate for its shell is exempted from brittle fracture concerns. This applies to a tank currently even if the tank is constructed of an unknown carbon steel. See the exemption curve in Fig. 3.7.

5. Hydrostatic testing has been accepted as a demonstration of fitness for service. Technically, this concept is weak. A hydrostatic test does not always develop stresses that are higher than those encountered in service. Also, the hydrostatic test temperature is usually

higher than the design metal temperature of the tank. An evaluation could be made that shows that the risk of brittle fracture could be significant at the lower temperatures. These issues are difficult to resolve because operation at low temperature involves flaw size and distribution and warm prestressing effects during hydrostatic testing that tend to confuse the issues. From a practical view, the code does accept the hydrostatic testing as proof of adequacy against brittle fracture, and all the tanks in service support this conclusion in that none has ever failed after being hydrostatically tested.

6. When the design metal temperature is above 60°F, a tank is exempted from consideration for brittle fracture. This is consistent with the general material dependence of toughness with temperature.

7. When the membrane stress is below 7 ksi, there is insufficient driving force to drive the brittle fracture mechanism. Although it is recognized that high local or residual stresses may initiate a fracture, the propagation of the fracture cannot occur without the average 7-ksi membrane stress. A failure occurring in a tank stressed to below the critical level would result in a leak-before-break condition without the probability of catastrophic failure.

Tanks that meet the following conditions should be suspect for the potential for a brittle failure:

1. Tanks built before 1960

2. Tanks operating below the transition temperature

3. Tanks that have been repaired and modified without subsequent hydrostatic testing

3.5.3 Mechanical assessment methods

Fracture mechanics is a branch of engineering that provides an understanding of the conditions that can contribute to mechanical failure. It has been used to develop the impact exemption curves in the ASME Boiler and Pressure Vessel Code as well as the API Standards. In the context of its applicability to storage tanks, fracture mechanics is particularly useful as a go/no-go indicator for susceptibility of a tank to brittle fracture.

Because fracture-mechanics-based analysis both is time-consuming and has a relatively high degree of uncertainty, it is used primarily as a tool of last resort. However, in cases where the API 653 methods cannot be used or when it is used and additional justification for putting a tank back into service is required, fracture mechanics can provide an additional perspective about the risks of failure.

References

1. James E. McLaughlin, "Brittle Fracture of Aboveground Storage Tanks—Basis for the Approach Incorporated into API 653," Exxon Research and Engineering Company, Florham Park, New Jersey.
2. API Recommended Practice 920, *Prevention of Brittle Fracture of Pressure Vessels,* American Petroleum Institute, Washington, D.C.
3. Materials Properties Council FS-17, draft 4, Consultants Phase 1 Report, New York, September 1993.
4. *Metals Handbook,* desk ed., American Society for Metals, Metals Park, Ohio, 1985.
5. *Metals Handbook,* 9th ed., American Society for Metals, 1989.
6. *The Making, Shaping and Treating of Steel,* 9th ed., United States Steel, 1972.
7. Clark and Varney, *Physical Metallurgy for Engineers,* 2d ed., Litton Education Publishing, 1962.
8. *Steel Tanks for Liquid Storage,* rev. ed., American Iron and Steel Institute, 1982.
9. Jutla and Gordon, Edison Welding Institute, "Structural Integrity of Aboveground Storage Tanks," presented at ASME PVP Conference, June 1992, New Orleans.

Corrosion and Corrosion Prevention

SECTION 4.1 BASIC TANK CORROSION MECHANISMS

4.1.1 Corrosion of Tanks

4.1.1.1 Descriptive nature of tank corrosion

A tank is a large structure that provides the following different environments for corrosion:

- An external surface exposed to the atmosphere
- An external surface under the tank bottom
- A vapor space
- An immersed liquid surface (when two phases are present such as a water layer under hydrocarbon)

Thus many different corrosion mechanisms and causes of corrosion can be at work on the same tank at the same time. To understand the susceptibility of tanks to corrosion, it is useful to categorize the basic types of corrosion into the following categories (also see Fig. 4.1.1 and Tables 4.1.1 and 4.1.2).

4.1.1.2 Atmospheric corrosion

This occurs on the roof and shell of the tank as a result of exposure to air. Warm locations near the sea have accelerated rates compared to other locations. The chloride levels in the atmosphere cause chloride levels to be in sufficient quantities to cause accelerated corrosion.

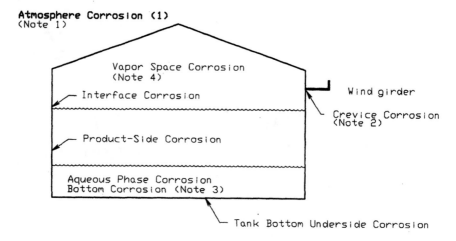

Atmosphere Corrosion (1)
(Note 1)

Vapor Space Corrosion
(Note 4)

Interface Corrosion

Wind girder

Crevice Corrosion
(Note 2)

Product-Side Corrosion

Aqueous Phase Corrosion
Bottom Corrosion (Note 3)

Tank Bottom Underside Corrosion

1. No Corrosion Allowance Usually Specified Since The Exterior Is Usually Coated.
2. Crevice Corrosion Can Be Prevented By Caulking or Seal Welding.
3. A Common Preventive Measure Is To Coat The Bottom and Up To 2'(Feet) or So On The Shell
4. Vapor Space Corrosion

Figure 4.1.1 Tank corrosion modes.

TABLE 4.1.1   Corrosion Prevention Methods

Corrosion type	Prevention measures	Results
Vapor space	Coating	■ Effective, but difficult to stop crevice corrosion above rafters and in lapped roof plates
	Design	■ Effective with coating to use butt-welded designs but costly
Product	Coating	■ Costly
	Corrosion allowance	■ Effective where pitting is not a problem
Bottom	Coating	■ Effective
	Corrosion allowance	■ Effective where pitting rate is not high
Exterior general	Coating	■ Effective
Exterior crevice	Seal weld attachments	■ Costly but effective
	Caulking attachments	■ May last only a few years

TABLE 4.1.2 Corrosion Control Methods

Type of corrosion	Control methods
Uniform corrosion	Inhibitors Protective coating Cathodic protection
Intergranular corrosion	Avoiding temperatures that can cause contaminant precipitation during heat treatment or welding
Pitting corrosion	Protective coating Allowing for corrosion in wall thickness
Stress corrosion cracking	Reducing residual or applied stresses Redistributing stresses Avoiding misalignment of sections joined by bolts, rivets, or welds Use of materials of similar expansion coefficients in one structure Protective coating Cathodic protection
Corrosion fatigue	Minimizing cyclic stresses and vibrations Reinforcing critical areas Redistributing stresses Avoiding rapid changes in load, temperature, or pressure Inducing compressive stresses through peening, swagging, rolling, vapor blasting, chain tumbling, etc.
Galvanic corrosion	Avoiding galvanic couples Completely insulating dissimilar metals (paint alone is insufficient) Using filler rods of same chemical composition as metal surface during welding Avoiding unfavorable area relationships Using unfavorable area relationships Cathodic protection Inhibitors
Thermogalvanic corrosion	Avoiding nonuniform heating and cooling Maintaining uniform coating or insulation thickness
Crevice corrosion; concentration cells	Minimizing sharp corners and other stagnant areas Minimizing crevices, especially in heat transfer areas and in aqueous environments containing inorganic solutions or dissolved oxygen Enveloping or sealing crevices Protective coating Removing dirt and mill scale during cleaning and surface preparation Welded butt joints with continuous welds instead of bolts or rivets Inhibitors

TABLE 4.1.2   Corrosion Control Methods (*Continued*)

Type of corrosion	Control methods
Erosion; impingement attack	Decreasing fluid stream velocity to approach laminar flow Minimizing abrupt changes in flow direction Streamlining flow where possible Installing replaceable impingement plates at critical points in flow lines Filters and steam traps to remove suspended solids and water vapor Protective coating Cathodic protection
Cavitation damage	Maintaining pressure above liquid vapor pressure Minimizing hydrodynamic pressure differences Protective coating Cathodic protection Injecting or generating larger bubbles
Fretting corrosion	Installing barriers which allow for slip between metals Increasing load to stop motion, but not above load capacity Porous protective coating Lubricant
Hydrogen embrittlement	Low-hydrogen welding electrodes Avoiding incorrect pickling, surface preparation, and treatment methods Inducing compressive stresses Baking metal at 200 to 300°F to remove hydrogen Impervious coating such as rubber or plastic
Stray-current corrosion	Providing good lubrication on electric cables and components Grounding exposed components or electrical equipment Draining off stray currents with another conducting material Electrically bonding metallic structures Cathodic protection
Differential-environment cells	Underlaying and backfill underground pipelines and tanks with the same material Avoiding partially buried structures Protective coating Cathodic protection

SOURCE: Adapted from Pludek, 1977.

Another form of atmospheric corrosion occurs by crevice corrosion. Many appurtenances such as wind girders and top angle rings are not seal-welded but are stitch-welded to the tank. The formation of rust causes a rust-streaked appearance. Sometimes the buildup of corrosion products in the space of the crevice can develop sufficient pressure to cause fastener failures and leaks.

4.1.1.3 Product side corrosion

This is the corrosion that occurs as a result of the stored liquid and its effect on the steel containing it. A tank material is usually selected so that the only significant form of corrosion is uniform or general corrosion.

4.1.1.4 Bottom corrosion

Because a water layer and sediment often lie beneath typical hydrocarbon products at the bottom, the bottom is usually corroding at the highest rate in the entire tank. Corrosion is frequently in the form of both general corrosion and pitting. For example, water that separates from crude oil usually contains a high level of chlorides which promotes high corrosion rates and pitting. Control of pitting is usually accomplished by coating the tank bottom and coating the shell up, off the bottom several feet to a point above which the high water level is never expected to rise. Sludges and deposits are usually greatest on the bottom, and this also promotes corrosion, particularly in the form of pitting.

4.1.1.5 Vapor space corrosion

In the vapor space of tanks, corrosion is accelerated by the presence of moisture condensing on the walls and roof as the temperature varies throughout the day and night. In the alternating wet and dry conditions, the concentrations of corrosive compounds are often increased. The rate of corrosion is often most severe at the interface between the vapor and the liquid.

4.1.1.6 Interface corrosion

At the interface of a liquid and gas in a tank, the corrosion rate is often accelerated because the oxygen or moisture concentration gradient at the interface varies sharply with depth into the liquid. For example, in sulfuric acid tanks where the level remains unchanged for a long time, the moisture from the atmosphere can dilute the acid to a concentration that is corrosive to the steel and can etch a deep groove into the tank wall. Unless the liquid level is relatively constant for long periods, this type of corrosion usually is not a problem.

4.1.1.7 Bottom underside corrosion

Because of the widely varied conditions, tank underside corrosion varies significantly based upon the site, design, foundation, drainage conditions, and many other factors.

When the water table is close to the grade level, the moisture that condenses on the underside can start bottom side pitting to occur. In the Gulf Coast region of the United States, bottom side pitting is a severe problem, and the life expectancy of tank bottoms is much less than that in other areas. The salinity in the groundwater, as well as the exposure to the saline waters, during flood conditions aggravates the bottom side pitting reaction.

4.1.2 Corrosion Control and Prevention

Corrosion control and prevention can take many forms. It may take the form of a design detail such as the application of a corrosion allowance to sophisticated lining systems and cathodic protection devices. Some of the most common methods of corrosion control and prevention relate to

- Linings
- Corrosion allowance
- Design (avoidance of dissimilar metals, galvanic couples, improper materials, high fluid velocities in inappropriate places, caulking or seal welding of areas susceptible to crevice corrosion, roof design, etc.)
- Sacrificial anodic systems
- Impressed current cathodic protection
- Use of high-alloy materials

Table 4.1.2 shows some corrosion control methods with respect to typical types of corrosion encountered in tanks.

4.1.3 Designing for Corrosion in Tanks

If the concept of preventing leaks before they occur is adopted by a tank owner, then a rational system can be designed to prevent leaks. Various private company operating practices, as well as API Standard 653 and API Recommended Practice 12R1, use this concept. In simple terms, the tank is removed from service periodically and at an interval that is short enough that the corrosion does not make a hole through the tank.

If a *design corrosion rate* can be estimated or determined, then it is possible to estimate the time between putting the tank into service and a hole through, whereupon leakage occurs. Design corrosion rates can often be obtained by comparison with historical data based upon tanks that have been in similar service. Often it is sufficient to know simply what will be stored in a tank to estimate the design leakage rate. However, the most accurate data can be obtained by performing

TABLE 4.1.3 Typical Stock Side Corrosion Rates

	Stock side*	Vapor space†	Bottom pitting
Black oil, crude oil	1	1	20–30
Light oily >500°F	1	1	4–8
Light oily 370–500°F	1	1.5	4–8
Light nonoily	2	2	4–8

*General or uniform corrosion rate applies to shell and bottom.
†Only applies to cone-roof tanks.

surveys that establish corrosion rates. Often this information is available in the literature or through information that vendors or cooperative companies have. Table 4.1.3 is a tabulation of some typical design corrosion rate values for various services that can be used if better data are not available.

Since there is uncertainty about the actual rate of corrosion, the leak prevention concepts and tank integrity programs usually limit the interval between inspections (removal of the tank from service to determine the extent of corrosion) such that the calculated remaining wall thickness is never less than some minimum amount. API 653 uses 0.1 in on the shell and bottom.

The *corrosion allowance* is the extra thickness that is specified to be supplied with the required thickness for a component. Corrosion allowance is an effective corrosion control tool for general corrosion in that it extends the life of the tank. However, it is usually not effective for pitting-type corrosion.

When the general corrosion rate is less than 10 mils/yr, then the addition of the corrosion allowance is often economical and effective. For higher rates, a coating is most economical.

The technicalities of specifying the corrosion allowance can be significant at times. Often, for shells and bottoms, the corrosion allowance is relatively easy to add. However, it may not always be applied to off-the-shelf blind flanges, nozzles, and other appurtenances, because it would result in nonstandard component thickness that would require special fabrication or in excess component thickness or because it is not necessary. When the corrosion allowance is less than a small value (say, $\frac{1}{16}$ in), it may be most efficient to specify that all wetted surfaces have some specific corrosion allowance. When larger corrosion allowances must be specified, then each surface and component must be investigated as to the applicability of the corrosion allowance. Internal piping that has both internal and external corrosion allowances may be required or may result in excessively thick components. Often trim such as cables and bolting is specified to be some material that either does not require or requires only some minimal corrosion allowance.

4.1.4   Specific Storage Tank Corrosion Service Problems (Petroleum Products)

4.1.4.1   Crude oil tanks

Crude oil has salt water in it that separates from the oil. Because it is saline the corrosion rates can be high, and pitting is usually a problem. Crude oil tanks are often lined on the bottom and up several feet on the shell to extend the life of the tank. Since crude and black oil tanks are often heated, the higher temperatures tend to accelerate corrosion. For the shell of crude oil tanks, the application of some corrosion allowance is one of the best methods of corrosion control. Sacrificial anodes used with coated crude tank bottoms are very effective in protecting the numerous holidays and surface imperfections that will exist in the bottom coating.

4.1.4.2   Refined hydrocarbon storage tanks

Because the water layer is usually cleaner than that for crude oil tanks and contains less oxygen, the bottom pitting and general corrosion are less severe. However, coatings are often used for product contamination reasons. Sacrificial anodes are another effective means of controlling corrosion.

For typical design corrosion rates see Table 4.1.3.

4.1.5   Special Problems

4.1.5.1   Mill scale

Mill scale is a heavy oxide layer that occurs on steel plate when it is exposed to the atmosphere while hot. Bare steel is anodic to the oxide layer. Wherever breaks in the film occur, accelerated corrosion in the form of pitting occurs. Removing mill scale from plates used on tank bottoms is often a way of extending the life of the tank bottom. Mill scale under the right conditions causes accelerated corrosion. Because it breaks away from the steel surface as it flexes or as moisture penetrates it, a galvanic cell is set up. Pitting is the usual result.

Where tanks have been placed on a concrete slab, it may not be worthwhile to use descaled plate. Where the plates rest upon clean, mineral-free, washed sand, or compacted material, and the moisture content is relatively low, it may be prudent to use descaled plate.

4.1.5.2   Corrosion of tanks with double bottoms

Tanks that have double bottoms are susceptible to corrosion from the material between the bottoms, as shown in Fig. 4.1.2. The new steel

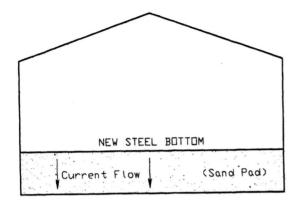

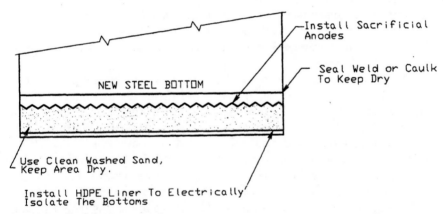

Figure 4.1.2 Tank bottom corrosion.

bottom is anodic to the old bottom. In many cases the new bottoms will only last 3 to 7 years unless the following precautions are taken:

- Install a high-density polyethylene liner or some other elastomer to electrically isolate the anode (new bottom) from the cathode (old bottom).

- Install a sacrificial anode in the spacer material if it is a granular material such as sand or gravel. When concrete is used, no sacrificial anodes are required.

- Use clean, washed materials so that the presence of minerals does not reduce the spacer material resistivity and form electrolytes and a foundation design that prevents the intrusion of groundwater into this space.

- Keep the space between the bottoms dry by caulking or seal welding. Another important aspect of keeping this area dry is to ensure that the foundation is up high enough off the ground to prevent floodwaters and impounded rainwater from rising above the tank bottom. When this is difficult or impossible, as in regions like the southern U.S. Gulf Coast area, then other solutions to double bottoms and leak detection may be required.

4.1.5.3 Corrosion of aluminum skin-and-pontoon internal floating roofs

Aluminum internal roofs are used frequently in all sectors of petroleum-related operations without many incidents or problems. Corrosion in the form of serious pitting is not usually considered a potential problem for crude oil, wastewater, or refined product service. However, in some cases the roof has been severely damaged by corrosion where a caustic wash carries over into finished storage or the product is contaminated in some way with chemicals that rapidly attack aluminum.

The aluminum alloys used in floating roofs vary by component, roof type, and manufacturer, but some generalities about material usage are possible. Probably the biggest factor affecting material selection is the roof type. There are two types of aluminum roofs: the full-contact type, or honeycomb roof, and the skin-and-pontoon type.

The full-contact type is made of sheet 3003 H14 aluminum in most cases, since it is constructed as a honeycomblike structure sealed by a top and bottom sheet. The 3003 series aluminum grade has about 0.12 percent copper alloyed into it. There are few cases where pitting has been a serious problem in these roofs in petroleum service. In other roof applications such as wastewater tanks or crude oil where there is a significant level of hydrogen sulfide, the 3003 aluminum stands up well and keeps the appearance of newness. The structural extrusions which support the 3003 sheets are usually made of 6061 aluminum extrusion, and these do show "white rust" and spot pitting. However, the pitting is usually acceptable from a mechanical integrity point of view.

Skin-and-pontoon roofs use a sheet deck of 3003 or 3004, and the actual buoyancy is provided for in the pontoons that are semisubmerged in the liquid. The skin is in the vapor space. The pontoons are often made of 6061 or 3004 or 3003. The aluminum roof vendors have consistently offered 3004 aluminum. The reason is that the 3004 series is readily available as irrigation tubing. Therefore, the bulk of the tanks use internal floating roofs with 3004 series pontoons in them.

The 6061 aluminum has 0.28 percent copper, whereas the 3004 has no copper in it and 3003 has 0.12 percent copper. The higher copper in 6061 does seem to cause spot pitting but not sufficiently to cause problems in normal refinery and marketing applications.

In sour services where there is a water layer at the tank bottom, severe corrosion of the roof legs occurs. This happens presumably because the H_2S reacts with the water to form sulfurous and/or sulfuric acid. In any case, the corrosion is so severe that most companies specify stainless steel legs for these services or insulating leg caps which reduce corrosion.

The skin-and-pontoon problems may be characterized as follows:

Crevice corrosion. This usually occurs on the ends of the pontoons where structural brackets that support the deck are welded to the endcaps on the pontoons. Another likely location for crevice corrosion to occur is where straps encircle the pontoons for attaching the deck. Although a white rust occurs, it has not caused failures or severe pitting is not considered a serious problem. However, the white rust can develop significant forces between plates or surfaces, causing failure of fasteners, or lead to cracking.

Pitting. This occurs and does cause failures at two places. Any submerged surface that can accumulate sediment causes pitting. A roof skirt that forms the periphery of the roof is often used which is a plate rolled into a channel section. The submerged leg of the channel is subject to rapid and severe pitting. Pitting in structural members is not serious provided that the strength is not reduced substantially.

Another area where pitting occurs is at the interface between the liquid and vapor on the pontoons (the 9 and 3 o'clock positions). A brown buildup that varies from a stain to a deposit approximately 1 to 2 in thick occurs here. One company has done elemental analysis of this and reported the presence of iron. This is presumably iron oxide whose presence forms a corrosion cell. Pitting does occur at the interface in the form of pinholes approximately 0.5 to 1.0 mm in diameter. They are very difficult to detect as the deposits tend to hide them, and they are not always visible to the naked eye. One technique they use to search for these pits is "feel." The pitting forms an elevated crater that may result from the iron oxide and aluminum oxide buildup within the pit, causing higher pressures and plastic deformation of the adjacent metal into a crater. These craters are detectable by running the hand against this area.

The problem of fuel entering the pontoons is serious for two reasons:

- The buoyancy of the floating roof depends on the vaportightness of the pontoons over the life of the pontoon. However, pitting can occur within a 6- to 8-year period.

- The presence of flammable liquids in the pontoons can be a fire and explosion hazard. One company has experienced an explosion that blew the endcaps off the pontoon when a welder inadvertently struck an arc against a pontoon while making repairs inside the tank.

Fabrication. Many pontoons leak because of porosity in the welds and because of weld failures. Welds frequently fail at the endcaps of the pontoon because of the loads transferred to it from the brackets supporting the deck. These cracks have allowed gasoline to enter the pontoon.

Design and Maintenance Considerations

1. In design, consider the alternative to skin-and-pontoon roofs: the full-contact, or honeycomb, roof. This design may be more costly, but it is more effective from a fire protection point of view. The roof has a good long-term service history. Although initial capital cost may be higher than that for the skin-and-pontoon design, the overall costs, given early inspections and repairs, may make it the economical choice.

2. Sniff and inspect pontoons for the presence of hydrocarbon on the interior when the opportunity presents itself or when any internal tank inspection or work is being conducted.

3. Assume that the pontoons have something flammable in them for the purposes of hot work or for inspection.

4. Monitor the buoyancy of internal floating roofs using the skin-and-pontoon design by checking the levelness of the deck during inspections and for tanks that have had these in service for more than 8 years.

5. Reduce the internal inspection period for tanks with internal floating roofs using the skin-and-pontoon design to a maximum interval of 10 years.

6. Use designs and specifications that address materials, fabrication and welding, and other causes of known failures.

4.1.5.4 Corrosion resulting from hydrostatic test water

In the past it was common to treat test water with chromate corrosion inhibitors, but this practice has died out because of the associated environmental problems. Currently, it is necessary to consider the disposal of the test water. One method of ensuring reduced corrosion resulting from hydrostatic testing is to raise the alkalinity of the test water to pH 10 to 12 by using soda ash. There is the possibility of residual alkalinity causing potential stress cracking problems, however. Probably the best and most reliable method is to drain the water from the tank as soon as possible and dry it out over a short time.

For tanks constructed of stainless steel, the chloride level should be limited to 50 ppm, or else the possibility of inducing stress corrosion cracking exists.

SECTION 4.2   CORROSION PREVENTION WITH LININGS

4.2.1   Basic Types of Lining

Tank liner types depend on whether they are applied to the external or internal surface of a tank. For external surfaces, liners must be able to withstand the effects of weather, ultraviolet light in sunlight, and industrial or marine atmospheres and often maintain an acceptable appearance. Internal coatings function primarily to prevent corrosion from exposure to both fumes and condensation in the vapor space, immersion exposure to chemicals or stored liquids, and usually an aqueous cuff layer at the bottom with sludge and deposits. However, another important and often overlooked internal lining function is to maintain product purity. For example, certain grades of glycols must be clear. If they are stored in steel tanks, enough iron will dissolve to discolor the product and degrade its worth. Another example is the storage of jet fuels. These tanks are internally lined to prevent corrosion products from entering the fuel in the form of a fine particulate of iron oxide called *rouge*. Often the internal bottom surface must be able to withstand abrasive effects of slurry movement caused by internal flow patterns, mixers, or inlet and outlet flows, or by mechanical action as by the movement of roof drain hoses lying on the tank bottom. Another unrecognized benefit of a bottom liner is that it reduces the cleaning effort substantially if the tank is removed from service for repairs or for inspection.

4.2.2   External Coating Types

The painting of the exposed external surfaces of a tank provides the following benefits:

- Corrosion protection
- Improved appearance
- Reduced evaporation loss

The coating type and selection for external surfaces depend on several critical factors. When the tank operating temperature is above 200°F, the possibility of a wet, corrosive environment is minimized and no coating is viable, because the corrosion rate is low and finding a coating that does not deteriorate is difficult and costly. Another factor affecting external coatings is whether the tank will be insulated. Because insulations contain minerals and salts that can contaminate the tank surface, a chemically resistant coating is required. Since it is covered by insulation, the coating need not be able to withstand the ultraviolet light in sunlight. A different coating system is typically applied to floating roofs because water stands on the deck surface for long periods after rains and the general weather exposure on the roof is much more severe than that on the shell.

Table 4.2.1 is an example of how some typical coating types and functions based upon the above variables might be selected. Note that this table includes alternative coating systems where the cost of higher-performance coating systems is justified. These conditions might be chemical plant environments where there are high ambient corrosives levels, coastal regions that are humid and warm, and where aesthetics are important as the high-performance options usually have

TABLE 4.2.1   Sample External Paint Type Selection Chart

Surface to be painted	Typical paint types
Tank shells and wind girders (uninsulated to 200°F)	Inhibited alkyd–alkyd enamel
	or
	Self-cured inorganic zinc–polyamide epoxy (High build)–aliphatic polyurethane*
Insulated, below 200°F and steamed out	Polyamide epoxy (primer only)
Insulated, over 200°F	Recommend not coating
Floating roof (uninsulated)	Self-cured inorganic zinc–polyamide epoxy (High build)
	or
	Self-cured inorganic zinc–polyamide epoxy (High build)–aliphatic polyurethane (We prefer galvanizing to coating.*) Inhibited alkyd–alkyd enamel (We prefer galvanizing to coating.)
	or
Stairways and railings	Self-cured inorganic zinc–polyamide epoxy (High build)–aliphatic polyurethane*

*Where high performance is required as in marine environments and industrial atmospheres or where the longest possible life is needed, this alternative should be selected.

higher cost. The high-performance alternatives are also more difficult to apply.

4.2.3 Internal Coatings

Internal coatings are specified for two reasons. First, they protect the tank from internal corrosion. Because it is considered essential to prevent a hole through the bottom before the next internal inspection or tank shutdown, the coatings can allow continued operation of tanks for reasonable periods (10 years or more for large tanks) in spite of high general corrosion and pitting rates.

Various methods of coating tanks internally have developed as a result of the location and nature of the corrosion:

4.2.3.1 Bottom coatings

When tank bottoms are coated with thin films in hydrocarbon service, it is most effective to coat not only the bottom but also a few feet up the shell sides. This is because many services have an additional phase—water, sludge, particulates, or bacteria—that tend to stay at the tank bottom and on the shell near the bottom. If the coating is extended up the shell a few feet, the corrosive effects of these cuffs, sludges, water layers, etc., are minimized.

4.2.3.2 Vapor space corrosion

When the roof must be coated, a number of problems arise. First, since fixed roofs have rafters, there are many crevices that cannot be properly prepared or coated. Tank operators often try to caulk these crevices, meeting with little success. Most often, to get a good coating job requires designing the structure for a smooth, cleanable, coatable surface.

Second, coatings are sometimes used to maintain product purity. Even at low corrosion rates, some corrosion products such as iron oxide or scale develop. These products contaminate stocks with particulates or dissolved corrosion products that produce an unacceptable product. A particularly common problem is the occurrence of rouge, or extremely small, reddish particulate that is generated on the tank interior by atmospheric exposure, which eventually ends up as a suspension in the product tank.

Internal coatings are often classified according to the final dry film coating thickness. Thin films range from 10 to 20 mils whereas thick-film linings are greater than 20 mils. Thin-film coatings are not reinforced whereas thick-film coatings have glass fiber reinforcement. Table 4.2.2 compares the basic attributes of the various internal coating systems.

TABLE 4.2.2   Internal Tank Coatings

| Attribute | Thin film | Thick film | |
		Glass flake	Laminate
Dry film thickness (mils)	10–20	30–40 (sprayed)	80–125
Application in tank	Roof, Shell, bottom	Shell, bottom	Bottom
Life expectancy, yr	10	10–15	20
Average installation time (50,000-bbl tank), weeks	2–3	5–6	6–8
Cost	Lowest	Medium	Highest
Good for corrosive service	Mild	Mild to severe	Severe
Abrasion resistance	None	Good	Excellent
Application	Spray 2+ coats	Spray to trowel	Hand lay up three layers of resin plus two layers of mat plus final coat gel
Chemistry	Epoxy resins, most common vinyls, inorganic zinc, elastomeric-urethanes		
Product purity protection	Good	Good	Good (requires coating on shell and roof as well)
Cathodic protection beneficial	Yes	No	No
Structural strength to bridge cracks and pits	No	No	Yes
Good for coating old pitted or corroded plate	No	Yes	Yes
Susceptibility to cracking	Low	Medium	High (at transition regions)
Inspection after coating	Easy	More difficult	Very difficult
Lining integrity test	Low-voltage wet-sponge detector	High-voltage spark tester	High-voltage spark tester

4.2.3.3   Thin-film linings

These linings contain no reinforcement fibers and are 10 to 20 mils thick. However, thin-film coating systems have the lowest cost and should be considered first for internal lining systems. Because they are thin, they are inadequate for abrasive service or where the substrate surface is severely corroded or pitted. Thin films are usually comprised of epoxy or epoxy copolymer coating systems. Table 4.2.3 lists some of the specific lining systems as well as typical services and service temperature limitations for thin-film linings.

TABLE 4.2.3   Thin-Film Tank Bottom Lining Systems

Lining system	Typical services	Temperature limitation, °F
Coal tar epoxy	Foul water services and crude oil	120–170
Epoxy phenolic	Light products, distillates, aromatics, high-purity water, sour products, crude, and gasoline	180–220
Epoxy amine	Water, light products, distillates, aromatics, crude, and gasoline	160–220
Epoxy amine adduct	Water, light products, distillates, crude, and gasoline	160–220
Epoxy polyamide	Water, distillates, crude, and gasoline	160–180
Epoxy polyamidoamine	Water, distillates, crude, and gasoline	160–180

Note: Generally applied over a white or near-white abrasive blast cleaning in two to three coats. No primer is required. Information related to the performance limitations of specific products with regard to chemical immersion and elevated temperatures should be obtained from the lining manufacturer.

For maintenance of product purity (where corrosion products could cause either iron compound or particulate contamination) thin films are the optimal linings. They are also good for mildly corrosive environments. All kinds of petroleum product tanks are often fully or partially coated with thin-film coating systems.

Thin-film systems always have holidays or imperfectly covered surfaces. Thin-film systems are sometimes made even more effective when they are used with internal cathodic protection. The expected life of a non-high-performance thin-film system is approximately 10 years. Thin-film systems can be applied by spray, brush, or roller in several coats.

4.2.3.4   Thick-film linings

These films are usually reinforced with glass flake, chopped glass fibers, glass mat, glass cloth, or organic fibers. Table 4.2.4 lists some specific thick-film lining systems, services, and service temperatures.

Within the thick-film lining category there are two types of coatings. The *glass-flake-reinforced* coatings can be used in lieu of thin-film coatings, but they cost approximately twice as much. They are normally selected for abrasive services or where highly corrosive services are involved. The *laminate* coatings are sufficiently thick and reinforced to provide some structural support. The laminate coating is a prime example of the only coating that will work to cover severely pitted or corroded tank bottoms that will be coated without repair. The laminate coatings have the ability to span cracks, pits, and holes

TABLE 4.2.4   Thick-Film Tank Bottom Lining Systems

Glass-reinforced lining systems*	Typical services	Temperature limitation, °F
Polyesters		
Isophthalic	Water, crude oil, distillates, and gasoline	140–160
Bisphenol-A	Water, crude oil, distillates, and gasoline	160–180
Vinyl ester	Water, crude oil, aromatics, solvents, alcohols, gasohol, oil, and chemicals	180–220
Epoxy	Water, crude oil, aromatic distillates, and gasoline	180

Note: All applied over a white or near-white abrasive blast cleaning in one to four coats. Primer frequently required. Dependent upon thickness, one or two body coats specified. Resin-rich topcoat or flood coat needed. Polyesters require wax addition to topcoats to ensure timely cure. Information related to the performance limitations of specific products with regard to chemical immersion and elevated temperatures may be obtained from the lining manufacturer.

*Glass reinforcement includes flake, chopped strand, mat, and roving.

even under the hydrostatic pressure of a full tank. However, they are susceptible to cracking wherever there is a transition such as at the shell to bottom joint or at lap patches. Because tank floors move as a result of settling, the highly localized loadings imposed by roof columns or roof legs, these areas are particularly prone to failure of the laminate coating by cracking. Because of cost, laminate coatings are only used on tank bottoms. Many companies will not use this coating system because of the cracking problem that occurs where subsequent corrosion under the laminate causes further bottom deterioration. The laminates are difficult to properly apply as well as to inspect later, and they also entail much greater effort to remove when the tank bottom will be recoated.

The thick-film coatings are less likely to have initial holidays in them, and, therefore, internal cathodic protection is usually not specified in conjunction with these coatings.

In general, the corrosion resistance of the coatings increases with increasing dry film thickness. However, as stated earlier, if a crack develops, which is likely in laminate-coated bottoms, the bottom may hole through rapidly.

Laminate coatings are sometimes specified where a very high underside pitting rate is anticipated due to soil conditions and the environment. The envisioned sequence of events is that once a hole-through occurs, the laminate coating will span the hole, allowing continued operation. While this is true, there is no evidence to suggest that the operable life of the tank without leaks is significantly increased.

4.2.3.5 Other factors affecting lining selection

Temperature is an important factor that affects coating selection. Tanks that are maintained between 160 and 200°F require special attention to coating selection.

When an immersion coating is chosen for service above 140°F, there are factors to consider in addition to the type of product contained and the service temperature. Premature failures could occur if the effects of cathodic protection currents and/or temperature gradients are not considered. Amine-cured epoxies are resistant to cathodic protection currents; however, some manufacturers add phenolics to the epoxies to make them immersion-resistant. Most of these epoxy phenolic coatings are not resistant to cathodic protection currents. Therefore immersion service coatings to be used above 140°F require special consideration, testing, or discussion with the supplier.

Recoatability is often a factor in coating selection. Coatings that cure by solvent evaporation are easily recoated since the new coat starts to dissolve the old coat, providing a good adhesion to the old layers. For epoxies, urethanes, and phenolics that cure to a hard, solvent resistance, shell-like finish, the surface may have to be sweepblasted to provide a sufficient profile for the new coat to adhere.

Weather often affects coating selection. For coating jobs that need to be installed in the winter or where the ambient temperatures are low, the cure time can become a significant problem or the coating may not even be able to fully cure.

Local regulations may limit both the surface preparation methods and the type of coatings that can be used. Since solvent emissions are a major contributor to pollution, many areas are regulated by volatile organic compound (VOC) rules. These rules specify the maximum grams per liter or pounds per gallon of solvent emission.

Long-term aesthetics may contribute to external tank coating selection. Since the ultraviolet light in sunlight degrades coatings by causing them to chalk or to yellow, the topcoat should be resistant to these modes of failure. Aliphatic urethanes have good weathering characteristics. Alkyds yellow and chalk with time. Epoxies chalk quickly; however, the corrosion resistance is unaffected.

Another important external coating selection factor is color. Because the tank color affects volatile emissions, the color selection may impact the overall emissions from a facility. All high-temperature service coatings should use an inorganic zinc primer, which is gray. If some other color is needed, a topcoat will be required. However, there are no VOC compliant, high-temperature coatings so this can become a problem.

4.2.4 Surface Preparation

The key to a long-lasting, successful coating job is surface preparation. Inadequate surface preparation is a prime reason for at least 70 to 90 percent of coating failures. Surfaces to be coated must first be free of any organic grease or oil films, and the surface anchor pattern must be appropriate for the coating. The *anchor pattern* is the surface profile. It is based upon the peak-to-valley height of the microscopic roughness that allows the primer or coating to grip the metal surface.

4.2.4.1 Precleaning

Before blasting to achieve the desired anchor pattern, it is necessary to remove films of any kind from the surface. Tar, oil, grease, salt, and other contaminants can be removed by solvent cleaning [Steel Structures Painting Council standard (SSPC-SP1)] or in combination with other methods such as high-pressure water washing or steam cleaning. Cleaning is followed by a freshwater wash to ensure complete removal of mineral contaminants and chemicals.

4.2.4.2 Abrasive blasting

Sand blasting or abrasive blasting produces the best surface preparation for coating adhesion and is required by many coating system manufacturers' instructions. There are numerous degrees of abrasive blast finish that are defined by the National Association of Corrosion Engineers (NACE) and the Steel Structures Painting Council (SSPC). The common terms defining the finish are *white metal, near white metal, commercial,* and *brush-off blast.* Table 4.2.5 gives the actual specification numbers and a brief description of each finish. Visual standards are also an excellent method of specifying the finish. The Society of Naval Architects and Marine Engineers (SNAME) publishes color pictorial standards for abrasive blasting. NACE also has visual standards in the form of plastic-coated samples of steel with the various finishes. Table 4.2.6 gives the average height of profile in mils that can be economically achieved by various abrasive materials as well as the amount of dust formed.

Abrasive blasting produces large amounts of particulates suspended in air. Since some air management districts restrict the amount of particulates in the air, abrasive blasting may be prohibited. Abrasive blasting should not be done when the temperature of the steel surface is less than 5°F above the dew point or if the relative humidity is above 80 percent. The abrasive must be free of contaminants such as salts, dirt, clay, oil, and grease that could cause surface contamination.

TABLE 4.2.5 Surface Preparation Standards

Steel Structures Painting Council	National Association of Corrosion Engineers	Description	Government of Canada	Swedish standard	British standard
SSPC-SP1		*Solvent cleaning*: Removal of oil, grease, dirt, soil, salts, and contaminants by cleaning with solvent, vapor, alkali, emulsion, or steam			
SSPC-SP2		*Hand tool cleaning*: Removal of loose rust, loose mill scale, and loose paint to a degree specified, by hand chipping, scraping, sanding, and wire brushing	31 GP-401	St. 2 (approx.)	
SSPC-SP3		*Power tool cleaning*: Removal of loose rust, loose mill scale, and loose paint to degree specified, by power tool chipping, descaling, sanding, wire brushing, and grinding	31 GP-402	St. 3	
SSPC-SP5	NACE 1	*White metal blast cleaning*: Removal of all visible rust, mill scale, paint, and foreign matter by blast cleaning by wheel or nozzle (dry or wet) using sand, grit, or shot (for very corrosive atmosphere where high cost of cleaning is warranted)	31 GP-404, type 1	Sa. 3	BS 4232 first quality
SSPC-SP10	NACE 2	*Near-white blast cleaning*: Blast cleaning nearly to white metal cleanliness, until at least 95% of each element of surface area is free of all visible residues (for high humidity, chemical atmosphere, marine, or other corrosive environment)		Sa. 2½	BS 4232 second quality
SSPC-SP6	NACE 3	*Commercial blast cleaning*: Blast cleaning until at least two-thirds of each element of surface area is free of all visible residues (for rather severe conditions of exposure)	31 GP-404, type 2	Sa. 2	BS 4232 third quality
SSPC-SP7	NACE 4	*Brush-off blast cleaning*: Blast cleaning of all except tightly adhering residues of mill scale, rust, and coatings, exposing numerous evenly distributed flecks of underlying metal	31 GP-404, type 3	Sa. 1	Light blast to brush-off
SSPC-SP8		*Pickling*: Complete removal of rust and mill scale by acid pickling, duplex pickling, or electrolytic pickling. May pacify surface			

TABLE 4.2.6 **Properties of Several Abrasives Used in Air Blast Equipment**

Abrasive	Dust factor	Free-silica content, %	Abrasive mesh, NBS sizes	Average height of profile, mils
Sand, very fine	High	>90	20/40	1.5
Sand, fine	High	>90	16/30	1.9
Sand, medium	High	>90	12/25	2.5
Sand, large	High	>90	10/20	2.8
Steel grit no. G-801	Very low	None	40	1.3
Iron grit no. 501	Very low	None	25	3.3
Iron grit no. 401	Very low	None	18	3.6
Iron grit no. 251	Very low	None	16	4.0
Iron grit no. 161	Very low	None	12	8.0
Steel shot no. S-170*	Very low	None	20	1.8
Iron shot no. S-230†	Very low	·None	18	3.0
Iron shot no. 330†	Very low	None	16	3.3
Iron shot no. 3902†	Very low	None	14	3.6
Flint sand	Moderate	>90	8/30	3.4
Granite sand*	Moderate	<5	12/40	3.0
Garnet sand*	Moderate	<1	12/40	3.3
Slag	Moderate	<1	8/40	3.6
Slag	Moderate	<1	10/50	3.5
Slag	Moderate	<1	16/30	3.8
Slag	Moderate	<1	20/40	2.5
Slag	Moderate	<1	16/50	1.5

*Only used in blast rooms and cabinets so abrasive can be contained, recycled, and reused.
†Generally used in automatic blast cleaning facilities using centrifugal wheels.

The typical anchor pattern for tank bottom linings is 1.5 to 4 mils and increases with increasing thickness of lining.

4.2.4.3 Other surface preparation methods

Hand tool cleaning for surface preparation is significantly less effective than abrasive blasting but is also much less costly and does not require the degree of concern for air pollution regulations. These methods include rotary cleaners, sanders, power wire brushing, and grinding. Using these methods for surface preparation may reduce the coating life by at least a factor of 2.

Water blasting depends on high-pressure jets of water to remove surface grit and dirt. However, it cannot produce an anchor pattern that is required for a good coating job. The high-pressure water stream (3000 to 30,000 psi) can be dangerous to anyone in the vicinity. Corrosion inhibitors are usually added to prevent rusting while the water dries.

4.2.4.4 Substrate repairs

Prior to sandblasting, all weld repairs should be performed in accordance with API 653. Typical repairs include bottom patches, weld-up of pits and cracks, removal of arc strikes, weld spatter, sharp weld crests, undercutting, and rough welds to ensure a uniformly smooth surface for the coating. Chipping followed by grinding can be used to remove sharp edges.

4.2.4.5 Lining application

For thick-film linings, the lining supplier may require that the prime coat be less than the thickness of the anchor pattern. This is important to adhere to because this allows the following coats to properly adhere, preventing premature delamination.

SSPC-PA 1 and NACE 6F164 are good practices for coating applications. Application of coatings should be done when the relative humidity is less than 80 percent and the dew point is at least 5°F above the metal surface temperature.

Improper curing is a major cause of early coating failures. Prior to placing the tank in service, the curing should be complete. If necessary, warm air, force-circulated through the tank interior, may accomplish the proper drying and curing.

4.2.5 Economics

True costs for coatings, like most problems associated with optimizing plant and equipment, are extremely elusive and complex. However, some simplified approaches can make the problem easier to understand.

Initial costs have little to do with true, overall, long-term costs. However, they are an important starting point. Variables such as the cost per percent of solids, surface preparation costs, and downtime or delays in schedules resulting from the coating can be looked at. Because many factors impinge upon the true costs, it might be considered axiomatic that the initial cost of the coating system is much less important than the coating's applicability, workability, longevity, and serviceability or than the projected life of the equipment.

To estimate initial costs, some important considerations should be examined:

Task	Things to consider
Materials selection	■ Cost per square foot at specified thickness (varies with percentage of solids, specified dry film thickness, and cost per gallon) ■ Paint loss (typically 15% for flat surfaces to 30% or more for complex shapes)
Surface preparation	■ Cleanliness versus life ■ Wheelabrators may take only single pieces of pipe which then need to be welded and reblasted before painting

Task	Things to consider
Application	■ Complexity of the paint (e.g., single or multiple component?) ■ Complexity of the shape (e.g., flat tank surfaces versus small piping) ■ Cost of access for final coats and maintenance (on ground, in the air, or offshore?) ■ Impact of curing time on schedule

4.2.6   Inspection of Linings

Inspection of coatings ensures that each step will provide the quality that is needed so the overall durability and coating life are achieved. All inspectors should either be NACE-certified or knowledgeable in coating and lining practices. Equipment needed for coating inspections is covered in NACE RP0288. Inspection procedures are covered in NACE RP0288. The surface should be verified when cleaned to meet the blast requirements specified. SSPC-VIS 1 and NAXW TM0175 provide visual means to compare the work performed to the standards specified. NACE RP0287 provides a method of measuring the anchor pattern. After the films are in place, the thickness can be verified by ASTM D 4414 or SSPC PA 2. If the hardness of coatings is required, ASTM D 2583, ASTM D 2240, ASTM D 3363, and the solvent wipe test can be used. Holiday testing is done according to procedures outlined in NACE RP 0188.

4.2.7   Lining Repairs

NACE RP 0184 provides guidelines for lining repairs. Spot repairs are made when the damage to the lining is minimal. They are often done on new coating jobs where subsequent work caused a small amount of damage. Topcoating is done to existing linings where the adhesion of the existing lining is still good. Complete relining is done where the lining is beyond repair.

4.2.8   Safety

The dangers of internal coating projects are much more significant than those of external coating projects. For internal coating projects, API Publication 2015 which covers guidelines for safe entry of tanks should be followed as well as the appropriate federal and state rules. Information about the proper precautions and procedures may also be found in the OSHA Standard for Abrasive Blasting, SSPC PA 3, and NACE 6D163. Manufacturer's material safety data sheets (MSDSs) are available to inform employees and contractors about the hazards of the coating components to be used. All applicable MSDSs should be reviewed prior to the start of any work.

Coatings contain a number of hazards that should be addressed in the planning and execution phases of any coating project. Most paint components are highly flammable and explosive in certain concentrations.

When various abrasives are used, there is an emission of dust or particulate pollution. Some areas in the United States require that specific criteria for dust be complied with.

Abrasive blasting is a source of static electricity. By grounding the blasting equipment and the material being cleaned, the possibility of an ignition in the presence of flammable vapors or substances is greatly reduced.

Abrasive blast may be a hazardous waste material and may have to be disposed of according to the Resource Conservation and Recovery Act (RCRA) rules. RCRA specifies the transportation, disposal, and recordkeeping requirements for solid hazardous wastes. It also has a method of determining whether the waste is hazardous through a simple laboratory test.

4.2.9 Specific Lining Applications

Linings should be specified for three reasons:

1. A corrosion rate that causes a hole through before the next scheduled tank inspection

2. Product purity

3. Ease of future tank cleaning, entry, and inspection

Some specific applications are covered below.

4.2.9.1 Potable water tanks

The U.S. Food and Drug Administration regulates the coatings that are acceptable for lining potable water tanks.

4.2.9.2 Crude oil tanks

Since crude oil causes bottom pitting rates on the order of 20 to 45 mils/yr, a hole through would occur in an unreasonably short period. Therefore, they are almost always coated on the bottom and up several feet on the shell (to a point that is always above the water phase in the bottom) with a thin-film lining.

4.2.9.3 Diesel and fuel oil tanks

Because these tanks are subject to sulfur-reducing bacteria and the related corrosion pitting that occurs, they are often coated with a thin film on the bottom and a few feet up the sides of the shell.

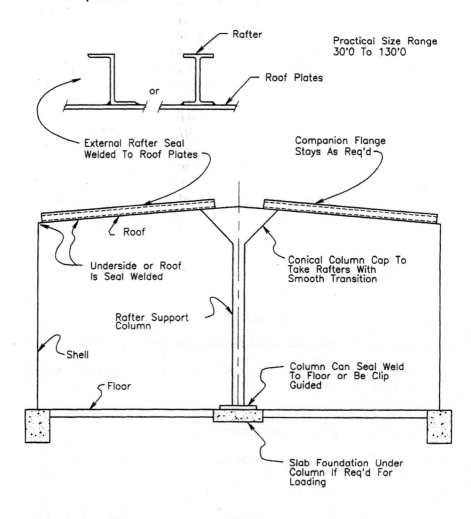

These tank designs have been used frequently.
Where the roof interior is to be coated with
Phenolic Epoxy where Hot–Wet H_2S Gas is present.

Figure 4.2.1   External rafter roof tank.

4.2.9.4   Aviation gas

These are typically coated on the bottom. Sometimes the shell is coated for floating-roof tanks. Some operators do not coat these tanks.

4.2.9.5   Jet fuel tanks

Because of the need to maintain product purity (water and particulate), these tanks are often coated on the entire interior surface. However, many operators do not require coating these tanks.

4.2.9.6 Motor gas

Motor fuel tanks are often coated on the bottom only because of the water-phase-induced corrosion that generates pitting and product purity problems. However, many operators do not coat these tanks.

4.2.9.7 Methyl Tertiary Butyl Alcohol (MTBE) or Other Oxygenates

Since MTBE is noncorrosive to steel, the purpose of coating these tanks is for product purity. Some companies will coat an external floating-roof tank storing MTBE. For fixed-roof tanks, it provides no benefit to coat the bottom unless there is a water phase present.

SECTION 4.3 CORROSION PREVENTION WITH CATHODIC PROTECTION

4.3.1 Electrolytic and Galvanic Corrosion

To understand cathodic protection, it is instructive to know about two corrosion mechanisms known as *electrolytic (stray)* and *galvanic corrosion*. In electrolytic corrosion (stray current), corrosion occurs where stray current enters a structure that happens to be in the electric flow path. Corrosion occurs where current leaves the structure. An example of this type of corrosion is shown in Fig. 4.3.1.

The other corrosion mechanism is galvanic corrosion, pictured in Fig. 4.3.2. When two metals are electrically connected in the presence of an electrolyte, corrosion of the more active metal occurs and is called galvanic corrosion. Table 4.3.1 shows some typical metals in the galvanic series. The important thing to remember about this table is that the metal higher up on the table is the more active metal and will corrode preferentially to any metal below it. Note that for steel there is a wide variance of potential differences in the galvanic scale. Galvanic corrosion is not limited to corrosion of different metals. The variable activity of the metal surface of a steel plate may cause local galvanic corrosion to occur, as shown in Fig. 4.3.3.

The elements required for galvanic corrosion are as follows:

- The presence of an electrolyte. The underside of tank bottoms provide this condition.

- An electrical path. This is the tank bottom itself.

- A more active metal (anode site). Because of the presence of any number of conditions such as oxygen concentration, rust, scale, etc., small areas may become anodic (higher up on the galvanic series)

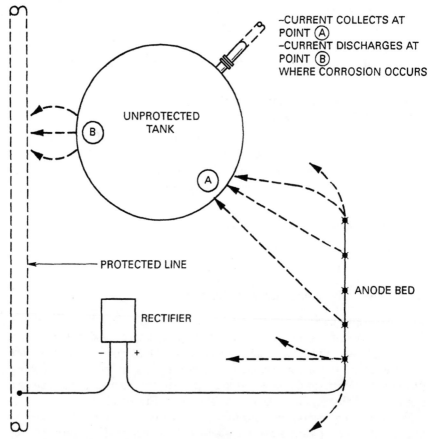

Figure 4.3.1 Stray current corrosion of an unprotected tank bottom. (*Source: American Petroleum Institute.*)

with respect to the rest of the tank bottom and therefore the site of pitting.

- A less active metal. This is the adjacent area near the site of pitting.

On the underside of tank bottoms, both general corrosion and pitting occur and are the result of the corrosion cell described above. In general corrosion, there are so many evenly distributed but shifting corrosion cells that the metal loss appears relatively uniform.

In both electrolytic and galvanic corrosion, the anode or area corrosion occurs where the current leaves the steel.

4.3.2 Cathodic Protection

Although the above description is a vast oversimplification, it can be appreciated that if the current could be reversed, then corrosion at

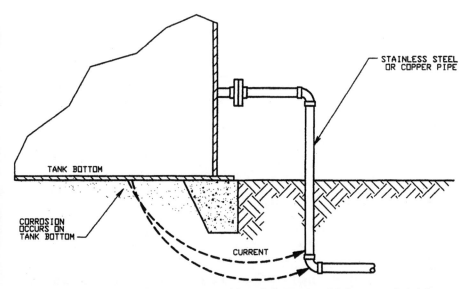

Figure 4.3.2 Galvanic corrosion. (Occurs with tank bottoms and belowground stainless steel or copper pipe.) (*Source: American Petroleum Institute.*)

TABLE 4.3.1   Partial Galvanic Series

	Metal	Volts*
↑	Commercially pure magnesium	−1.75
Active or anodic	Magnesium alloy (6% Al, 3% Zn, 0.15% Mn)	−1.6
	Zinc	−1.1
	Aluminum alloy (5% Zn)	−1.0
	Commercially pure aluminum	−0.8
	Mild steel (clean and shiny)	−0.5 to −0.8
	Mild steel (rusted)	−0.2 to −0.5
	Cast iron	−0.5
	Lead	−0.5
	Mild steel in concrete	−0.2
	Copper, brass, bronze	−0.2
	High-silicon cast iron	−0.2
Noble or cathodic	Mill scale on steel	−0.2
↓	Carbon, graphite, coke	+ 0.3

Note: Data apply in environments consisting of neutral soils and water.
*Voltages are referenced to copper/copper sulfate.

the anode might cease because the ionization of metals due to current flow is the reason for corrosion. If fact, this is the principle of cathodic protection. There are two methods of causing current flow reversal.

In the first method, called *galvanic,* or *sacrificial,* cathodic protection, a material that is higher up in the galvanic series table is put into electrical proximity to the structure being protected. This forces

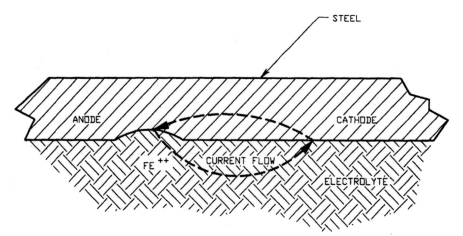

Figure 4.3.3 Electrochemical corrosion cell. (*Source: American Petroleum Institute.*)

the current to flow from the anode to the cathode. The anode protects the structure since it provides a steady source of current into the protected structure.

In the second method, called *impressed current* cathodic protection, a direct current is forced through anodes to the protected structure. A source of direct current of sufficient potential is required.

Both types of systems are shown in Figs. 4.3.4 and 4.3.5.

Properly designed cathodic protection systems in the right conditions are a proven method of protecting tanks from the corrosive effects of tank bottoms in contact with underlying corrosive soils. Cathodic protection is also useful for protecting double-bottom tanks and for internal corrosion caused by immersion conditions as well.

4.3.3 Polarization

Figure 4.3.6 shows the corrosion of steel in water or an electrolyte. At the cathode (where current enters the steel) there is a buildup of hydrogen ion concentration due to the half-cell reaction. The hydrogen ion film reduces the free ionic flow by blanketing the metal surfaces with an excess of ions and is called a *polarization film*. The rate of corrosion is reduced by this film because it repels the ionic flow (current flow). Anything that tends to remove the film increases the corrosion rates. The presence of oxygen can remove the film by supplying oxygen at the cathode which can react to become water instead of hydrogen gas. Another reaction that occurs to reduce the polarization film is the combining of hydrogen ions to form hydrogen gas. This usually requires a higher potential called the *hydrogen overvoltage* to make this reaction occur in neutral solutions. This reaction occurs

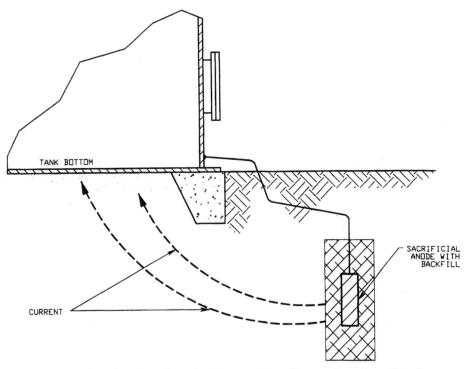

Figure 4.3.4 Sacrificial anode cathodic protection. (*Source: American Petroleum Institute.*)

freely in acidic environments. In neutral solutions, the oxygen level usually controls the polarization film.

When a structure such as a tank bottom is cathodically protected, it may take several days to months to become polarized. The initial current requirements are high and then slowly drop to the steady-state value after the structure is polarized.

4.3.4 Electrical Potential Measurement

The potential measurements that can be made on a tank bottom are used to evaluate the corrosion problem before the cathodic protection is installed and to assess the degree of protection after it is installed. The standard way of taking potential measurements is to use a copper–copper sulfate reference cell (Fig. 4.3.7) and to hook it up as shown in Fig. 4.3.8. These measurements are also useful for indicating if stray current corrosion is harming the tank. If the measurements on one side of the tank are significantly different from those on the opposite side, there is a strong possibility of stray current. The topic of potential measurements is covered in greater detail later.

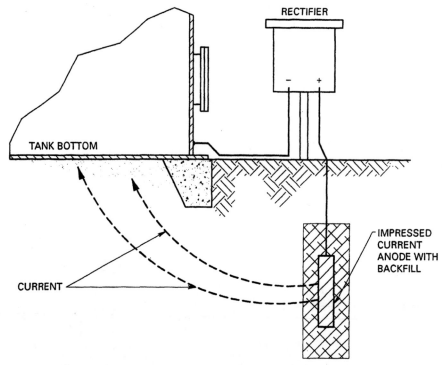

Figure 4.3.5 Impressed current cathodic protection. (*Source: American Petroleum Institute.*)

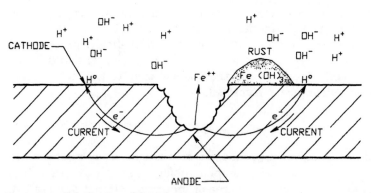

Figure 4.3.6 Polarization of steel during corrosion.

There are two primary criteria for determining the adequacy of cathodic protection. One is called the *instant-off criterion* potential. This potential is really the tank-to-soil potential. Because voltage drop as a result of electric current flow may cause erroneous readings, the power supply is shut off, and tank-to-soil readings are immediate-

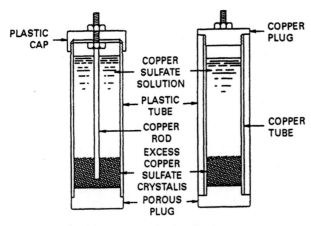

Figure 4.3.7 Typical copper sulfate electrodes.

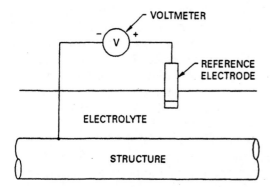

Figure 4.3.8 Structure potential measurement.

ly made. This is the true tank-to-soil potential. The other criterion is the *potential shift criterion.*

4.3.5 Current Requirements

The current required to protect a tank is directly proportional to the bare metal surface area. Therefore, coatings reduce the amount of current required for protection. Other factors affect current requirements:

- Acidic environments require more cathodic protection current.

- Increased temperature increases the required current.

- Increased oxygenation of the electrolyte requires more current.

- Soils with lower resistivities require more current. Since the presence of salts lowers soil resistivities, more current is required.

- Where there is fluid velocity such as inside a tank, more current is required.

4.3.6 Internal versus External Cathodic Protection

It should be understood that a cathodic protection system protects only one surface of a structure. The inside of a tank can be protected by an internal cathodic protection system. However, if it is necessary to protect the underside of the tank bottom, then an additional external cathodic protection system needs to be installed.

4.3.7 Basic Types of Cathodic Protection Systems

4.3.7.1 Sacrificial

(See Fig. 4.3.4.) A sacrificial anode cathodic protection system depends on coupling a more active metal on the galvanic series with the protected metal or steel. This type of protection is applicable to small-diameter tanks (40 ft or less) where they are isolated from other underground structures and where the soil resistivity is less than 5000 $\Omega \cdot$cm. This system is limited because the amount of current and the potential generated by the sacrificial anodes are limited.

The sacrificial anode system is particularly suited and almost always used to provide internal cathodic protection such as for crude oil storage tanks. A schematic diagram of an internal sacrificial cathodic protection system is shown in Fig. 4.3.9.

The most common sacrificial anode material for tanks is magnesium, zinc, and aluminum alloys. They are usually in the form of castings or ribbon. They are distributed around the perimeter of the tank or buried beneath the tank. The depth can vary from 5 to 15 ft.

4.3.7.2 Impressed current

The impressed current system is more versatile than the sacrificial system in that a wide range of current and potential levels can be controlled by the system design. A schematic diagram of this system is shown in Fig. 4.3.5. Since large amounts of current may be required to protect a large tank, the impressed current system is used. Direct current is required so that most installations that use impressed current cathodic protection take standard alternating current and reduce its voltage and rectify it. Power for these systems, however, has been supplied by other means as well:

- Batteries
- Thermoelectric generators
- Wind-driven generators

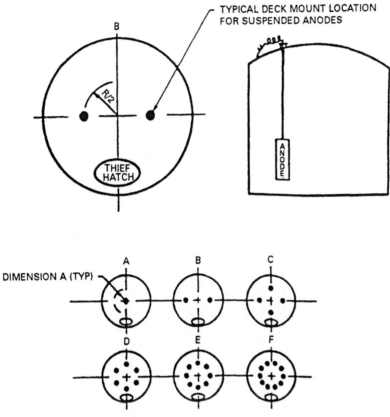

Figure 4.3.9 Suspended internal anodes.

- Turbine-driven generators
- Solar cells

Table 4.3.2 compares some of the generic advantages and disadvantages of the sacrificial anode system and the impressed current system.

4.3.8 Applications

4.3.8.1 Internal cathodic protection

Internal cathodic protection almost always uses the sacrificial cathodic protection system. The reason is the low maintenance and sufficient life. The system can be designed to protect the interior of the tank until the next scheduled shutdown.

Internal cathodic protection may be applied to tank bottoms that have a water layer in them or are filled with an electrolyte. Good candidates are crude oil tanks, knockout tanks, or brine tanks. The inter-

TABLE 4.3.2 Comparison of Sacrificial and Impressed Current Cathodic Protection Systems

Sacrificial anode cathodic protection systems	Impressed current cathodic protection systems
Advantages	*Advantages*
■ No external power required	■ Can be designed for wide range of voltage and current
■ No regulation required	
■ Easy to install	■ High ampere-year output available from single ground bed
■ Minimum of cathodic interference	
■ Anodes can be readily added	■ Large areas can be protected by single installation
■ Minimum of maintenance	
■ Uniform distribution of current around periphery of tank	■ Variable voltage and current output
	■ Applicable in high-resistivity environments
■ Installation can be inexpensive (if done during construction)	■ Effective in protecting uncoated and poorly coated structures
■ Minimum right-of-way and easement costs	■ Can be routinely monitored for effectiveness
■ Efficient use of protective current	
Disadvantages	
■ Limited driving potential	*Disadvantages*
■ Lower/limited current output (typically 1–2 A maximum)	■ Can cause cathodic interference problems
■ Installation can be expensive (if done after construction)	■ Subject to power failure and vandalism
■ Can be ineffective in high-resistivity environments (generally greater than 5000 Ω•cm)	■ Requires periodic inspection and maintenance
	■ Requires external power
■ Practical only for small tanks (typically less than 40 ft in diameter)	■ Monthly power costs
	■ Overprotection can cause coating damage
■ Useful only for protecting one tank at a time	
■ Difficult to monitor effectiveness	

nal surfaces of tank bottoms are particularly vulnerable to corrosive effects because water, sludge, and solid mater provide the ideal conditions for corrosion to occur. There is usually the presence of water because it settles out of the product, comes in through external floating-roof seals, or is created from atmospheric condensation on the inside vapor surfaces of fixed-roof tanks.

Figure 4.3.9 shows an example of a suspended sacrificial anode system. When there is a water layer in the bottom with noncorrosive material above such as hydrocarbon, the cable on the suspended sacrificial anodes can be lengthened to the point where the anodes lie flat on the tank bottom. The suspension method is good on tanks up to about 30 ft in diameter. In protecting larger-diameter tanks where there is a water layer, the anodes are often bolted or welded to the bottom. If the interior of the tank is coated, internal cathodic protection is less frequently used; however, it can be effective in protecting the tank where holidays in the coating exist.

4.3.8.2 External cathodic protection

The term *external cathodic protection* is synonymous with the protection of the underside of a tank bottom. Either the galvanic or impressed current method may be used. The galvanic system is limited to small tank diameters and where there is no galvanic interaction between other tanks. This system is maintenance-free and lasts up to 30 years. Where these criteria cannot be met, then impressed current systems are used. This is discussed in greater detail in the following sections.

4.3.8.3 Liners and secondary containment

Since cathodic protection current cannot pass through dielectric liners such as polyethylene or hypalon, there are severe limitations on the applicability of cathodic protection systems. In all these cases, the cathodic protection may be applied if there is sufficient space between the tank bottom and the liner. Both sacrificial and impressed current methods are used in this application of cathodic protection.

The use of a thick-film lining should not be considered justification for eliminating underside cathodic protection measures.

4.3.8.4 Double bottoms

Both impressed current and sacrificial anode systems have been used successfully in the space between the new bottom and the old bottom. Unless the spacer material is concrete, cathodic protection should be used in these cases.

4.3.9 External Cathodic Protection Design

External cathodic protection involves numerous factors that affect the selection and design of the system, some of which are

- Basic design considerations
- Soil resistivity
- Type of foundation
- Atmospheric conditions
- Type of tank bottom (single, double)
- Secondary containment
- Tank operating conditions (temperature, fill-empty rate, etc.)

4.3.9.1 Basic design considerations

The process of designing a cathodic protection system includes these steps:

1. Calculate the current requirements of the structure, using the design rules of thumb.

2. Select an anode type, size, and weight.

3. Calculate the resistance of the anode to the electrolyte.

4. Calculate the current output of a single anode by using the resistance (from step 3) and the driving voltage between the anode and the desired protected potential from Ohm's law ($I = V/R$).

5. Calculate the total number of anodes needed to meet the current requirement by dividing the total current requirement by the individual anode current output.

6. Calculate the life of the anode system, using the anode alloy current capacity.

7. Repeat steps 2 to 6 with different size and weight anodes until a reasonable number and system life are achieved.

Since the art of cathodic protection involves so many different variables, it is recommended that tank systems be designed and installed by personnel knowledgeable in both the design and installation of these systems.

4.3.9.2 Soil resistivity

External cathodic protection systems are applicable to tanks of any size where the soil corrosivity is sufficient to reduce the tank bottom life to an unacceptably short period. One of the key indicators of corrosivity is the soil resistance. Soil resistivity is used not only to evaluate corrosivity, but also to design the anode ground bed. Table 4.3.3 gives some typical soil resistivities and corresponding corrosivities. For existing tanks, soil resistivity measurements should be taken at a minimum of four locations around the perimeter at depths of 5 to 10 ft.

TABLE 4.3.3 Soil Resistivity versus Corrosivity

Resistivity range, $\Omega \cdot cm$	Potential corrosion activity
<500	Very corrosive
500–1,000	Corrosive
1,000–2,000	Moderately corrosive
2,000–10,000	Mildly corrosive
>10,000	Progressively less corrosive

SOURCE: A. W. Peabody, *Control of Pipeline Corrosion.*

4.3.9.3 Criteria for effective cathodic protection

The primary method of determining the effectiveness of a cathodic protection system is to measure the soil-to-tank potential. For tank bottoms, a measurement can be made using a reference cell such as the copper–copper sulfate cell in Fig. 4.3.7. The natural potential is about -0.5 V when measured against a reference copper sulfate electrode (CSE). More positive values represent steel that has undergone corrosion, and more negative values are indicative of new, well-coated structures or structure with cathodic protection operating. If a tank has a potential of -0.85 V with respect to the copper–copper-sulfate cell, then it is usually considered *protected*.

The problem with the measurements is that tank bottoms are large-surface structures and that taking uniformly spaced readings across the bottom is rarely possible. Frequently, the only measurement that can be gotten is a measurement near the perimeter. Experience shows that the measured potential can vary significantly from the center of the tank bottom to the perimeter. A common variance may be as much as a 0.3-V difference between the center and edge of the tank. Under normal conditions an allowance of 1 mV/ft is considered a reasonable voltage gradient for design purposes.

The reason for this variance is that there is an IR (current times soil resistance) drop in potential. As the current requirements increase, this IR drop becomes more important. For the underside of tanks the rules of thumb are approximate and may not apply to the actual field conditions, because the soil resistivity may vary under the tank. Generally, while the perimeter may be protected at -0.85 V, the center may not be protected due to the variance of the current field.

These criteria are stated in API RP 652 for achieving effective protection:

1. A minimum of -0.85 V measured across the structure-electrolyte interface relative to a saturated copper–copper-sulfate reference electrode.

2. A minimum of 0.100 V of cathodic polarization measured between the structure surface and a stable reference electrode contacting the electrolyte. The formation or decay of this polarization can be used in this criterion.

For new installations a permanent reference cell should be placed under the center of the tank.

4.3.9.4 Current requirements

In general, as the corrosivity of the soil increases, so does the protection current required. Bare steel bottoms typically require between

1.0 and 2.0 mA/ft^2 of protected underside surface. For optimal design, a current requirement test can be performed on existing tanks. This test uses temporary anodes and a source of direct current. A known amount of current is forced through the temporary anode bed to the tank bottom. When the specified potential is achieved, the known current establishes the design current requirements for the tank.

4.3.9.5 Foundation

Foundations constructed of concrete or asphalt or the presence of leak prevention liners or hard rock may prevent cathodic protection from being effective or needed. Flat-bottom tanks may be installed on many different types of underlying material including concrete foundations, compacted soil or gravel foundations, or native soil foundations.

Concrete foundations are highly effective in reducing not only the corrosion rates because of their passivating effect, but also the need for cathodic protection. Although concrete has a much higher resistance than typical soils do, the cathodic protection current can be made to traverse the foundation. Consideration must be given to the presence of reinforcing steel. For most cases, the use of cathodic protection on tanks which have continuous concrete foundations would not be necessary. For cathodic protection to work on a tank bottom, the bottom plates must rest in intimate contact with the concrete, because the tank really rests on low points and it is unknown what percentage would really be protected. However, where there is much moisture near the periphery of the tank, the possibility of accelerated corrosion due to the moisture and oxygen might provide sufficient reason to consider cathodic protection.

Tanks constructed on soil foundations are subject to high rates of corrosion where the soil is corrosive. The presence of chlorides and sulfates and low pH increase the probability of bottom underside corrosion (see Table 4.3.4). Cathodic protection should be considered where these conditions occur. Even if clean backfill material is used, the presence of minerals can often be found. In a seacoast environment, the salt in the atmosphere carried as a mist settles on the tank roof and shell and eventually washes down the sides and into the foundation near the periphery of the tank bottom. If the foundation is

TABLE 4.3.4 Chemical Corrosivity

Constituent	Corrosive	Very corrosive
pH	5.0–6.5	<5.0
Chlorides	300–1000 ppm	>1000 ppm
Sulfates	1000–5000 ppm	>5000 ppm

subject to floodwaters or to impounded secondary containment waters, there is a good chance that the foundation material will be infiltrated with corrosive minerals.

4.3.9.6 Double-bottom tanks

Figure 4.3.10 shows a double-bottom tank with an old and new bottom utilizing a granular material (sand) spacer. Even if the sand is clean and mineral-free, the new bottom will experience an unsually high rate of corrosion. The problems with this detail are that the new steel bottom is anodic to the old bottom and that accelerated corrosion can occur. It is almost impossible to completely seal the space from the two bottoms from atmospheric moisture infiltration. The girth joint caused by installing the new bottom is not leakproof unless a great deal of effort and cost are applied to this joint by seal-welding it and inspecting it. Even if this joint could be sealed, the old bottom often has holes or cracks in it. Because of the thermal changes occurring in the new tank bottom, the space between the bottoms breathes. This breathing pumps moisture in where it condenses and starts the corrosion process. Groundwater infiltrating from the old bottom or floodwater entering the double-bottom girth joint often brings in the necessary minerals to create an effective and corrosive electrolyte.

To correct the problem, a dieletric (electrically insulating) liner material which usually doubles as a release prevention barrier can be installed as shown. The use of a liner blocks current flow between the old and new bottoms, thus eliminating galvanic corrosion between the old and new bottoms. However, if a sand or granular material is used as the spacer, then corrosion can still occur locally on the new bottom underside. It is prudent to place a sacrificial anode into the spacer material. Sacrificial systems using zinc or magnesium ribbon have been effectively used; however, there are limitations on life. Typically these are approximately 20 years. Impressed current systems are also used for these spaces. Titanium ribbon with a TIR 2000 mixed metal oxide coating, $\frac{1}{4}$ in wide laid in a grid pattern, is effective. Ribbon spacing is usually at 5-ft intervals. Zinc ribbon has also been used. Table 4.3.5 is a cost and life-cycle comparison of typically used systems.

Generally for double-bottom tanks, the most effective system is the impressed current cathodic protection system, because of the limited space, high soil resistivities, and large current requirements for bare steel bottoms. However, galvanic systems have no operational or maintenance considerations associated with them.

4.3.9.7 Secondary containment areas

Secondary containment provides a means of collecting, impounding, and storing leaks and spills from storage tanks. As a result, some

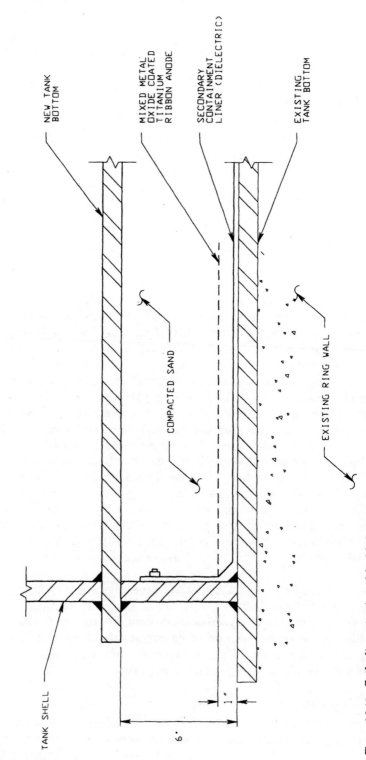

Figure 4.3.10 Cathodic protection of double bottoms.

TANK SHELL

NEW TANK BOTTOM

MIXED METAL OXIDE COATED TITANIUM RIBBON ANODE

SECONDARY CONTAINMENT LINER (DIELECTRIC)

EXISTING TANK BOTTOM

COMPACTED SAND

EXISTING RING WALL

6'

1"

TABLE 4.3.5   Cathodic Protection Cost Comparison (120-ft-diameter Tank Bottom)

System	Life, yr	Installation cost, $/ft^2	Life-cycle cost, $/(ft^2·yr)
Titanium grid	50	1.17	0.023
Zinc at 4.0 ft	11	1.28	0.116
Zinc at 2.5 ft	18	1.60	0.089

companies and regulatory bodies require that the secondary containment area be lined with an impervious material such as clay or impervious membranes. Since dielectric elastomeric liners are impervious to cathodic protection current, special consideration needs to be given to the design of both double-bottom tanks or tanks that are constructed using dielectric liners. Figure 4.3.11 shows the effect of the liner on a cathodic protection system. Because the cathodic protection system must be placed between the tank bottom and the liner, it must be a shallow system. For existing tanks, it may be impossible to place anodes under the tank because of the inadequate clearance between the liner and the tank bottom. Where clay has been used, cathodic protection can operate through the clay lining, provided it remains moist. For higher-temperature tanks, the use of cathodic protection may be ineffective due to drying of the clay.

The service conditions often dictate the use of sacrificial or impressed current systems. Where a long life is needed, the impressed current system is preferred. However, it has higher maintenance, inspection, and operating costs. The amount of current is another important design variable. Current requirements are often proportional to the surface area to be protected. Where large surfaces

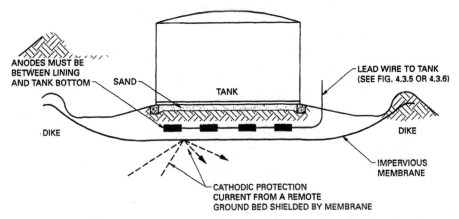

Figure 4.3.11 Impervious membrane beneath storage tanks. (*Source: American Petroleum Institute.*)

are involved, the relatively high amounts of current necessitate an impressed current system. One way to determine the current requirement is to run a current requirement test. Experience in the amounts of current required for other similar tanks is also valuable information that can be used to estimate current requirements.

Once the current requirements are known and the resistivity of the electrolyte, soil, or medium is known, then the number and weight of anodes can be computed.

4.3.9.8   Anode selection

Sacrificial system.   The most commonly used anode materials are zinc, aluminum, and magnesium. Each metal has unique properties that give it a best fit for the application:

Magnesium has the highest potential but a low efficiency. Its application is limited to more resistive (fresher) waters where a higher potential is required.

Zinc is applicable where sparks cannot be tolerated if the anode is dropped, but is not widely used.

Aluminum is the primary metal used for sacrificial internal anodes. It is doped with indium to prevent oxide films from passivating the anode.

Table 4.3.6 shows some of these characteristics.

Impressed current anodes.   Impressed current anodes are classified as to whether they are decaying or nondecaying. As the name implies, decaying anodes are used up with time. For tanks, by far the most common type used is the decaying anode. It may be

- High-silicon-chromium cast iron

- Lead silver

- Treated graphite

TABLE 4.3.6   Driving Potential of Three Common Anode Materials

Metal	Potential, V	Capacity, A•h•lb	Density, g/cm^3
Zinc	1.1	335	6.6
Magnesium	1.7	500	1.6
Aluminum	1.1	1100	2.4

Values are approximate for seawater at ambient temperature.

TABLE 4.3.7 Consumption Rates for Various Anodes,* lb/(A•yr)

Material	Theoretical	Actual	Efficiency, %	Driving voltage ΔE, V
Galvanic				
Magnesium	8.75	17.5	50	0.9†
				0.7‡
Zinc	23.5	25.8	90	0.25
Aluminum	6.46	7.0–8.0	90	0.25
Impressed				
Durichlor 51		0.25–1.0		
Graphite		0.5–2.0		

*These values are for anodes surrounded by select backfill materials.
†Galvomag alloy.
‡H-1 alloy.

The nondecaying types are

- Platinum
- Titanium dioxide

Table 4.3.7 shows the consumption rates for anodes of various materials. Because design life is a function of consumption rates, these data are important for optimizing the anode material selection and setting a reasonable design life.

4.3.10 Installation

4.3.10.1 Typical configurations

The most common three configurations for cathodic protection are as follows:

Distributed-anode system. Figure 4.3.12 shows this configuration. Anodes are evenly spaced around the perimeter of the tank. The farthest the current must flow is to the center of the tank. The current flow will be greater near the periphery than at the center. Since this is often where the greatest moisture and thus corrosion occur, this system is effective. This configuration is used with either the impressed current or the sacrificial cathodic protection system.

Deep-anode system. The concept of the deep-anode system is illustrated in Fig. 4.3.13. By placing an anode deep in the ground, fewer anode wells need to be installed. Uniformity of cathodic protection current flow is provided by the depth of the anode. This system is usually applicable where many structures or large tanks are to be protected. The design of the system is complex and should be based upon the experience of knowledgeable designers or contractors. This configura-

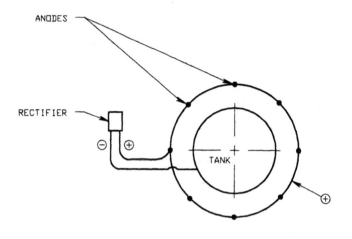

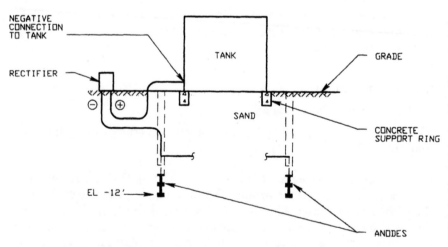

Figure 4.3.12 Distributed anode system.

tion applies only to the impressed current system where sufficient voltages and currents may be applied to make the system effective.

Angle-drilled anode system. In this system (Fig. 4.3.14), the anodes are drilled in at an angle to the vertical. This angle provides anodes that can be located under the tank and thus provide better distribution of current flow to the center of the tank bottom as well as less potential variation. The advantage of this configuration is that the variation in potential between the outside edge of the tank and the

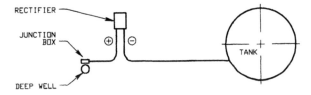

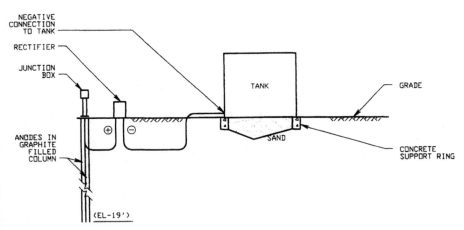

Figure 4.3.13 Deep anode system.

center of the tank is reduced. Because of the typical current and potential requirements, this system usually uses the impressed current system of cathodic protection.

4.3.10.2 Anode installation

Internal applications. For internal cathodic protection, anodes may be mounted by suspension from cables. This method is convenient for diameters up to about 30 ft. For larger tanks the most effective method is to use the flush-mounted anodes.

External applications. For external cathodic protection using impressed current methods, the anodes may be made of graphite, steel, high-silicon cast iron, or mixed metal oxides on titanium. The anodes are buried in a coke breeze backfill to extend the life and reduce anode resistance. The anode installation details for a sacrificial system are similar. Figure 4.3.15 shows an example of an anode installation.

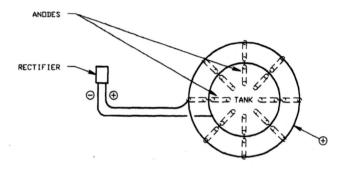

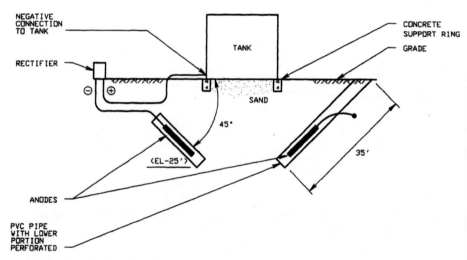

Figure 4.3.14 Angle drilled anode system.

4.3.11 Maintenance

Periodic surveillance and recordkeeping or current measurements and potentials are necessary for long-term effective operation. This is particularly true for the impressed current systems. It is most important that an initial survey be conducted just after the initial installation to benchmark the future surveys. Because complete polarization of a structure may take some time, the periodic surveys will reveal when adequate protection has been attained or if additional anodes are required.

During dry seasons, it may be useful to water down the anodes to

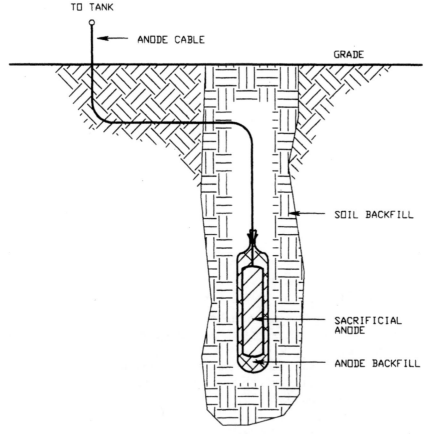

Figure 4.3.15 Typical sacrificial anode installation. (*Source: American Petroleum Institute.*)

maintain an acceptable flow of current by decreasing the anode resistance. A decrease in anode current in successive cathodic protection surveys may indicate that the anode is consumed and needs replacing or that the electrical connections are deteriorating.

Impressed current anodes will fail if the wire-to-wire connection contacts brine or short-circuits to the tank.

4.3.11.1 Anode watering

The cathodic protection current may be reduced when the top layer of the ground dries out because of the increase in anode resistance. Watering the anode and saturating the immediate area until the entire anode area is moist will ensure that sufficient current continues to protect the tank.

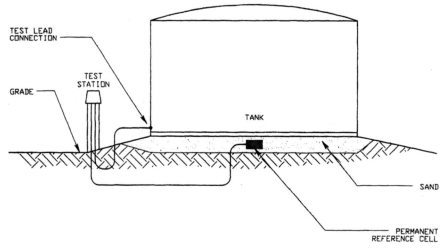

Figure 4.3.16 Permanently installed reference cell and test station.

4.3.11.2 Monitoring

A problem with monitoring cathodic protection systems is the inability to obtain tank-to-soil potentials near the center. For new tank construction, the problem is solved by installing a permanent reference cell (Fig. 4.3.16). Another useful method is to install a perforated pipe under the tank through which a reference electrode may be inserted for future measurements.

References

Section 4.1

1. L. C. Case, *Water Problems in Oil Production: An Operator's Manual,* 1st ed., 1977.
2. H. Uhlig, *Corrosion and Corrosion Control,* Wiley, New York, 1971.
3. Munger, *Corrosion Prevention by Protective Coatings,* NACE, Houston, 1984.
4. Nace, *Corrosion Basics—An Introduction,* 1984.
5. Uhlig, ed., *Corrosion Handbook,* Wiley, New York, 1948.

Section 4.2

1. API:
   Spec. 12B Specification for Bolted Tanks for Storage of Production Liquids.
   Spec. 12D Specification for Field Welded Tanks for Storage of Production Liquids.
   Spec. 12F Specification for Shop Welded Tanks for Storage of Production Liquids.
   500A-C *Classification of Areas for Electrical Installations.*
   STD 620 *Design and Construction of Large, Welded Low Pressure Storage Tanks.*
   STD 650 *Welded Steel Tanks for Oil Storage.*
   RP 652 *Tank Interior and Bottom Lining.*
   STD 653 *Tank Inspection and Repair.*
   RP 1615 *Installation of Underground Petroleum Storage Systems.*
   RP 1621 *Bulk Liquid Stock Control at Retail Outlets.*

RP 1632 *Cathodic Protection of Underground Petroleum Storage Tanks and Piping Systems.*

Manual *Guide for Inspection of Refinery Equipment,* chap. 13, Atmospheric and Low Pressure Storage Tanks.

ASTM:*

G-57 *Method for Field Measurement of Soil Resistivity Using the Wenner Four-Electrode Method.*

NACE:†

RP0169-83 *Control of External Corrosion on Underground or Submerged Metallic Piping Systems.*

RP0177-83 *Mitigation of Alternating Current and Lightning Effects on Metallic Structures and Corrosion Control Systems.*

RP0285-85 *Control of External Corrosion on Metallic Buried, Partially Buried, or Submerged Liquid Storage Systems.*

RP0388-88 *Impressed Current Cathodic Protection of Internal Submerged Surfaces of Steel Water Storage Tanks.*

RP0572-85 *Design, Installation, Operation, and Maintenance of Impressed Current Deep Groundbeds.*

RP0572-75 *Design, Installation, Operation, and Maintenance of Internal Cathodic Protection Systems in Oil Treating Vessels.*

TPC:

Pub. 11 *A Guide to the Organization of Underground Corrosion Control Coordinating Committees.*

Control of Pipeline Corrosion, A. W. Peabody.

NFPA:‡

30 *Flammable and Combustible Liquids Code.*

70 *National Electrical Code.*

PEI:§

2. Underwriters Laboratories of Canada, ULC-S603.1-M, *Standard for Galvanic Corrosion Protection Systems for Steel Underground Tanks for Flammable and Combustible Liquids.*

3. Steel Tank Institute Standard *Specification and Manual for Corrosion Protection of Underground Steel Storage Tanks.*

4. J. H. Morgan, *Cathodic Protection—Its Theory and Practice in the Prevention of Corrosion,* Leonard Hill [Books] Limited, 1959.

5. "How to Size Cathodic Protection Cable for Least Cost," Paper no. HC-23, Harco Corporation.

6. Marshall E. Parker and E. G. Peattie, *Pipeline Corrosion and Cathodic Protection,* 3d ed., Gulf Publishing Company, 1984.

7. Department of Transportation, *Regulations for Transportation of Liquids by Pipeline,* 49 CFR 195, Bureau of National Affairs, Inc.

8. Department of Transportation, *Regulation for Transportation of Natural and Other Gas by Pipeline,* 49 CFR 192 Minimal Federal Safety Standards, Bureau of National Affairs, Inc.

9. Department of Transportation, *Regulations for Transportation of Natural and Other Gas by Pipeline,* 49 CFR 191, Annual Reports, Incident Reports, and Safety-Related Reports, Bureau of National Affairs, Inc.

10. John Morgan, *Cathodic Protection,* 2d ed., National Association of Corrosion Engineers, Houston.

11. *Control of External Corrosion on Underground or Submerged Metallic Piping System,* NACE RP-01-69, National Association of Corrosion Engineers, Houston.

---

*American Society for Testing and Materials, 1916 Race Street, Philadelphia, PA 19103.

†National Association of Corrosion Engineers, P.O. Box 218340, Houston, TX 77218.

‡National Fire Protection Assocation, Batterymarch Park, Quincy, MA 02259-9990.

§Petroleum Equipment Institute, P.O. Box 2380, Tulsa, OK 74101.

12. *Corrosion Control,* Air Force Manual no. AFM88-9, U.S. Air Force, 1962.
13. H. B. Dwight, "Calculations of Resistance to Ground," *AISS Transactions,* **55:** 1319–1328 (1939).
14. E. D. Sunde, *Earth Conduction Effects in Transmission Systems,* D. Van Nordstrand, New York, 1949.
15. M. J. Szeliga, "Understanding Corrosion," *Maintenance Technology,* October 1988.
16. P. O. Gartland, E. Bardal, R. E. Anderson, and R. Johnsen, "Effects of Flow on the Cathodic Protection of a Steel Cylinder in Seawater," *Materials Performance,* **40**(3), March 1984.
17. *Pipeline and Gas Journal* staff, *Corrosion Control Report,* February 1988.
18. O. Osborn, C. F. Schrieber, W. B. Brooks, R. C. Jorgensen, and B. Douglas, "Cathodic Protection Handbook," *Hydrocarbon Processing,* June 1957.
19. *Cathodic Protection* (theory and data interpretation), National Association of Corrosion Engineers, Houston, 1986.
20. L. P. Sudrabin and F. W. Ringer, "Some Observations on Cathodic Protection Criteria," *Corrosion,* **13**(5), May 1957.
21. F. N. Speller, *Corrosion—Causes and Prevention,* 3d ed., McGraw-Hill, New York, 1951.
22. H. H. Uhlig, *Corrosion Handbook,* Wiley, New York, 1948.
23. David H. Kroon, "Tank Bottom Cathodic Protection with Secondary Containment," Corrpro Companies, Inc., Spring, TX.
24. S. R. Rials and J. H. Kiefer, "Evaluation of Corrosion Prevention Methods for Aboveground Storage Tank Bottoms," Conoco, Inc., Ponca City, OK, January 1993.
25. J. A. Hanck, "Cathodic Protection Interference—Detection and Correction," *Pipe Line News,* Pacific Gas & Electric Co., Emeryville, CA.
26. Underground pipelines are subject to costly corrosion which can be mitigated by proper design and protection. M. G. Fontana, *Corrosion: A Compilation,* Columbus, OH, 1957, pp. 223–224.
27. B. Husock, "Use of Pipe-to-Soil Potential in Analyzing Underground Corrosion Problems," *Corrosion,* **17**(8): 97–101, August 1961.
28. "Statement of Minimum Requirements for Protection of Buried Pipelines," prepared by NACE Technical Group Committee T-2 on Pipeline Corrosion, *Corrosion,* **12**(12): 19, October 1957.
29. B. Husock, "Cathodic Protection of Water Storage Tanks," *Proceedings of the Sixth Annual Appalachian Underground Corrosion Short Course,* West Virginia University, Morgantown, WV, 1961, pp. 279–287.
30. *Corrosion Prevention for Gravity Tanks. Approved Equipment for Industrial Fire Protection.* Associated Factory Mutual Fire Insurance Companies, 1957.
31. Cathodic Protection Investigation (Interim Report), *Civil Works Investigation No. CW 311.* Rock Island District, U.S. Army Corps of Engineers, Rock Island, IL, 1954.
32. J. Segall, "Cathodic Protection for Coal Unloading Facilities for Thos. H. Allen Steam Electric Station," AIEE Paper CP61-424, presented at AIEE Winter General Meeting, Jan. 29–Feb. 3, 1961.
33. B. Husock, "Fundamentals of Cathodic Protection," Harco Corporation, Cleveland, OH, 1990.
34. M. Romanoff, "Underground Corrosion," *National Bureau of Standards Circular 579,* April 1957.
35. H. H. Uhlig, *Corrosion and Corrosion Control,* Wiley, New York, 1963.
36. Environmental Protection Agency, Title 40, CFR, Parts 280 and 281, Sept. 23, 1988.
37. Department of Transportation, Title 49, CFR, Parts 192 and 195, as amended June 30, 1971.
38. National Association of Corrosion Engineers, Recommended Practice RP-01-69, *Control of External Corrosion on Underground or Submerged Metallic Piping Systems,* revised in 1983.
39. D. H. Kroon, "Tank Bottom Cathodic Protection with Secondary Containment," Paper 529, presented at NACE Corrosion/91, Cincinnati, OH, March 1991.

Tank Foundations

5.1 Introduction to Tank Foundations

This section applies to vertical, cylindrical tanks with uniformly supported flat bottoms. Although tank foundations in many respects are similar to other foundations, they have evolved into a relatively specialized form with specific requirements. Some of the requirements are covered below.

This section also provides the basis for anchorage requirements due to wind loads, internal pressure, and seismic loads.

5.1.1 Preliminary studies

In the early phases of tank foundation engineering, several types of information should be considered and evaluated:

- Site conditions
- Condition and settlement of similar equipment in nearby areas
- Soil conditions including soil bearing capacity and soil characteristics
- Acceptable amounts and rate of settlement
- Earth pressures, pore water pressures, and dewatering quantities
- Applicable codes and standards

5.1.2 Codes and standards

There are a handful of codes and standards that are frequently used by the tank designer. One of the most useful is the American Concrete Institute (ACI) Standard ACI 318 which covers the requirements of steel-reinforced concrete structures. The Uniform Building Code

(UBC) also provides useful guidance and principles for foundation design. Another useful reference is found in API Standard 650, Appendix B, as well as API Standard 620, which provides useful foundation design provisions specific to storage tanks.

5.1.3 Soil investigations

Although many tanks have been built without the advantages of soils reports, the advantages of having these reports available are substantial. Foundation design usually begins with specifications provided in the soils report. The soils report addresses subsurface conditions, soil bearing capacity, and potential settlement. These are determined from soil borings, load tests, sampling, and laboratory testing. In addition, one of the most important bits of information available to the tank foundation designer is the experience with similar structures available from a soils engineer familiar with the area.

Although there have been numerous tanks built at sites judged to be adequate by visual assessment, some tanks have settled so much as to be damaged during hydrostatic testing. There is at least one case where a tank settled during the hydrostatic test and collapsed as a result of settlement.[1]

When the subgrade conditions are inadequate to support the tank without undue settlement, some of the following approaches have been used:

- Remove inadequate material and replace with suitable compacted material.

- Preload the area of the proposed foundation by building up sufficient backfill.

- Design the foundation to reduce the bearing pressures by the use of extended foundation areas and/or piles.

- Use vibrocompaction or dynamic compaction methods to stabilize the subgrade.

Appendix B of API 650 has more details on these methods.

5.2 Important Elements to Consider in Foundation Design

5.2.1 Foundation elevation

Not only is the basic foundation elevation important for adequate tank life span, but also the slope across the top of the foundation impacts costs as well as life and constructability. Some of the factors that affect foundation design are as follows:

5.2.1.1 Finished foundation elevation. A significant cause of bottom failures is corrosion from the tank bottom underside. Several factors contribute to this:

- The final elevation of the tank bottom is important for several reasons. After settlement, the tank shell may sit in moist conditions because water is unable to drain away adequately. This results in accelerated corrosion or pitting and a reduced tank bottom life. Tanks should be designed to be at least 8 to 12 in above the surrounding grade level, if possible, after settlement. Foundations which are too low or settle to an elevation that is too low relative to surrounding finished grade elevations are particularly prone to bottom side corrosion. Foundations constructed of compacted soil are particularly subject to this problem also.

- Use of a material which contains mineral salts or is contaminated with organic matter that can decompose, induce microbial corrosion, or turn acidic can accelerate underside corrosion. The solution is to use appropriate backfill. When a granular material such as sand is used, it should be cleaned and washed to minimize the presence of salts and minerals.

Another consideration for setting tank foundations is the possibility of buoyancy of the tank due to submergence in water. The most likely cause of this is rainfall or firewater filling the secondary containment area. Since typical tanks require less than 1 ft of submergence to float off of the foundation, the probability of this happening while the tanks are empty is relatively high.

The tank base elevation should be made high enough to prevent uplift due to buoyant forces acting on the tank when it is empty and the secondary containment area is filled with some designated level of water, such as 6 in. The maximum liquid level in the secondary containment area is also a consideration for setting the final tank elevation. Numerous considerations and details need to be assessed in this case, such as anchorage of the tank or the probability of the tank's being empty.

5.2.1.2 Foundation profile and surface geometry. Although cursory visual examination of tank foundation surfaces indicates that they are constructed as a flat, plane surface, they usually are not. Instead, their shallow profiles put them into the following categories:

- *Cone up.* This is the most common profile. A typical slope of 1 to 2 in per 10 ft of horizontal run is used. This pattern prevents and minimizes intrusion of rainwater from the outside periphery of the tank bottom. It also allows water layers, or water bottoms, inside the tank to migrate to the tank water draw for removal.

- *Cone down.* The bottom slopes toward the center of the tank, and an internal sump is usually included here for water bottom removal. The rate of slope is the same as for a cone-up tank.

- *Planar sloped bottom or "shovel bottom".* In this pattern, the bottom is constructed as a plane, but it is tilted to one side. This allows for better removal of water than a cone-up design, but has the advantage that it is easier to construct. However, it is slightly more complex for the shell construction to accommodate this pattern. This configuration is usually limited to tanks under about 120 ft in diameter, as the total elevation change across the tank diameter becomes excessive.

- *Plane flat bottom.* For small tanks it is not worthwhile to provide a sloped bottom for services where water removal is not required. It is also not necessary to use sloped bottom tanks.

5.2.2 Drainage

Drainage plays a key role in the useful life of the tank foundation. Not only does lack of adequate drainage cause accelerated bottom side corrosion and pitting, but also the soil stability and bearing capacity can be affected by inadequate drainage. In some cases, salt soils when secondary containment areas have not been quickly drained after rainfall; the saturated soil has become destabilized, resulting in edge settlement.

5.2.3 Oil sand under tank bottom

Although past practice has been to use an oiled sand cushion under the tank, this is becoming less popular. It has not been demonstrated that the oiled sand reduces bottom side corrosion. In fact, some companies claim that it causes accelerated bottom side pitting in certain areas. Oil sand is also an organic contaminant that might not be an environmentally acceptable medium to place on soil.

5.3 Tank Foundation Types

While it is difficult to classify all possible foundation types for storage tanks, some general types have proved to be most common for specific applications. Foundation types may be broken into several classifications in generally increasing order of costs:

- Compacted soil
- Crushed-stone ringwall
- Concrete ringwall

- Slab
- Pile-supported

A brief discussion of each type of foundation follows. Table 5.1 summarizes the characteristics of various foundations by type.

5.3.1 Concrete ringwall foundation

The concrete ringwall foundation is so called because of its appearance, shown in Fig 5.1. It is used on foundations for tanks of a diameter of at least 30 ft or more. In the large-diameter tanks is usually the most cost-effective reinforced concrete foundation. It has many advantages that have made its use widespread. Characteristics of the concrete foundation are as follows:

- Where suitable soil bearing capacities do not exist at the surface but are at some depth below it, a ringwall can be used to provide an adequate foundation by preventing lateral squeezing of the soil.

- It provides a level working surface for tank construction to begin. Checking flatness for construction tolerances is made relatively easy during construction.

- It virtually eliminates the phenomenon of edge settlement. It reduces differential shell settlement and the resulting distortion of the shell. This is important, particularly for floating-roof tanks which must have relatively round shells for proper operation of the roof.

- The concrete ringwall provides a good means of sealing the leak containment liners under tanks where they are used.

- The design criteria for ringwall foundations should be carefully planned and thought out because of the numerous types of loading conditions that can be applied to them, particularly in anchored tank designs. More information concerning ringwall design is given in the section on design loads.

- The potential for catastrophic and sudden tank content loss is reduced because of the probability that rapid erosion of the soil under the tank is reduced in the event of a bottom leak.

- The potential for erosion of the tank support at the chime area due to wind, rain, and running surface waters is reduced.

- Ringwalls should be proportioned so that the pressure under the base of the ringwall bearing on the soil is approximately the same as the hydrostatic pressure acting on the tank bottom. This can be accomplished with a few assumptions. The weight acting on the ringwall is the sum of the tank shell weight and the portion of the

TABLE 5.1 Foundation Characteristics by Type

Foundation type		
Concrete ringwall	Applications	■ Usually used for tank diameters exceeding 30 ft. ■ Requires good to medium soils, properly prepared. ■ Used where design for uplift required since it provides for anchorage.
	Advantages	■ Provides a plane base for tank construction. ■ Provides foundation for anchorage. ■ Minimizes edge and differential shell settlement. ■ Minimizes entry of water to the tank bottom, thus corrosion problems. ■ Prevents erosion of supporting soil under tank. ■ Distributes concentrated shell loads well. ■ Provides a controlled variable force resisting uplift by adjustment of ring width.
	Disadvantages	■ Relatively costly. ■ Ineffective or costly where soils are poor.
	Other	■ Requires reinforcing steel to be effective. ■ Requires design for hoop stress caused by active soil pressures. ■ Cathodic protection can be used under the tank bottom.
Crushed-stone ringwall	Applications	■ May be used on any size tank. ■ Can be used in poor soils.
	Advantages	■ Low cost. ■ It allows for more uniform soil loadings at the expense of slightly larger tank deformations.
	Disadvantage	■ Subject to edge settlement and differential shell settlement.

Compacted soil	Application	▪ Can be used on any size tank provided the soil is good.
	Advantage	▪ Low cost.
	Disadvantages	▪ Subject to local edge and differential settlement. ▪ Higher potential for bottom side pitting corrosion. ▪ Unexpectedly high local settlement may occur.
Concrete slab	Application	▪ Limited to relatively small diameters less than about 30 ft.
	Advantages	▪ Provides a plane base for tank construction. ▪ Can provide anchorage foundation. ▪ Minimizes edge and differential shell settlement although this is not usually a consideration for small-diameter tanks. ▪ Minimizes entry of water to the tank bottom, thus corrosion problems. ▪ Prevents erosion of supporting soil under tank. ▪ Can act as a leak detection liner.
	Disadvantages	▪ High cost. ▪ Slab subject to cracking on shifting or settling soil. ▪ Cannot use cathodic protection.
Pile-supported	Applications	▪ Used where soil bearing pressures are very low. ▪ Applicable to any size tank where other foundation types cannot be used.
	Advantages	▪ Minimizes most kinds of settling. ▪ Provides for anchorage for uplift forces.
	Disadvantages	▪ Very costly. ▪ Requires good soils information and soil reports. ▪ Design procedures are involved and require special procedures.

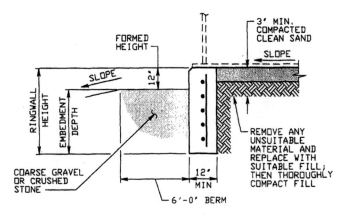

VIEW A-A

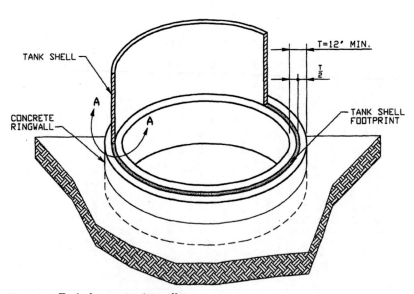

Figure 5.1 Typical concrete ringwall.

roof supported by it, the liquid weight projecting vertically on the ringwall, and the weight of the ringwall itself. The formula for this is

$$t = \frac{W}{P(1 - f) - 150h} \geq 12 \text{ in} \qquad \text{minimum recommended ringwall width}$$

where t = width of ringwall, ft

P = hydrostatic pressure on tank bottom, pounds per square foot (psf)

f = fraction of foundation width with liquid projecting vertically onto it

h = height of ringwall, ft

By proportioning the ringwall this way, settlement will usually be more uniform than if this is not the case.

- For very soft clays or soils, the difference in settlement between the ringwall and the tank bottom can introduce bottom sag and bottom stresses, resulting in bottom failure.

5.3.2   Crushed-stone ringwall foundations

This type of foundation can be considered the "poor man's ringwall." A typical crushed-stone ringwall foundation is shown in Fig. 5.2. This design happens to incorporate a leak detection system. While it costs less than the concrete ringwall, it has many of the advantages of the concrete ringwall. These are some of its key characteristics:

- It can be significantly less costly than a concrete ringwall.

- It provides uniform support of the tank bottom by dissipating concentrated loads in a gradual pattern.

- The crushed-rock ringwall does not provide as good a construction base to work from as does a concrete ringwall.

- A crushed-rock ringwall allows moisture infiltration under the periphery of the tank base. Unless it is elevated sufficiently after settling, moisture can cause accelerated corrosion. Where the soil conditions are corrosive such as in acidic or saline areas, the corrosion will be highly accelerated. In addition, high point pressures induced by the steel on gravel points can cause pitting corrosion. The solution is to blanket the top layer with a clean, granular material such as sand so that the tank does not rest directly on the gravel. However, the sand is subject to washout, and care must be taken to prevent this with placement of geotextile fabrics or to use progressively smaller particle sizes from the base of the ringwall up to the surface. If this is not done, the sand cushion tends to percolate into the depths of the coarse material.

- The crushed-stone ringwall is more appropriate for very soft soils than the concrete ringwall, but edge settlement is a potential problem in these cases.

- Catastrophic failure of the bottom is possible if a small leak starts and washes out the underlying support.

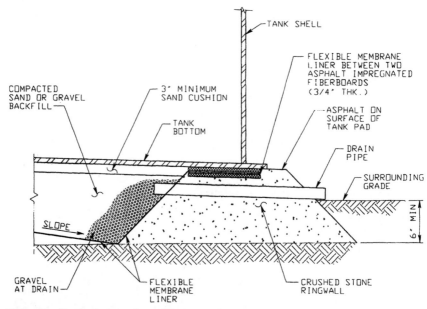

Figure 5.2 Crushed-stone ringwall.

5.3.3 Compacted soil foundations

These foundations can be used where the soil quality and bearing capacity are good. Generally, the top 3 to 6 in will be removed and replaced with a sand or granular backfill. These are often called *sand pad foundations,* laid directly on earth. It is important to remove all organic material and loam. This should be replaced with compacted fill or gravel. The advantage of this type of foundation is the relatively low cost. Some characteristics are as follows:

- This type is not usually suitable where the tank must be designed for uplift because anchorage cannot be provided easily. However, helical soil anchors may be used in special cases with appropriate design considerations.

- These foundations are prone to washout during heavy rainfalls and storms.

- Underside tank bottom corrosion is more unpredictable than with other types of material. Mixtures of various types of soils with slightly different chemistries often lead to accelerated corrosion areas on the underside of the tank bottom.

- Catastrophic failure of the bottom is possible if a small leak starts and washes out the underlying support.

Erosion control is accomplished by using at least a 5-ft berm or shoulder, sloping the surface away from the tank. This prevents washout of the foundation under the perimeter of the shell and prevents moisture from infiltrating the area under the tank bottom. On compacted soil foundations, the bottom profile is usually crown up with a 1- to 2-in slope for each 10 ft of horizontal run. In large-diameter tanks, the center portion is often flat to keep the total rise below 6 in over the diameter of the tank.

5.3.4 Slab foundations

An example of a concrete slab is shown in Fig. 5.3. The concrete slab foundation has the advantages of the concrete ringwall but is usually limited to tanks with diameters less than 30 ft. Often the edge of the slab will be thickened sufficiently to provide for anchorage or to provide some of the characteristics of a ringwall. A slab foundation is very versatile, but its high cost limits it to use in small tanks. The slab provides a level and plane working surface that facilitates rapid field erection or placement of shop-fabricated tanks. It can be used for all types of tanks including those requiring design for uplift. Sometimes leak detection grooves are put at the surface of the foundation.

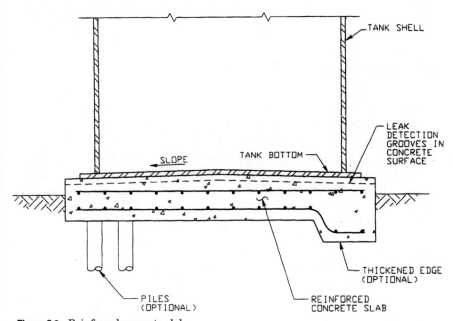

Figure 5.3 Reinforced-concrete slab.

5.3.5 Pile-supported foundations

The pile-supported foundation is usually found where the soil bearing pressures are very low. Examples might be river deltas and land adjacent to bays. These foundations are characteristically found in the Gulf Coast region of the United States where soft, claylike soils extend over wide areas. They are also used where high foundation uplift forces are encountered resulting from internal pressure or seismic loading.

In spite of being the lowest-cost type of foundation in the applications involving them, they still are expensive. Therefore, it is important to economize the design by securing as much accurate soils information as possible. In addition, a civil engineer experienced in pile foundation design should be active in the design process.

5.4 Design Principles

The basic principles of foundation design for tanks are provided in this section. However, nothing precludes the possibility of alternative procedures or methods. Generally, foundation design builds from the principles and requirements outlined in the Uniform Building Code. However, critical structures that would have significantly greater than usual hazards to the environment or public would require that more stringent and very customized design criteria and requirements be applied.

5.4.1 Design procedure

The design starts with a review of the applicable tank codes and standards as well as local building code requirements, tank drawings, vendor drawings, and soil reports. To establish a final design, the relationship between loads and soil bearing capacities arises from factors of safety. Typically, dead loads use a factor of safety of 2.0, and for dynamic loads (wind and earthquakes) it is 1.5. Other factors such as setting the foundation depth below the front line should also be considered.

5.4.2 Design loading conditions

In addition to considerations for integrating alternate functions of the foundation such as leak detection, the primary purpose of the foundation is to satisfactorily carry both static and dynamic loads adequately over the life of the tank. Some of the following load conditions and considerations must be addressed:

5.4.2.1 Dead load. The dead load comprises the weight of the steel used in the tank including the bottom, shell, roof internal, roof supporting structures and appurtenances, and connected piping loads. API 650 assumes a fixed-roof dead load of about 20 psf in roof computations, and this should prove to be conservative for preliminary computations when the actual roof dead load is not known. This value corresponds to the maximum $\frac{1}{2}$-in thickness of roof plates allowed. Remember that although the dead loads are usually averaged out over the area of the tank, local areas under the shell and column bases are subject to much higher loads and are not averaged out due to the lack of stiffness of the bottom plates. Failure to consider this often leads to local soil bearing pressure failures, resulting in extremely localized settlement (e.g., edge settlement).

Column loads. The comments in this section apply to any roof columns within the tank. Generally, the most significantly loaded column will be the center column of a fixed-roof tank. The foundation supporting it must be designed to carry the load transferred to it by the portion of the roof supported by it.

There are basically two ways to support the load from the column. One method is to install a concrete pad under the tank floor. The other is to use a thickened plate that spreads the load out so that the soil loadings are within acceptable limits.

A common problem that arises for design engineers is to determine what soil bearing allowable capacities should be used under columns. If the hydrostatic load caused by the liquid level is deducted from the allowable soil bearing capacity, then this approach often leads to large foundations or bearing plates because the net soil bearing capacity is much reduced. There are many installations where no deduction has been used that are adequately supported. Although some form of multiplier may be used to reduce the soil bearing capacity due to the presence of the hydrostatic load, a reduction is most often not used in practice, and the results have been acceptable (Table 5.2).

In high-seismicity areas, columns are subjected to not only vertical loads but also lateral loading caused by liquid sloshing; and lateral acceleration of the column can produce significant bending stresses. Combined axial compression and bending stresses will often lead to roof column buckling.

5.4.2.2 Live loads

Hydrostatic load. The allowable solid bearing pressure may limit the height of the tank. For example, a 50-ft-high tank storing process water has a live load of 3120 psf. If the allowable soil bearing capacity is 3000 psf, then the design height of the tank must be reduced.

TABLE 5.2   Allowable Foundation and Lateral Pressure

Class of materials	Allowable foundation pressure, psf	Lateral bearing, psf per ft of depth below natural grade	Lateral sliding	
			Coefficient s	Resistance, psf
1. Massive crystalline bedrock	4000	1200	0.70	130
2. Sedimentary and foliated rock	2000	400	0.35	
3. Sandy gravel and/or gravel (GW and GP)	2000	200	0.35	
4. Sand, silty sand, clayey sand, silty gravel, and clayey gravel (SW, SP, SM, SC, GM, and GC)	1500	150	0.25	
5. Clay, sandy clay, silty clay, and clayey silt (CL, ML, MH, and CH)	1000	100		

During earthquakes, there is a sloshing wave generated at the liquid surface. The effect is to increase the hydrodynamic pressures at the bottom of the tank. For the purposes of foundation design, these increases may often be neglected, as the allowable increase of soil pressures and foundation materials can be increased by a factor of one-third to cover transient forces generated on foundations. However, these loads may be computed according to the formulas given in the Seismic Loads section.

Snow load. Snow loading can be a significant portion of the load in certain parts of the country that must be considered. Several references provide methods of establishing snow loads. One is the ANSI/ASCE Publication 7-88, *Minimum Design Loads for Buildings and Other Structures,* and the other is the 1989 Uniform Building Code.

For remote mountain locations, it is prudent to ensure adequate snow load capability because the snow loading can vary significantly from one locale to another separated by a distance of only a few miles. Probably the best source of information is the local building inspection officials. Some tank owners have specified roof snow loads as high as 100 psf.

Thermal loads. Significant load conditions may be introduced into slab and ringwall foundations from thermal conditions, particularly in hot tanks.

Overturning loads and uplift loads. Overturning loads can be induced by either of two possibilities:

Wind loads. Overturning moments induced by wind on empty tanks with a large height-to-diameter ratio can be significant. Although often specified by the purchaser of a tank, wind loads can be assumed to exert pressures of 18 psf on the projected area of a tank shell and 15 psf on the projected area of tank roofs. This is based on a 100 mi/h wind speed. For different wind speeds, these pressures can be factored by $(V/100)^2$, where V is the wind speed in miles per hour.

To maintain a factor of safety of 1.5, for unanchored tanks, the overturning moment should not exceed two-thirds of the dead load resisting moment according to this equation:

$$M_W = \frac{2}{3}\left(\frac{WD}{2}\right)$$

where M_W = wind load overturning moment, ft·lb
   W = weight of tank shell and portion of roof supported by it, lb
   D = diameter of tank, ft

When any combination of wind or internal pressure causes net uplift or exceeds the code requirements for anchorage, then the tank must be anchored.

For anchored tanks the following values for anchor bolt loading should be used:

$$t_b = \frac{4M}{Nd} - \frac{W}{N}$$

where t_b = design tension load per anchor, lb
d = diameter of anchor circle, ft
N = number of anchors

Seismic loads. Tanks can overturn or tip and slide laterally as a result of seismic activity. In high-seismicity zones, the wind load conditions will not govern either the foundation or anchorage requirements. Instead, overturning due to seismic activity will establish the design conditions.

In addition to rigid-body overturning loading, the effect of sloshing waves in the tank actually increases the foundation pressures due to the increased depth of liquid caused by the wave and the acceleration of the liquid. These effects are often overlooked in the design of the foundation.

Although API 650 is the most widely used practice for establishing the overturning moment, other methods have been developed that could change the foundation loading conditions.

Figure 5.4 shows the forces that are generated on a ringwall foundation that must be considered in the design of tank foundations in seismic areas (this diagram also applies to wind load conditions).

The following discussion applies to unanchored tanks.

In establishing load conditions for a foundation subject to seismic overturning, three cases should be considered (see Fig. 5.5).

Case 1: Between 0 seismic moment and some value of an earthquake-induced overturning moment, the weight of the tank shell is sufficient to prevent a net upward force at the edge of the tank, causing uplift of the tank shell. The basic equation for the compressive load on the foundation is

$$b = W_t + \frac{M}{\pi R^2}$$

where b = axial compressive force from tank shell per foot of circumference
W_t = weight to tank shell and portion of roof supported by it
M = seismic overturning moment

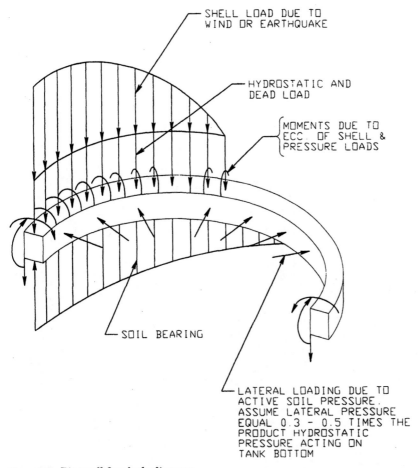

SHELL LOAD DUE TO
WIND OR EARTHQUAKE

HYDROSTATIC AND
DEAD LOAD

MOMENTS DUE TO
ECC. OF SHELL &
PRESSURE LOADS

SOIL BEARING

LATERAL LOADING DUE TO
ACTIVE SOIL PRESSURE.
ASSUME LATERAL PRESSURE
EQUAL 0.3 - 0.5 TIMES THE
PRODUCT HYDROSTATIC
PRESSURE ACTING ON
TANK BOTTOM

Figure 5.4 Ringwall free-body diagram.

R = tank radius
D = tank diameter

In API, this equation is written as

$$b = W_t + \frac{1.273M}{D^2}$$

See Fig. 5.5, case 1.

Case 2: When the overturning moment produces a force that exceeds the deadweight of the shell, the shell tends to lift up from the foundations, but the liquid contents resist this uplift. Many of the seismic theory variations that still exist today involve the nature of the liquid resistance to uplift. However, all theories assume that a column of liquid in the region adjacent to the shell can be used to

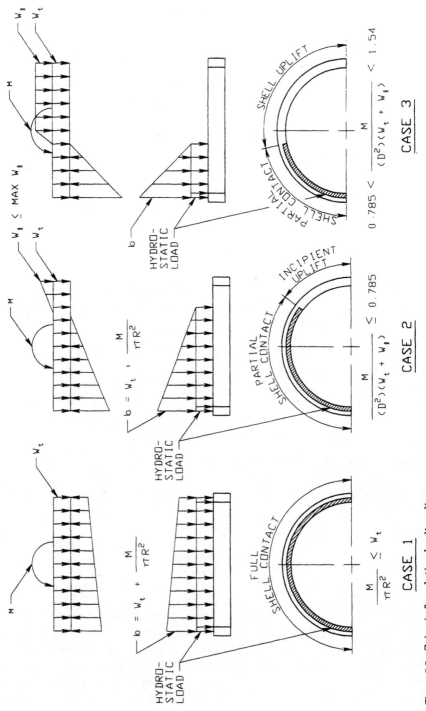

Figure 5.5 Seismic foundation loading diagrams.

resist uplift. As the seismic moment increases, the liquid that resists uplift increases up to a certain value that represents the maximum liquid weight that can be used to resist uplift, as long as the overturning coefficient $M/[D^2(W_t + W_l)] < 0.785$.

The value of the overturning coefficient is determined by setting the uplift force equal to the shell dead load plus the maximum annular hold-down force:

$$\frac{4M}{\pi D^2} = W_t + W_l$$

$$\frac{M}{D^2(W_t + W_l)} = \frac{\pi}{4} = 0.785$$

where W_l = weight of liquid near the tank shell tending to hold the tank down.

Case 3: When the seismic overturning coefficient exceeds 0.785, then the maximum hold-down force is acting to resist overturning for the portion of the tank shell that is uplifted. The models that are the basis for API 650 and AWWA D-100 assume that the limit of stability is based on restricting the eccentricity of the load to a point within the footprint of the shell. AWWA limits the overturning coefficient to a more conservative value of 1.54 than the API limit of 1.57. Setting the vertical load P equal to the hold-down force $\pi D(W_t + W_l)$ and setting the eccentricity API solves for equilibrium under these conditions and uses the following equations:

$$\frac{M}{P} \leq \frac{D}{2}$$

$$\frac{M}{D^2(W_t + W_l)} < \frac{\pi}{2} = 1.571$$

$$b = \frac{1.49(W_t + W_l)}{1 - 0.637M/[D^2(W_t + W_l)]} - W_l$$

where P = vertical load. As the seismic moment increases beyond stability, then additional measures are required to stabilize the tank such as anchorage, changes of the annular plate dimensions or material, and changing the aspect ratio of the tank.

Internal pressure loads.   Since most tank roofs are designed for a minimum live load of 25 psf and a typical roof of $\frac{3}{16}$-in plate weighs about 7 psf, a net uplift can occur at 32 psf, or 0.22 psi. Tanks built in accordance with API 650, Appendix F, may be pressurized up to 2.5 psig, and tanks built in accordance with API 620 may be pressurized up to

15 psig. Because of the relatively large areas over which the internal pressure acts, large uplift forces tending to pull the shell up away from the bottom are involved.

Sufficiently high internal pressure inside a flat-bottom tank produces an uplift force in the shell that must be held down with anchorage. In designing for uplift ringwall foundations, slab foundations are sometimes used. The uplift force on the anchors is counterbalanced by the following forces:

- Weight of the shell and portion of roof supported by the shell
- Weight of the concrete ringwall foundation
- Effective soil weight carried by the foundation
- Friction between the foundation and the earth
- Weight of the column of liquid projecting over the foundation
- Forces generated in the soil by piles, if used

These forces are shown in Fig. 5.6. Generally the uplift is resisted by the foundation; however, piles can be placed under the ringwall or slab to further increase resistance to uplift.

The appropriate combination of counterbalancing forces resisting uplift and the operating conditions is essential. For example, if there is always a minimum liquid head in the bottom of a tank that is never pressurized when empty, then this liquid weight may be used to counterbalance the internal pressure.

5.4.3 Anchorage

Anchorage is used to hold the tank shell down against uplift resulting from the wind overturning moment, seismic overturning moment, or internal pressure. The link that ties the uplift forces produced by seismic, wind, or internal pressure from the tank to the foundation is the anchorage system. The system comprises the foundation-to-anchor bond, the anchor, and the shell-to-anchor bond, or anchor chair. When this system is extended beyond its design limits, a number of different failure modes may occur. Some of these may be acceptable during, e.g., a large seismic event, and others may not. Any condition which leads to loss of tank contents is probably considered unacceptable.

These failure modes in the anchorage system can be briefly categorized by

- Foundation problems

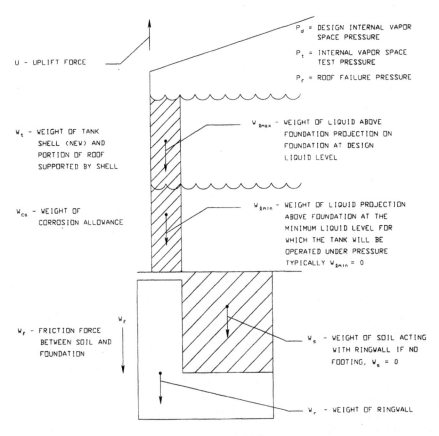

U - UPLIFT FORCE

P_d = DESIGN INTERNAL VAPOR SPACE PRESSURE

P_t = INTERNAL VAPOR SPACE TEST PRESSURE

P_f = ROOF FAILURE PRESSURE

W_t - WEIGHT OF TANK SHELL (NEW) AND PORTION OF ROOF SUPPORTED BY SHELL

$W_{\ell max}$ - WEIGHT OF LIQUID ABOVE FOUNDATION PROJECTION ON FOUNDATION AT DESIGN LIQUID LEVEL

W_{ca} - WEIGHT OF CORROSION ALLOWANCE

$W_{\ell min}$ - WEIGHT OF LIQUID PROJECTION ABOVE FOUNDATION AT THE MINIMUM LIQUID LEVEL FOR WHICH THE TANK WILL BE OPERATED UNDER PRESSURE TYPICALLY $W_{\ell min} = 0$

W_f - FRICTION FORCE BETWEEN SOIL AND FOUNDATION

W_f

W_s - WEIGHT OF SOIL ACTING WITH RINGWALL IF NO FOOTING, $W_s = 0$

W_r - WEIGHT OF RINGWALL

DESIGN CRITERIA

$$\left.\begin{array}{c} U_w + U_p \\ \text{or} \\ U_s + U_p \end{array}\right\} \leq 1.5\ (W_t + W_{\ell min} + W_r + W_f + W_s - W_{ca})$$

$$U_t \leq 1.25\ (W_t + W_r + W_f + W_s)$$

$$U_f \leq 1.5\ (W_t + W_{\ell max} + W_r + W_f + W_s)$$

WHERE:

U_w = UPLIFT DUE TO WIND FORCE
U_s = UPLIFT DUE TO SEISMIC
U_p = UPLIFT DUE TO INTERNAL PRESSURE
U_t = UPLIFT DUE TO P_t, INTERNAL TESTING PRESSURE
U_f = UPLIFT DUE TO P_f, CALCULATED FAILURE PRESSURE OF ROOF TO SHELL FAILURE

Figure 5.6 Uplift counterbalancing weight.

- Pullout of anchor
- Failure of foundation in shear (the pulling out of a shear cone encasing the embedded anchor)
- Tank and chair failures
- Tearing off of the anchor from the shell (weld failure)
- Top-of-chair failure
- Tearing of the tank shell

- Bolt failures
- Necking and yielding of the bolt
- Pullout of the bolt

Anchor bolts should be designed so that in a failure of the anchorage system, the bolts provide ductility. This minimizes the chance of tearing of the shell or loss of anchorage altogether. The chair should be designed so that the bolt yields before the anchor chair or its attachment yields. This can be ensured by designing the chair for some safety factor times the bolt yield strength. The bolt must also yield before the bolt pulls out of the foundation. Several general design criteria listed in the various design codes can be summarized as follows:

1. The design of anchorage and its attachment to the tank shell are a matter of agreement between the purchaser and the manufacturer.

2. Typically, a minimum diameter of 1-in anchor bolts is required by the various codes and standards. When corrosion allowance is specified, then a $\frac{1}{4}$-in minimum is added to the diameter (effective corrosion allowance equals $\frac{1}{8}$ in). Note that adding corrosion allowance to bolts can have a ripple effect in the design of other elements and in the cost of the anchorage system as well. For example, if the diameter of an anchor bolt is increased from 1 to $1\frac{1}{4}$ in for corrosion allowance, the area of the bolt is increased by over 50 percent. This means that the anchor chair must be designed 50 percent larger and that the embedment of the anchor in the foundation must also be designed for this increased load. Since the anchors must fail in a ductile mode in either the new or the corroded condition, the components of the anchorage system should be designed for the condition of the uncorroded bolt. Unless there is a severe corrosion problem introduced by proximity to chemical atmospheres or seacoast environments, it may be more effective to paint the anchors or to consider the effects of corrosion than to arbitrarily specify an extra corrosion allowance.

3. Anchors should not be placed farther than 10 ft apart.

4. Thermal expansion should be accounted for when the tank temperature exceeds 200°F.

5. The number and size of anchor bolts can be determined from the information provided in Table 5.3.

5.4.4 Anchor chair design

TABLE 5.3 Anchor Bolts

Materials	Yield strength, ksi	Tensile strength, ksi	Cited in code
ASTM A 36	36	58	API 620
A 307		55	API 620, API 650, AWWA
A 193	105	125	API 620, API 650
A 325			AWWA
A 320	105	125	API 620
Allowable stresses for condition of uplift due to:			
Internal pressure	650	620	AWWA
	σ_a	σ_a	
Internal pressure + wind	$(4/3)\sigma_a$	Min. $(4/3)\sigma_a/(0.8\sigma_y)$	NA
Pressure + seismic	$(4/3)\sigma_a$	Min. $(4/3)\sigma_a/(0.8\sigma_y)$	$(4/3)\sigma_a$
Internal test pressure	σ_y	Min. $(4/3)\sigma_a/(0.8\sigma_y)$	NA
Failure pressure			NA

	σ_a for:		
	API 620	API 650	AWWA
A 36	15.3	15 (all materials)	15 (all materials)
A 193 B7	24.0		
A 307	15.0		
A 320	24.0		

163

Anchor chair design is based on the procedure described in Ref. 11. If the chair functions as designed, the bolt will yield before the tank shell or chair yields. However, it is anticipated that the tank shell will experience some local yielding before the bolt fails. Since the maximum stress in the shell diminishes quickly as the distance from the chair is increased, this local yielding is not considered a problem.

Allowable shell stresses for the chair design are taken from API 650, chapter 3, which for most shell materials is 20 kilopounds per square inch (ksi). With the one-third increase for seismic loads, the allowable stress becomes 26.6 ksi. Allowable chair stresses are $0.6(1.33)F_y$, which, for A 36 material, equals an allowable stress of 28.8 ksi.

Figure 5.7 shows loads and design criteria that can be used in accordance with Ref. 1 to design anchor chairs.

5.4.5 Allowable soil pressure

Most codes accept increasing allowable soil bearing pressures due to wind or earthquake loadings by a factor of one-third more than for normal service. The stability ratio of resisting moment to either seismic or wind-induced moments should be equal to or greater than 1.5.

Foundation soil bearing pressures for wind or seismic loading conditions should be based upon sound engineering principles and judgment, taking into account the nature of the subsoil and load distribution. In the absence of better data, the maximum soil bearing pressure should not exceed that permitted under normal loading conditions when the subsoils are cohesionless and may be up to 2 times that permitted under normal conditions when the subsoils are cohesive.

5.4.6 Final designs

After preliminary foundation cases are compared and estimated, an agreement between the purchaser and the designer can be reached about the desired way to go. Often, this is specified in the bid documents, but it can often be influenced by the knowledge of the designer. The designer should point out the advantages and disadvantages of the various designs, taking into account that foundations often serve multiple purposes such as leak detection and containment.

References

1. Irving E. Boberg, *Oil Storage Tank Foundation,* The Water Tower, Chicago, Ill., March 1951.
2. Joseph E. Bowles, *Foundation Analysis and Design,* 4th ed., McGraw-Hill, New York, 1988.

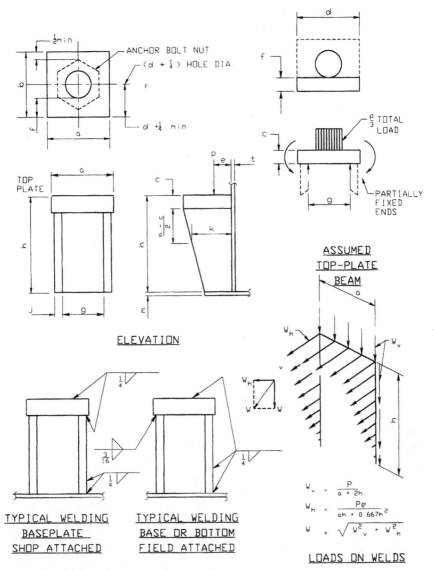

TOP PLATE

$\frac{1}{2}$min

ANCHOR BOLT NUT

(d + $\frac{1}{4}$) HOLE DIA.

d +$\frac{1}{4}$ min

$\frac{P}{3}$ TOTAL LOAD

PARTIALLY FIXED ENDS

ELEVATION

ASSUMED TOP-PLATE BEAM

TYPICAL WELDING BASEPLATE SHOP ATTACHED

TYPICAL WELDING BASE OR BOTTOM FIELD ATTACHED

LOADS ON WELDS

$$W_v = \frac{P}{a + 2h}$$

$$W_h = \frac{Pe}{ah + 0.667h^2}$$

$$W = \sqrt{W_v^2 + W_h^2}$$

Figure 5.7 Anchorage design.

3. Wang and Salmon, *Reinforced Concrete Design,* 4th ed., Harper & Row, New York, 1985.
4. Lockheed Aircraft Corporation and Holmes and Narver, Inc., *Nuclear Reactors and Earthquakes,* Chap. 6 and App. F, ERDA TID 7024, August 1963, pp. 183–195, 367–390.
5. R. S. Wozniak and W. W. Michell, "Basis of Seismic Design Provisions for Welded Steel Oil Storage Tanks," presented at the 43d midyear meeting of the American

Petroleum Institute, May 1978.
6. American Water Works Association, *AWWA Standard for Welded Steel Tanks for Water Storage*, ANSI/AWWA D-100-84 (AWS D5.2-84).
7. American Petroleum Institute, *Welded Steel Tanks for Oil Storage*, API Standard 650, 9th ed.
8. American Petroleum Institute, *Design and Construction of Large, Welded, Low-Pressure Storage Tanks*, API Standard 620, 8th ed.
9. W. C. Young, *Roark's Formulas for Stress and Strain*, 6th ed., McGraw-Hill, New York.
10. M. H. Jawad and J. R. Farr, *Structural Analysis and Design of Process Equipment*, 2d ed., Wiley, New York, 1989.
11. American Iron and Steel Institute, *Steel Plate Engineering Data*, vol. 2: *Useful Information on the Design of Plate Structures*, Washington, D.C.

6

Seismic Considerations

SECTION 6.1   SEISMIC DESIGN

6.1.1   Overview

In this chapter the primary focus is on the provisions of API 650, Appendix E. While the provisions of Appendix E are not mandatory, they are the most widely used methods for assessing seismic vulnerability for both existing and new tanks.

Although tanks are considered quite flexible structures, there is a limit to the amount of ground shaking or seismic activity to which they may be subjected without damage and/or spillage of contents. Figures 6.2.3 to 6.2.7 give examples of the type of damage that occurs during severe seismic activity. These photos show examples of damage that occurred to a water storage tank near the Los Angeles basin during the Northridge earthquake of 1992.

Notice in all these examples that the damage occurs near the base of the tank as a result of either uplift of the tank base from the foundation and tearing or breaking of anchorage and/or shell as well as from crushing of the lower portion of the bottom course which tears shell components such as attached piping and reinforcing pads.

The problem of understanding the seismic considerations of API 650, Appendix E can be divided into four broad categories:

- Risk assessment
- Fluid hydrodynamics
- Buckling of thin shells
- API methodology and assumptions

6.1.2 Risk Assessment

See the discussion of risk assessment in Sec. 6.2.1.

Fortunately, API 650, Appendix E has made the task of providing reasonable safety and economy associated with seismicity of tank design by providing a cookbook approach. By dividing the geographical regions of the United States into areas of differing seismicity as well as establishing an importance factor, the problem of setting reasonable risk becomes simply a matter of completing the calculation methodology used in Appendix E. Because of the simplicity and elegance of API 650, Appendix E it has survived pretty much unchanged since 1978 even though more accurate methodologies have been developed.

6.1.3 Fluid Hydrodynamics

The fundamental design problem confronting the design engineer after having established a design basis earthquake is to find the forces acting on the tank shell and foundation so that seismic capacity may be confirmed. To do this is fundamentally a problem of establishing the hydrodynamic fluid pressures exerted against the tank wall and bottom. Once the pressures are known, they can be integrated to find total forces and moments.

Critical to the tank designer are:

- Hoop tension, axial compression and tension, and end shear and moments in the tank wall as a result of hydrodynamic effects.
- Sloshing wave effects.
- Moments and forces transmitted to the foundation and soil.
- Other variables such as bending moments in upper shell courses may be important in some applications.

6.1.3.1 Basic model

Figure 6.1.1a shows a tank that has been subjected to horizontal acceleration. Because waves are created by the seismic forces there are really two components that are dealt with in the simplified model used to derive the forces and moments acting on the tank. These are the *impulsive* and *convective* components. The model assumes that part of the tank contents acts like a rigid solid that moves exactly in unison with the ground motion and tank shell, which is assumed to be rigid. The part of the model that represents the sloshing waves is modeled as a mass attached to the rigid tank walls by springs. While the assumption of a rigid tank leads to some error it is not significant

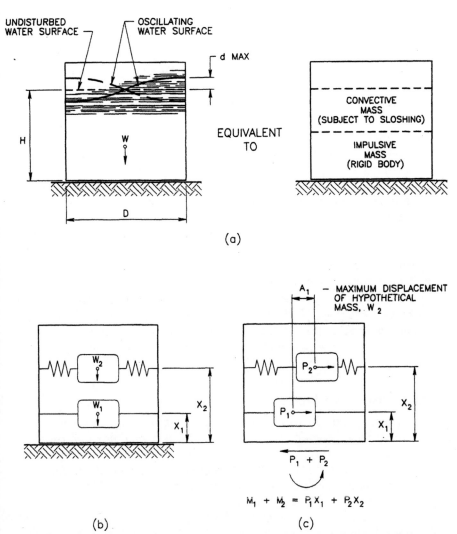

Figure 6.1.1 Simplified model for seismic effects on ASTs. (*a*) Fluid motion in tank; (*b*) dynamic model; (*c*) dynamic equilibrium of horizontal forces.

for either open top tanks with wind girder stiffeners or fixed roof tanks which are stiffened by the roof. The impulsive and convective components of the model are shown in Figs. 6.1.1*b* and *c*.

Impulsive component. Below some depth, the liquid acts as if it were a solid body, moving in unison with the tank and ground motion. This mode is the *impulsive mode* and is modeled as a rigid body accelerated uniformly at the same rate as the ground motion. The force P_1 is simply proportional to the acceleration and it acts through the centroid of the impulsive portion of the tank liquid mass W_1 at a distance X_1 above the ground producing the portion of the overturning moment on the tank due to the impulsive component. W is the total mass of the tank and liquid. The subscripts 1 and 2 refer to the impulsive and convective components respectively.

Convective component. Near the surface the liquid is free to accelerate vertically forming sloshing waves. This can be represented by a mass connected to springs which are tied to the tank walls. The period of such waves for large tanks is typically 6 to 10 s for large tanks. Some models consider the effects of the flexibility of the tank walls including multiple masses with spring attachments of varying stiffnesses to achieve better accuracy. Because the first sloshing mode frequency is usually so much lower than the motions associated with dominant seismic excitation frequencies, the assumption of a rigid tank does not usually produce a significant error. When the tank is assumed to be a rigid structure, a single mass-spring system is used as shown in Fig. 6.1.1b. The force P_2 is the force acting on the convective component of the tank liquid mass which acts convectively, W_2. This force acts through the centroid of the convective mass at a distance X_2 above the ground.

6.1.3.2 Simplified approach

The latest approaches to establishing the hydrodynamic pressures associated with seismic activity tend to be very complex and difficult to use. The advantage of using the original approach, worked out by Housner[1] and adapted for API by Wozniak and Mitchell,[2] is that the models illustrated above are sufficiently accurate for engineering purposes, and they are simple to use.

The result of computing the hydrodynamic pressures using any of these methods is similar to Fig. 6.1.2. The horizontal acceleration acting on a tank produces a resultant force P which is the sum total hydrodynamic pressures acting on the wall of the tank. This force tends to translate the position of the tank, while at the same time tending to cause overturning. It varies with cos ϕ reaching a peak value at $\phi = 0$, where ϕ is the angle between the direction of the ground motion and the direction under consideration. The base shear just above the bottom of the tank is numerically equal to P. This also induces a moment M at a section just above the base of the tank.

As shown in Figure 6.1.2 the dynamic pressures acting on the tank bottom membrane also tend to introduce overturning into the foundation. The methods in API 650 are aimed at addressing the weakest component under consideration, the tank, and therefore do not esti-

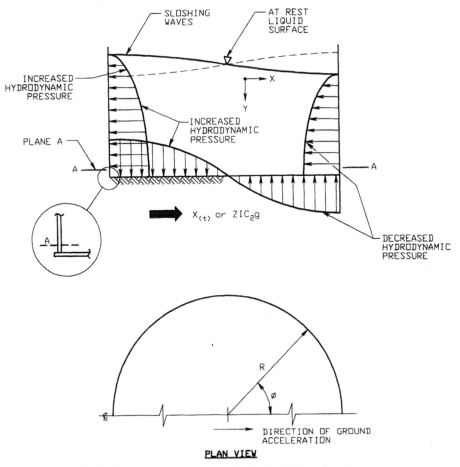

Figure 6.1.2 Hydrodynamic pressures caused by horizontal acceleration.

mate the increased load on the tank foundation resulting from the hydrodynamic pressures acting on the tank bottom. These loads can be significant, and procedures used to determine the foundation loads are given later in this section.

6.1.3.3 Tank aspect ratio

The model described above includes one other important feature that is illustrated by Fig. 6.1.3. Whenever the tank has an aspect ratio (height-to-diameter ratio) greater than 3:4 it is a *tall tank*. That is, any liquid beneath this depth is always treated as a rigid body (impulsively). Since the diameter-to-height ratio is a constant for this condition as shown by the dotted line in Figure 6.1.3 for tall tanks, a constant ratio may be used to determine the convective liquid portion

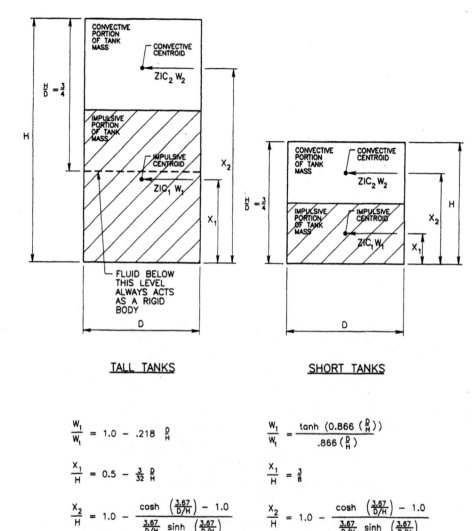

Figure 6.1.3 Tall and short tank models.

of the tank contents. This leads to the different formulas for both W_1 and X_1 for the tall versus short tanks as shown in the figure. Figure 6.1.3 is the basis for the formulas used to compute pressures, centroid locations, and impulsive and convective masses given in the equations below.

6.1.3.4 Horizontal forces

When designing structures the seismic engineer will want to know the maximum ground shaking that will occur during the life of the structure. This data is not usually available, so what is often used is an estimate of the future probabilities of ground shaking magnitude based upon seismological and geological data from the local area. This also is difficult to ascertain because of the short time frame over which data have been collected and the number of seismographs in use. However, some seismic methods in use involve the application of a design spectrum to determine components of acceleration and response. For example, the New Zealand Code requires that the engineer start with charts of annual probabilities of exceedence of design earthquakes. Then earthquake design spectra are required. These are based on limited data based on extrapolation of seismicity models and the adoption of wave attenuation expressions derived from overseas data. The code then goes on to provide recommended earthquake zones, geographic coefficients, acceleration spectra, and probability factors. In all, the procedure is complex and hard to understand.

The API codes have used the method of the Uniform Building Code because of its simplicity and the ease with which the components of acceleration are applied to the structures. The method of API is very easy to apply because it is based on designs to resist seismic forces based upon equivalent lateral forces under static conditions. That is, the structure is designed to resist a statically applied horizontal force instead of actual dynamic inertial forces resulting from ground shaking. However, for API tank seismic design there has been controversy about the lateral force coefficients for sloshing modes, levels of damping, estimation of compressive forces for a tank, and treatment of tank flexibility and buckling.

For tanks, the equivalent static design force is ZIC_1W_1 for the impulsive portion of the tank contents or ZIC_2W_2 for the convective portion of the tank contents applied to the respective centroids of the respective masses as shown in Fig. 6.1.3.

6.1.3.5 Wall and bottom pressures

The total pressure at any point on the tank wall is given by

$$P_T = P_1 + P_2 \qquad (6.1.1)$$

where P_1 is the impulsive component of total hydrodynamic pressure and P_2 is the convection component of total hydrodynamic pressure.

The impulsive tank wall pressure at any depth Y below the liquid surface in the direction of ground shaking on the increased pressure side is given by

$$P_1 = 108GZIC_1H\left[\left(\frac{Y}{H}\right) - \frac{1}{2}\left(\frac{Y}{H}\right)^2\right]\tanh\left(0.866\frac{D}{H}\right)$$

$$\text{for short tanks } (H/D < \tfrac{3}{4}) \quad (6.1.2)$$

$$P_1 = 66.4 GZIC_1 D \left[\left(\frac{Y}{0.15D} \right) - \frac{1}{2} \left(\frac{Y}{0.75D} \right)^2 \right]$$

for tall tanks ($H/D > \frac{3}{4}$), where Y is less than $\frac{3}{4}D$ (6.1.3)

and

$$P_1 = 33.2 GZIC_1 D^2 \qquad \text{where } Y \text{ is greater than } \frac{3}{4}D \qquad (6.1.4)$$

The convective wall pressure at any depth Y is given by

$$P_2 = 23.4 GZIC_2 D \frac{\cosh\left(3.67 \dfrac{H-Y}{D} \right)}{\cosh\left(3.67 \dfrac{H}{D} \right)} \qquad (6.1.5)$$

Figure 6.1.4 shows the wall pressures developed by the effective masses acting impulsively and convectively. The pressures are in the direction of acceleration on the high-pressure side. On the opposite side the pressures are negative or reduced. It should be understood that for very short tanks the main contribution to horizontal force is from the convective effect and for very tall tanks it is from the impulsive effect.

Both impulsive and convective components of the model give rise to pressures acting on the bottom of the tank. These are shown in Fig. 6.1.5 for a tank with a $H/D = 0.75$. For considering the integrity of the tank itself, these are not important. However, for considering the effect of the overturning moment on the foundation they have a considerable effect. This is because the bottom is considered to be a flexible membrane, and the pressure on the bottom does not impact the forces acting on the tank shell itself. While this is not strictly true, for the purpose of analysis this assumption can be made. In fact, the basis for other seismic models does include the membrane forces acting in the tank bottom during uplift to provide additional resistance to overturning. So in this sense, API 650 is conservative.

6.1.3.6 Effective portion of mass that is impulsive or convective

The expression for the fraction of total liquid weight that contributes to the impulsive component of horizontal force is

$$\frac{W_1}{W} = \frac{\tanh(0.866D/H)}{0.866D/H} \qquad \text{for short tanks } (H/D < \frac{3}{4}) \qquad (6.1.6)$$

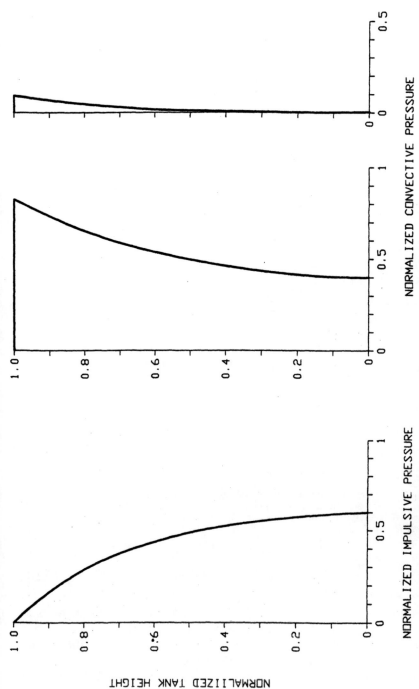

NORMALIZED IMPULSIVE PRESSURE

NORMALIZED CONVECTIVE PRESSURE

NORMALIZED TANK HEIGHT

Figure 6.1.4 Impulsive and convective shell pressures. Dimensionless functions for impulsive and convective components of hydrodynamic wall pressure for tanks with $H/R = 0.75$.

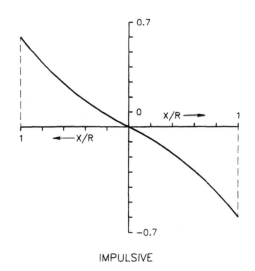

IMPULSIVE

CONVECTIVE

Figure 6.1.5 Impulsive and convective bottom pressures. Dimensionless functions for impulsive and convective components of hydrodynamic base pressure for tanks with $H/R = 0.75$.

$$\frac{W_1}{W} = 1.0 - 0.22\frac{D}{H} \qquad \text{for tall tanks } (H/D > \tfrac{3}{4}) \qquad (6.1.7)$$

Figure 6.1.6 is a plot of the effective impulsive versus convective mass normalized by dividing by the total mass. Notice that the convective mass is predominant as the tank becomes shallower but the impulsive mass is predominant for the tall tanks.

To exert an equivalent moment to that exerted by the fluid pressure on the walls only, the horizontal force component acts at a distance above the base of the tank of

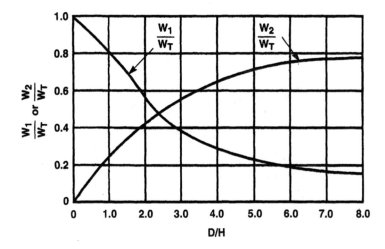

Where

$$\frac{D}{H} > \frac{4}{3} \qquad \frac{W_1}{W_T} = \frac{\tanh (0.866 \, D/H)}{0.866 \, D/H}$$

$$\frac{D}{H} < \frac{4}{3} \qquad \frac{W_1}{W_T} = 1.0 - 0.218 \, (D/H)$$

For All D/H $\quad W_2 = 0.230 \left(\frac{D}{H}\right) \tanh \left(\frac{3.67}{D/H}\right)$

Figure 6.1.6 Effective mass of seismic impulsive and convective components. (API 650, Fig. E-2.) (*Courtesy of American Water Works Association.*)

$$\frac{X_1}{H} = \frac{3}{8} \qquad \text{for short tanks } (H/D < \tfrac{3}{4}) \tag{6.1.8}$$

$$\frac{X_1}{H} = 0.5 - \frac{3}{32}\frac{D}{H} \qquad \text{for tall tanks } (H/D > \tfrac{3}{4}) \tag{6.1.9}$$

Figure 6.1.7 is a plot of the height of the centroids of the impulsive and convective portions of the tank contents. This is the same as the plot found in API 650, Appendix E. Since the horizontal force acts at the centroid of the horizontal pressure wall, pressure which is assumed to be caused by a rigid body, the centroid is at a height of one-half of the elevation of the portion of liquid contents that acts impulsively.

It is important to note here that the effects of the bottom pressures generated by seismic acceleration are not included in these expressions. In other words, the moment determined from the expressions above are equivalent to the moment applied to the tank at a plane *A-A* just above the tank bottom as shown in Fig. 6.1.2. This moment is the

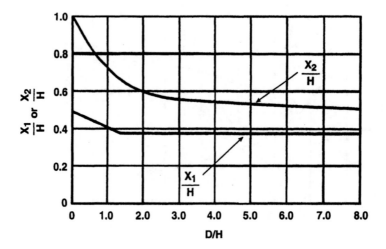

For $\dfrac{D}{H} > \dfrac{4}{3}$ $\dfrac{X_1}{H} = 0.375$

$\dfrac{D}{H} < \dfrac{4}{3}$ $\dfrac{X_1}{H} = 0.5 - 0.094\ (D/H)$

For All $\dfrac{D}{H}$ $\dfrac{X_2}{H} = 1.0 - \dfrac{\cosh\left(\dfrac{3.67}{D/H}\right) - 1}{\left(\dfrac{3.67}{D/H}\right)\ \sinh\left(\dfrac{3.67}{D/H}\right)}$

Figure 6.1.7 Centroid location of effective seismic impulsive and convective masses. (API-650, Fig. E-3.) (*Courtesy of American Water Works Association.*)

one that is used to determine both the uplift forces and the compressive forces acting on the tank shell near the base.

For foundation design, the effects of bottom pressure must be considered as the overturning moment below plane *A-A* of Fig. 6.1.2 can be substantially higher than the overturning moment applied to the tank. In order to account for the increased moment on the tank foundation by the hydrodynamic fluid pressures acting on the tank bottom, the height at which the concentrated resultant wall pressure acts above the tank bottom may be increased to produce an equivalent moment as follows.

For determination of the height of the centroid of the impulsive component of the tank mass use

$$\frac{X_1'}{H} = \frac{3}{8}\left[1.0 + \frac{4}{3}\left(\frac{0.866D/H}{\tanh 0.866D/H} - 1.0\right)\right] \qquad \text{for short tanks } (H/D < \tfrac{3}{4})$$

$$(6.1.10)$$

$$\frac{X_1'}{H} = 0.5 + 0.06\frac{D}{H} \qquad \text{for tall tanks } (H/D > \tfrac{3}{4}) \qquad (6.1.11)$$

The convective response is modeled by a single mass attached to rigid tank walls with springs as shown in Fig. 6.1.1c. The response of the convective portion of the liquid is free oscillations of the fluid surface in the fundamental mode. Integration of the resultant wall pressures leads to the total horizontal force acting on the wall:

$$\frac{W_2}{W} = 0.23\frac{D}{H} \tanh\left(\frac{3.67H}{D}\right) \qquad (6.1.12)$$

For determination of the height of the centroid of the convective component of tank mass use

$$\frac{X_2}{H} = 1.0 - \frac{\cosh(3.67H/D) - 1.0}{(3.67H/D)\sinh(3.67H/D)} \qquad (6.1.13)$$

If the pressure acting on the bottom of the tank is accounted for by increasing the height of the moment arm to

$$\frac{X_2'}{H} = 1.0 - \frac{\cosh(3.67H/D) - 31/16}{(3.67H/D)\sinh(3.67H/D)} \qquad (6.1.14)$$

then the convective force is the equivalent of a mass of weight W_2 attached to the tank walls and vibrating according to

$$X = A_1 \sin wt \qquad (6.1.15)$$

where

$$W_2 = \frac{3.67g}{D} \tanh\left(3.67\frac{H}{D}\right) \qquad (6.1.16)$$

$$A_1 = 0.27\Theta_h \frac{1}{\tanh(3.67H/D)} \qquad (6.1.17)$$

$$\tau = \frac{2\pi}{W} = \frac{2\pi\sqrt{D}}{\sqrt{3.67g \tanh(3.67H/D)}} \qquad (6.1.18)$$

The sloshing wave that is generated by the seismic accelerations can be estimated by

$$d_{max} = \frac{3}{4}\sqrt{\frac{27}{8}}A \tanh\left(4.77\sqrt{\frac{H}{D}}\right)$$

$$= 1.124ZIC_2\, \tau^2 \tanh\left(4.77\sqrt{\frac{H}{D}}\right) \qquad (6.1.19)$$

6.1.3.7 Sloshing modes

The acceleration coefficients C_1 and C_2 required for seismic calculations in accordance with API 650, Appendix E are computed as follows:

$$
C_2 = \begin{cases} \dfrac{0.75S}{T} & \text{when } T \le 4.5 \\[2ex] \dfrac{3.375S}{T^2} & \text{when } T > 4.5 \end{cases}
$$

where S = site coefficient from Table 6.1.1 and T = natural period of the first sloshing mode, in seconds.

The work above is based on an infinitely rigid tank wall. Tanks typically have natural periods of vibration on the order of 0.10 to 0.25 s. Further work shows that the rigid tank concept is reasonable if the maximum ground acceleration is replaced with the spectral value of the pseudo acceleration corresponding to the fundamental natural frequency of the tank fluid system. A constant value for this of 0.6 is used which is the value of C_1. The value of 0.6 is consistent with the Uniform Building Code.

The period of the first sloshing must be determined to calculate C_2. The formula for C_2 is based on a maximum spectral velocity of 1.5 to 2.3 ft/s and a maximum spectral displacement of 1.1 to 1.65 ft depending on soil type. The period of the first sloshing mode can be determined from $T = kD^{1/2}$, where k is determined from Figure 6.1.8 and the formulas

$$
T = \frac{2\pi\sqrt{D}}{3.67g \tanh\left(\dfrac{3.67}{D/H}\right)} \quad \text{and} \quad k = \frac{0.578}{\sqrt{\tanh \dfrac{3.67}{(D/H)}}}
$$

TABLE 6.1.1   Site Coefficients

Soil type	Description	S factor
S_1	A soil profile with either (a) a rocklike material characterized by a shear wave velocity greater than 2500 ft/s or by other suitable means of classification or (b) stiff or dense soil conditions where the soil depth is less than 200 ft	1.0
S_2	A soil profile with stiff or dense soil conditions where the soil depth exceeds 200 ft	1.2
S_3	A soil profile 40 ft or more in depth containing more than 20 ft of soft to medium stiff clay but more than 40 ft of soft clay	2.0 1.5
S_4	A soil profile containing more than 40 ft of soft clay	2.0

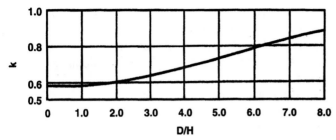

Figure 6.1.8 Factor k.

6.1.4 Buckling of Thin Shells

6.1.4.1 Basis for allowable buckling stresses

If a cylindrical shell as shown in Fig. 6.1.6 is subjected to axial compression, the theoretical buckling stress is given by

$$\sigma_e = \frac{Et}{\nu\sqrt{3(1-\nu^2)}} \tag{6.1.20}$$

where σ_e = elastic buckling stressing
E = modulus of elasticity
t = shell wall thickness
r = radius
ν = Poisson's ratio

Substituting Poisson's ratio, $\nu = 0.3$, the expression becomes

$$\sigma_{\text{Cl}} = 0.6\frac{Et}{r} \tag{6.1.21}$$

The expression above, called the classical buckling stress formula, only applies to materials within the proportional limit which occurs if the value of t/r is small (i.e., thin shells). When the thickness-to-radius ratio is greater, then adjustments must be made to the formula to account for the nonlinearity of the stress-strain curve. Even thicker shells will fail by plastic collapse or yielding rather than elastic buckling. This phenomenon is called inelastic or plastic buckling.

After theoretical foundations were established for buckling, a number of tests were conducted to verify the results. In all cases the actual buckling stresses were much less than the expected values. In addition, the discrepancy between experiment and theoretical values increased for thinner shells.

In early work on buckling of cylindrical shells it was suggested that actual buckling stresses are approximately one-third of the theoretical buckling stresses because the actual fabricated shells had imperfections in the geometry that significantly affected their capacity for buckling loads. Theoretical work has confirmed that small imperfections indeed have the ability to drastically reduce the buckling capacity of cylindrical shells.

API Bulletin 2U gives a series of buckling formulas for buckling of unstiffened, ring-stiffened, and bay-stiffened cylinders. The formulas given below are based on local buckling of unstiffened or ring-stiffened cylinders and are classified as to whether the buckling mode is elastic or plastic. Further, it is assumed that the axial buckling stress, whether caused by uniform axial compression or as a result of bending (as in seismic), is the same.

API Bulletin 2U shows for elastic buckling stresses (σ_e)

$$\sigma_e = \frac{\alpha C E t}{r} \tag{6.1.22}$$

where $\alpha = 0.207$ for shells where r/t is greater than 610 and $C = 0.605$ is based upon typical tank dimensions.

The equation for buckling of thin shells then becomes

$$\alpha_e = 0.125 E \left(\frac{t}{R} \right) \tag{6.1.23}$$

When shells are thicker than about $r/t = 300$, then the inelastic buckling stress is limited to one-half of yield.

6.1.4.2 API 650 allowable compressive stresses

For very thin shells (most tanks) API uses Eq. (6.1.23) to determine the elastic buckling stress, but it has an overall limit of one-half of the yield strength to cover small values of r/t. If a safety factor of 1.5 is applied, then the allowable stress F_a may be written

$$F_a = \frac{0.125(29 \times 10^6)}{1.5} \left(\frac{t}{r} \right) = 2.416 \times 10^6 \left(\frac{t}{r} \right) \cong 400000 \frac{t}{D} \tag{6.1.24}$$

where D is in feet.

However, theoretical work on shell buckling has also showed that the stability ultimately increases. Experiments and theoretical work showed that buckling pressure increases according to a curve similar to that shown in Fig. 6.1.9. The abscissa plots $\bar{p}$ which basically represents pressure versus the buckling coefficient which we designate as C_p. The parameter $\bar{p}$ is defined as

$$\bar{p} = \frac{P}{E} \left(\frac{R}{t} \right)^2$$

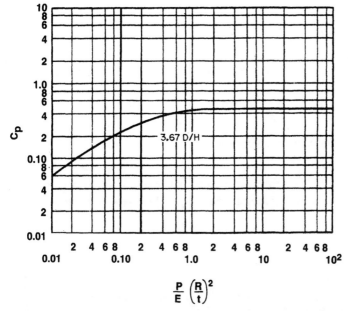

$$\frac{P}{E}\left(\frac{R}{t}\right)^2$$

Figure 6.1.9 Increase in buckling strength of cylinders with internal pressure. (*Source: Baher et al., Structural Analysis of Shells, Krieger Publishing Co., Malabar, Fla., 1972.*)

Converting this to the terminology of API 650 by defining $\xi = GHD^2/t^2$

$$\bar{p} = 0.538\times10^{-6}\xi$$

By rewriting Eq. (6.1.23) to include pressure we have

$$F_a = \frac{(0.125 + C_p)}{FS}\, E\!\left(\frac{t}{R}\right) = \frac{(0.125 + C_p)}{1.5}\, E\!\left(\frac{t}{6D}\right) \quad (6.1.23)$$

It was recognized that a square root function approximated the values in Figure 6.1.9, but a factor of safety of 2 on the chart was used as follows:

$$C_p = k\sqrt{\bar{p}}$$

Using a midrange value of 0.6 for the abscissa and 0.4 for ordinate the value of k is determined to be 0.26. This allows us to write the allowable stress as

$$F_a = \frac{(0.125 + 0.26\sqrt{\bar{p}})}{1.5}\, E\!\left(\frac{t}{6D}\right)$$

Substituting and rounding numbers

$$F_a = 400{,}000\,\frac{t}{D} + 600\sqrt{\frac{GHD^2}{t^2}}\left(\frac{t}{D}\right) = 400{,}000\,\frac{t}{D} + 600\sqrt{GH}$$

This is the equation found in API 650, Appendix E. However, this equation is limited to values of $\zeta \leq 10^6$. API caps the allowable stress $F_a = 10^6 t/D$ for values of $\zeta > 10^6$ and has an ultimate limit on $F_a = \frac{1}{2}F_y$, where F_y is the yield strength of the material.

6.1.5 Seismic Demand as Specified by API Codes

Figure 6.1.10 shows the factors on which the resistance to overturning as a result of liquid loading are based. The assumption is made

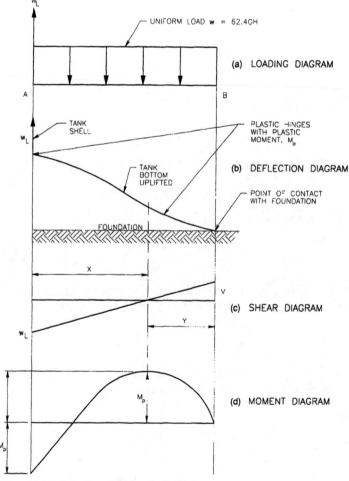

Figure 6.1.10 Seismic uplift free-body diagram.

that plastic hinges develop in the tank bottom as shown. Plastic moments M_p occur at the shell-to-bottom junction as well as a point at a distance x from the shell.

The calculations are based on small deflection theory meaning that no membrane stresses act in the bottom plates to further resist uplift. Using static equilibrium conditions, we sum the moments at point A and get

$$\Sigma M_A: \qquad 2M_p = \frac{wx^2}{2} \tag{6.1.33}$$

The uplift force on the shell is the weight of liquid acting on the left side of the plastic hinge:

$$w_L = wx \tag{6.1.34}$$

Combining Eqs. (6.1.33) and (6.1.34),

$$w_L = 2\sqrt{wM_p} \tag{6.1.35}$$

The next step is to sum movements at point B:

$$\Sigma M_B: \qquad M_B = w_L - \frac{w_L^2}{2} \tag{6.1.36}$$

Substituting Eq. (6.1.35) into Eq. (6.1.36) and simplifying,

$$L = 1.707 \frac{w_L}{w} \tag{6.1.37}$$

Using the bending stress in a plate subject to a moment with a safety factor of 1.5, we find w_L with the following relations:

$$M_p = \frac{F_{by} t_b^2}{4} \qquad \text{and} \qquad w = 62.4GH \tag{6.1.38}$$

Substituting these into (6.1.35) we have

$$w_L = 7.9 t_b \sqrt{f_{by} GH} \tag{6.1.39}$$

This equation is used in the API standards for the resistance to overturning that may be attributed to the liquid contents on the tank. The liquid resistance to uplift is, however, arbitrarily limited to uplift length L to less than about 7 percent of the tank radius. This is accomplished by setting

$$w_L \leq 1.25GHD \tag{6.1.40}$$

Substitution into Eq. (6.1.37) gives

$$L = 0.068R \tag{6.1.41}$$

Having established a basis of the hold-down force resulting from the liquid weight w_L, it is possible to derive the shell compression formulas that are used in API Standards 650 and 620.

For tanks without uplift or tanks that are anchored, the formulas that are found in basic strength-of-material textbooks can be used assuming the tank is a single cylindrical section subjected to a bending movement:

$$N = \frac{M}{\pi R^2} = \frac{4M}{\pi D^2} = \frac{1.273M}{D^2} \qquad (6.1.42)$$

where N = the unit axial shell force in pounds per foot.

On the compression side of a tank with a seismically induced movement, the maximum unit force on the shell or foundation is

$$b = w_t + \frac{1.273M}{D^2} \qquad (6.1.43)$$

where w_t = the weight of the tank shell and a portion of the roof weight supported by the shell in pounds per foot of shell circumference.

For anchored tanks, the maximum uplift force is found by subtracting the weight of the tank:

$$U = \frac{1.273M}{D^2} - w_t \qquad (6.1.44)$$

For unanchored tanks that experience uplift, the following derivation is the basis for the API Standards. Assume that uplift has occurred as a result of a seismic overturning movement as shown in Fig. 6.1.11a. The loading diagram shows that the shell weight w_T (including the portion of the roof supported by it), and the weight of liquid w_L that is near the shell and carried by the annular region during uplift are balanced by compression forces generated on the portion of shell in contact with the foundation.

By a simple trick, which is to add w_L to b approximating the liquid hold-down force as shown in Fig. 6.1.11b, a simplification can be made that allows for determination of the compression forces. Summing all vertical forces in Fig. 6.1.11b to equate to zero:

$$2 \int_0^\beta f_b R \, d\theta = 2\pi R(w_T + w_L) \qquad (6.1.45)$$

By proportioning the shell compression force according to the diagram in Fig. 6.1.11b:

$$f_b = (b + w_L)\frac{\cos \theta - \cos \beta}{1 - \cos \beta} \qquad (6.1.46)$$

and substituting and integrating:

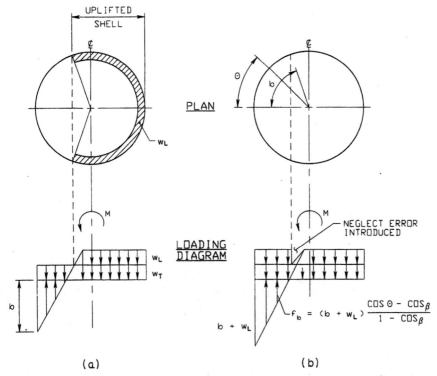

Figure 6.1.11 Seismic uplift load diagrams. (a) Assumed load distribution; (b) equivalent load distribution.

$$\frac{b + w_L}{w_T + w_L} = \frac{\theta(1 - \cos \beta)}{\sin \beta - \beta \cos \beta} \tag{6.1.47}$$

By summing up the moments we obtain

$$M = 2R \int_0^\beta f_B R^2 \cos \theta \, d\theta = \frac{D^2(b + w_L)}{2} \frac{\frac{1}{2}\beta - \frac{1}{4} \sin 2\beta}{1 - \cos \beta} \tag{6.1.48}$$

and by substituting

$$b + w_L = (w_T + w_L) \frac{\pi(1 - \cos \beta)}{\sin \beta - \beta \cos \beta} \tag{6.1.49}$$

The following relation is obtained:

$$\frac{M}{D^2(w_T + w_L)} = \frac{\pi}{8} \frac{2\beta - \sin 2\beta}{\sin \beta - \beta \cos \beta} \tag{6.1.50}$$

Substitution of values of b into the expression for vertical forces and moments (6.1.47) and (6.1.50) allows for the determination of the relationship between the terms:

$$\frac{M}{D^2(w_T + w_L)} \quad \text{and} \quad \frac{b + w_L}{w_T + w_L} \tag{6.1.51}$$

This is plotted in Fig. 6.1.12. This is the relationship used in API 650, E-5. To improve estimation of the value of

$$\frac{b + w_T}{w_T + w_L} \tag{6.1.52}$$

when

$$\frac{M}{D^2(w_T + w_L)} \tag{6.1.53}$$

is greater than 1.5, the following approximate relation may be used:

$$\frac{b + w_T}{w_T + w_L} \cong \frac{\dfrac{3\pi}{2\sqrt{10}}}{\sqrt{1 - \dfrac{2}{\pi}\dfrac{M}{D^2(w_T + w_L)}}} \tag{6.1.54}$$

This equation is found in API 650, E.5.1.

6.1.6   API Methodology

Having discussed all of the factors that go into the seismic considerations above, it is now possible to use a simple procedure to determine the capability of a tank for seismic forces. API 650 loads the tank with a horizontal acceleration (vertical acceleration is not considered) according to the impulsive and convective components as follows:

$$M = ZI(C_1 W_s X_s + C_1 W_t H_t + C_1 W_1 X_1 + C_2 W_x X_2)$$

where M = overturning moment applied to the base of the tank shell
$\quad\quad Z$ = seismic zone factor from Table 6.1.2
$\quad\quad I$ = importance factor (see discussion below)
$\quad C_1, C_2$ = lateral earthquake force coefficients
$\quad\quad W_s$ = total weight of tank shell, lb
$\quad\quad X_s$ = height from tank bottom to center of gravity of tank shell, ft
$\quad\quad W_r$ = total weight of tank roof plus a portion of snow load, if any, specified by purchaser
$\quad\quad H_t$ = total height of tank shell, ft
$\quad\quad W_1$ = weight of effective mass of the tank contents that move in unison with the tank shell

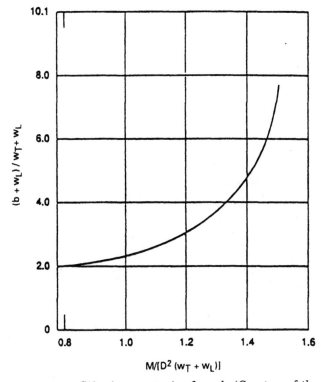

Figure 6.1.12 Seismic compressive force b. (*Courtesy of the American Petroleum Institute.*)

TABLE 6.1.2   Seismic Zone Factors for Horizontal Acceleration

Seismic factor	Seismic zone factor (horizontal acceleration)
1	0.075
2A	0.15
2B	0.20
3	0.30
4	0.40

W_2 = weight of effective mass of the tank contents that move in the first sloshing mode

X_2 = height from the bottom of the tank shell to the centroid of lateral seismic force applied to W_2, ft

The seismic factor or zone is determined from Fig. 6.1.13 which is used with Table 6.1.2 to obtain the seismic zone factors for various geographical areas of the United States.

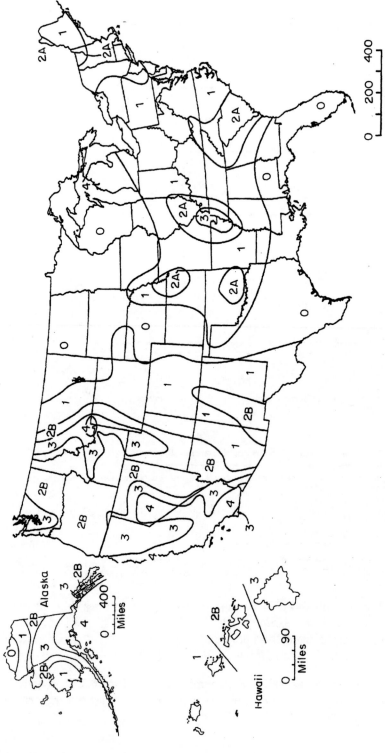

Figure 6.1.13 Seismic zones of the United States. (*Used by permission of the International Conference of Building Officials; map courtesy of Structural Engineering Society of California.*)

The importance factor I should always be set to 1.0 unless the tank is to serve in postearthquake duty in services such as emergency water or fuel supplies. If this is the case, the I should be set equal to 1.25. Typical petroleum storage tanks for refineries and terminals have values of $I = 1.0$.

6.1.7 Nomenclature

b	Maximum longitudinal shell compression force, lb/ft of shell circumference
C_1, C_2	Lateral earthquake coefficients for impulsive and convective forces, respectively
d	Height of sloshing wave above mean depth, ft
D	Tank diameter, ft
E	Modulus of elasticity, psi
F_a	Maximum allowable longitudinal compressive stress in tank shell, psi
F_{by}, F_{ty}	Minimum specified yield strength of bottom annular ring and tank shell, respectively, psi
g	Acceleration due to gravity = 32.2 ft/s^2
G	Specific gravity (1.0 for water)
H	Maximum filling height of tank, ft
H_t	Total height of tank shell, ft
I	Essential facilities factor
k	Parameter for calculating T, (s^2/ft)$^{1/2}$
L	Bottom uplift length, ft
M	Overturning moment applied to bottom of tank shell, ft·lb
M_p	Plastic bending moment in bottom annular ring, in·lb/in
p	Internal pressure, psi
P_1, P_2, P_E	Increased hoop tension in tank shell due to impulsive, convective, and total earthquake forces, respectively, lb/in
R	Tank radius, in
S	Site amplification factor
t	Thickness of cylindrical shell, in (when used in formulas for tank design applies to thickness of bottom shell course excluding corrosion allowance)
t_b	Thickness of bottom annular ring, in
T	Sloshing wave period, s
w	Unit weight on tank bottom, lb/ft^2
W_L	Maximum weight of tank contents which may be utilized to resist shell overturning moment, lb/ft of shell circumference

w_t	Weight of tank shell, lb/ft of shell circumference
W_r	Total weight of tank roof plus portion of snow load, if any, lb
W_s	Total weight of tank shell, lb
W_T	Total weight of tank contents, lb
W_1, W_2	Weight of effective masses of tank contents for determining impulsive and convective lateral earthquake forces, lb
X_s	Height from bottom of tank shell to center of gravity of shell, ft
X_1, X_2	Height from bottom of tank shell to centroids of impulsive and convective lateral earthquake forces, respectively, for computing M, ft
X_1', X_2'	Height from bottom of tank shell to centroids of impulsive and convective lateral earthquake forces, respectively, for computing total overturning moment on foundation, ft
Y	Vertical distance from liquid surface to point on shell being analyzed for hoop tension, ft
Z	Seismic zone coefficient
β	Central angle between axis of tank in the direction of earthquake ground motion and point on circumference

Design Example A petroleum storage terminal tank 80 ft in diameter and 40 ft high will be designed according to API 650, Appendix 4. The specific gravity of the stored fluid will be $G = 1.0$ and the corrosion allowance $c = 0$. The tank shell weighs approximately 131,400 lb, the roof weighs 50,000 lb, and one-half of its weight is supported by the shell. The tank will be located in Denver, Colo. The tank is constructed on a stiff soil greater than 200 ft deep. Consider what options are available to the tank designer—anchorage, increased annular plate thickness, or lowering liquid level.

solution According to our model this tank is a "short tank" since $H/D = \frac{1}{2} < \frac{3}{4}$. Using the refinery location and referring to Fig. 6.1.13 or to API 650, Fig. E-1, the seismic zone factor $Z = 1$. Since the tank is an ordinary petroleum storage tank the importance factor $I = 1.0$. Using API 650, Table E-3 we determine $S = 1.2$. From API 650, Appendix A, Table A-4 we have a listing of the shell wall thicknesses that are used to construct the tank:

Course number	Course height, ft	Course thickness, in
1	8	0.46
2	8	0.37
3	8	0.27
4	8	0.25
5	8	0.25

The unit shell weight is $W_s = 131,000$ lb and the weight of shell plus roof weight supported by shell is

$$w_t = \frac{131,000 + (50,000/2)}{\pi(80)} = 620 \text{ lb/ft of circumference}$$

The given roof weight is $W_r = 50,000$ lb and the weight of contents is

$$W_T = \frac{\pi D^2 H}{4} \times 62.4G = \frac{\pi(80)^2(40)}{4}(62.4)(1.0) = 12.55 \times 10^6 \text{ lb}$$

$$\frac{W_1}{W_T} = 0.54 \qquad \text{from Fig. E-2 for } D/H = 2.0, \text{ or Eq. (6.1.6)}$$

$$W_1 = (0.54)(12.55 \times 10^6) = 6.78 \times 10^6 \text{ lb}$$

$$\frac{X_1}{H} = 0.375 \qquad \text{from Fig. E-3 or Eq. (6.1.8)}$$

$$X_1 = (0.375)(40) = 15.0 \text{ ft}$$

$$\frac{W_2}{W_T} = 0.44 \qquad \text{from Fig. E-2 or Eq. (6.1.12)}$$

$$W_2 = (0.44)(12.55 \times 10^6) = 5.52 \times 10^6 \text{ lb}$$

$$\frac{X_2}{H} = 0.61 \qquad \text{from Fig. E-3 or Eq. (6.1.13)}$$

$$X_2 = (0.61)(40) = 24.4 \text{ ft}$$

$$K = 0.59 \qquad T = 0.59\sqrt{80} = 5.3 \text{ s} \qquad \text{(from Fig. 6.1.8)}$$

$$C_2 = \frac{1.35S}{T^2} = \frac{(1.35)(1.2)}{(5.3)^2} = 0.058 \qquad \text{for } T > 4.5 \text{ s}$$

$$M = ZI[C_1 W_s X_s + C_1 W_T H + C_1 W_1 X_1 + C_2 W_2 X_2]$$
$$= (1.0)(1.0)[(0.24)(0.131 \times 10^6)(17.3) + (0.24)(0.05 \times 10^6)(40)$$
$$+ (0.24)(6.78 \times 10^6)(15.0) + (0.058)(5.52 \times 10^6)(24.4)]$$
$$= 0.54 \times 10^6 + 0.48 \times 10^6 + 24.4 \times 10^6 + 7.8 \times 10^6 = 33.2 \times 10^6 \text{ft} \cdot \text{lb}$$

Try a tank with a normal $\frac{1}{4}$-in bottom thickness. Then $t_b = 0.25$ ft, $F_{by} = 30,000$ psi (ASTM A283C36).
The weight of contents resisting overturning is

$$w_L = 7.9t_b \sqrt{F_{by}GH} = 7.9(0.25)(\sqrt{30,000)(1)(40)} = 2160 \text{ lb/ft}$$

The limit of the liquid weight that may be used to resist uplift is

$$1.25GHD = (1.25)(1)(40)(80) = 4000 > 2160$$

$$\frac{M}{D^2(w_t + w_L)} = \frac{33.2 \times 10^6}{(80)^2(520 + 2160)} = 1.936 > 1.57 \text{ N.G.—unstable}$$

One of the options is to anchor the tank. For anchored tanks the maximum longitudinal compressive force at the bottom of the shell is

$$b = w_t + \frac{1.273M}{D^2} = 620 + \frac{1.273 \times 33.2 \times 10^6}{80}{}^2 = 7223 \text{ lb/ft}$$

This compressive force results in longitudinal shell compressive stress of $b/12t$ = 1308 psi. The maximum allowable shell compression to resist buckling must also be within the allowable stress limits to prevent shell buckling. When GHD^2/t^2 is greater than 10^6, then $F_a = 10^6 t/D$. When it is less than 106, then

$$F_a = \frac{10^6 t}{2.5D} + 600\sqrt{GH}$$

Another limit is that F_a shall not exceed $0.5F_y$. First the parameter GHD^2/t^2 is determined to be

$$\frac{(1.0)(40)(80^2)}{0.46^2} = 1.2 \times 10^6$$

which is greater than 10^6 so that the allowable stress is

$$F_a = \frac{10^6 t}{D} = \frac{(10^6)(0.46)}{80} = 5750 \text{ psi.}$$

A check is made for $0.5 \times 30,000 = 15,000$ psi. Since the compressive stress 1308 is below all of these limits the tank may be anchored. If this option is chosen, then the anchorage must be designed to provide anchorage resistance to uplift of

$$\frac{1.273M}{D^2} - w_t = \frac{(1.273)(33.2 \times 10^6)}{80^2} - 620 = 5983 \text{ lb/ft}$$

When rigid foundations such as reinforced mats are used, it is necessary to determine the seismic moment which is significantly higher than the tank overturning moment acting on the foundation. This is due to the hydrodynamic bottom pressure that acts on the tank bottom which is not included in the tank overturning moment calculations. This is done by using the basic moment equation in API 650 and substituting X_1' for X_1 and X_2' for X_2 using Eq. (6.1.10), (6.1.11), or (6.1.14) depending on whether the tank is tall or short.

In the current case for the short tank model

$$\frac{X_1'}{H} = \frac{3}{8}\left[1.0 + \frac{4}{3}\left(\frac{0.866D/H}{\tanh 0.866D/H} - 1.0\right)\right]$$

$$\frac{X_1'}{H} = \frac{3}{8}\left[1.0 + \frac{4}{3}\left(\frac{0.866 \times 2}{\tanh 0.866 \times 2 - 1.0}\right)\right] = 0.797$$

$$\frac{X_2'}{H} = 1.0 - \frac{\cosh(3.67H/D) - (31/16)}{(3.67H/D) \sinh(3.67H/D)} \qquad \frac{X_2'}{H} = 1.0 - \frac{\cosh(3.67/2) - (31/16)}{(3.67/2) \sinh(3.67/2)}$$

$$= 0.772$$

$$\begin{aligned}
M_{\text{found}} &= ZI[C_1 W_s X_s + C_1 W_T H + C_1 W_1 X_1' + C_2 W_2 X_2'] \\
&= (1.0)(1.0)[(0.24)(131,000)(17.3) + (0.24)(50,000)(40) \\
&\quad + (0.24)(0.54)(12,550,000)(0.797)(40) \\
&\quad + (0.058)(0.437)(12,550,000)(0.772)(40)] \\
&= 62.7 \times 10^6 \text{ ft} \cdot \text{lb}
\end{aligned}$$

Note that this is significantly above the moment tank overturning moment of 33.2×10^6 ft·lb. However, anchoring a tank is usually the least desirable option in that it is costly, it imposes high point loads on the tank during seismic activity, and other options are simpler and easier to use in general. Often, when the overturning stability is not too much greater than 1.57, then increasing the annular plate thickness is an effective means of providing increased liquid hold down resistance to overturning. Try increasing the standard $\frac{1}{4}$-in-thick bottom to a $\frac{3}{8}$-in annular plate thickness: $t_b = 0.375$ in; $F_{by} = 30,000$ psi

$$w_L = 7.9(0.375)\sqrt{(30,000)(1)(40)} = 3250 \text{ lb/ft} < 4000 \text{ lb/ft}$$

$$\frac{M}{D^2(w_t + w_L)} = \frac{33.2 \times 10^6}{(80)^2(520 + 3250)} = 1.376 < 1.57$$

The tank is therefore structurally stable.

$$\frac{b + w_L}{w_t + w_L} = 4.17 \text{ (from Fig. E-5 or from the discussion in the note below)}$$

$$b = 4.17(520 + 3250) - 3250 = 12,470 \text{ lb/ft}$$

$$f_a = b/12t = 12,470/(12 \times 0.46) = 2260 \text{ psi}$$

$$GHD^2/t^2 = (1)(40)(80)^2/(0.46)^2 = 1.21 \times 10^6 > 10^6$$

$$F_{a'} = \frac{10^6 t}{D} = 1 \frac{0^6(0.46)}{80} = 5750 \text{ psi} > 2260 \text{ psi} \qquad \text{so OK.}$$

Use a $\frac{3}{8}$-in annular plate:

$$\text{Minimum width} = 0.0274 \, \frac{w_L}{GH} = 0.0274 \left[\frac{3250}{(1)(40)} \right] = 2.231 \text{ ft}$$

Another option is to lower the maximum design liquid level such that the tank does not need to be anchored. By trial and error, using the same physical tank but reducing the liquid level to 36 ft, the tank has a stability ratio of about 1.5 which is acceptable for an unanchored tank. For this case we have the following

Liquid contents	11,291
First sloshing mode	5.36 s
W_1/W_T	0.498
W_2/W_T	0.815
Overturning moment	25.65×10^6 ft·lb
Foundation moment	52.6×10^6 ft·lb
Annular ring thickness	0.25
Liquid contents resisting uplift	2052
Liquid contents limit	3600
Buckling force	16,629 lb/ft
Max compressive stress	3012 psi
Allowable stress	5750 psi
Anchorage force	4479 lb/ft (if used)
Stability ratio	1.49

To determine the height of the sloshing waves generated by the seismic event we have

$$d_{max} = \frac{3}{4}\sqrt{\frac{27}{8}}A\,\tanh\!\left(4.77\frac{H}{D}\right)$$

$$= 1.124\,ZIC_2\,T^2\,\tanh\!\left(4.77\sqrt{\frac{H}{D}}\right)$$

$$= 1.124(1.0)(1.0)(.058)(5.3^2)\tanh(4.77\sqrt{1/2})$$

$$= 50\text{ ft}$$

By an iterative procedure, the value from API 650, Appendix E, Fig. E-5 of the ratio $(b + w)/(w_t + w_L)$ for the values of $M/[D_2(w_t + w_L)]$ between 0.785 and 1.57 can be determined as follows. Determine β in radians by trial and error solution of the formulas:

$$\frac{M}{D(w_t + w_L)} = \frac{\pi}{8}\frac{2\beta - \sin 2\beta}{8\sin\beta - \beta\cos\beta}$$

Then using this value of β solve the formulas:

$$\frac{b + w_L}{(w_T + w_L)} = \frac{\pi(1 - \cos\beta)}{\sin\beta - \beta\cos\beta}$$

Solution of this is facilitated by using the initial trial value of β:

$$\beta = \pi - 2 - \frac{\pi}{2\,D^2}\frac{(w_t + w_L)}{M} \qquad \text{for } M/D^2(w_T + w_L) < 1.0$$

$$\beta = \pi - 10\left[1 - \frac{\pi}{2}\frac{D^2(w_t + w_L)}{M}\right] \qquad \text{for } M/D^2(w_T + w_L) > 1.0$$

Convergence in determining β can be quickly found using a computer program or programmable calculator using the first derivative of

$$\frac{M}{D(w_t + w_L)} = \frac{\pi}{8}\frac{2\beta - \sin 2\beta}{\sin\beta - \beta\cos\beta}$$

SECTION 6.2   SEISMIC EVALUATION AND RETROFITTING OF EXISTING TANKS

6.2.1   Introduction

Assessing overall risk is complex and difficult task when compared to the relatively easy task of determining the seismic vulnerability of a specific tank. Facility risk entails the evaluation of innumerable variables such as the environmental sensitivity of an event occurring at a particular facility and its impact. For example, a facility located in Hawaii has great potential for environmental and irreparable damage should flammable and/or toxic liquids be released. Not only could the release prove immediately hazardous to life and health, but also delicate marine ecosystems could be adversely affected or even destroyed.

Establishing risk criteria associated with seismic damage to existing facilities including storage tanks is also a complex and difficult task. A facility design philosophy often adopts the UBC design provisions and the seismic zone maps, and related hazard levels can be used. A design-basis earthquake (DBE) can be based on ground motion with a 10 percent probability of being exceeded over a 50-year period. This corresponds to an earthquake with a 475-year recurrence interval and is usually accepted for many structures. When facilities are located at faults or other specific conditions exist such as the possibility of soil liquefaction, then site-specific hazard levels and design bases should be established.

Because the location of an earthquake and its epicenter as well as the frequency and magnitude of the event are of a probabalistic nature, the risks cannot be eliminated altogether. In addition, for existing equipment the acceptable level of integrity for equipment against seismic failure may be made at a lower degree of ground shaking than that for new equipment, which is usually set by the applicable building codes.

The reason that varying degrees of mechanical integrity are possible for new versus existing equipment depends on the underlying performance objectives. This philosophy requires maintaining varying degrees of structural integrity dependent on the facility use and consequent hazard. A table showing an example of a matrix for establishing such criteria is Table 6.2.1. Experience has shown that even when shell collapse occurs, only approximately 10 percent of these tanks lose their contents. It would seem, then, that a reasonable design criterion would be the prevention of shell collapse. However, there have been many cases where seismic forces were well below the level sufficient to produce shell collapse and loss of contents occurred for other reasons.

TABLE 6.2.1   Seismic Performance Objectives for ASTs

Usage categories*	Performance objectives			
	Expect shell or roof damage	Maintain confinement of liquid	Maintain storage functions	Maintain storage seismic risk
Low-risk facilities	Yes	No†	No	Low
General facilities	Yes	Yes	No	Medium
Essential facilities	Minor	Yes	Yes	Low

*Low-risk facilities are those that are remotely situated and, should a major spill or fire develop, would not pose serious health or life endangerment. General risk facilities are all others. Essential facilities are those that are used by public works departments during a general emergency such as fire water tanks or potable water tanks for public emergencies.

†In this context, it is assumed that secondary containment is provided and would provide a localized confinement of the tank contents should it rupture, preventing an environmental risk.

Although there are many approaches to considering the problem of seismic activity on existing tanks, only one is presented here:

- Evaluate the site- and equipment-specific risk to determine the seismic criteria to establish.

- Evaluate the seismic capacity of existing tanks based on API seismic criteria.

- Evaluate the seismic capacity of existing tanks based on an alternative method when the API method fails the tank. (API 650 may be overly conservative in many cases.)

- Evaluate mitigation possibilities such as reducing the maximum fill height, adding hold-down anchorage, increasing flexibility of attachments, modifying the tank to include an annular plate, etc.

6.2.2   Ranking of Exposure and Risk

A first step in a seismic assessment program is to identify tanks with high vulnerability and risk. Here are a few of the most likely factors:

6.2.2.1   Age

Most aboveground storage tanks have been designed to API Standard 650. The seismic considerations contained in Appendix E of this standard are a relatively recent addition (1978). Although the general theory was developed earlier, few tanks were designed for earthquakes in accordance with Appendix E before 1978. Since 1978 there have been several technical advancements beyond API 650, Appendix E. However, no change has been incorporated into an existing code.

Because only recently constructed tanks have been designed to resist earthquakes, there may be several potentially vulnerable tanks in any given tank population. To limit the seismic exposure, these tanks should be identified and possibly retrofitted. In addition, although API 653 does not require an upgrade of existing tanks, a facility may wish to upgrade to limit risk and exposure for vulnerable tanks.

Prioritizing tanks for seismic vulnerability begins with the simplest classification. As previously mentioned, tanks that were built prior to 1978 were constructed without the benefit of seismic requirements. Therefore, tanks in high-seismicity zones (zones 4, 3, and 2) that were constructed prior to 1978 should be evaluated for seismic stability.

6.2.2.2   Aspect ratio

A tall, narrow tank is more vulnerable to seismic events than one that is broad and flat. The shell compressive stresses are higher, which

TABLE 6.2.2 Aspect Screen Table

Tank diameter, ft	Height/diameter ratio
D 60	0.50
60>D 95	0.40
95>D 175	0.30
175>D	0.25

Assumptions made in table: API 650, Appendix E calculation method, seismic zone 4, specific gravity 1.0, annular ring thickness of ¼ in, S_3 soil site.
D = tank diameter, ft
S_3 = A soil profile 40 ft or more in depth containing more than 20 ft of soft to medium stiff clay but not more than 40 ft of soft clay

tends to cause elephant's foot, and the possibility of complete overturning is greater. In addition, tall tanks move more at the top than broad, short tanks do. Table 6.2.2 provides a guide to which tanks are probably able to pass the API 650 criteria based upon height-to-diameter ratio in the worst seismic zone. This type of table can be used to screen for tanks needing a more detailed analysis where there are many tanks to consider. Similar tables can be developed for other seismic zones. However, for any particular case, the API 650 computation can easily be performed to determine whether the tank meets the criteria.

To be seismically safe, API 650, Appendix E evaluates the compressive stresses in the shell. When the ratio $M/[D^2(w_t + w_L)]$, called the *stability ratio,* approaches 1.57, the tank is at an imminent state of toppling over. The shell compressive stresses become excessive, and the tank is unstable. The tank's compressive stress should also be checked at a stability ratio slightly less than 1.57. In accordance with API 650 Appendix E, the equation for calculating the overturning moment is

$$M = ZI(C_1 w_S X_S + C_1 w_r H_t + C_1 w_1 X_1 + C_2 w_2 X_2)$$

where Z = seismic zone factor = seismic zone
I = importance factor = 1.0 for all normal tanks
C_1 = impulsive mass coefficient
C_2 = convective mass coefficient
w_S = weight of tank shell
X_S = height of center of gravity of shell
w_r = weight of roof
H_t = height of shell
w_1 = impulsive mass
w_2 = convective mass
X_1 = height to centroid of impulsive mass

X_2 = height to centroid of convective mass
D = diameter of tank
w_t = unit circumferential weight of tank
w_L = weight of liquid that may be used to resist overturning

In a liquid storage tank, the liquid accelerated by ground motion moves in impulsive and convective components. The impulsive component is at the bottom of the tank and is confined by the liquid above so that it acts as a rigid body. The convective component is free to slosh around and is at the top of the tank. The C_1 coefficient is applicable to the impulsive component, and the C_2 coefficient is applicable to the convective component. The rest of the equation can be broken down into four parts. The first two parts $C_1 w_S X_S$ and $C_1 w_r H_t$ reflect the contributions of the tank shell and tank roof, respectively, to the overturning forces. The last two parts $C_1 w_1 X_1$ and $C_2 w_2 X_2$ reflect the contributions of the liquid moving rigidly with the tank and the sloshing liquid to the overturning moment. The last two terms account for most of the overturning moment. Resistance to the overturning forces comes from the roof and shell weight as well as the liquid resting on the uplifted bottom plate.

6.2.2.3 Construction details

Tanks which are riveted are more vulnerable to failure than welded tanks. Very old tanks may be lap-welded or have butt welds with incomplete penetration and fusion. These joints are much more likely to fail during severe seismic activity than modern butt-welded construction. Old tanks also have a high probability that the steel is notch-toughness-sensitive at ambient temperatures, meaning that a rapid strain rate caused by a seismic event could lead to brittle fracture failure. The basic rule of thumb is that the older the tank, the less likely it will stand up to a severe seismic event. This does not mean that facility owners should eliminate all old tanks meeting the above description, but that an assessment should be performed. Analysis can be used to show that such a tank is, in fact, adequate to meet the basic acceptable level or risk or that it meets the code requirements.

6.2.2.4 Attachment details

Most failures during seismic events are caused by the tearing loose of piping, ladders, platforms, or other equipment which is sufficiently rigid that the movement in the tank wall causes it to tear the tank wall or attachment. Piping that rises out of the ground immediately adjacent to the tank shell is a common example of excessive rigidity. Another example is a ladder that is bolted not only to the tank but also to a concrete pad. When the tank sways during an event, the ladder will tear out at the weakest point which can be its attachment to the tank shell. More details are given under Tank Damage Mechanisms below. Figure 6.2.1 relates these tank failures associated with attachment details and the likelihood of loss of tank contents.

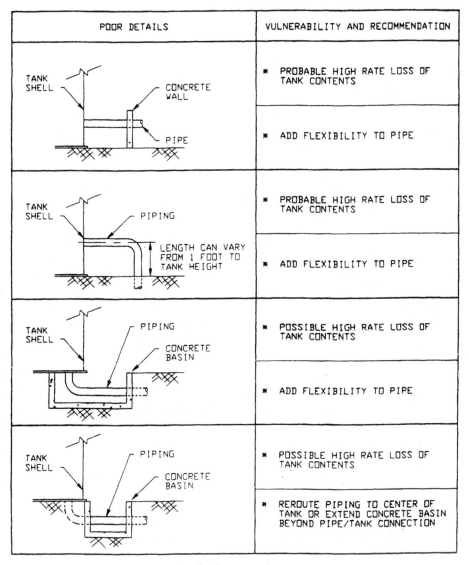

Figure 6.2.1 Poor tank, seismic details of construction.

POOR DETAILS	VULNERABILITY AND RECOMMENDATION
TANK SHELL — PIPE — TANK SHELL	* POSSIBLE HIGH RATE LOSS OF TANK CONTENTS
	* INCREASE FLEXIBILITY BY PROVIDING HORIZONTAL OR VERTICAL BENDS
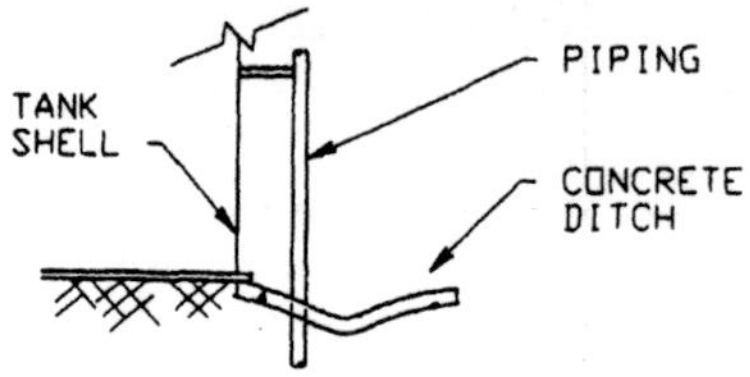 TANK SHELL — PIPING — CONCRETE DITCH	* POSSIBLE LOW RATE LOSS OF TANK CONTENTS
	* ANCHOR PIPE AT SHELL ROOF CONNECTION
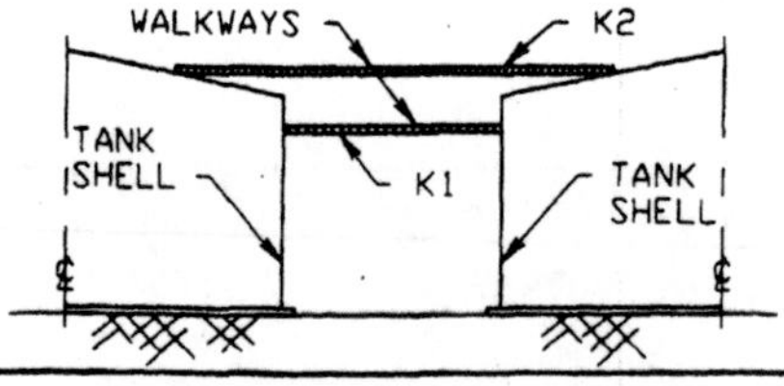 WALKWAYS — K2 — TANK SHELL — K1 — TANK SHELL	* UNLIKELY LOSS OF TANK CONTENTS
	* INCREASE WALKWAY FLEXIBILITY TO ACCOMODATE RELATIVE DISPLACEMENTS
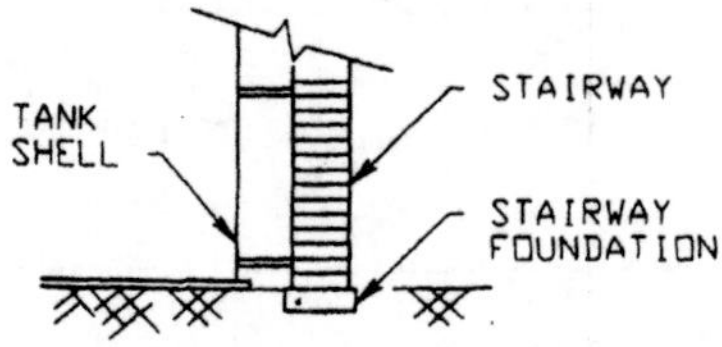 TANK SHELL — STAIRWAY — STAIRWAY FOUNDATION	* PROBABLE LOW RATE LOSS OF TANK CONTENTS
	* SUPPORT STAIRWAY EXCLUSIVELY ON TANK SHELL
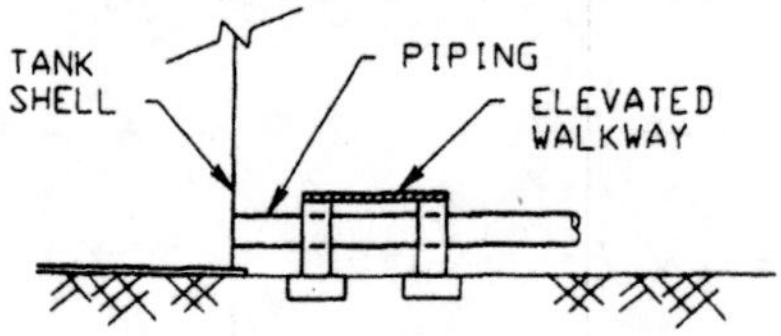 TANK SHELL — PIPING — ELEVATED WALKWAY	* POSSIBLE HIGH RATE LOSS OF TANK CONTENTS
	* INCREASE PIPING FLEXIBILITY, ATTACH WALKWAY EXCLUSIVELY TO TANK SHELL, OR PROVIDE MORE PIPING CLEARANCE

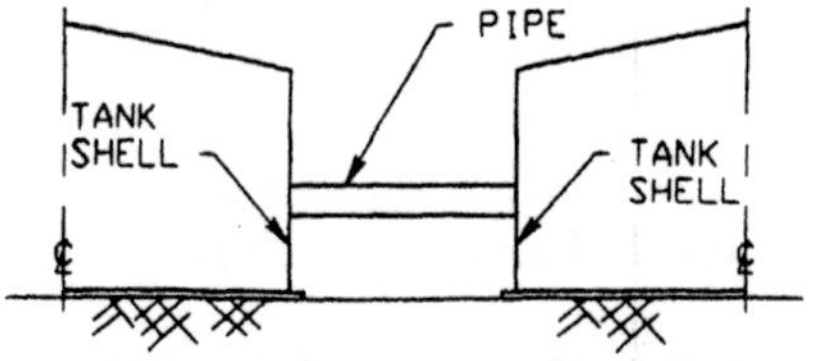

Figure 6.2.1 (*Continued*)

6.2.2.5 Anchored tanks

If a tank is anchored, the compressive shell stresses decrease significantly compared to a similar unanchored tank. Also, anchoring limits uplift displacements, and failure of rigidly attached components and piping is reduced. However, inadequate anchorage or foundation details can lead to tank failure. These systems should be evaluated as part of the overall seismic assessment.

6.2.3 Tank Damage Mechanisms

Figures 6.2.2 to 6.2.7 show examples of a tank that was near the epicenter of the 1994 Northridge, California, earthquake. This tank held a municipal water supply. Although it was designed to AWWA D-100 standards, these are almost identical to the API 650 seismic requirements. It is believed to have failed in spite of its being a "code tank" because the local acceleration was much greater than that of a design-basis earthquake.

One should have a basic understanding of seismic behavior and tank damage modes when assessing existing tanks for suitability for service.

Seismic forces can occur in any direction; however, one can generally assume that the majority of ground acceleration is in the horizontal plane. In fact, the API approach to seismic design assumes that the seismic forces are strictly horizontal. Since seismic forces are proportional to the mass of the structure involved, most of the force generated by seismic activity on a tank results from the liquid contents.

0°
View

Figure 6.2.2 Tank T-3 location plan.

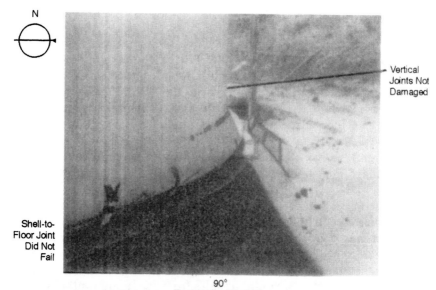

Vertical
Joints Not
Damaged

Shell-to-
Floor Joint
Did Not
Fail

90°
Elephant Foot Bocking

Figure 6.2.3   Tank T-3 elephant's foot.

Failure of
Rigidly
Attached
Piping

Seperation of
Reinforcement
to Shell

Paint Failure
Due to
Compressive
Strains on
Shell

Figure 6.2.4   Tank T-3 piping failure.

Figure 6.2.5 Tank T-3 anchorage failures.

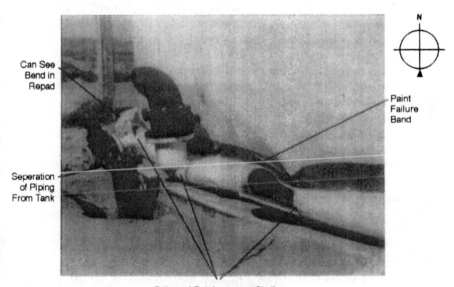

Failure of Reinforcement Shell

Figure 6.2.6 Tank T-3 reinforcing pad and piping failure.

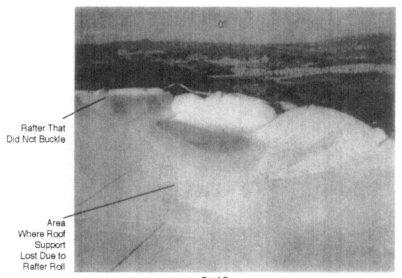

Rafter That
Did Not Buckle

Area
Where Roof
Support
Lost Due to
Rafter Roll

Roof Damage

Figure 6.2.7 Tank T-3 roof damage.

However, the hydrodynamic action of the tank's liquid content can further be classified into two behaviors. One behavior is that the liquid acts as a solid. Liquid that is confined to the bottom by liquid above it behaves as though it were a solid. This makes computing the seismic forces for this behavior quite simple. The liquid near the surface, however, is free to move, creating surface waves. Both behaviors produce a force that wants to tip the tank over, called an *overturning moment.*

The response of the tank structure to overturning is to develop hydrodynamic wall pressures on the tank shell and bottom. With sufficient acceleration the tank starts to tip. The weight of the tank contributes to the resistance to overturning. However, if the overturning moment is great enough, then a portion of the tank starts to lift off the foundation. Uplift is a well-known and documented occurrence. In one case a 100-ft-diameter by 30-ft-high tank was reported to have uplifted about 14 in during the 1971 San Fernando earthquake. In the Loma Prieta earthquake of 1989 in the San Francisco Bay Area, there were reports of uplift of approximately 6 to 8 in. Uplift by itself is not so damaging to the tank as the effect on rigidly attached piping, ladders, or appurtenances. During uplift, any rigidly attached piping or equipment will usually tear away from the tank, perhaps opening the tank and causing spills.

Tanks are surprisingly flexible structures, and because of this they have stood up well to earthquakes. However, *two* very specific modes of failure occur that could be considered catastrophic, which in this context means loss of liquid contents:

1. Various appurtenances, clips, railings, piping, or other attachments can be pulled off the tank and can tear the shell, causing spills. These events can result from either uplift or relative motion between tanks which are connected via piping or walkways.

2. If the bottom of the tank collapses, causing elephant's foot, the connected piping will tear out of the shell.

Tanks can fail in many ways during a seismic event. Typical damage occurs due to the following categories:

- Sloshing waves are generated, and these may cause roof damage. Sometimes the waves twist the rafters sideways, and this allows the roof to collapse or sag. Greater impact by sloshing waves can cause the roof to separate from the shell. The sloshing wave motion can also cause roof support columns to be knocked from their support points at the base and can cause roof failure. Since roof supports must not be connected to the tank bottom so that settlement does not pull the roof down, the design of guides should include at least 8 in of free vertical motion without lateral support. During an inspection, this should be checked to ensure that the roof can lift 8 in without the column's pulling out of its guide.

- Any attachments to the tank such as ladders, platforms, piping, clips, or other equipment can tear loose from the tank due to differential movement between the tank and the connecting structure.

- Another cause of tank damage is due to excessive compressive loading of the shell caused by the seismic overturning moment. This usually produces buckling in the shell and, at worst, completely collapses the the shell at the tank base. This is referred to as *elephant's foot*. When the tank is very thin and tall, as in the case of stainless steel chemical tanks, elastic buckling can initiate a diamond buckling pattern. While this does not usually cause loss of tank contents or loss of structural integrity, it is often repaired.

Various tank damage mechanisms that are based on attachments and piping connections are shown in Fig. 6.2.1. For existing and new tanks, the degree of damage to the tank during an earthquake is not so much of a concern as its loss of contents. In fact, during a severe earthquake it is anticipated that damage will occur. The purpose of the minimum design standards in the codes is to reduce the damage to an acceptable level. Certainly, shell buckles and roof distortion are not serious problems that would endanger lives or the environment.

Loss of tank contents, however, can pose immediate threats to those near the tank and entails the possibility of increased risk due to flammable or toxic releases affecting other structures, equipment, or tanks, providing the potential for a chain reaction of increasing risk.

A good definition of tank failure in the context of seismic events is tank damage that results in loss of integrity of the shell with subsequent loss of content.

6.2.4   Seismic Assessment Program

6.2.4.1   Establishing criteria for acceptance

It is important to understand that the codes do not and cannot prevent catastrophes from occurring. The codes can only provide good practice that has a reasonable chance of offering a measure of protection against earthquakes. They do not address risk issues, vulnerability, or other issues associated with facilities at all. Simple practices such as flexibility in piping attached to the tank are not really addressed because the amount that a tank will uplift is not really known with the current state of the technology. However, the owner/operator has the opportunity to intuitively understand the mechanisms of damage and to provide some mitigating measures in the design and construction. API and most alternative methods of seismic analysis only address the tank shell buckling failure mode, and not failure due to movement of attachments relative to the tank or other modes of failure.

The overall approach to determining seismic stability is shown in Fig. 6.2.8. It is patterned after the following sequence.

6.2.4.2   A model approach

Screen tank on aspect ratio. Tall tanks are more likely to fail than short tanks. By walking around a facility with Table 6.2.2 in hand, you can readily determine which tanks should be subjected to a seismic analysis.

Screen tanks using tank age. Tanks that do not pass the aspect ratio test should be divided into category by age. All riveted tanks and tanks built prior to 1978 should be put on the list for analysis.

Perform an API 650 seismic analysis on the candidate tanks. The tank should be subjected to the API 650 Appendix E analysis. When the tank has a stability ratio greater than 1.57, it is unstable. The tank owner/operator should then decide whether to lower the safe maxi-

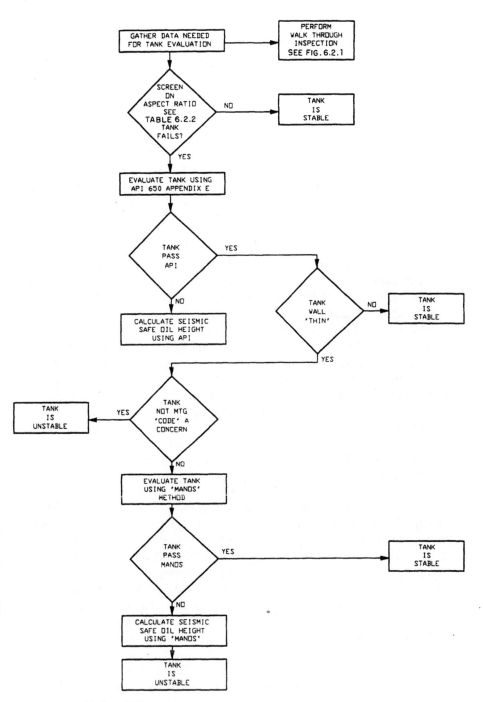

Figure 6.2.8 Tank stability assessment.

mum operating liquid level or to conduct an alternative analysis. The safe maximum operating level can be determined by running the Appendix E analysis with different liquid levels until the stability ratio drops to below 1.57. A computer program is useful for this purpose.

Perform an alternative analysis.   For many tank configurations API 650 Appendix E conservatively estimates their seismic performance. This approach is presently the most widely accepted method for seismic analysis in spite of better understanding of seismic performance of tanks and its somewhat outdated technology. Alternative methods such as the one used by the New Zeland Tank Code are probably more accurate and utilize the latest understanding of seismic dynamics and performance, but may be extremely complex and cumbersome to use. Another useful and simple alternative method that has been used for seismic analysis is a method developed by George C. Manos. This method may predict a tank's performance in earthquakes more accurately than the API 650 method for many configurations and has some experimental and field data to back it up. One of the main differences that allows a tank to pass when API does not is the uplift mechanism. API arbitrarily limits the uplift dimension and does not take into account bottom membrane stresses which help to resist overturning. This is shown in Fig. 6.2.9. However, acceptance of an alternative method by the jurisdiction having authority should be considered. The Manos approach is much more sensitive to tank shell thickness than the API 650. Therefore, API 650 may not be conservative when one is assessing tanks with shells that are thinner than normal, usually found on small-diameter, tall tanks.

The Manos approach calculates the overturning moment required to cause buckling in the tank shell. An empirical compressive stress distribution is assumed that peaks at 75 percent of the classical buckling stress or yield stress, whichever is less. The integration of the compressive stress is balanced by the uplifted portion of the tank bottom and the liquid projecting over it. The moment caused by an earthquake in terms of acceleration is then calculated; however, the convective component of the tank liquid mass is ignored for the sake of simplicity. The assumption justifying this is that the phase of the convective components is not synchronous with the ground acceleration, especially for tall tanks, for which the Manos method is ideally suited. In addition, a somewhat conservative moment arm is used for the centroid of the impulsive component of the liquid mass. By taking the ratio of the two, the acceleration can be solved for. This acceleration is then compared to the maximum acceleration that the tank is expected to experience during an earthquake. The tank's acceleration is expected to be 3 to 4

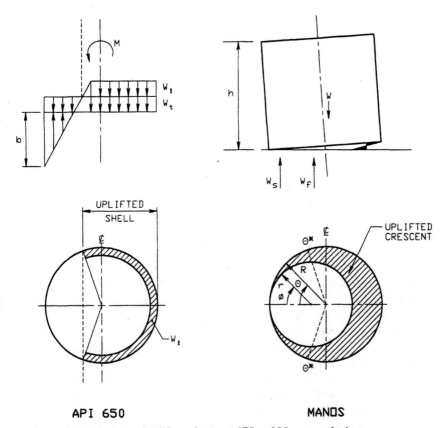

Figure 6.2.9 Comparison of uplift mechanism, API and Manos methods.

times the ground acceleration. If the acceleration needed to cause tank buckling is greater than that found on the response spectra, then the tank has adequate resistance to earthquakes.

The method presented here is modified from the original version presented by Manos:

- Whenever the tank H/D ratio exceeds 2, then the tank should be anchored, which is believed to be the upper limit of applicability to the Manos method.

- The allowable stress values can be increased from 75 to 90 percent of the classical buckling stress as needed. Experimental and observation data seem to support this. The compressive stress should not be allowed to exceed the yield stress.

- The sum of the Manos compressive stresses around the perimeter of the tank should not exceed the total weight of the tank and contents, which puts an upper bound on the resisting moment.

The Manos method does not consider the effect of nonlinearity in the stress-strain curve (e.g., strain beyond the proportional limit), and this may make it somewhat unconservative under elastic buckling conditions. Foundation stiffness can have a considerable effect on tank performance. Since a flexible foundation allows for more rigid-body motion, it has more uplift, radial displacement, and penetration. The compressive stresses in the tank shell are decreased as the foundation becomes more flexible. In the Manos method there is a soil factor to account for the decreased stresses. In API 650 there is none. Tanks that have thinner-than-average shells may pass API 650 criteria but fail the Manos method. This is because the Manos method is more sensitive to the tank's shell thickness than API 650. The difference becomes more pronounced in lower seismic zones. Details on computing the seismic capacity based on the Manos method are given in App. 6A.

Review attachment details. A walk-through should be conducted to review the flexibility of piping connected to the tank, attached ladders and platforms, handrails, or any other structures connected to tanks. Decisions should be made on how to modify them so that they do not cause failure of the tank even with uplift or swaying of the tank top. Although the dynamics and degree of uplift are not really fully understood at this time, experience shows that provisions for 10 to 12 in of vertical uplift will conservatively ensure that the tank contents are not lost. Horizontal sliding of the tank on foundations is not common, but there was a case in the 1989 Loma Prieta earthquake in the San Francisco area where a jet fuel tank 90 ft in diameter slid about 6 in on its foundation. Fortunately, flexibility in the piping which was 12 in in diameter prevented loss of contents. Piping that is less than 6 in in diameter can usually be considered flexible unless it is restrained within 5 to 10 ft of the connection to the shell. Ladders or stairs that are connected to the tank shell with clips and also rigidly attached to the ringwall or to concrete foundations pose a serious hazard. It is often possible to eliminate this problem simply by unbolting or unfastening the connection to the foundation or even trimming it off. When tanks are connected at their tops via interconnecting platforms, the relative motion between them caused by a seismic event can tear out sections of the tank. Although these failures may not be serious because they are above the liquid level, it may be possible that when one end of the platform falls, it damages the shell below the liquid level. These factors should be considered in the walk-through seismic examination.

Implement a plan. The final step is to form a strategic plan that addresses the worst hazards first and systematically addresses all tanks over time. All this work should be documented so that a change in management or personnel does not undermine the overall plan, which could span a long time. For tanks failing to pass both the API 650 and the alternative analysis for suitability for service, consider the following measures:

- Lower the maximum operating liquid height.

- Anchor the tank.

- Modify the tank and foundation.

Lower operating liquid level. Lowering the safe maximum liquid level is often an appropriate choice. Care should be exercised in computing the seismic capacity of existing tanks that the shell height is not the height used in the computations. Rarely will a tank be filled to the top of the shell. For floating-roof tanks, the roof occupies the top several feet, and therefore this should be accounted for. For fixed-roof tanks, there may be overflows that limit the maximum liquid level. In addition, freeboard should be left in the tank for sloshing waves. Without freeboard, the sloshing waves can collapse a roof by rolling the rafters over or by knocking columns over.

Another approach is to consider the probability that the tank is above its safe maximum liquid height during a seismic event. Many tanks are rarely filled to the top but are operating at about some medium level. If data can be used to show that the probability of the liquid level's being above the safe fill height is comparable to the level of acceptable risk, then nothing may need to be done in spite of the tank's not meeting the seismic criteria specified in the codes. This is because the codes are based on an absolute liquid level. They do not consider the probability that the shell may rarely be filled to the top rim angle.

Anchor tank. A tank can be anchored to its foundation. If a tank is anchored, generally its full operating height can be obtained. If the seismically safe liquid height is compared to the maximum operating height, a cost analysis can be performed to assist in the decision to anchor a tank using the incrementally increased capacity.

During the walk-through inspection, care should be exercized to excuse tanks solely on the basis that they are already anchored. The anchorage should be examined to ensure that they are of the chair design. Anchor straps cannot be considered adequate.

Modify the tank. Depending on specific conditions it may be possible to upgrade the tank by replacing a lower shell course, causing the tank to meet the seismic analysis criteria; to add or thicken the annular plate, which can increase the seismic resistance according to API; or to stiffen the shell by using vertical stiffeners. Most of the time these methods will be economically prohibitive.

Another method is to add a new bottom approximately 6 to 12 in above the old bottom and to fill the space with a heavy material such as concrete to increase the overturning resistance. This method is discussed in detail below. Still other methods modify the foundation to improve resistance to overturning. Some of these are discussed as well.

6.2.5 Design Considerations For Retrofit Approaches

The decision to retrofit involves not only the tank and appurtenances but also the foundation. Resistance to the overturning forces must be developed back down to the soil. There are several methods to do this. The most economical method for an individual tank will depend on the tank's existing conditions. A flowchart for foundation retrofitting decisions is shown in Fig. 6.2.10.

6.2.5.1 Tank foundations on grade

Aboveground tanks are placed on compacted soil foundations or concrete ringwalls or slabs or combinations. Tanks which rest directly on grade are placed on sand, gravel, asphalt, or crushed rock. It is difficult to anchor these tanks. Three methods are usually considered for anchoring tanks with soil foundations.

For small tanks, either the tank can be lifted and a new slab foundation can be built under the tank, or the foundation can be built in parts by excavating portions of the tank foundation and working from the periphery of the tank. When this type of retrofit is done, the anchor bolts should be cast into the new foundation. Retrofit concrete anchors should not be used. In the case of this type of retrofit, the soil does not need to take the uplift load as the foundation, for this is accomplished through the anchors. Allowable bearing stresses should be checked as in normal foundation design for seismic events as well as for the adequacy of the foundation to carry the anchorage loads.

Larger tanks are much more difficult to move without damaging them and the costs associated with this operation are prohibitive. It is, however, possible to retrofit a ringwall under the tank's shell. In most cases the weight of the ringwall would be insufficient to provide the required resistance to uplift. Helical piles or caissons may need to be retrofitted.

A seismic retrofit often occurs in conjunction with a tank bottom replacement. This happens because old tanks, when inspected, often need new bottoms, but they were also built prior to 1978 and so may not meet current seismic standards. Replacing a bottom lends itself to increasing the resistance to uplift by increasing the weight and, hence, resistance to uplift, as described later.

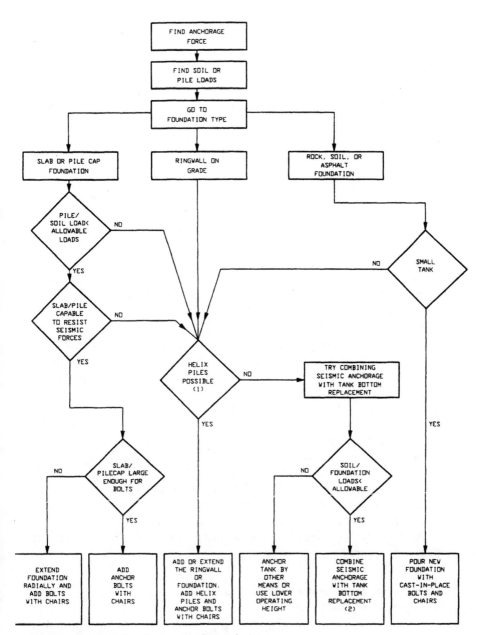

Figure 6.2.10 Tank foundation retrofitting decisions.

6.2.5.2 Slabs

Tanks with slabs under them usually require minimal modifications. However, the foundation may be inadequate to resist the overturning forces without modification. This should be checked. The tank can be anchored directly to the slab. If the edge distance between the anchor bolt and the foundation is small, the foundation may have to be extended radially for adequate anchor bolt confinement. Thin slabs may require that the bolt pass through the slab. In these cases large washers or plates under the slab have been used to distribute the anchor load over wider areas. As with all retrofits, the soil bearing capacities should be checked and the slab checked for its ability to carry the seismic overturning forces generated. If inadequate, then modifications to the foundation must be considered.

6.2.5.3 Ringwalls

When a tank is anchored to its ringwall for earthquakes, the ringwall must be designed for the additional loads caused by the earthquake. The ringwall must be designed for the shear and moment caused by anchoring the tank. The ringwall is analyzed as a continuous beam with supports at pile locations. The moment can be positive or negative depending on the location in the ringwall and the load direction. Load direction may be from the tension loading on the bolts or the compressive stresses from the tank shell.

The eccentricity between the bolts and the piles will cause a torsion in the ringwall which must also be designed for. To resist the uplift forces, the piles must have uplift connections that are strong enough to resist the full design loads.

Adding piles under the ringwall makes it much more resistant to vertical loads than the tank's center, which sits on the soil. If the soil is still settling the differential settlement between the ringwall and the tank's center should be accounted for. For existing tanks with ringwalls, anchor chairs can be attached to the shell and the shell anchored to the ringwall. The ringwall may require modifications to accommodate the overturning forces. One approach to these types of modifications is to extend the ringwall outward radially, increasing the resisting weight. Usually, the addition of helical piles or caissons will also be needed.

6.2.5.4 Seismic anchorage combined with tank bottom replacement

For normal tank bottom replacements using the double-bottom methods, the section below the new bottom is not normally designed for strength and is often not even connected to the new bottom and tank which rests directly atop. By increasing the thickness of the space between the old and new bottom, significant weight can be applied to resisting overturning moments. However, the design details required to join the old and new bottoms together with sufficient strength become important. The reason is that as the tank lifts up, it must lift not only the new bottom and liquid contents but also the old bottom and the spacer material (usually concrete). Using concrete increases the section modulus of the bottom and produces stiffness, which reduces uplift. Again, the soil bearing pressure should be checked. This method does reduce the tank's effective capacity because the bottom must typically be placed 6 to 12 in above the old bottom to be effective.

To implement this method, find the radial uplifted length. The uplifted bottom plate length is found by summing the moments about the far compression side, which equal zero. The forcing moment is the everturning moment, and the resisting moment is due to the fluid resting on the uplifted concrete slab. To account for the uncertainty of the uplifted length, the overturning moment is multiplied by 1.5.

6.2.6   Using Piles

This section will give some background information on helix piles and compare them to other pile types. Detailed designs will not be covered.

To be earthquake-resistant, the tank must be able to bring the earthquake forces back down into the soil. The earthquake usually causes uplift on one side of the foundation which can be resisted by piles in tension. Typical piles that can resist tension are

- Driven piles
- Caissons
- Helix piles

Because of their economy, helix piles can be one of the best choices. Helical piles are basically large screws using plate that has been wrapped around a shaft into a helical pattern. They are usually driven by screwing them into place. The equipment to do this can be much smaller than that for other kinds of piles and is usually a driver device mounted on a truck.

Resistance against pullout is provided by the soil itself, and for shallow depths the failure is a shear cone projecting upward from the bottom helical plate. Shallow embedment is usually considered to be less than about 5 turns of the helix, and deep embedment is greater than this. For deep embedments the failure surface is a plug about the diameter of the anchor. A deep anchor is preferred because it will usually have a ductile failure mode instead of pulling out of the soil. Anchor diameters range from 8 to 14 in. The helix pile must be embedded into the foundation so that it does not pull out. This is accomplished by adding a plate on the helix pile's end that is embedded into the concrete. An example of helical piles is shown in Fig. 6.2.11. The plate and the surrounding concrete must be checked for the following:

- Concrete bearing pressures under the plate
- The bending and shear stresses in the plate
- The bearing plate's pulling out of the concrete

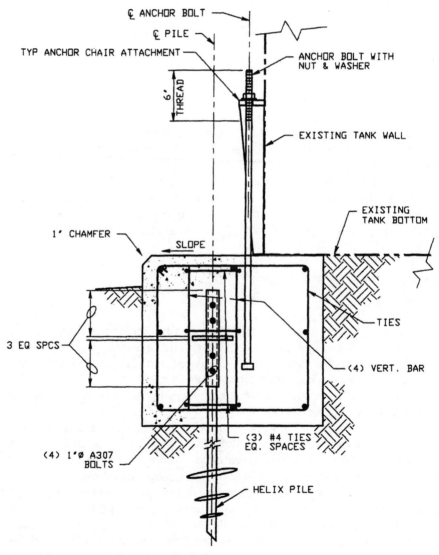

Figure 6.2.11 Helical piles used with ringwall foundation.

Since the helix pile can be in tension or compression, the above three items should be checked in both directions.

Since the piles cannot be anchored to rock, the soil must be deep enough to qualify the pile as having a deep embedment (5 diameters). Also, since helix piles are torqued into place, the soil cannot be too dense or have large boulders.

Certain spacing considerations also apply:

- Minimum spacing between piles is 3 diameters.
- Minimum spacing to an adjacent structure is 5 diameters.

If the above limitations cannot be met, an alternate pile type should be used. Helical piles may cost less than caissons, but soil conditions could be the determining factor in an economical selection. Careful investigation of the soil conditions beneath the tank is required to ensure adequate anchorage. In addition, the life of the helical piles should be considered in conjunction with the corrosion rate anticipated. If necessary, cathodic protection could be applied to increase the life of the helical anchors, or anchors sufficiently large that there is an effective corrosion allowance could be considered.

The soil may not be able to resist the pullout forces. It may not be deep enough, or the soil may be especially poor. In some instances it may be more economical to use piles or caissons instead of helix piles. In these cases, the method of combining seismic anchorage with the tank bottom replacement may have to be used. This method uses the tank's own mass to anchor it and may be applied to any foundation system as long as the soil loads are less than the allowable loads. This method may not be the most economical.

6.2.7 Anchorage Design of Existing Tanks

In an anchored tank, it is important that the anchor bolts fail ductilly before the anchor chair, tank shell, or foundation attachment fails, in order to prevent the loss of tank contents. One way to ensure this is to design the anchor chair for some factor (usually 1.25) times the bolt yield strength.

Designing to prevent pullout of the anchor bolt is beyond the scope of this section, but certain critical factors should be considered. To develop a ductile anchorage system, the steel component capacities must be less than the capacities determined by the concrete. A design factor of safety of 4 is recommended, and this accounts for the variations in steel and concrete properties encountered in the field. It ensures that the bolt will yield before the concrete fails and covers a variation in the assumed failure cone shape that occurs in concrete.

This factor of safety is consistent with the anchor bolt specification outlined in the Uniform Building Code. Careful consideration needs to be given to inadequate edge spaces which could cause the concrete to fail at a lower than computed value.

6.2.7.1 Types of anchors

The most common anchors used, in order of preference, are

1. Cast-in-place anchors
2. Capsule anchors
3. Grouted-in-place A-307 bolts
4. Through bolts which penetrate the foundation thickness

6.2.7.2 Cast-in-place anchors

These are the preferred anchor because the concrete is cast around the anchor and they develop the best connection. They can be used only with foundation retrofits that involve the pouring of new concrete and are, therefore, narrowly limited in application.

6.2.7.3 Capsule anchors

These anchors use an epoxy adhesive to develop the bond between the anchor and the concrete. They provide a convenient means of anchorage as they require little embedment length and can be placed close to the tank. A ductile design requires that the epoxy bond develop a strength greater than that of the bolt. Anchor manufacturer recommendations should be used. A rule of thumb is that 2 times the minimum embedment depth is sufficient to develop a ductile anchorage system. For hot tank foundations, then, care must be exercised to ensure that the epoxy retains its strength. It should be assumed that the anchor is at the same temperature as the heated tank; otherwise, a detailed thermal analysis should be performed to determine the worst-case steady-state temperature of the anchor at its deepest point of embedment.

6.2.7.4 Grouted-in-place A-307 bolts

Where the above conditions cannot be met due to spacing or edge distance concerns, grouted bolts may work. The advantages of grouted bolts are that

- Small edge distance is required.
- Anchors may be placed as deep as the minimum concrete cover will allow.

The disadvantages are

- Large eccentricities due to the large hole that needs to be drilled
- Higher factor of safety required due to the greater variability in concrete cone capacity

6.2.7.5 Steel through bolts

For slabs or foundations which are not very thick, the anchor can penetrate the foundation. A stainless steel bolt is recommended since corrosion is likely to attack carbon steel anchors.

Appendix 6.2A. Manos Method of Evaluation of Existing Tanks

The equation for determining the tank's seismic resistance is

$$C_{eq} = \frac{0.372}{\delta_w} \frac{SEt_s^2}{GRH^2} \frac{M_t}{M_1} \left(\frac{R}{H}\right)^n \left(\frac{t_s}{t_p}\right)^{0.1} \qquad (6.2.1)$$

where E = Young's modulus, psi
  G = specific gravity of contents
  H = liquid height, ft
  R = tank radius, ft
  M_t = total liquid mass, lb
  M_1 = impulsive liquid mass, lb
  $n = 0.1 + 0.2H/R \le 0.25$
  t_p = bottom plate thickness, in
  t_s = tank wall thickness, in
  δ_w = unit weight of water, lb/ft^3

This equation only accounts for the impulsive liquid for the earthquake forces. The convective liquid forces have been compensated by increasing the moment arm X_1/H for the impulsive liquid. Also, the method does not account for the usually neglectible shell and roof weight and forces, which constitutes only a small inaccuracy.

Procedure using the Manos approach

It is recommended that the valid H/D range for the Manos method be 0.4 to 2.0, which was supported by experimental data. For tanks with $H/D > 2.0$ this method will always fail. Therefore, this sets the upper bound. When $H/D < 0.4$, then the tank can be computed as though it had a value of 0.4, using the actual diameter. This will give conservative results.

The peak buckling stress is arbitrarily limited to 75 percent of the classical buckling stress to account for the fact that elastic buckling can occur at values well below the classical value. On small tanks, however, the buckling stress may exceed the yield stress. To limit the method when this occurs, a constant α is defined:

$$\sigma_{buckle} = \frac{0.454Et_s}{R} \qquad (6.2.2)$$

$$\alpha = \frac{\sigma_{buckle}}{\sigma_{yield}} \geq 1.0$$

The constant α is used as a divisor later in the procedure.

The weight of the tank's contents is

$$W_t = \frac{62.4\rho(\pi HD^2)}{4000} \qquad (6.2.3)$$

Next, calculate the shell's total compressive force.

$$F_c = 0.38SEt_s^2\left(\frac{R}{H}\right)^n\left(\frac{t_s}{t_p}\right)^{0.1} \qquad (6.2.4)$$

The Manos method also sums the total shell compression, assuming that the peak stress level is 75 percent of the classical buckling stress as indicated above. Again, certain conditions can cause the total compressive stress to be greater than the total tank weight including contents. For this purpose a constant is defined:

$$\beta = \frac{F_c}{W_t} \geq 1.0$$

Next, the soil rigidity factor S can be selected from the Table 6.2.3. The applicable soil type should be the soil immediately under the tank shell. Soil types 2 and 3 are defined in API 650, Appendix E.

The ratio W_1/W_t is determined by using fig. E-2 in API 650 Appendix E. Then the value of C_{eq} can be computed from Eq. (6.2.1). However, it must be divided by the larger of α or β to get the modified C_{eq}. A comparison between the modified C_{eq} with the peak spectral values in Table 6.2.4 defines the acceptability under this method. If a site-specific response spectrum is available, C_{eq} can be compared to the spectral peak at 2 percent damping. It must be less than C_{eq} for the tank to be stable.

Unstable tanks

A tank that has a C_{eq} less than the peak acceleration which the tank would be expected to see in an earthquake is unstable and should

TABLE 6.2.3   Foundation Type

Concrete slab, asphalt pad, ringwall, or rock	Crushed rock, plank, or type 2 soil	Type 3 soil
—		
$S = 1.0$	$S = 1.2$	$S = 1.5$

TABLE 6.2.4   Peak Spectral Value

UBC zone	Soil type		
	S_1	S_2	S_3
1	0.194	0.232	0.292
2A	0.387	0.464	0.581
2B	0.516	0.619	0.774
3	0.774	0.929	1.161
4	1.032	1.238	1.548

have its safe seismic fill height calculated. The safe seismic fill height can be compared to the operating height for an economic analysis. The example problem describes how to perform a seismic analysis by the Manos method.

Example: Seismic Evaluation of an Existing Tank. The tank described below has been targeted as high risk and an analysis is to be performed to determine its suitability for service. A first step is to gather the required information which is:

Tank diameter	30 ft
Tank height	30 ft
Maximum fill height	28 ft
Seismic zone	4
Tank fluid	Hydrocarbon
Specific gravity	0.80
Tank material	A36, $F_y = 36$ ksi
Cone roof weight	5.4 kip
Percent supported by shell	50%
Importance factor	1.0
Site amplification factor, S	1.0

Shell courses	Height, ft	Thickness, in
1	6	0.25
2	6	0.25
3	6	0.25
1	6	0.1875
2	6	0.1875
Bottom plate		0.25

Use Table 6.2.2 to get an idea if the tank will meet API 650. $H/D = 1.0$ so an analysis should be performed.

Compute the overturning moment according to API 650, Appendix E:

$$M = ZI(C_1 W_s X_s + C_1 W_r H_t + C_1 W_1 X_1 + C_2 W_2 X_2)$$

where Z = seismic zone factor
    = 0.4 for zone 4
    I = importance factor (1.0 for all normal tanks)
    C_1 = 0.60
    $C_2 = 0.75S/kD^{1/2} = (0.75)(1.5)(0.578 \times 5.48) = 0.355$
    W_s = 25,976 lb
    X_s = height of center of gravity of shell = 14 ft
    W_r = weight of roof = 5411 lb
    H_t = height of shell = 30 ft
    W_t = weight of tank contents = 987,840 kip
    W_1 = impulsive mass = $(W_1/W_t)W_t$ = (0.77)(988) = 760 kip, where (W_1/W_t) is read from the graph in API 650 using the actual maximum liquid level, giving H/D = 28/30 = 0.93
    W_2 = convective mass = $(W_1/W_t)W_t$ = (0.24)(988) = 237 kip
    X_1 = height of centroid of impulsive mass = 0.4 × 28 = 11.2 ft
    X_2 = height to centroid of convective mass = 0.73 × 28 = 20.4 ft

Inserting the above values, compute the moment:

$$M = ZI(C_1 W_s X_s + C_1 W_r H_t + C_1 W_1 X_1 + C_2 W_2 X_2)$$

$$= (0.4)(1.0)[(0.6 \times 25976 \times 14 + 0.6 \times 5411 \times 30 + 0.6 \times 760000 \times 11.2 + 0.355 \times + 237000 \times 20.4)]$$

$$= 2.9 \times 10^6 \text{ ft·lb}$$

Next, the resistance to overturning is computed by the minimum of the expressions

$$w_L = \sqrt{F_{by} GH} \qquad 1.25GHD$$

The controlling value is $1.25GHD$ = 840 lb/ft. Next we find

$$w_t = \frac{25,976 + 0.5 \times 5411}{\pi \times 30} = 304 \text{ lb/ft}$$

The overturning ratio is then

$$\frac{M}{D^2(w_t + w_L)} = \frac{2.8 \times 10^6}{30^2(304 + 840)} = 2.72$$

Since this is greater than 1.57, the tank is unstable. The choices going from the easiest to the most difficult are

1. Lower the safe fill height.
2. Perform an alternative analysis to see if it passes this criteria.
3. Anchor the tank.
4. Thicken the annular plate.
5. Thicken the shell.
6. Modify the foundation.

The facility decides that lowering the operating height is not feasible. Therefore, it is decided to perform an analysis according to the Manos method. The Manos analysis proceeds as follows. Since H/D = 28/30 = 0.93, it is within the acceptable H/D range of 0.4 to 2.0 and the analysis is valid. Since the foundation is concrete, a foundation rigidity factor of 1.0 is selected. Since the soil type is type A and the seismic zone is 4, from Table 6.2.4, the peak spectral acceleration is 1.032. A maximum buckling stress will be set at 75 percent of the classical buckling stress:

$$\sigma_{classical} = \frac{0.453E_t}{R} = 18.3 \text{ ksi}$$

The tank shell compressive force is

$$F_c = 0.38(1.032)(29 \times 10^3)(0.25^2)\left(\frac{15}{28}\right)^{0.25}\left(\frac{0.25}{0.25}\right)^{0.25} = 608 \text{ kip}$$

Sometimes the Manos method for small tanks can give a total compressive stress that is greater than the total weight of the tank and contents. Computing β takes this into account. The total weight of the tank contents computed above was

$$W_t = 988 \text{ kip}$$

Therefore, if $W_t < F_c$, then calculate

$$\beta = \frac{F_c}{W_t} = \frac{608}{988} \quad \text{since less than 1,} \quad \beta = 1.0$$

$$C_{eq} = \frac{0.372}{62.4} \frac{(1.032)(29 \times 10^6)(0.25)^2}{(0.8)(15)(28)^2}\left(\frac{988}{760}\right)\left(\frac{15}{28}\right)^{0.25}\left(\frac{0.25}{0.25}\right)^{0.1} = 1.32$$

Since the Manos limiting acceleration 1.32 is greater than the peak spectral acceleration of 1.03, the tank is adequate and does not need to be anchored.

In spite of the tank failing the API 650, Appendix E criteria, the alternative method that more realistically takes into account the uplift resistance provides a reasonable argument for accepting this tank for service without retrofitting it with anchors. If the tank was computed with a site amplification factor of 1.5, then the tank would have to be anchored by either methodology.

API 650		Manos	
Instability ratio	2.77	Acceleration ratio (peak spectral 2%/Manos limiting)	1.55/1.28 = 1.21
Safe oil height (unanchored), ft	18.4	26	

Different combinations of shell height, wall thickness, as well as other variables will give substantially differing results. A trial and error analysis can be used to determine the maximum safe oil height for both the API and Manos methods to preclude the possibility of installing anchors. When the modified Manos method fails, then anchorage needs to be applied or other modifications made if the safe oil height is not reduced.

To complete the analysis, the forces acting on the foundation must be computed. These will be based on a liquid height of 30 ft. Seismic overturning moment may produce very high bearing loads under the shell. For sandy, noncohesive or soft soils new tanks should use ringwalls or slab foundations. When small tanks will be retrofitted to be supported on ringwalls or matts with piles, the foundations should be designed for the overturning moment applied to the bottom of the foundation, which is greater than the moment applied to the tank just above the bottom.

Moments acting on foundations are greater than the moments computed using the method of API 650. This is because hydrodynamic pressures develop on the tank bottom. To account for this a lever arm that is greater than X_1 and X_2 is used for the foundation. The overturning moment can be found by substituting values of X_1' and X_2' for X_1 and X_2. Figure 6.2.12 is a plot of X_1' and X_2' for various values of H/D. Using the figure for our example, $H/D = 1$ and $X_1' = 0.56 \times 30 = 16.8$ and $X_2' = 0.75 \times 30 = 22.5$.

The moment acting at the top of the foundation is found by substituting these values into the basic moment equation as follows:

$$M = ZI(C_1 W_s X_s + C_1 W_r H_t + C_1 W_1 X_1' + C_2 W_2 X_2')$$

$$= (0.4)(1.0)[(0.6 \times 25976 \times 14 + 0.6 \times 5411 \times 30 + 0.6 \times 760000 \times 16.8 + 0.355 \times 237000 \times 22.5)]$$

$$= 4.0 \times 10^6 \text{ ft·lb}$$

This is considerably greater than the moment applied to the tank shell computed above (2.9×10^6 ft·lb). The moment acting at the pile or soil level the base shear must be taken into account using

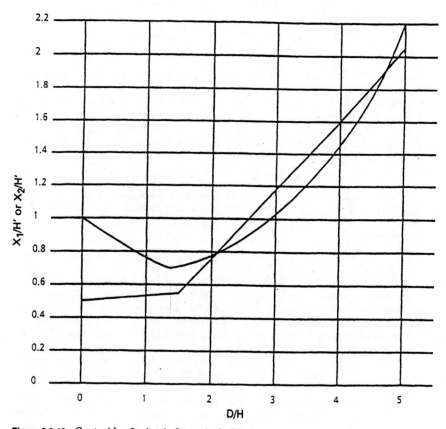

Figure 6.2.12 Centroids of seismic forces including base pressure.

$$V = ZI(C_1 W_s + C_1 W_r + C_1 W_1 + C_2 W_2)$$

$$= (0.4)(1.0)(0.6 \times 25{,}976 + 0.6 \times 5411 + 0.6 \times 760{,}000 + 0.355 \times 237{,}000)$$

$$= 224 \text{ kip}$$

The moment on the soil is $M + V \times t$, where t is the slab thickness. The moment on the bottom of the 12-in-thick foundation is

$$M = 4.0 \times 10^6 + \frac{224 \times 10^3 \times 12}{12}$$

$$= 4224 \text{ ft} \cdot \text{lb}$$

Use this moment to check the soil and/or pile loads.

Now we will check the freeboard at the top of the tank to determine if the roof may be damaged by sloshing seismic waves. The basic equation for determining sloshing wave height is

$$d = \frac{1.124Z}{C_2 T^2} \tanh\left(4.77\sqrt{\frac{H}{D}}\right) = 1.2 \text{ ft}$$

Since the liquid level is 2 ft below the top angle, there is sufficient freeboard to prevent tank damage on the roof.

In the walk-through inspection it is advisable to check ladders and piping. Actual stresses, loads, and moments should be analyzed with a piping flexibility analysis to determine just what is acceptable and what is not. Defining the acceptable motion for uplift and for translation is a judgment call on the part of the tank owner/operator.

References

1. G. W. Housner, "Earthquake Pressures on Fluid Containers," A Report on Research Conducted under Contract with the Office of Naval Research, California Institute of Technology, Pasadena, Earthquake Research Laboratory, August 1954.
2. R. S. Wozniak and W. W. Mitchell, "Basis of Seismic Design Provisions for Welded Steel Oil Storage Tanks," presented at the Session on Advances in Storage Tank Design, American Petroleum Institute, Refining, 43d Midyear Meeting, Toronto, Canada, May 9, 1978.
3. J. E. Rinne, "Oil Storage Tanks, Alaska Earthquake of 1964," *The Prince William Sound, Alaska, Earthquake of 1964*, vol. II-A, U.S. Department of Commerce, Coast and Geodetic Survey, 1967.
4. R. D. Hanson, "Behavior of Liquid Storage Tanks," *The Great Alaska Earthquake of 1964*, Engineering, National Academy of Sciences, Washington, 1973.
5. P. C. Jennings, "Damage of Storage Tanks," *Engineering Features of the San Fernando Earthquake, February 9, 1971*, Earthquake Engineering Research Laboratory Report 71-02, Cal. Tech., June 1971.
6. R. Husid, A. F. Espinosa, and J. de las Casas, "The Lima Earthquake of October 3, 1974: Damage Distribution," *Bulletin of the Seismological Society of America*, vol. 67, no. 5, pp. 1441–1472, October 1977.
7. G. W. Housner, "Dynamic Pressures on Accelerated Fluid Containers," *Bulletin of the Seismological Society of America*, vol. 47, pp. 15–35, January 1957.
8. Lockheed Aircraft Corporation and Holmes & Narver, Inc., *Nuclear Reactors and Earthquakes*, chap. 6 and App. F, ERDA TID 7024, August 1963, pp. 183–195 and 367–390.
9. A. S. Veletsos and J. Y. Yang, "Earthquake Response of Liquid-Storage Tanks," *Advances in Civil Engineering through Engineering Mechanics, Proceedings of the 2d Annual Engineering Mechanics Division Specialty Conference*, American Society of Civil Engineers, pp. 1–24, May 1977.

10. R. S. Wozniak, "Lateral Seismic Loads on Flat Bottomed Tanks," Chicago Bridge & Iron Company, The Water Tower, November 1971.
11. D. P. Clough, "Experimental Evaluation of Seismic Design Methods for Broad Cylindrical Tanks," University of California Earthquake Engineering Research Center, Report no. UCB/EERC/77/10, May 1977.
12. C. D. Miller, "Buckling of Axially Compressed Cylinders," *Journal of the Structural Division, ASCE,* vol. 103, no. ST3, *Proceedings* paper 12823, pp. 695–721, March 1977.
13. H. Lo, H. Crate, and E. B. Schwartz, "Buckling of Thin-Walled Cylinders under Axial Compression and Internal Pressure," NACA TN 2021, 1950.
14. D. C. Ma, J. Tani, S. S. Chen, and W. K. Liu (eds.), "Sloshing and Fluid Structure Vibration—1989," *PVRC,* vol. 157, American Society of Mechanical Engineers, New York, presented at the 1989 ASME Pressure Vessels and Piping Conference, Honolulu, HI, July 23–27, 1989 (copyright 1989).
15. D. C. Ma and T. C. Su (eds.), "Fluid-Structure Vibration and Liquid Sloshing," *PVP,* vol. 128, American Society of Mechanical Engineers, New York, presented at the 1987 Pressure Vessels and Piping Conference, San Diego, CA, June 28–July 2, 1987 (copyright 1987).
16. S. T. Algermissen, *An Introduction to the Seismicity of the United States,* Earthquake Engineering Research Institute.
17. N. M. Newmark and W. J. Hall, "Earthquake Spectra and Design," Department of Civil Engineering, University of Illinois at Urbana-Champaign, Earthquake Engineering Research Institute.
18. G. W. Housner and P. C. Jennings, "Earthquake Design Criteria," Division of Engineering and Applied Science, California Institute of Technology, Earthquake Engineering Research Institute.
19. A. K. Chopra, "Dynamics of Structures—A Primer," University of California, Berkeley, Earthquake Engineering Research Institute.
20. M. A. Haroun and E. A. Abdel-Hafiz, "A Simplified Seismic Analysis of Rigid Base Liquid Storage Tanks under Vertical Excitation with Soil-Structure Interaction," *Soil Dynamics Earthquake Engineering,* vol. 5, pp. 217–225, 1986.
21. M. A. Haroun and G. W. Housner, "Seismic Design of Liquid Storage Tanks," *Journal of Technical Councils, ASCE,* vol. 107, no. 1, pp. 191–207, 1981.
22. Y. Tang, R. A. Uras, and Y. W. Change, "Effect of Viscosity on Seismic Response of Waste Storage Tanks," Argonne National Laboratory, Reactor Engineering Division, March 1992.
23. A. S. Veletsos, "Seismic Response and Design of Liquid Storage Tanks," *Guidelines for the Seismic Design of Oil and Gas Pipeline Systems,* Technical Council on Lifeline Earthquake Engineering, American Society of Civil Engineers, New York, 1984, pp. 255–370 and 443–461.
24. A. S. Veletsos, "On Influence of Liquid Viscosity on Dynamic Response of Tanks," Report presented to BNL/DOE Seismic Experts Panel, Rockville MD, April 10, 1992.
25. A. S. Veletsos and Y. Tang, "Dynamics of Vertically Excited Liquid Storage Tanks," *Journal of Structural Division, ASCE,* vol. 112, pp. 1228–1246, 1986.
26. A. S. Veletsos and Y. Tang, "Interaction Effects in Vertically Excited Steel Tanks," *Dynamic Response of Structures,* G. C. Hart and R. B. Nelson (eds.), American Society of Civil Engineers, New York, 1986, pp. 636–643.
27. A. S. Veletsos and Y. Tang, "Rocking Response of Liquid Storage Tanks," *Journal of Engineering Mechanics, ASCE,* vol. 113, pp. 1774–1792, November 1987.
28. A. S. Veletsos and Y. Tang, "Soil-Structure Interaction Effects for Laterally Excited Liquid Storage Tanks," *Journal of Earthquake Engineering and Structural Dynamics,* vol. 19, pp. 473–496, 1989.
29. A. S. Veletsos, P. ShivaKumar, Y. Tang, and H. T. Tang, "Seismic Response of Anchored Steel Tanks," *Proceedings of the 3d Symposium on Current Issues Related to Nuclear Plant Structures, Equipment, and Piping,* A. J. Gupta (ed.), North Carolina State University, Raleigh, 1990, pp. X/2-2 to X/2-15.
30. A. S. Veletsos and Y. Tang, "Earthquake Response of Liquid Storage Tanks," *Advances in Civil Engineering through Engineering Mechanics, Proceedings of the ASCE/EMD Specialty Conference,* Raleigh, NC, 1977, pp. 1–24.

31. M. J. N. Priestly, B. J. Davidson, G. D. Honey, D. C. Hopkins, R. J. Martin, G. Ramsay, J. V. Vessey, and J. H. Wood (eds.), "Seismic Design of Storage Tanks," Recommendations of a Study Group of the New Zealand National Society for Earthquake Engineering, December 1986.

32. *Welded Steel Tanks for Oil Storage,* Appendix E, *Seismic Design of Storage Tanks,* API 650-E, American Petroleum Institute, Washington, November 1980.

33. "Manufacture of Vertical Steel Welded Storage Tanks with Butt-Welded Shells for the Petroleum Industry," BS 2654, British Standards Institution, London, 1984, p. 83.

34. H. F. Winterkorn and H. Y. Fang, *Foundation Engineering Handbook,* van Nostrand-Reinhold, New York, 1975.

35. J. M. Duncan and T. B. D'Orazio, "Stability of Steel Oil Storage Tanks," *ASCE Journal of Geotechnical Engineering,* vol. 110, no. 9, September 1984.

36. D. M. Penman, "Soil Structure Interaction and Deformation Problems with Large Oil Tanks," *Ground Engineering.*

37. G. M. Harris, "Foundations and Earthworks for Cylindrical Steel Storage Tanks," *Ground Engineering,* July 1976, pp. 24–31.

38. J. M. Duncan, P. C. Lucia, and R. A. Bell, "Simplified Procedures for Estimating Settlement and Bearing Capacity of Ringwalls on Sand," American Society of Mechanical Engineers.

39. R. E. Hunt, *Geotechnical Engineering Investigation Manual,* McGraw-Hill, New York, 1984.

40. "Site Investigation (Subsurface)," MWD Civil Engineering Division Publication CDP 813/B, 1982.

41. J. S. Sherard, L. P. Dunnigan, and R. S. Deeker, "Identification and Nature of Dispersive Soils," *ASCE Geotechnical Engineering Division,* vol. 102, no. GT4, 1976.

42. H. B. Seed and I. M. Idress, "Ground Motions and Soil Liquefaction during Earthquakes," Monograph published by Earthquake Engineering Research Institute.

43. J. B. Berrill and R. O. Davis, "Energy Dissipation and Seismic Liquefaction of Sands: Revised Model," *Soils and Foundations,* vol. 25, no. 2, pp. 106–118, June 1985.

44. A. T. Moore, "The Response of Cylindrical Liquid Storage Tanks to Earthquakes," *Proceedings of the International Conference on Large Earthquakes,* Misc. Series no. 5, Royal Society of New Zealand, 1981.

45. G. V. Berg and J. L. Stratta, "Anchorage and the Alaska Earthquake of March 27, 1964," American Iron and Steel Institute, New York.

46. K. V. Steinbrugge and R. Flores, "Engineering Report on the Chilean Earthquakes of May 1960—A Structural Engineering Viewpoint," *B.S.S.A.,* vol. 53, no. 2, 1963.

47. H. Kawasumi (ed.), "General Report on the Nigata Earthquake of 1964," Tokyo Electrical Engineering College Press.

48. G. W. Housner, "The Dynamic Behavior of Water Tanks," *B.S.S.A.,* vol. 53, no. 2, 1963.

49. "Code of Practice for General Structural Design and Design Loadings for Buildings," NZS 4203:1984, Standards Association of New Zealand, Wellington, 1984.

50. "Seismic Design of Petrochemical Plants," Ministry of Works and Development Civil Division Publication, Wellington, New Zealand, 1981.

51. *AWWA Standard for Welded Steel Tanks for Water Storage,* AWWA-D100, American Waterworks Association, Denver, CO.

52. "Guidelines for the Seismic Design of Oil and Gas Pipeline Systems," American Society of Civil Engineers, 1984, p. 466.

53. G. W. Housner and D. E. Hudson, *Applied Mechanics—Dynamics,* D. Van Nostrand, New York, 1957, p. 260.

54. Uniform Building Code.

55. *Bulletin on Stability Design of Cylindrical Shells,* API Bulletin 2U, 1st ed., American Petroleum Institute, Washington, May 1, 1987.

56. S. P. Timoshenko and J. M. Gere, *Theory of Elastic Stability,* McGraw-Hill, New York, 1961.

57. T. Von Karman and H. S. Tsien, "The Buckling of Thin Cylindrical Shells under Axial Compression," *Journal of Aeronautical Science,* vol. 8, no. 8, p. 303, 1941.
58. H. Lo, H. Crate, and E. B. Schwartz, "Buckling of Thin-Walled Cylinders under Axial Compression and Internal Pressure," NACA TN 2021, NACA, 1950.
59. AWWA D-100, *Theory of Elastic Stability,* McGraw-Hill, New York, 1961.
60. J. M. Rotter, "Buckling of Ground-Supported Cylindrical Steel Bins under Vertical Compressive Wall Loads," Melbourne, Australia, May 23–24, 1985.
61. K. Bandyopadhyay, A. Cornell, C. Constantino, R. Kennedy, C. Miller, and A. Veletsos, "Seismic Design and Evaluation Guidelines for the Department of Energy High-Level Waste Storage Tanks and Appurtenances," Department of Nuclear Energy, Brookhaven National Laboratory, Associated Universities, Inc. (U.S. DOE contract no. DE-AC02-76CH00016), January 1993.

7

General Design of Tanks

7.1 General Design Considerations

This section applies primarily to atmospheric, flat-bottom steel tanks; however, many of the principles can be extended to other designs and to other materials.

7.1.1 Standards

An early consideration in the design of tanks is to establish the appropriate codes and standards that may cover the proposed tank. Small tanks are covered by Steel Tank Institute standards or UL-142 as well as the API 12 Series or API 650, Appendix F standard. Larger vertical, cylindrical flat-bottom tanks are covered by API 650, API 620, and AWWA D-100 for water. Tanks of double curvature are covered by API 620. Pressure vessels and spheres are covered by the ASME Boiler and Pressure Vessel Code, section VIII. Complete coverage of codes and standards is given in Chap. 3.

7.1.2 Site and process data

A compilation of relevant data should be made prior to detailed tank design. This includes any existing site-specific geotechnical reports; meteorological data; required loading conditions for wind, snow, rain, or other loads; collection of physical properties for the range of liquids under consideration; flow rates into and out of the tank; any special hazards associated with the stored liquid; and other process and local data. The design life of the plant is another key item for design consideration. The design life of a tank is often different from the design life of other components in the plant. Utility costs may be important for establishing tank heater sizing and insulation requirements.

7.1.3 Materials

Corrosion and material compatibility are major factors in establishing what materials are acceptable for use in the tank. In addition, cost factors and establishment of design life play key roles in determining which thicknesses may be used. A carbon steel tank with a lining may or may not be more costly than a stainless steel tank, depending on the assumptions made in the cost analysis.

Mixing steels. The question often arises about mixing various steels in the same tank. The tank designer is confronted with many issues at the outset involving material selection and design thickness requirements. Although a single shell material might be selected, it is often more economical to select shell materials by course. These economies are usually only important in larger-diameter tanks because of the increased thicknesses. For these tanks the variable design point method of computing shell course thicknesses is often used. For multiple calculations it is efficient to set up a computer program. After computing the tank courses in one material, the designer can select other materials for various shell courses. The mixing of shell course steels usually occurs for the following reasons:

The economic optimization of material usage and costs is primary. For example, an A573-70 plate costs about 5 percent more than A36 plate but has over a 20 percent (28 versus 23.2 ksi) increase in design allowable stresses. This fact alone tends to promote the use of higher-strength steels when other factors are not at issue. However, in the upper shell courses the required thickness may be set by the minimum thickness for fabrication or because of the corrosion allowance. It would therefore be more economical to use the lower strength-to-cost steels in the upper courses as the thickness is not governed by design hydrostatic stress levels but by other factors such as these:

- The influence of material exemption curves for impact toughness testing can often cause the designer to switch to a more economical steel in the upper courses as a result of the decreasing thickness. The exemption curves of API 650, fig. 2-1 minimum permissible design metal temperature for plates used in tank shells without impact testing, is a good example.

- If the thickness of the shell is based on wind load conditions rather than hydrostatic design stresses, then it will usually be economical to use the lower-cost steel in the upper courses as opposed to the higher-cost steels used in the lower courses.

- Sometimes the thickness of the top course can be increased to meet the minimum thickness required for wind loading and to eliminate the need for a wind girder. In this case, the steel with the lowest cost is selected.

- Sometimes thermal stress relief requirements of the standards used can cause the material selection to vary by shell course. For example, in API 650 thermal stress relieving is required when the shell material is group I, II, III, or IIIA greater than 1 in thick; and for group IV, IVA, V, or VI the requirement applies when the thickness is $\frac{1}{2}$ in. The cost of thermally stress-relieving the shell plates may outweigh the higher cost of the thicker material on the bottom courses while not affecting the upper courses, which may govern.

- Availability of steels often plays a role in mixing of shell course steels.

The material of thickened insert plates is also often different from the shell course material. Since the thickened insert plate must meet the toughness impact testing requirements of Table 2-1 in API 650, tank fabricators will switch to a material that is not required to be impact-tested at its installed thickness. Another reason is that many of the materials are limited in thickness. For example, tank shell courses fabricated from A573-70 may be used up to $1\frac{1}{2}$-in maximum thickness including insert plates. If the shell is over $\frac{3}{4}$ in thick, this material cannot be used.

Annular plate materials. The material of the annular plate does not need to be the same as the steel used in the first course of a tank. In fact, for large tanks they are often different materials. For API 650 tanks, the annular plate material should be selected on the basis of the material exemption curve API 650, Fig. 2-1.

7.1.4 Operations

The operational needs tend to drive a primary design variable such as tank capacity. It is important to thoroughly understand the nature of the operations because they tend to be somewhat vague at times. For example, a tank may be used mostly for one stock but may be used for some other stock in the future. The stock change may affect the corrosivity or specific gravity and hence the storage capacity. A further example is transfer rates. The changes in filling and withdrawing may require that larger venting devices be selected so that the tank is not damaged during unusual operations. Operations will also dictate the requirements for access including ladders and platforms, means of sampling, use of floating suction lines, instrumentation requirements, and numerous additional items. One less obvious operating requirement may be product purity. For example, glycol is normally stored in steel tanks since steel has a low corrosion rate in this service. However, minute concentrations of dissolved iron tend to discolor the product, and for high-purity glycol, which must be clear and transparent, the storage tank must be coated with a vinyl or phenolic resin as a minimum. Cleaning the tank also becomes an important operating and, therefore, design issue. Design to provide maximum cleanliness and the ability to clean the tank becomes important for this case. For glycol tanks which must be heated, the coils have to be made of stainless steel to prevent product discoloration.

7.1.5   Liquid properties

Each stored liquid has unique problems associated with its storage. In addition to collection of physical properties and consideration of these properties in storage, research into the unique problems should be done. This often involves discussing these problems with others who manufacture the same compounds—suppliers or pilot plant personnel who worked with these liquids. An example of the design problems that arise from insufficient research into hard-to-handle liquids can be seen in molten sulfur. Many liquid sulfur tanks have suffered a total collapse due to internal vacuum. This resulted because the pressure vacuum (PV) valves became stuck. Although the design calls for steam-traced vent lines and PV valves, experience shows that this is inadequate. Fully steam-jacked PV valves and vent lines are usually required. Another important variable is vapor pressure. This physical property will almost solely govern whether a fixed-roof or a floating-roof tank is used for volatile organic liquids because these requirements are specified in the Clean Air Act. Flammability and flash point not only determine basic tank design but also drive spacing and layout requirements.

7.1.6   Sizing considerations

Establishing the optimal size of a storage tank may not be an easy task. Generally, the design capacity should be made approximately equal to the maximum desired inventory plus the unusable volume that remains in the bottom of the tank. Inventory represents working capital plus operating expense which both reduce net profitability. However, unexpected outages in tank volumes that limit operations can quickly wipe out the advantages of minimizing design storage capacity. Some studies can be performed to optimize the design capacity required:

- Compare raw material storage costs to risked cost of plant downtime.

- Compare raw material and inventory capital costs to reordering costs, considering volume discounts and bulk shipment savings.

- Consider the process and probabilities of accepting shipments to partially empty tanks with insufficient capacity to accept full loads and the consequent demurrage costs.

- Stored product capacity and inventory costs must be balanced against the possible lost costs, market share, and goodwill due to insufficient capacity or unexpected shutdowns.

Although it might appear that one large tank for all material of the same kind should be used, there are several reasons to consider several tanks for a given fluid:

- The tank size may limit the use of one tank. Although the largest steel tanks are approximately 1 million bbl, the limits for plastic tanks, aluminum tanks, and other kinds of tanks may be much smaller.

- With a sufficient number of tanks on a site, the probability that at least one is out-of-service for inspection or other reasons favors installation of extra "day" or "swing" tanks to cover the out-of-service inventory capacity.

- Check tanks may needed to ensure quality control.

- Dedicated tanks may be required.

- Risk assessment may indicate that the probability of loss warrants storage of a given material in several separated locations within the plant. Examples are fire loss, degradation, polymerization, contamination, and security.

Once the required design capacity is determined, the mechanical considerations for tank capacity become important. Some definitions are in order at this point:

The *nominal capacity* is the total volume of the shell to the very top. This is also called the *gross capacity*.

The *operating capacity* is the usable volume of the tank, which is limited at the top by the safe oil height or maximum operating level and by the low pump-out height.

The *safe oil height* or *maximum operating capacity* is the level above which the liquid is not allowed to rise. For cone-roof tanks, this is normally set at 6 to 12 in below the top rim or below the level above which an overflow would occur or the lowest roof component, such as a rafter or girder. A value of 6 in is recommended in the absence of other restraints. In internal floating-roof tanks, the maximum liquid level is often set by how high the roof may travel without impinging on the fixed roof, which would cause damage. In external floating-roof tanks, the maximum level can also be determined by how high the roof can travel without the seal's losing contact with the shell. In either internal or external floating roofs with secondary seals, the safe oil height should always be such that the top edge of the secondary seal does not go beyond the rim of the tank. Several tanks have been damaged because the roofs were run up higher than the safe oil height, the rim seal caught at the rim, and when the liquid level was reduced, the seal and/or roof was damaged. In earthquake areas, the effect of sloshing waves should be considered in order to provide freeboard for determining the maximum safe oil height. Generally, in the most severe earthquake regions, a freeboard of 2 ft has been found to be adequate to prevent major damage to the tank. The probability of the tank's being at the safe oil height during an earthquake should also be considered in determining how much, if any, freeboard should be used. In general, tanks which are kept full for long periods, such as seasonal tanks, should be provided with freeboard, whereas it may not be justified for tanks which cycle through high to low very frequency. Equations for sloshing wave height are given in API 620, Appendix L.

The *unavailable inventory* is stock which cannot be pumped out of the tank without special changes to the operations of the tank. Low pump-out is often set by the nozzle elevations. Attempting to pump below specified levels entrains air and/or causes the pump to cavitate. The low pump-out level should be at least 6 in above the top surfaces of fill line deflectors or suction line vortex breakers or above any point which would cause the pump to inspire air or to cavitate. For floating-roof tanks, the low landed leg position gives the generally unavailable inventory in the tank. The low roof level is in turn set by appurtenances such as mixers and floating suction lines or roof drain piping below which the roof may not go. Much effort has gone into attempting to reduce the unavailable inventory such as by designing special bottoms, using cone-down bottoms, or using special auxiliary piping, but these systems have generally not proved useful.

7.1.7 Venting

For flammable liquids, the use of API Standard 2000 is almost universal. It has provisions for handling the displacement due to liquid transfer into and out of a tank as well as the vaporization that occurs due to normal thermal changes, called *tank breathing*. It also has provisions for emergency venting requirements. For normal and emergency venting it breaks the categories of liquids into those with flash points over 100°F and those with flash points under 100°F. The liquids with flash points under 100°F have slightly higher venting capacities. The emergency venting requirements are based on heat flux into the tank as a result of a nearby ground fire. However, API 2000 does not cover internal pressure generated as a result of explosions, deflagrations, exothermic reactions, decomposition, and similar events. These must be considered by the design engineer and provisions made for safely relieving this pressure or preventing it. The design for vacuum must also be carefully considered because most flat-bottom tanks can withstand very little vacuum. For nonflammable liquids, a modified analysis similar to that presented in API Standard 2000 should be made.

A typical sample of venting devices follows:

Tank service	Minimum venting requirements
Gasoline, crude oil, etc., with flash point <100°F	Pressure vacuum vent valve
Fuel oils, furnace oils with flash point >100°F	Weather hood
Asphalt and unfinished lubes	Weather hood
Finished lube oil	Pressure vacuum vent valve
Transformer oil	Special moistureproof breathing units
Sulfur tanks	Steam jacketed pressure vent valves

7.1.8   Life span

The design life of a tank will influence the design in the areas of corrosion allowance and the use of linings and coatings and cathodic protection.

7.2   Welding and Fabrication

7.2.1   Welding

The vast majority of tanks are fabricated using the arc welding process. Although residual stresses are significant within and adjacent to the welds, they are considered to be acceptable when they are statically loaded. Analysis shows that local stresses in places such as the shell-to-bottom joint can exceed yield stresses. When this occurs, a small amount of yielding occurs that redistributes the stress fields, particularly in ductile materials. To prevent loss of ductility in the heat-affected zones, carbon steels with less than 0.35 percent carbon content are usually used. For heavy sections or where significant restraint is present, such as in the welds of a flush cleanout type of fitting, extra requirements such as postweld heat treating may be beneficial.

Welding processes used.   API 650 does not specify the welding processes to be used except as limited by the ASME. Some typical welding processes used on tanks are SMAW, SAW, GMAW, and FCAW. The gas-shielded processes such as GTAW, GMAW, and FCAW may be adversely impacted by air drafts or winds over 5 mi/h. Short-circuiting GMAW or short-circuiting FCAW should not be used because of the risk of nonfusion and cold lap defects in the fill passes. Table 7.1 indicates some of the characteristics of the various welding processes.

Weld joint design. The API standard provides the required joint design, and typical weld joints are shown in the tank standards for vertical cylindrical flat-bottom tanks. The required weld joint design results from empirical, theoretical, and practical considerations. For example, single fillet welds are allowed on the roofs and bottoms of atmospheric tanks because they are not subject to high membrane stresses. However, the shells are required to be double butt-welded with full penetration and fusion. API 620 makes an exception by allowing fillet-welded shells which are rarely used by tank manufacturers. When tanks have fillet-welded shells, then a substantial penalty in allowable stress (joint efficiency) is paid. When conditions warrant, higher-quality joints than specified by the codes may be used. An example would be a tank storing concentrated sulfuric acid. The National Association of Corrosion Engineers (NACE) Standard T-5A-18 requires the use of butt-welded tank bottoms for this service.

General welding requirements for tanks constructed to API 650. The minimum fillet weld size is for plates with $\frac{3}{16}$-in or less full-fillet welds and not less than one-third the thickness of the thinner plate and at least $\frac{3}{16}$ in. Single fillet welds are permissible only on the roof and bottom joints. Full fusion and penetration is required for the shell. API 620 tanks may use lap-welded shells with an appropriate joint efficiency. Horizontal shell welds shall be full fusion and full penetration. Top angles may be attached to the top course with a double fillet-welded lap joint.

Tank bottom welds. Because tank bottoms are usually accessible from only one side for both fabrication and inspection, they are more susceptible to cracks and the resulting leaks are more difficult to detect than those for the shell and the roof. Although API 650 allows single-pass fillet welds, many companies require that the bottom welds of these tanks use a minimum of two passes. This is a requirement for API 620 flat-bottom tanks. The dual pass on bottom fillet welds significantly reduces the chance of leaks due to porosity, cracks, or flaws. Table 7.2 shows the advantages and disadvantages of dual-pass welding. When higher quality is required, butt-welded bottoms can be specified, but the costs are significantly greater and it is practical only for limited diameters. The use of annular plates reduces many of

TABLE 7.1 Comparison of Various Welding Techniques

Welding process	Characteristics	Advantages	Disadvantages	Application on tanks
SMAW (shielded metal arc welding)	■ Called *stick welding* ■ Manual process ■ Flux-covered consumable electrode	■ Simple ■ Versatile ■ All positions ■ Quick and easy setup	■ Low rate of deposition	■ Used widely on almost all field-erected tanks when thickness <½ in ■ Used for repairs and maintenance
GTAW (gas tungsten arc welding)	■ Called *TIG* or *heliarc* ■ Uses inert gas flux, argon, or helium	■ High-quality weld ■ High-quality root pass even from one side	■ Low rate	■ Stainless steel or aluminum
GMAW (gas metal arc welding)	■ Called *MIG* ■ Three types: spray transfer, short-circuiting, and pulsed arc ■ Uses spooled electrode wire ■ Inert gas flux	■ Used semiautomatically or automatically ■ Little smoke and spatter	■ Penetration and fusion problems ■ Expensive equipment ■ Difficult setup ■ Quality ■ Subject to air currents	
FCAW (flux core arc welding)	■ Similar to GMAW but uses flux-containing electrode ■ Two types: self-shielded and gas-shielded	■ Higher penetrations and deposition rates than SMAW	■ Weld quality subject to air currents ■ Not low-hydrogen process ■ Can have poor notch toughness	

TABLE 7.1 Comparison of Various Welding Techniques (Continued)

Welding process	Characteristics	Advantages	Disadvantages	Application on tanks
SAW (submerged arc welding)		■ Low-hydrogen process ■ Deep penetration ■ Easy visual inspection	■ Handling and setup time ■ Notch toughness dependent on wire-flux combinations	■ Used for field-welded tanks and spheres
EGW (electrogas welding)	■ For vertical positions or shell joints	■ Useful in field ■ Simple joint preparation ■ Uses large, open gap	■ Degradation of properties in HAZ	■ Can be used for vertical tank joints where design metal temperature is relatively warm

When the FCAW process is used, the following requirements improve the weld quality:
  Use E6XT-1, E6XT-5, E7XT-1, or E7XT-5 electrodes.
  Only T-1 electrodes $\frac{5}{64}$ in or smaller should be used for all positions.
  Larger T-1 and T-5 electrodes up to $\frac{3}{32}$ in are acceptable for welding in the flat position.
  Short-circuiting transfer FCAW should not be used.
  Vertical FCAW is used only in the uphill progression.
Weld joint design: API 650 does not specify this, but it is usually best to align the shell courses in a vertical profile such that the inside surfaces are straight. This prevents shelves where the lower thicker course joints to the thinner upper course which can trap sediment and cause accelerated corrosion.
Welding electrodes: API 650 requires AWS A5.5 E60 or E70 classification electrodes to be used on materials with tensile strengths less than 80 ksi and E80XX-CX electrodes for tensile strengths greater than this.

TABLE 7.2 Typical Tank Weld Joints

Advantages	Disadvantages

Single-Pass Weld

Advantages	Disadvantages
High productivity Low cost	Requires an experienced welder. Usually two passes are required to cover undercuts and leaks.
Allowed by API	Steel must be cleaned either by sandblast or powerbrushing
Most used in area	Low productivity if repairs are needed.
	Difficult to get a good washout, usually leaves a sharp edge on plate.
	More QA/QC required (visual inspection).
	Possible cracking in both welds and heat affected zone, however type of material in tank may also cause the same problem.

Two-Pass Weld

Advantages	Disadvantages
Covers first pass leaks	Higher cost
Tempers weld and heat-affected zone. Less chance of cracking.	Pay contractor for sloppy work.
Best assurance of a good bottom.	
Less inspection required.	
If E6010 is used as a root pass, no cleaning of the plate is required (sandblasting, etc.).	

Weld Rod Alternatives

E6010: Good root weld. Will burn out impurities, can operate at high temperatures, and can be welded at a fast rate. Can be subject to cracking depending on service conditions.

E6012: Old tanks welded with this wire. Some cracking and poor fusion can be present in these welds. Most contractors no longer use it. As listed in the Lincoln catalog, this rod may be used for sheet metal. It has a fast fill, can handle poor fit up, and has low penetration qualities. Welding instruction is to move as fast as possible.

E6012A: Special made Hobart E6012A rod which overcomes cracking problems and works well.

E7024: Iron powder, high speed, maximum deposition rate, easy slag removal, and smooth finish.

---

the problems that occur in the vicinity of the shell-to-bottom joint, by reducing the weld flaws in this region and by reducing the subsequent deformation due to settling or other load conditions that act on welds in this region. Bottom annular plates must be butt-welded. They may be either double butt-welded or welded using a backup bar.

When the tank bottom is less than $\frac{1}{4}$ in thick, then it is advisable not to use E6010 weld rod for the following reasons:

- For fillet welds, the edge of the plate will be burned off because of the penetration capability of the rod and will produce spatter, uneven contour, and a weld less than the nominal plate thickness.
- It results in a convex weld cap that is subject to high stresses and potential cracking.
- There is a greater chance of burn-through.

Since E6010 is not a low-hydrogen rod (it is cellulosic), it can cause cracking on thicker sections. API prohibits the use of low-hydrogen rods for the shell-to-bottom joints where the shell is greater than $\frac{1}{2}$ in or is constructed of higher-strength materials.

If the bottom is greater than $\frac{1}{4}$ in thick and will be welded with two or more passes, then the 6010 root pass followed with a 7018 or 7028 cover pass is good practice.

Shell-to-bottom weld. This weld is the most critical weld for a vertical cylindrical storage tank. It undergoes the most severe local stress in the tank. The size of the weld is controlled by the thickness of the thinner plate (usually the bottom on larger tanks) and is limited to a maximum of $\frac{1}{2}$ in. When annular plates are greater than $\frac{1}{2}$ in, then the attachment weld is sized so that either the leg of the fillet weld or the groove depth plus the leg of the fillet weld is equal to the thickness of the annular plate.

Shell welds. Almost all tanks use full penetrations and fusion, double butt welds. For the SMAW or FCAW process, square butt joints are practical up to about $\frac{5}{16}$-in thickness. For EGW no beveling is necessary up to about 1 in. The square joints are easy to prepare because no beveling is required. For thicker sections from about $\frac{3}{4}$ in thick, backgouging the root after welding from the first side is necessary to obtain complete penetration.

From a fabrication standpoint, a low-hydrogen weld rod should not be used unless it is required for environmental corrosion reasons or required by the API standards. For example, on moderate-strength steels (API groups I through III) for practical reasons E6010 can and should be used because it can be welded in the downhill position, putting little heat into the shell but penetrating the square-edge joint on shell thicknesses below about $\frac{3}{8}$ in. If a low-hydrogen rod were used, it would have to be used in the uphill position and would cause not only increased joint preparation time but much more heat input and warpage of the shell plates.

Thickness variations. Most codes require that differing thicknesses of plate joined by butt welding be made by tapering the thicker plate to the thickness of the thinner plate at a definite rate. This requirement limits the effect of stress concentrations.

The two most common places where shell thickness varies are

- At the transition between shell courses which decrease in thickness from bottom to top
- At thickened insert plates

For girth or roundabout seams, the API standards do not require beveling or tapering the thicker section to the thinner one. This is because the stresses in the vertical direction are much lower than the hoop stresses and because the ratio of course thicknesses is not very large. However, for thickened insert plates, large stresses occur, and the thickened plate is often twice as thick as the shell material. For this reason the API standards require a 4-to-1 transition taper which keeps the stress concentration factor below 5 percent.

Preheat. Preheating is done to prevent cracking of welds, reduce residual stress levels, improve notch toughness, and improve the properties of the heat-affected zone (HAZ). Cracking mostly occurs in the HAZ soon after weld completions due to hydrogen, which is called delayed cracking, cold cracking, or underbead cracking. It can result from inadequate preheat and use of steels with high hardenability. Underbead cracking is not usually a problem with low and plain carbon steels. To occur, three factors must be present:

- Hydrogen
- High stresses
- A base steel of sufficient hardness

The hydrogen content is controlled by using low-hydrogen electrodes. In addition, preheat allows the hydrogen sufficient time to diffuse out of the HAZ quickly enough to avoid becoming a problem.

Preheating is necessary when the temperature of the base metal is below 50°F. This essentially requires the steel to be welded to be warm to the touch. This prevents delayed cracking problems.

Postweld heat treatment (PWHT). Postweld heat treatment is used in the API codes to accomplish one or more of the following:

- Reduce residual stresses
- Reduce weld and HAZ hardness
- Improve toughness
- Outgas hydrogen from welds
- Increase dimensional stability

The primary reason for postweld heat treating carbon steel tank components is to reduce the residual stresses resulting from the restraint of the weld and weld shrinkage. Since it is impractical to postweld heat-treat the entire tank, those parts with the most welds and weld restraint are subject to postweld heat treatment. Examples are nozzles and flush cleanouts. These fittings are postweld heat-treated as assemblies, and the entire assembly can then be welded into the shell with a weld which is not required to be postweld heat-treated.

The higher-strength materials have more stringent postweld heat treatment requirements. Moderate-strength materials specified in API 650 are group I through group III materials, and the detailed specifications for these are included in the standard. Higher-strength materials are group IV to group VI materials. PWHT is required for the following cases:

- All flush cleanouts. These have an unusual amount of welding per unit volume of steel, and the restraint of the welds during cooling produces high residual weld stresses. These are reduced by the PWHT requirements.

- When the moderate-strength materials are used (groups I to III) for nozzles or manways 12 in or larger in shells or insert plates greater than 1 in thick, the fitting must be prefabricated, postweld heat-treated at a temperature of 1100 to 1200°F per hour per inch of thickness, and the assembly welded into the shell.

- When the higher-strength materials are used (groups IV to VI), then any fitting requiring reinforcement and greater than $\frac{1}{2}$ in thick must be postweld heat-treated.

Weld hardness. Weld hardness is an overall control limit that is an indication of the maximum stresses in the weld and HAZ. The deposited weld metal for steel tanks constructed of moderate-strength steels should generally not exceed 95-ksi tensile strength, and this is related to weld hardness. Excessive hardness can cause cracking in some services. Weld hardness measurements are often used to determine the effectiveness of PWHT where required by the API standards. Hardness is typically taken with a portable unit such as a Telebrineller, because it is easy to use, simple to operate, and accurate. A typical hardness limit specified is 200 BHN maximum for carbon steel tanks. Hardness tests are also required for test plates during weld procedure qualification tests. Welds made manually with a low-strength stick rod such as E60XX or E70XX need not be tested.

A requirement to check welds is often specified. They are made by automatic weld processes because the weld chemistry and thus material properties can be adversely affected by improper flux-wire combinations.

Hardness testing of bottom welds is normally not required, but in sour service it may be beneficial.

Weld space requirements. The intent of the weld spacing requirements is to prevent the interaction of residual welding stresses from compounding and leading to potential fracture. Since these requirements were developed before the significant advances in materials sciences and finite elemental analysis, the weld spacing requirements in the 8th edition of API 650 may be conservative.

Note that neither API 620 nor the ASME pressure vessel codes have requirements for minimum weld-to-weld spacing.

API 650 has no weld spacing requirements for the roof because a leak will not result in loss of contents. Although the bottom plates of tanks are not considered pressure-containing membranes subject to membrane stresses, the nature of bottom construction makes the likelihood of flaws much greater. The requirements for spacing of these plate laps are shown in Fig. 7.1.

API 650 also has weld space requirements that are applicable to shell plates. It requires that the vertical seams in adjacent courses be staggered or offset by at least 5 times the thickness of the lower course. This would prevent propagation of cracks into adjacent courses. In addition, API 650 has requirements for spacing of welds that involve thickened insert plates or reinforcement pads.

Tank proportioning. Most atmospheric tanks range from 10 to 300 ft in diameter and 6 to 72 ft high. The aspect ratio or the height-to-diameter ratio is set by a number of factors that are optimized. Some of these are as follows:

- *Seismic considerations.* Since it is preferred to use unanchored tanks when possible, tank proportions in areas of high seismicity favor low height-to-diameter ratios. Table 7.3 shows the maximum typical height-to-diameter ratio of unanchored tanks by zone. This table should not be used for design since detailed calculations in accordance with some accepted method such as that in Appendix E of API 650 should be used.

- High land costs or limited space favors tanks with a greater height-to-diameter ratio.

- Process considerations often affect optimal height-to-diameter ratios. For example, for mixing variation in the height can cause different levels of mixing and power requirements.

- A low height-to-diameter ratio tends to increase the percentage of unusable inventory.

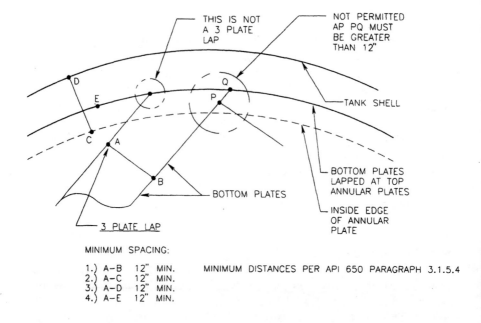

TANKS WITH ANNULAR PLATES

MINIMUM SPACING:

1.) A—B 12" MIN.
2.) A—C 12" MIN.
3.) A—D 12" MIN.
4.) A—E 12" MIN.

MINIMUM DISTANCES PER API 650 PARAGRAPH 3.1.5.4

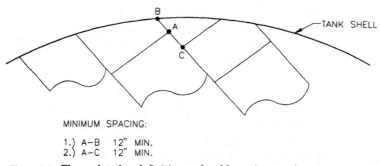

TANKS WITHOUT ANNULAR PLATES

MINIMUM SPACING:

1.) A—B 12" MIN.
2.) A—C 12" MIN.

Figure 7.1 Three-plate lap definition and weld spacing requirements.

- Low soil bearing pressures will limit the tank height.

- Use of gravity flow may determine the required height of the tank.

- Designs for tanks connected by platforms may use height as a fundamental consideration to efficiently connect them or to maintain OSHA-approved stairs or platforms.

- Costs and material utilization are relevant.

TABLE 7.3 Tank Height-to-Diameter Ratio, by Seismic Zone

Seismic zone	Maximum tank height-to-diameter ratio
1	2.4:1
2	1.25:1
3	0.67:1
4	0.5:1 (small tanks, soft soil)
4	0.6:1 (large tanks, firm soil)

7.3 Design of Roofs for Vertical Cylindrical Tanks

Roof design starts with the establishment of roof dead and live loads. The Uniform Building Code or the ANSI/ASCE 7-88 Standards provide a basis for establishing the methodology of combining dead and live loads as well as wind and seismic loads. The typical load combination is the most severe of

- Dead load
- Dead load plus live load or snow load
- Dead load plus wind load or earthquake load
- Dead load plus live load or snow load plus wind or earthquake load

Earthquake loadings for roofs tend not to be significant in the most severe of the roof load combinations. Wind and earthquake are also not assumed to act simultaneously. Concentrated loads imposed by equipment located on top of the tank should be individually considered.

The two most important live loads to consider are loads due to vacuum inside the tank and snow loads. Roofs designed according to API 650 include provisions for live plus dead load of 45 psf. Since it requires that the minimum live load be set at 25 psf, it can be inferred that the remaining load covers the deadweight of the roof structure. The thickest roof allowed by API 650 is ½ in, and this is about 20 psf. Presumably the 25-psf load must cover loads resulting from external pressure and/or snow loads. The minimum 25-psf live load has been found adequate to cover the majority of worst load combinations of vacuum loading, snow loading, rain load, or activities that occur on roofs, including personnel walking on the roof. However, for unusual conditions the roof live load should be increased. The single most likely cause of requiring increased live load is the effect of vacuum.

It may be safely assumed that a tank roof designed in accordance with API 620 may be subjected to a vacuum not exceeding 1 ounce per square inch (osi) or 9 psf. API 650 tanks may be considered for vacuum levels of up to 1 in water column without further analysis. When the vacuum level exceeds these levels, then consideration should be given to increasing the specified live load and performing specific computations for ensuring the integrity of the roof.

7.3.1   Type of roof

For cylindrical tanks there are basically two categories of roof design:

- Self-supported dome or umbrella or cone roofs
- Column-supported cone roofs

The computations for thickness of self-supported roofs can be established by treating the roofs as simple shells subjected to various load conditions.

Dome or umbrella roof

Required minimum thickness.   The equations used by API to set the thickness of self-supporting dome or umbrella roofs are based on the elastic stability of spheres under external pressure. Figure 7.2 shows the free-body diagram of a spherical or dome roof. The stress in a spherical shell is

$$\sigma = \frac{Pr}{2t} \tag{7.1}$$

Experimental work has shown that the critical buckling stress for spheres subjected to external pressure is

$$\sigma_{\mathrm{Cr}} = 0.125 \frac{Et}{r} \tag{7.2}$$

This equation is the basis taken by ASME section VIII, division I, for spherical pressure vessels subject to external pressure. It assumes no reduction for tolerance, theory versus tests, or other unknown variables.

By substitution of Eqs. (7.1) and (7.2) we have

$$P_{\mathrm{Cr}} = 0.25E\left(\frac{t}{r}\right)^2 \tag{7.3}$$

Using a factor of safety of $4t/r$ assuming a critical buckling pressure of 45 psf, and defining r in feet and t in inches, we get the following equation:

$$t = \frac{R}{200} \tag{7.4}$$

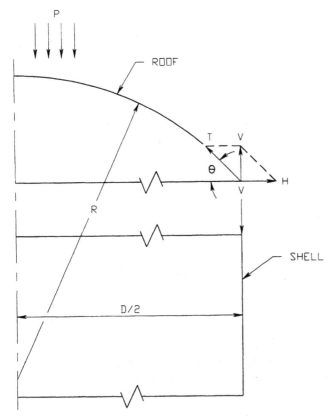

Figure 7.2 Dome-roof free-body diagram.

This equation gives the required thickness of a dome or umbrella roof.

Since for a given pressure and wall thickness a spherical shell has twice the strength of a cylindrical shell, API decided to make the radius of the dome twice as large as the radius of the shell, which will allow a constant thickness between the top course and the dome. The API standards allow a 20 percent variation in either direction, and

$$\text{Dome or umbrella radius} = R \pm 0.2D \qquad (7.5)$$

$$\text{or } 0.8D - 1.2D$$

Dome roof compression area requirements. Although the thickness of the roof was determined by external pressure and the elastic stability of a sphere, the internal pressure is limited by the compressive forces that develop at the periphery of the dome. Figure 7.2 is a free-body diagram that illustrates the unbalanced forces that are resolved by the compression region at the periphery of the dome.

Balancing the radial force H to the horizontal roof force component gives

$$H = T \cos \left(\theta \, \frac{PR}{2} \right) \cos \theta \qquad (7.6)$$

The area required to resist H is

$$A = \frac{H(D/2)}{\sigma} = \frac{PR \cos \theta}{2} \frac{D}{2\sigma} = \frac{DR}{4\sigma/(P \cos \theta)} \qquad (7.7)$$

API 650 uses a compression region allowable stress of 15 ksi.

Substituting the value of 45 psf, the maximum value of $\cos \theta$ for $R = 0.8D$ is 0.909. Expressing R and D in feet and A in square inches, the required area is

$$A \cong D \frac{R}{1500} \qquad (7.8)$$

This is the API 650 equation for the dome roofs.

Cone-roof tanks

Required minimum thickness

Thickness of self-supported fixed-cone roofs. Like the dome roof, the thickness of self-supported cone roofs is based on the elastic stability of cones under external pressure. The classical buckling stress for curved plates is

$$\sigma_{\mathrm{Cr}} = 0.6E \frac{t}{r} \qquad (7.9)$$

Experimental investigations have shown that the actual safe compression stress is about one-twelfth of the theoretical value. Using $E = 29 \times 10^6$ psi gives

$$f_{\mathrm{allow}} = 1.5 \times 10^6 \frac{t}{r} \leq \frac{1}{3} \sigma_y \qquad (7.10)$$

To adapt this criterion to cone roofs, refer to Fig. 7.3. By substitution of

$$r = \frac{6D}{\sin \theta}$$

with r in inches and D in feet

$$f_{\mathrm{allow}} = 1.5 \times 10^6 \frac{t \sin \theta}{6D} = 250{,}000 \frac{t \sin \theta}{D} \qquad (7.11)$$

The stress in shallow cones subject to internal pressure is

$$f_s = \frac{PD}{2t \sin \theta}$$

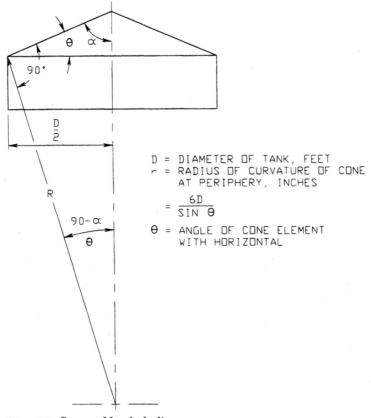

D = DIAMETER OF TANK, FEET
r = RADIUS OF CURVATURE OF CONE
    AT PERIPHERY, INCHES

$$= \frac{6D}{SIN\ \Theta}$$

Θ = ANGLE OF CONE ELEMENT
    WITH HORIZONTAL

Figure 7.3 Cone-roof free-body diagram.

By substitution

$$\sin \theta = \frac{D}{1000t} \sqrt{P/6} \tag{7.12}$$

Choosing a live load of 25 psf and a dead load of 7.65 lb/ft^2 (which is equivalent to typical $\frac{3}{16}$ in roof plate), we get

$$t = \frac{D}{430 \sin \theta} \tag{7.13}$$

API 650 uses

$$t = \frac{D}{400 \sin \theta} \tag{7.14}$$

Required compression area. There are several distinct load conditions due to internal or external pressures acting on the fixed-cone roofs of tanks. When the load is external, two factors limit the pressure. The first is the elastic stability of the roof. The second is the so-called participating compression region that, in the case of external loading, is in tension. This region is an effective or equivalent area that acts as a structural ring at the location of the roof-to-shell joint. This ring goes into tension for external loading and compression for internal loading.

When the external pressure is at or below the weight of the roof plates, the roof does not lift up and no compression is generated in the compression region. When the internal pressure exceeds the weights of the roof plates but is less than the weight of the roof plus shell, compressive forces act in the participating compression area, but there is no tendency of the tank to lift up. A special design case is included by API 650 when the internal pressure is equal to the weight of the roof plus shell. If the participating compression area is set to a small enough area to yield, then a frangible roof can be used.

The frangible roof concept is to cause a rupture at the roof-to-shell junction due to buckling of the compression region at an internal pressure that occurs at incipient uplift in an empty tank. When the internal pressure exceeds the weight of the roof plus shell, then uplift will occur. API 650 requires that tanks for this condition be anchored and limited to those with shells less than $\frac{1}{2}$ in thick and that the roof-to-shell junction be designed in accordance with API 620. To derive the basic equations for the compression region of fixed-cone roofs, see Fig. 7.4. The force in the compression region must be balanced against the radial thrust developed by the pressure acting on the roof, resulting in

$$A = \frac{HD}{2\sigma} \tag{7.15}$$

where A = participating compression area
$\quad H$ = radial thrust
$\quad D$ = tank diameter
$\quad \sigma$ = stress in compression region

Also a force balance requires that

$$HD(H \sin \theta) = \left(\frac{\pi D^2}{4}\right) P \quad \text{or} \quad H = \frac{PD}{4 \sin \theta} \tag{7.16}$$

Combining Eqs. (7.15) and (7.16) gives

$$A = \frac{HD}{2\sigma} = \frac{PD^2}{8\sigma \sin \theta} \tag{7.17}$$

API uses an allowable stress of 4000 psi. If the pressure is set to $P = 0.315$ psi (45 psf), then Eq. (7.17) becomes

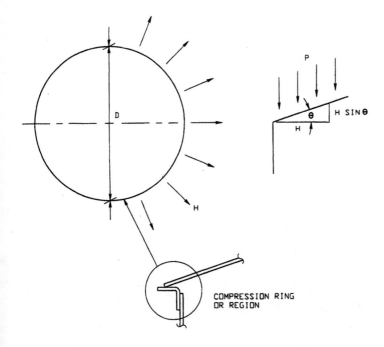

H = UNIT RADIAL FORCE RESULTING FROM INTERNAL PRESSURE

A = AREA OF PARTICIPATING COMPRESSION

Figure 7.4   Roof compression area requirements.

$$A = \frac{0.315D^2(144 \text{ in}^2/\text{ft}^2)}{8.15000 \sin \theta} = \frac{D^2}{2645 \sin \theta} \approx \frac{D^2}{3000 \sin \theta} \quad (7.18)$$

The required compression area must be at least equal to A, using the specific fabrication details of API 650.

When the compression area is to be determined for a frangible roof, the compression area is set to yield at the point of incipient uplift:

$$W = P\pi r^2 \quad (7.19)$$

where r = tank radius
P = internal pressure
W = weight of tank roof plus shell

Substituting the yield strength, which API assumes to be 32,000 psi for the stress σ, into Eqs. (7.17) and (7.19) gives

$$A = \frac{0.159W}{32,000 \tan \theta} = \frac{0.153W}{30,800 \tan \theta} \tag{7.20}$$

This is the area below which the participating compression region must be to qualify for a frangible roof joint.

To derive the allowable design pressure where a tank has been built, a safety factor η is chosen. Recognizing that steel has a specific gravity of 8 times that of water and subtracting it from the internal pressure to find the net upward force, we have

$$(P - 8t)\frac{\pi D^2}{4} = \frac{W}{\eta} \tag{7.21}$$

where t = thickness of roof
P = internal pressure, in WC
η = safety factor

Using consistent units, we get

$$\left(0.03606 \ \frac{\text{lb/in}^2}{\text{in WC}}\right)\left(\frac{144 \ \text{in}^2}{\text{ft}^2}\right)\left(\frac{\pi D^2}{\text{ft}^2}\right)(P - 8t) = \frac{W}{\eta} \tag{7.22}$$

Rearranging gives

$$\frac{W}{\eta} = 4.0738D^2(P - 8t) \tag{7.23}$$

Using Eq. (7.17) yields

$$A = \frac{W}{2\pi\sigma \tan \theta} = \frac{\eta(4.0783)D^2(P - 8t)}{2\pi\sigma \tan \theta} \tag{7.24}$$

Setting a safety factor against yield of $\eta = 1.6$ (an allowable stress of 20,000 psi) gives

$$A = 1.0385 \ \frac{D^2(P - 8t)}{\sigma \tan \theta} \tag{7.25}$$

Rearranging yields

$$P = \frac{A\sigma_y \tan \theta}{1.0385D^2} + 8t = \frac{30,800A \tan \theta}{D^2} + 8t \tag{7.26}$$

This is API's equation for allowable internal pressure.

To prevent uplift, another check is required. Using Eq. (7.23) and setting $h = 1.0$, we have

$$P = \frac{W}{4.0738D^2} + 8t = \frac{0.245W}{D^2} + 8t \tag{7.27}$$

API uses the equation as a check to limit the internal pressure to a value which does not exceed uplift.

If the safety factor is removed by setting $\eta = 1.0$ in Eq. (7.21), the failing pressure of a tank can be determined.

For design, Eq. (7.21) is written including the safety factor $\eta = 1.6$:

$$1.6PD - 12.8t = \frac{4W}{\pi D^2} \tag{7.28}$$

For failure, $\eta = 1.0$ and

$$P_f - 8t = \frac{4W}{\pi D^2} \tag{7.29}$$

This is the equation API uses to determine the failure pressure.

Figure 7.5 is a flowchart illustrating the use of Appendix F of API 650.

7.3.2 Design of tank shells

The discussion below focuses on the design of vertical cylindrical storage tanks. For large tanks of double curvature, the methods of API 620 or finite element modeling should be used to determine acceptable shell thicknesses. Because hydrostatic stresses for small tanks are usually not high enough to impose stresses approximating the material allowable stresses, a completely different set of criteria affects the shell designs for small tanks. For large tanks, hoop stress becomes important and limiting in the selection of shell thickness. For large low-pressure tanks, biaxial stress conditions may govern the required shell thickness.

For most small tanks the shell thickness is not governed by the requirements to maintain the hoop tensile stresses below allowable values but by fabricability and minimum thickness requirements. For example, the required design thickness of an API 650 tank storing water in an A36 tank with a design liquid level of 20 ft and a diameter of 20 ft is less than 0.05 in. However, this is too thin to work with and does not have sufficient capacity to handle local loadings from ladders and handrails or from wind or seismic loads.

Minimum wall thickness. For most tanks, experience has shown minimum thicknesses are acceptable (see Table 7.4). These are the minimum thicknesses required by both API 620 and API 650.

In general, the small tank will have a wall of a single thickness whereas large tanks will step down the thickness with higher shell courses. Because hydrostatics does not play a role in the thickness requirements of small tanks, several standards use this advantageously to provide cookbook sizes and designs for tanks that require little or no calculations. An example is API 12F, *Specification for Shop Welded Tanks for Storage of Production Liquids.*

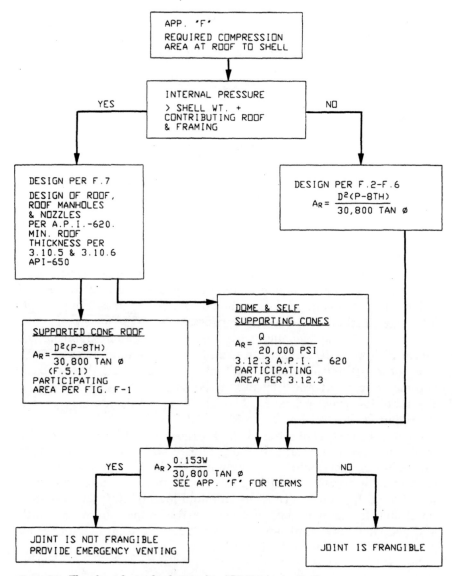

Figure 7.5 Flowchart for tanks designed to API 650 Appendix F.

TABLE 7.4 Minimum Shell Wall Thicknesses

Nominal tank diameter, ft	Min. shell plate thickness, in
<50	$\frac{3}{16}$
>50, <120	$\frac{1}{4}$
>120, <200	$\frac{5}{16}$
>200	$\frac{3}{8}$

Shell design of large vertical cylindrical tanks. Large tanks have significant stresses of several kinds:

- *Hoop stress* is the stress developed in the circumferential direction in the shell as a result of hydrostatic pressure. It is the most important as it usually determines the wall thickness of the tank. It is sometimes called *membrane stress.*

- Longitudinal or axial stresses arise from the weight of the shell, live and dead load transferred from the roof to the shell, or internal pressure. These stresses generally have no effect on design shell thicknesses.

- Residual stresses result from welding, heat-treating, or localized heating.

- Thermal stresses result from thermal changes in the stored product or from ambient diurnal conditions.

- Local stresses result from imposed loads from mixers or equipment which may be static or dynamic.

- Dynamic stresses result from earthquake and wind. The API Standards specifically address these kinds of stresses for design.

- Local discontinuity stresses result from restraint of components. These stresses are usually referred to as *edge stresses* for cylindrical shells in stress analysis or as *local stresses.* For example, the bottom course expands radially as hydrostatic pressure increases but is restrained by both the joint to the bottom and the shell course above it. Another example of this kind of stress is found at the roof-to-shell junction for tanks with internal pressure.

- Still another stress is the membrane stress found in tank bottoms that have settled.

Because the vast majority of shell thicknesses are determined by the hoop stresses resulting from the product hydrostatic stress, this is covered in detail below.

Classical hoop stress. Hoop stress usually sets the required thickness for vertical cylindrical storage tank design. Although axial stresses, local stresses, and discontinuity stresses are important, they usually do not set the wall thickness of storage tanks. However, they have been considered and incorporated through design and experience in the existing codes and standards that are available to the tank consumer and fabricator. These load conditions will be considered as appropriate in greater detail later.

For an allowable hoop stress of f psi, the required thickness of a cylindrical shell under internal pressure at depth H, from the classical hoop stress formula, is

$$t = \frac{2.6DHG}{f} \tag{7.30}$$

However, in practice, it is usually modified by two factors: joint efficiency and corrosion allowance.

Joint efficiency is simply an arbitrary factor that is used to account for the fact that the welded or riveted joint may not be as strong as the parent material and the actual practices and qualities as well as inspection found in the field. For lap-welded tanks, significant local stresses are induced as a result of local bending, and the fillet weld has a built-in tendency to crack from the root which must be accounted for in the design joint efficiency. The codes and standards have set criteria for setting joint efficiency which have been based on experience. The joint efficiency is always less than or equal to unity. In the design of most modern tanks using API Standards, it is 1.0. It is still used in API 650, Appendix A tanks and for API 620 tanks which allow for lap-welded shells.

The modified formula that includes the joint efficiency e is then

$$t = \frac{2.6DHG}{fe} \tag{7.31}$$

Another critical factor in the shell wall thickness requirements is the specification of corrosion allowance. At the end of the design life given uniform corrosion, the wall thickness in the equation above is for the corroded condition. Therefore, an extra thickness of steel is added as corrosion allowance c, and the resulting formula is

$$t = \frac{2.6DHG}{fe} + c \tag{7.32}$$

The equation results in a required thickness that is proportional to the depth. Since plate is not available in tapered thicknesses, the bottom course is thickest with upper courses becoming thinner.

Since a shell course might be 8 to 10 ft high, a question arises as to what hydraulic depth to use for the shell course thickness. If the pressure at the bottom of the shell course is selected, then the thicknesses at all other elevations in the shell course are excessive. To use an average depth between the top and bottom of the shell course would lead to hoop stress in excess of the allowable stress below the center of the shell course and below the allowable stress above it.

The 1-ft method. It turns out that an empirical average depth is satisfactory. This can be seen by referring to Fig. 7.6. An unrestrained ring can grow radially outward, and the stresses will approximate those of the classical hoop stress formula derived earlier. However, each shell course is restrained at its bottom edge and cannot grow. Since hoop stress is proportional to radial strain, the net result is that the stresses

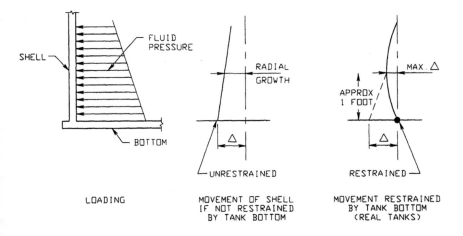

$\triangle$ = RADIAL SHELL DEFLECTION DUE TO INTERNAL PRESSURE

Figure 7.6 Strain in shell at bottom.

are actually lower than the classical hoop stress equation would predict for the bottom of the shell course. A simplified empirical method used by API 650 bases the hydraulic pressure on a distance 1 ft above the bottom of the shell course. This gives rise to the equation used by API 650

$$t = \frac{2.6DH(H-1)G}{Fe} \tag{7.33}$$

This method is called the *1-ft method*. While this method has been successfully used for many years, later finite element analysis and advanced stress analysis methods have shown that it is slightly conservative. To include this better understanding of shell stresses, API 650 included the option to use the *variable point method* to reduce the required shell thickness, using the same allowable stresses.

Location of variable design point. The location of the design point on each shell course for the hydrostatic pressure, to be used in calculating the plate thickness, can be developed through consideration of the elastic movement, radial and rotational, of the plate edges at each joint.

Bottom shell course. Computer calculations have shown that the design point 1 ft above the bottom results in a thickness and an actual stress reasonably close to the stress used to calculate the thickness. The actual stress does exceed the assumed design stress, however, for the larger tanks. A formula to determine the bottom course thickness more accurately has been developed based upon a shell of uniform thickness. All bottom courses, however, are influenced by a thinner second course when the height of the bottom shell course is less than

$$2.625\sqrt{rt_1} \qquad (7.34)$$

where r = normal radius of tank shell, in inches, and t_1 = thickness of bottom course, in inches. The bottom course thickness, therefore, can be obtained from the following expression, using the symbols of API Standard 650:

$$t_1 = \left(1.06 - 0.463\,\frac{D}{H}\,\sqrt{\frac{HG}{s}}\right)\frac{2.6HDG}{s} \qquad (7.35)$$

where D = nominal tank diameter, ft
$\quad H$ = height from bottom of shell to top angle or to bottom of overflow, ft
$\quad G$ = design specific gravity of liquid
$\quad s$ = allowable design stress for calculating plate thicknesses, psi

The above formula includes adjustment because of the effect of the second course and becomes conservative when the height of the bottom course is greater than $2.625rt_1$. In such cases, the bottom course thickness need not exceed the thickness calculated using the 1-ft design point.

Second shell course. The theoretical location of the design point to determine the thickness of the second course is complicated because the restraint of the tank bottom raises the location of the maximum stress in the lower shell of larger tanks to the vicinity of the girth joint between the first two shell courses.

The theoretical thickness of the second course is dependent upon the height of the bottom course and the radius-thickness values of the bottom course, in addition to the hydrostatic head on the second course. The three governing conditions for calculating the second course thickness are as follows:

$$\text{When} \quad \frac{h_1}{\sqrt{rt_1}} = 1.375 \qquad t_2 = t_1 \qquad (7.36)$$

$$\text{When} \quad \frac{h_1}{\sqrt{rt_1}} = 2.625 \qquad t_2 = t_{2a} \qquad (7.37)$$

$$\text{When} \quad 1.375 < \frac{h_1}{\sqrt{rt_1}} < 2.625$$

$$t_2 = t_{2a} + (t_1 - t_{2a})\left(2.1 - \frac{h_1}{1.25\sqrt{rt_1}}\right) \qquad (7.38)$$

where h_1 = height of bottom shell course, in

t_2 = final thickness of second shell course, in

t_2 = thickness of second course calculated as described for the upper shell courses section below, in

Upper shell courses.   The theoretical thickness of an upper shell course is a function of the two thicknesses at the girth joint at the lower edge of the shell course. The elastic expansion and rotation at the girth joint must result in common values since the two plate edges are connected to provide continuity at the joint. Figure 7.7 illustrates the elastic movement of two upper shell courses at the girth joint between them.

For a design where the thickness of each course is determined by a common stress, the theoretical location of the design point is at a variable distance above the bottom of the course. This distance is the lowest value obtained from the following three expressions:

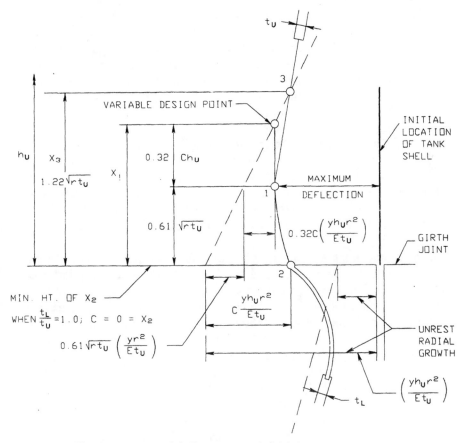

Figure 7.7   Elastic movement of shell courses at girth joint.

$$X_1 = 0.61\sqrt{rt_U} + 0.32Ch_U \tag{7.39}$$

$$X_2 = Ch_U \tag{7.40}$$

$$X_3 = 1.22\sqrt{rt_U} \tag{7.41}$$

where t_U = thickness of upper course at joint, in
$\quad t_L$ = thickness of lower course at joint, in

$$C = \frac{\sqrt{K(K-1)}}{1 + K\sqrt{K}}$$

K = thickness lower course at joint/thickness upper course at joint = t_L/t_U
h_U = height from bottom of course under consideration to top angle or to bottom of overflow, in

Figure 7.7 illustrates the location of the X_1, X_2, and X_3 distances from the girth seam.

The preceding calculations for the design point require an estimated thickness for the upper course, the thickness of the lower course having been previously calculated. Since the thicknesses will be reasonably proportional to the total pressure existing at the bottom of each shell course, or, more closely, to the total pressure existing at least 1 ft up from the bottom of the shell course, the thickness obtained by the usual 1-ft design method can be used as the first approximation. A further refinement in the design can be made by another calculation using the thicknesses resulting from the previous design.

Another reasonably accurate approximation of the design point location in the upper courses, which can be used for the preliminary thickness calculation, is the distance $1.0/(rt_L)$ above the girth joint. Obviously, the latter approximation is only valid when the thicknesses at the joint are determined by a common stress.

Design of flat tank bottoms. Flat bottoms for vertical cylindrical tanks include cone-up, cone-down, and single-slope as well as truly flat bottoms. The original design basis for tank bottoms was to assume that they were membranes for liquid containment with no real structural requirement. Indeed, Fig. 7.8 shows the construction details for tank bottoms without annular plates. Although this design has proved adequate in many cases, it does not recognize the very high local stresses that occur in the shell-to-bottom region. This design is limited to tanks using group I through group III materials. When the higher-strength materials of group IV, IVA, V, or VI are used, butt-welded annular plates are required.

Bottoms of tanks are subject to a number of load conditions:

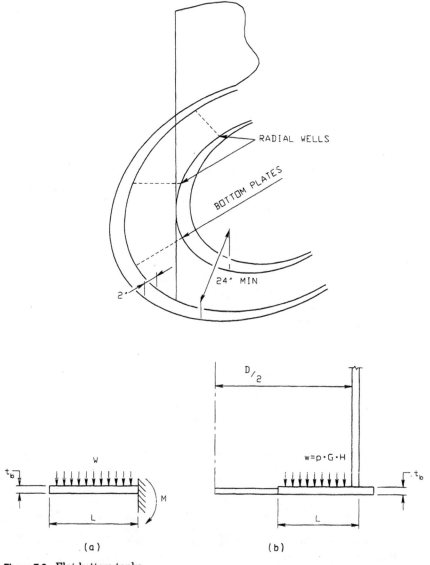

Figure 7.8 Flat-bottom tanks.

- Primary membrane stresses resulting from global or local dish settling
- High local stresses resulting from edge settlement
- High bending stresses from the restraint of the shell at the bottom corner joint
- Dynamic stresses such as occur during uplift in seismic activity

- Stresses resulting from washout or uneven settlement of areas under the shell

Some companies require the use of annular plates on tanks over 100,000 bbl regardless of the material groups or stresses used. Annular plates reduce edge settlement and reduce the possibility of failure during seismic events.

Design basis for annular plates. Figure 7.8 shows how an annular plate is constructed. Annular plates must be at least 24 in between the inside bottom plates and the shell. They must project outside the shell by at least 2 in. They range from ¼ to ¾ in thick. The bottom plate in an API Standard must be a butt-welded annular plate whose thickness varies between 0.25 and 0.75 in and is a function of the stress and thickness of the first shell course. The width of the annular plate must be adequate to support the column of water on top of it in case of a foundation settlement. The plastic moment of a beam is

$$M_p = \sigma_y \frac{(\text{width})(\text{depth})^2}{4}$$

Treating the annular plate as a beam of unit width gives

$$\sigma_y = \frac{4M}{t^2} \tag{7.42}$$

or

$$L = \frac{\sqrt{\sigma_y t^2}}{2\rho GH} \tag{7.43}$$

For $\rho = 62.4 \text{ ft}^3$, $\sigma = 33,000$ psi, H in feet, and t_6 in inches, the equation becomes

$$L = \frac{195t_h}{\sqrt{GH}} \tag{7.44}$$

API 650 uses a factor of safety of 2 for the length. The length of the annular plate is thus expressed as

$$L = \frac{390t_h}{\sqrt{GH}} \quad \text{(but not less than 24 in)} \tag{7.45}$$

API 620 tanks. The shells of most tanks storing liquids are shells of revolution about a single vertical axis. Therefore, any horizontal plane cut through the shell is a circle. The simplest kind of vessel or tank is cylindrical. However, other shapes such as spheres, elliptical vessels, or nodded spheres are used. These more complex shapes have two radii of curvature, one in the meridional direction and one in the latitudinal, or hoop, direction.

Membrane theory. To analyze this, a branch of stress analysis called *membrane theory* is used. The derivation of the equations can be found in Ref. 4. Membrane theory assumes all forces acting are in place, that is no shear can be transmitted through the tank's walls. If both radii are large compared to the shell thickness, this assumption works well for large steel plate structures.

From the basic equation of statics

$$\frac{\sigma_1}{r_1} + \frac{\sigma_2}{r_2} = \frac{p}{t} \tag{7.46}$$

where σ_1 = tensile stress in medridional direction, or meridional stress
σ_2 = tensile stress along parallel circle, or hoop stress
t = uniform thickness of shell
r_1 = meridional radius of curvature
r_2 = radius of curvature of a section taken perpendicular to meridian
p = pressure

The signs of both radii are positive if they point toward the center of the vessel.

Special cases

Sphere. The simplest case of a doubly curved tank is a sphere under internal pressure. Since both radii are equal ($r_1 = r_2 = r$ and $\sigma_1 = \sigma_2 = \sigma$), Eq. (7.46) becomes

$$\sigma = \frac{pr}{2t} \tag{7.47}$$

Cylinder. Vertical cylindrical tanks have the meridional radius $r_1 = \infty$ and the hoop radius $r_2 = r$, and Eq. (7.46) becomes

$$\frac{\sigma_1}{\infty} + \frac{\sigma_2}{r} = \frac{p}{t}$$

or

$$\sigma_2 = \frac{pr}{t} \tag{7.48}$$

Conical sections. As in the case of the cylinder, $r_1 = \infty$, but $r_2 = r/\cos \infty$, where ∞ is the half angle of the cone. Equation (7.46) results in

$$\sigma_2 = \frac{pr}{t \cos d} \qquad (7.49)$$

API 620 allowable stresses. The allowable stress basis for API 620 is somewhat complex because it includes provisions for several biaxial stress combinations:

1. Axial compressive stress with no concurrent hoop stress
2. Compressive stress in meridional and hoop direction of equal magnitude
3. Compressive stress in meridional and hoop direction of unequal magnitude
4. Compressive stress in one direction and tensile stress at 90° to the other direction

Because the radius-to-thickness ratios are large, the effect of critical buckling stress must be included. In fact, API 620 covers the elastic region, the elastoplastic, and fully plastic stress regions and sets reduction factors to the allowable stresses for simple tension and compression for all three regions:

1. *Axial compressive stress with no concurrent hoop stress.* API Bulletin 2U specifies that elastic buckling can be expected when $\sigma = 0.125Et/r = 3.625 \times 10^6 \, t/r$. API 620 uses a factor of safety of 2, and for $E = 29 \times 10^6$ the allowable buckling stress is $\sigma = 1.8 \times 10^6 t/r$. This is shown on segment OA in Fig. 7.9. Segment BC is the limiting

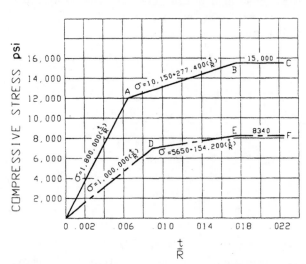

Figure 7.9 Biaxial allowable stresses for tanks of double curvature. (*Courtesy of the American Petroleum Institute.*)

value of stress for structural components in compression and is limited to 15,000 psi. Segment *AB* represents a transition between the elastic and plastic regions.

2. *Compressive stress in meridional and hoop direction of equal magnitude.* Collapse tests have shown that spheres buckle at about one-half the value of cylinders for the same t/r and external pressure. API 620 allows an elastic buckling stress of $1,000,000t/r$ for $t/r < 0.0067$. This is plotted as segment *OD* in Fig. 7.9. Dividing the limiting plastic allowable stress of 15,000 psi by 1.8 gives the value of 8333, which is plotted as segment *EF*. In a similar manner, the elasto-plastic transition is plotted as line *DE*.

3. *Compressive stress is meridional and hoop direction of unequal magnitude.* When the magnitudes of the coincident compressive stresses are unequal, the stress state is viewed as a superposition of stresses experienced as a sphere plus the additional stress (the stress difference) that would act as a cylinder. Figure 7.10 shows this concept. Using these assumptions, we find that

$$\frac{\text{Smaller stress}}{\substack{\text{Max. allowable stress for} \\ \text{sphere based on R} \\ \text{associated with} \\ \text{greater stress}}} + \frac{\text{Difference in stresses}}{\substack{\text{Max. allowable stress for} \\ \text{cylinder based on R} \\ \text{associated with} \\ \text{smaller stress}}} \leq 1.0 \quad (7.50)$$

and

$$\frac{\text{Smaller stress}}{\text{Max. allowable stress for sphere}} \leq 1.0 \quad (7.51)$$

Remembering the allowable stress ratio for spheres versus that of cylinders for the same t/r, we can rewrite the above equation as

$$\frac{(\text{Larger stress}) + 0.8(\text{Smaller stress})}{\substack{\text{Stress determined from OABC in} \\ \text{Fig. 7.10 using R for larger force}}} \leq 1.0 \quad (7.52)$$

$$\frac{1.8(\text{Smaller stress})}{\substack{\text{Stress determined from OABC in} \\ \text{Fig. 7.10 using R for smaller force}}} \leq 1.0 \quad (7.53)$$

4. *Compressive stress in one direction and tensile stress at 90° to these compressive direction.* The basis for setting allowable stresses for the cases is these Von Mises–Hinckley theory or the distortion energy theory of failure of ductile materials. The failure stress in this theory for the biaxial stress state can be written

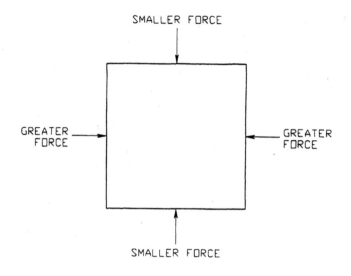

EQUIVALENT TO

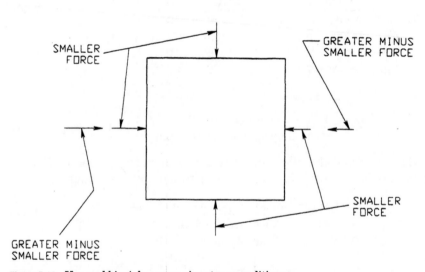

Figure 7.10 Unequal biaxial compressive stress conditions.

$$\left(\frac{\sigma_1}{\sigma_y}\right)^2 - \frac{\sigma_1}{\sigma_y}\frac{\sigma_2}{\sigma_y} + \left(\frac{\sigma_2}{\sigma_y}\right)^2 = 1 \tag{7.54}$$

where σ_1 and σ_2 are principal stresses and σ_y is the Von Mises failure stress. By redefining the stresses as follows:

σ_t = tensile stress, actual or limiting

σ_c = compressive stress, actual or limiting

σ_{ta} = allowable stress for simple tension

σ_{tc} = maximum allowable stress for longitudinal compression, psi, for a cylinder subjected to axial compression with no concurrent stresses in hoop direction

we can rewrite this as

$$\left(\frac{\sigma_t}{\sigma_{ta}}\right)^2 + \frac{\sigma_t}{\sigma_{ta}}\frac{\sigma_c}{\sigma_{tc}} + \left(\frac{\sigma_c}{\sigma_{ca}}\right)^2 = 1 \qquad (7.55)$$

This concept is graphically illustrated in Fig. 7.11.

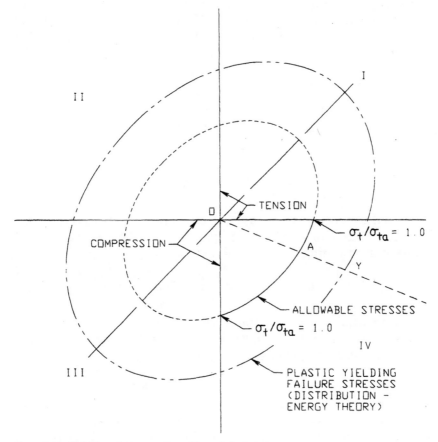

Figure 7.11 Relations between allowable and design stresses.

By choosing

$$N = \frac{\sigma_t}{\sigma_{ta}} = \frac{\text{actual tensile stress}}{\text{allowable tensile stress}}$$

$$M = \frac{\sigma_c}{\sigma_{ca}} = \frac{\text{actual compressive stress}}{\text{allowable compressive stress}}$$

we get

$$N^2 + MN + M^2 = 1 \qquad (7.56)$$

This is plotted in Fig. 7.12.

Reinforcement of shell openings. The discussion in this section is based upon API 650, but similar principles apply to the methods used in other codes.

Stress concentrations cause high stresses at openings. Circular holes in plates subject to tension cause high stresses, called *stress concentrations*, that develop around the opening. The stress at any point near a hole in a plate subject to tensile stresses is given by the following equations:

$$\sigma_r = \frac{\sigma}{2}\left(\frac{1+a}{r^2}\right) + \frac{\sigma}{2}\left(1 + \frac{3\sigma^4}{4} - \frac{4a^2}{r^2}\right)\cos 2\theta \qquad (7.57)$$

$$\sigma_t = \frac{\sigma}{2}\left(\frac{1+a}{r^2}\right) - \frac{\sigma}{2}\left(\frac{1+3\sigma^4}{r^4}\right)\cos 2\theta \qquad (7.58)$$

$$\sigma_{rt} = \frac{\sigma}{2}\left(1 + \frac{3a^4}{r^4} + 2\frac{a^2}{r^2}\right)\sin 2\theta \qquad (7.59)$$

The nomenclature and directions for stresses are illustrated in Fig. 7.13.

At the circumference of the hole, the maximum stresses develop, as found by substituting

$$r = a \quad \text{and} \quad \partial = \frac{\pi}{2} \quad \text{or} \quad 3\frac{\pi}{2}$$

The tangential stress is 3 times the applied stress. A plot of the stresses is shown in Fig. 7.14, where the stress concentration factor K is the ratio of the actual stress to the applied stress. It can be considered a normalizing factor. By solving Eqs. (7.1), (7.2), and (7.3) for stresses along the diametrical section *nn,* it is found that a compressive stress, tangent to the hole, equal to the applied tensile stress, develops. It damps out very rapidly, as shown in Fig. 7.14. By using the principle of superposition, stresses for applied biaxial stresses can be determined. The following are the most common cases for tanks.

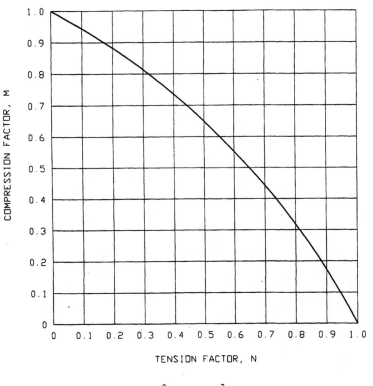

$$N^2 + MN + M^2 = 1$$

$$(\sigma_t/\sigma_{ta})^2 + (\sigma_t/\sigma_{ta})(\sigma_c/\sigma_{ca}) + (\sigma_c/\sigma_{ca})^2 = 1$$

WHERE:

N = σ_t/σ_{ta}

σ_t = TENSILE STRESS, IN POUNDS PER SQUARE INCH, AT THE POINT UNDER CONSIDERATION.

σ_{ta} = MAXIMUM ALLOWABLE STRESS FOR SIMPLE TENSION, IN POUNDS PER SQUARE INCH, AS GIVEN IN TABLE 3-1.

M = σ_c/σ_{ca}

σ_c = COMPRESSIVE STRESS, IN POUNDS PER SQUARE INCH, AT THE POINT UNDER CONSIDERATION.

σ_{ca} = MAXIMUM ALLOWABLE LONGITUDINAL COMPRESSIVE STRESS, IN POUNDS PER SQUARE INCH, FOR A CYLINDRICAL WALL ACTED UPON BY AN AXIAL LOAD WITH NEITHER A TENSILE NOR A COMPRESSIVE FORCE ACTING CONCURRENTLY IN A CIRCUMFERENTIAL DIRECTION.

Figure 7.12 Reduction of biaxial design stresses of opposite sign.

For a cylindrical tank with an opening stressed with internal pressure as shown in Fig. 7.15, the longitudinal stress is one-half the hoop stress. Using the principle of superposition, the maximum stress occurs at plane mm with a value of

$$\sigma_{max} = 3\sigma_y - \sigma_x = 3\sigma_y - \tfrac{1}{2}\sigma_y = 2.5\sigma_y \qquad (7.60)$$

where the subscripts are defined in Fig. 7.15.

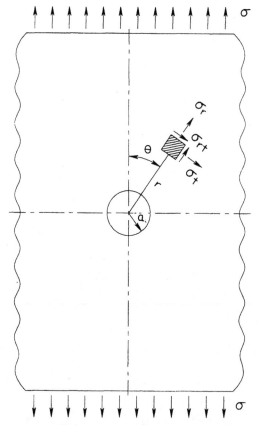

Figure 7.13 Hole in a plate subjected to tension.

For spherical tanks, $\sigma_x = \sigma_y = \sigma$. So

$$3\sigma_x - \sigma_y = 3\sigma - \sigma = 2\sigma \qquad (7.61)$$

For spheres subject to internal pressure, the stress concentration is two times the applied stresses for circular holes in the shell.

Table 7.5 summarizes the stress concentration factors for the basic types of tank shapes.

Basic reinforcement concepts. As shown in the previous section, high local stress concentrations occur as a result of openings cut into the shell of tanks. The idea of reinforcing the opening is to replace the amount of material removed adjacent to the opening through which the stresses may flow. The addition of material, however, must be placed within definite limits of the opening, as it is ineffective beyond

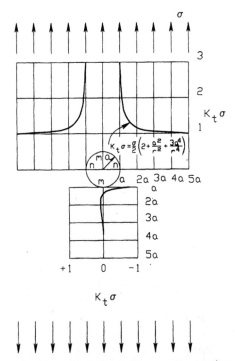

Figure 7.14 Variation in stress concentration in a plate containing a circular hole and subjected to uniform tension.

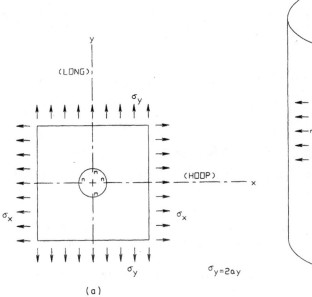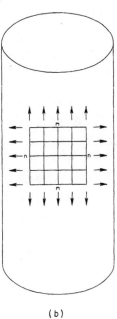

Figure 7.15 Hole in plate subjected to biaxial stress.

TABLE 7.5   Stress Intensity Factors for Various Ratios of Applied Stress

Stress ratio	Stress concentration factor K
1:0 (axial only)	3.00
2:1 (cylinder)	2.50
1:1 (sphere)	2.00

these limits. Since the reinforcement area is equal to the removal material, the API Standards assume that the same load-carrying capability is reestablished and that the primary stresses have been returned to a value nearly the same as the unperforated shell.

The limits of effective reinforcement are shown in Fig. 7.16.

Amount of reinforcement required.   Small openings less than 2-in nominal pipe size do not need to be reinforced because the amount of material removed has been shown by experience to be not significant. However, for openings larger than 2 in, reinforcement must be provided. There are two methods of determining reinforcement:

- The area of nominal shell plate thickness times the diameter of the opening when viewed through the vertical plane, as shown by method 1 of Fig. 7.16.

- The area of the required shell plate thickness times the diameter of the opening when calculations are made to determine tr, as shown by method 2 of Fig. 7.16.

The material that may be included as reinforcing area within the limits of reinforcing, shown in Fig. 7.16, is

- The reinforcing plate
- The extra material in the shell beyond the required thickness
- The nozzle neck thickness within $4t_n$ of the outside surface of the shell, as shown in Fig. 7.16.

When the standard nozzle and manway details of API are used, the amount of reinforcing actually available is usually conservative. In addition, the weld strength requirements that apply to the welding of the reinforcing pad to the shell and the nozzle to the shell and reinforcing pad are automatically covered and are also conservative.

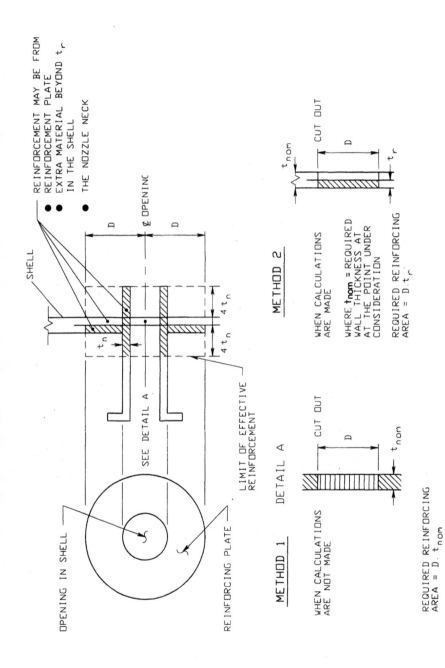

Figure 7.16 Limits of reinforcement.

OPENING IN SHELL

SHELL

REINFORCEMENT MAY BE FROM
- REINFORCEMENT PLATE
- EXTRA MATERIAL BEYOND t_r IN THE SHELL
- THE NOZZLE NECK

D

¢ OPENING

D

t_n

$4 t_n$

$4 t_n$

SEE DETAIL A

LIMIT OF EFFECTIVE
REINFORCEMENT

REINFORCING PLATE

METHOD 1   DETAIL A

WHEN CALCULATIONS
ARE NOT MADE

CUT OUT

D

t_{nom}

REQUIRED REINFORCING
AREA = $D \cdot t_{nom}$

METHOD 2

WHEN CALCULATIONS
ARE MADE

WHERE t_{nom} = REQUIRED
WALL THICKNESS AT
AT THE POINT UNDER
CONSIDERATION

REQUIRED REINFORCING
AREA = $D \cdot t_r$

t_{nom}

CUT OUT

D

t_r

REFERENCES

ANSI/ASCE 7-88, "Minimum Design Loads for Buildings and Other Structures," November 27, 1990.

NACE Standard T-5A-18, National Association of Corrosion Engineers, Houston.

Uniform Building Code, vol. 1, 1994 edition, Administrative, Fire- and Life-Safety, and Field Inspection Provisions.

Uniform Building Code, vol. 2, 1994 edition, Structural Engineering Design Provisions.

1. L. S. Beedle et al., *Structural Steel Design,* Ronald Press, New York, 1964.
2. J. E. Shigley and C. R. Mischke, *Mechanical Engineering Design,* 5th ed., McGraw-Hill, New York, 1989.
3. H. H. Bednar, *Pressure Vessel Design Handbook,* 2d ed., Krieger Publishing Company, Malabar, Fla., 1991.
4. J. F. Harvey, *Theory and Design of Pressure Vessels,* 2d ed., Van Nostrand Reinhold, New York, 1991.
5. J. J. Dvorak and R. V. McGrath, *Biaxial Stress Criteria for Large Low-Pressure Tanks.*
6. P. Newton, W. R. Von Tress, and J. S. Bridges, *Liquid Storage in the CPI.*

8

Tank Roofs

SECTION 8.1 FLOATING ROOFS

8.1.1 Introduction

The floating roof was developed in the 1920s to reduce product evaporation loss that occurred in the vapor space of fuels that were stored in fixed-roof tanks. It has not only proved effective for reducing emissions from the storage of volatile organic compounds when compared to fixed-roof tanks, but also helped to reduce the potential for vapor space explosions that regularly occur in fixed-roof tanks. The floating roof also virtually eliminates the possibility of a boilover phenomenon that occurs in fixed-roof tank fires where crude oils are stored. Because of these advantages the floating roof is now used extensively throughout the industry to store petroleum and petrochemical substances in large quantities.

Floating roofs can be used on both external (open-top) tanks and internal (fixed-roof) tanks. Because the external floating roof is subject to rainfall, snow, and wind loads, it is constructed of steel and is required to meet more stringent loading criteria. Appendix C of API 650 gives the minimum requirements that apply to external floating roofs. Since the internal floating roof is shielded from rain, snow, and wind, it can be of a lighter construction; Appendix H of API 650 gives the requirements for this type of roof. External floating roofs are usually specified for new tanks unless specific considerations require the use of an internal floating-roof tank. The following are a few uses of internal floating roofs:

- For storage of stocks which are sensitive to water contamination such as jet fuels and oxygenates (MTBE).

- In areas which are subject to very high snow or rain loadings. The fixed roof prevents excessive loads from developing on the roofs, which could cause them to capsize.
- When an existing fixed roof needs to be converted to a floating-roof tank, an aluminum floating roof can be inserted through manways as it is not usually economical to remove the fixed roof.

8.1.2 External Floating Roofs

8.1.2.1 Roof types

There are several types of floating roofs. Some basic configurations for external floating roofs are illustrated in Fig. 8.1.1.

Pan. The pan roof derives its buoyancy from the rim at the perimeter of the deck. Since it has a pan configuration, it has no inherent buoyancy. A single pinhole anywhere below the liquid surface can cause this type of roof to sink or capsize. Although these roofs are allowed by the codes, they are not recommended due to safety problems they create, and they should be considered incident-prone. There have been many cases of capsized roofs and subsequent fires as well as repeated problems after refloating these types of roofs. Most companies which understand the risks do not use the pan roof designs any longer.

Pontoon. The pontoon roof is the workhorse of the industry. It is the most commonly found type of floating-roof tank. It has inherent buoyancy which is derived from the outer pontoon compartments. API 650, Appendix C gives specific design requirements. The optimum economic diameter is from about 30 to about 200 ft. Beyond this diameter, wind can cause ripples to form in the deck section of the roof, causing fatigue cracking of the welds. In addition, it is harder to maintain good drainage when the diameter exceeds 200 ft because the roof has excessive flexibility. Because of the flexibility, vapor bubbles can form in the center of large pontoon roofs, which leads to vapor space corrosion problems and an excessive rise in the center deck.

Double-deck. This roof is the heaviest but most durable construction of all the roofs. It is most economical for small diameters up to about 30 ft and very large diameters from about 200 to over 300 ft. This roof maintains good rigidity under normal conditions, and therefore the drainage patterns can be adequately controlled in the very large diameters. Figure 8.1.2 shows the circumferential bulkheads. Radial bulkheads are also used. Since the double-deck roof has an air blanket between the upper and lower skins, it is much better insulated from solar heat gain, which tends to produce vapor under the roof in a pontoon-style

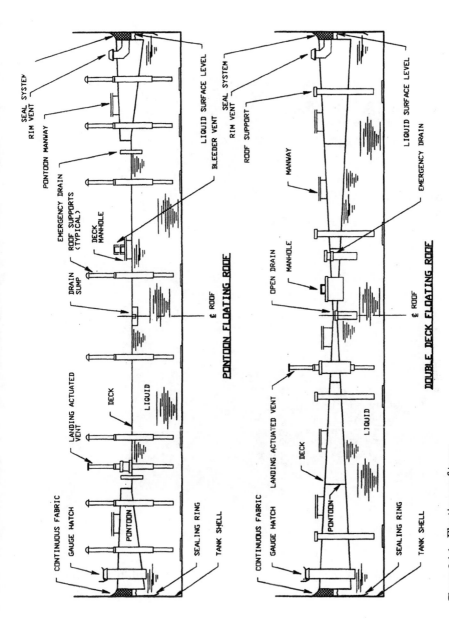

Figure 8.1.1 Floating-roof types.

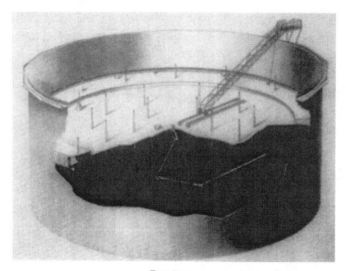

Pontoon

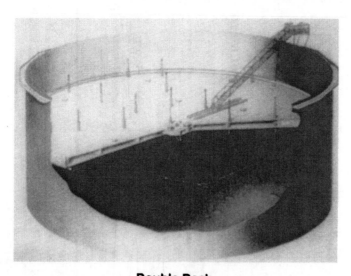

Double Deck

Figure 8.1.2   Pictorial view of external floating roofs.

roof. Because the roof is much heavier and less flexible than the pontoon-type roof, it is more suitable for a roof which needs to be insulated.

8.1.2.2 External roof design considerations

The design requirements for external floating roofs are given in API 650, Appendix C. The API Standards specify only the minimum requirements that directly affect safety or operability. The manufacturers have considerable latitude in implementing proprietary designs that address specific problems. Many of the specific details of construction must be a matter of agreement between the purchaser and the manufacturer.

Material. External floating roofs are constructed of steel with deck plates a minimum of $\frac{3}{16}$ in thick. All deck plates should be seal-welded on the top side and stitch-welded below within 12 in of any girders, support legs, or other rigid members. When crevice corrosion is anticipated, then seal welding of the bottom side of the roof can be one design countermeasure.

Load conditions

Landed condition and roof legs. When the roof is landed, API specifies that it shall be designed for a uniform live load of 25 psf. When roof drains freeze or plug, loads higher than this can damage roofs or collapse them when they are in the landed position. Figure 8.1.3 shows the operation of roof legs.

Floating condition. API specifies that the roofs shall maintain buoyancy under these conditions: 10 in of rainfall in 24 h with the primary drains inoperative or two adjacent pontoons punctured with no water accumulation. The pontoons must be designed so that permanent deformation does not occur when the roof is subjected to the design conditions listed above. Any penetration of the floating roof such as a gauge hatch should not allow product liquid to spill onto the roof under the conditions specified above.

Other design requirements

Bulkheaded compartments. Although it is not clear to many fabricators that API requires this, the top edge of all bulkhead compartments should be seal-welded using a single fillet weld, providing a liquidtight and vaportight compartment. There have been numerous cases where the compartments of a tipped roof allowed liquid to spill over the top of the bulkheads, leading to flooding of adjacent compartments and subsequent capsizing. Also, a leak in one compartment can allow flammable vapors to get into adjacent compartments. This improves the inspection and identification of leaking compartments when gas testing is used. The roof legs support the roof in its landed position. They should be designed to support at least twice the dead load plus the live load, because not all legs contact the bottom uniformly.

Support legs. The legs usually have two positions to allow the roof to rest at a high position or a low position. When maintenance work or cleaning will be done, the roof is landed on the high position which

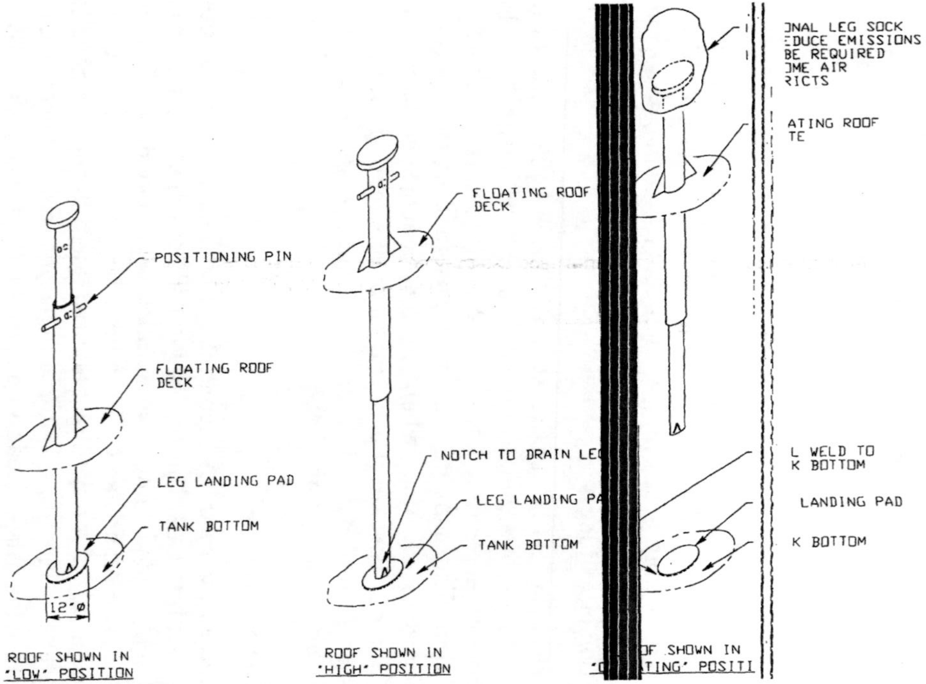
POSITIONING PIN
FLOATING ROOF DECK
LEG LANDING PAD
TANK BOTTOM
12"ø
ROOF SHOWN IN 'LOW' POSITION
FLOATING ROOF DECK
NOTCH TO DRAIN LE
LEG LANDING PA
TANK BOTTOM
ROOF SHOWN IN 'HIGH' POSITION
DNAL LEG SOCK
EDUCE EMISSIONS
BE REQUIRED
DME AIR
RICTS
ATING ROOF
TE
L WELD TO
K BOTTOM
LANDING PAD
K BOTTOM
OF SHOWN IN
TING' POSITI

Figure 8.1.3 Floating-roof support legs.

allows personnel to work in the space beneath the roof without bending over. When the working capacity of the tank is maximized, the roof is landed in the low position. The roof leg positions must be adjustable from above the roof while it is floating. Removable legs should be made from 3-in schedule 40 pipe because the heavier 4-in legs are too difficult to handle. When the legs are not removable, then 4-in schedule 80 pipe can be used. The legs must have notches or drain perforations to allow liquid that becomes trapped in the legs after it has landed to drain out.

Roof legs and their reinforcing pads are usually welded only on the top side of the decking with stitch welding below, as required by the API. Consideration should be given to welding the bottomside of each deck as well as to preventing potential cracking of the topside welds. This is more important when the roof leg loads are high, as is typical with large-diameter roofs or in services which are prone to developing weld cracks such as sour water service.

Landing pads are usually installed beneath the leg landing position to distribute the load. Usually $\frac{1}{4}$- or $\frac{3}{8}$-in-thick pads approximately 6 to 12 in across are used. When the leg loading is greater than about 10,000 lb, then special design considerations need to take into account the stability of the soil and bending stresses acting on the bottom plates and the welds. The landing pads should be seal-welded to the floor to prevent crevice corrosion from attacking the underside of the landing pads and to eliminate the need for inspecting the area beneath the pad for pits or corrosion.

Vents. The roof should have pressure and vacuum vents to allow for vapor expansion and contraction beneath the roof when it is landed. There are two basic types:

8.1.2.3 Landing-actuated vent

These are mechanically actuated so that they open when the roof lands. However, these are becoming barned by many air districts because the tank is freely vented to the atmosphere when its roof is landed.

8.1.2.4 Deck PV valve

The trend is to use PV valves to provide this venting capability. In addition, the vent valves relieve vapor pockets that build up from solar gain or from noncondensable gases or vapors that enter the tank via inflowing streams.

8.1.2.5 Drainage

The primary roof drain is required to be 3 in for tank diameters to 120 ft and 4 in for larger tanks. A check valve must be installed at the deck drain. In the event that a rupture of the roof drain piping or hose occurs, the check valve will reduce the likelihood of the deck's filling with product.

Ponding of rainwater for large pontoon roofs is more of a problem than it is for double-deck roofs. This is because the flexibility of the pontoon roofs in large sizes becomes significant, and water loading that can deflect the deck to increase ponding. The rolling-roof ladder also poses odds causing a deflected low point where water tends to accumulate.

8.1.3 Internal Floating Roofs

The various types of internal floating roofs are shown in Fig. 8.1.4. Note that all the external floating roofs are constructed of steel, but the internal floating roofs may be constructed of steel or aluminum or plastic.

8.1.3.1 Steel roofs

All the external floating-roof designs may be used for internal floating roofs including the pan roof, the pontoon roof, and the double-deck roof. In fact, many external roof tanks that were converted to internal floating-roof tanks were converted by simply adding a roof over the external floating-roof tank. Therefore, most of the steel internal roofs are the result of a conversion of the tank from external to internal roof.

As with the external floating roof, it is necessary to warn the user that the use of a pan roof even for an internal floating roof can lead to serious safety problems. A single pinhole in the pan can cause submergence or capsizing of the roof. Since floating roofs are infrequently inspected, even a slow leak can cause a failure of the roof over a long time. These roofs are not recommended as they have no inherent buoyancy.

There is an additional type of steel roof that is built exclusively for the internal application, called a *bulkhead roof*. It is shown in Fig. 8.1.4*d*. By bulkheading a pan roof, inherent buoyancy is obtained.

8.1.3.2 Aluminum roofs

General. Aluminum is subject to corrosion, and special consideration should be given to the service conditions. For example, some aluminum roofs have been used on slop water service. However, during upsets, caustic entered the slop tank, completely destroying the internal floating roof. Aluminum has been successfully used on a number of petroleum and organic stocks.

Aluminum roofs are much more delicate than steel roofs and are therefore used only for internal floating-roof applications. Because

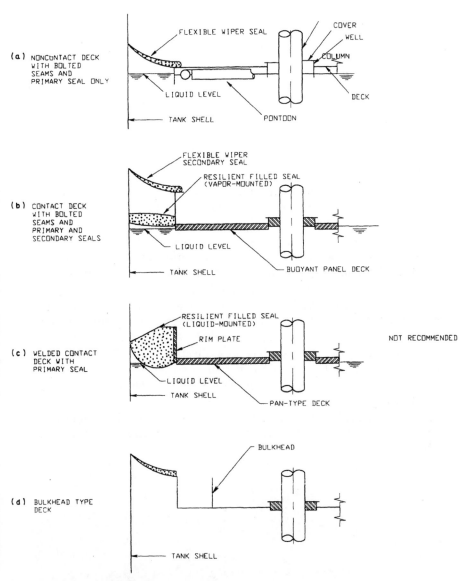

Figure 8.1.4 Internal floating-roof types.

the API Standard has very few structural requirements, manufacturers have produced a number of designs which are both adequate and inadequate depending on the context. Most designs have problems with mounting mechanical shoe seals or double seals. Some manufacturers have even produced mechanical shoes with smaller shoes to work better with internal roofs. In some cases, the friction generated by the use of dual seals has caused warpage of the roof deck and damage to deck seams as the roof changes direction. Manufacturers are constantly improving the designs, however, and the designs are entirely adequate in many cases.

Another consideration in the selection of aluminum internal roofs is the presence of a confined vapor space. A roof that has full contact has no vapor to breathe or expand. It cannot therefore burn or explode. Provided the vapor pressure is sufficient, the confined vapor space in a skin-and-pontoon roof will be too rich to burn. Not enough testing has been done to confirm the fire safety of one roof versus another at this time.

Skin-and-pontoon.   Figure 8.1.5 shows the construction of this type of roof. The buoyancy of this roof is derived from the pontoons which are simply closed-end aluminum pipes framed together so that coils of aluminum sheet can be rolled out into the deck. While there are many of these roofs in service, they do have some particular problems resulting from inadequate design details which cause cracks in the pontoons or the framing. Leaking pontoons are extremely dangerous to an unsuspecting welder or anyone else doing hot work during repairs. Pontoon corrosion is another problem. Manufacturers have developed various ways of making seams between the rows of aluminum sheeting that are more or less effective. Eventually, manufacturers will be able to obtain a certified emission loss factor for their particular roof designs which will help the user select the roof based upon emission losses.

Note that the skin-and-pontoon style of roof also has a vapor space of approximately 6 in between the stock liquid surface and the aluminum skin.

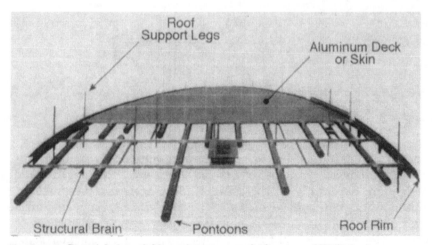

Figure 8.1.5   Pictorial view of skin-and-pontoon roof. (*Courtesy of HMT, Inc.*)

Figure 8.1.6 Aluminum honeycomb panel.

Honeycomb. Figure 8.1.6 shows a cross section of the main panels of the honeycomb floating roof. As long as the skin does not delaminate, this roof has inherent bouncy. In fact, numerous punctures cause by corrosion will not cause this roof to lose its buoyancy. This roof also maintains full contact with the liquid surface. Since the roof is only about 1 in thick, it does not have much excess buoyancy. When the seals on this roof do not move smoothly or have too much friction, the surface can be flooded with product. For this reason for very small tanks consideration should be given to this factor.

Appendix H of API 650 covers the basic design requirements for internal floating roofs. The primary major differences in design considerations between the internal and external floating roofs are as follows:

- Drainage is not required for the internal floating roof because the fixed roof prevents rainfall. However, should firewater be introduced into an internal floating roof via the inspection manways at the roof or from fixed or semifixed fire-fighting streams, then there is a significant danger of capsizing the internal roof without a drain. However, internal floating-roof fires are extremely rare.

- The buoyancy requirements are less stringent or do not apply depending on the roof type.

- In the landed condition, the roof is required to support a uniform live load of 12.5 psf. If the roof is designed so it cannot accumulate liquid above it, such as a honeycomb-style roof, then there is no live load requirement.

SECTION 8.2   RIM SEALS

8.2.1   Introduction

Both external floating-roof tanks (EFRTs) and internal floating-roof tanks (IFRTs) use rim seals to seal the annular space between the floating roof and the tank shell, as shown in Fig. 8.2.1. The use of these seals reduces emissions and evaporative losses, maintains product purity, reduces the amount of water that enters the product, keeps the interior shell walls clean, and keeps the roof centered in the tank. Seals are subjected to severe conditions and requirements. They must be able to

- Accommodate the irregularities of the tank shell such as weld irregularities, small buckles or imperfections in the shell and out-of-round diameter variations while allowing free movement of the roof as the liquid level changes
- Move up and down as the liquid level changes without exerting damaging forces on the roof, the shell, or the seal itself, even with buildup of waxes and corrosion products
- Withstand operating upsets such as gas entrained into the tank and exiting through the seals without damage

8.2.2   External Floating-Roof Seals

EFRTs use either one or two seals. These are called *primary* and *secondary* seals. The primary seal is the lower seal. If there is only one seal, then it is also called the primary seal. The secondary seal is always mounted above or on top of the primary seal. The function of the secondary seal is to provide additional reduction in emissions.

Primary seals can be classified as to type:

- Mechanical shoe seal
- Resilient toroid seal
- Flexible wiper

See Table 8.2.1 for a comparison of the fundamental characteristics of these seal types.

Additionally, the primary seal may be a *liquid-mounted* or a *vapor-mounted* seal. A liquid-mounted seal extends into the liquid surface. A vapor-mounted seal is above the liquid surface. It has a vapor space underneath it through which wind-driven air currents may circulate. The distinction is important in the regulation of acceptable seal types. Table 8.2.2 shows the three basic seal types as well as their function and mounting combinations.

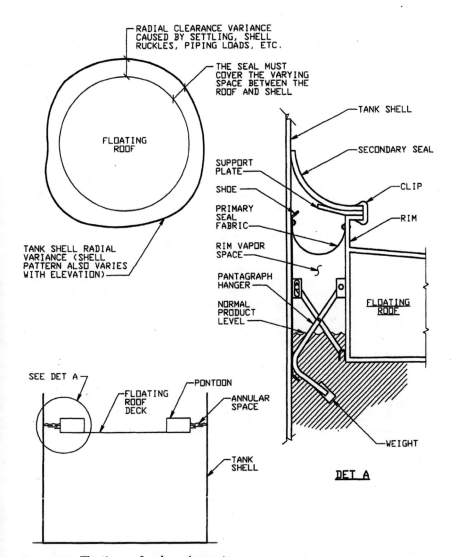

Figure 8.2.1 Floating-roof seal requirements.

Although the mechanical shoe seal is always given a separate classification in the federal regulations, it is and can be considered a liquid-mounted seal for this purpose. All secondary seals are rim-mounted except for shoe-mounted secondary seals. In this instance, a secondary seal may be mounted on a mechanical shoe seal. Today, the shoe-mounted seal is accepted by few, if any, regulatory authorities. Secondary seals are also always vapor-mounted seals.

Seal combinations are controlled by the nature of the hardware itself. Table 8.2.3 shows possible seal combinations.

TABLE 8.2.1   Seal Type versus Characteristics and Use

Seal type	Characteristics	Used for:
Mechanical Shoe	Rugged, long-lasting, most costly, details of attachment important for long life, accepted by all agencies, typical for EFRTs and EFRTs converted to IFRTs, tends to wear away internal shell coatings faster than other seal types, can attach wax scrapers	Primary only Liquid-mounted only
Resilient toroid	Good seal, low-cost, subject to tearing on rough surfaces, deflates when torn, poor in liquid-mounted service, better for vapor-mounted service, material compatibility critical for long life, typical for EFRTs and IFRTs, main failure mode is wear of outer protective cover, cannot attach scrapers	Primary or secondary Liquid-mounted or vapor-mounted
Flexible wiper	Low-cost, good seal,conforms to shell irregularities, best designs use compressor plates and vapor barrier fabrics, material selection and trim important, wiper tip subject to wear, subject to approval by air agency, typical for	Primary (not acceptable for EFRTs) Secondary only Vapor-mounted only

TABLE 8.2.2   Seal Function and Mounting

Seal type	Function	Mounting
Mechanical shoe	Primary only	Liquid only
Resilient toroid	Primary or secondary	Vapor or liquid
Wiper	Primary or secondary	Vapor only

TABLE 8.2.3   Seal Combinations

No.	Primary seal	Secondary seal	Weather shield
1	Mechanical shoe	Wiper	Not used
2	Resilient toroid Liquid-Mounted	Wiper	Not used
3	Resilient toroid Vapor-mounted	Not used	Can be used
4	Wiper	Not used	Not used

8.2.2.1 Regulations drive seal requirements

Currently, the requirements for the type and number of seals are largely regulated by EPA and the local regulatory enforcement agencies. Although there are many different and complex rules regarding the use, type, inspection, and maintenance of seals, some generalizations can be made, using the federal New Source Performance Standard (NSPS) as a minimum requirement for tanks.

By far the vast majority of tanks will meet regulatory requirements if they use seal combination 1 or 2 above. The reason is that the regulations require a liquid-mounted primary seal as well as a secondary seal. In addition to type of seal and type of seal mounting required, the regulations control the permissible seal gap requirements. The *seal gap* is the space that can be measured between the seal and the shell. Typically, rodlike probes of known diameter are inserted between the shell and the seal to determine compliance.

For very low-vapor-pressure stocks or those that are nontoxic and nonregulated, a single seal or even a weather seal alone is often used. However, in many of these cases a fixed-roof tank with no floating roof is used so that the roof seal is not needed. The purpose of seals when used in cases where they are not regulated is primarily to control water ingress and dust and dirt contamination as well as to help center the floating roof.

8.2.2.2 Mechanical shoe seals

Figures 8.2.2 and 8.2.3 show typical mechanical shoe seals. Mechanical shoes can be used on welded or riveted tanks of any diameter exceeding approximately 30 ft. The mechanical shoe seal is the industry workhorse. Although it may cost more than other types of seals, it is rugged and capable of lasting indefinitely with little or no maintenance. The concept is simple. It is a sheet metal band pressed to the inside tank wall by springlike mechanisms. It is sealed by a coated-sheet fabric called the *primary-seal fabric* that extends between the shoe and the rim of the floating roof, effectively blocking any path for vapors to escape. However, vapor can escape by the small gaps between the tank shell and the shoe. Also, as the roof moves down, the clingage of stored liquid to the shell results in emissions.

The most characteristic component of the mechanical shoe seal system is the shoe itself. It is typically constructed of 16 or 18 gauge galvanized or stainless steel sheet. It is made in sections approximately 12 ft long and has a vertical width of at least 3 or 4 ft. The vertical height of the shoe is governed by regulations that require it to extend into the liquid on the bottom and rise at least 24 in above the stored liquid surface. Experience has shown that this width of shoe provides

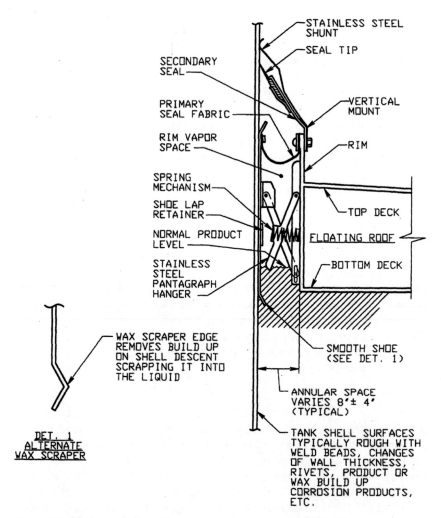

Figure 8.2.2 Mechanical shoe primary seal and rim-mounted secondary seal.

for adequate sealing of the annular space as well as mechanical stability. To get a continuous band, some manufacturers have bolted the shoe sections together whereas others allow the shoe sections to simply overlap one another by a few inches. The bolted configuration is stiffer and does not accommodate shell variations such as buckles and out-of-round as well as the lapped-section designs. Experience shows that lapped shoe plates that are free to slide past one another provide better sealing and accommodate the irregularities of the shell better than bolted shoes do. At both the top and bottom edges of the shoe, the sheet metal is bent inward slightly to prevent the shoe from catching on imperfections in the tank wall and being torn out. The

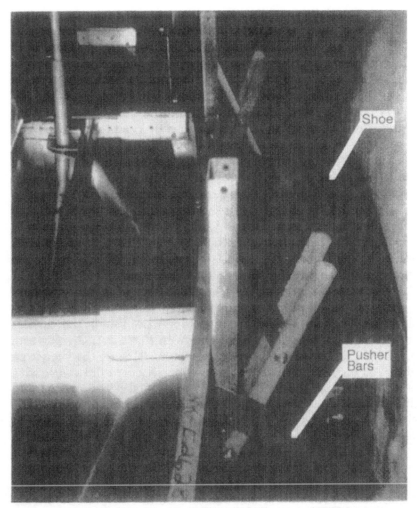

Figure 8.2.3   Pictorial view of mechanical shoe seal. (*Courtesy of HMT, Inc.*)

stainless steel shoe lasts longer from both corrosion and wear considerations, but a premium of approximately 10 to 20 percent of the
installed seal cost will result.

An often overlooked component of mechanical shoe seal selection is
consideration of the elevation of the top of the shoe above the liquid
surface. As stated earlier, the regulations for mechanical shoe seals
require that the top of the shoe be a least 24 in above the liquid surface. Because the shoe is mounted on a lever system or trapezoid
mounting system, it may drop in elevation slightly relative to the roof
rim elevation as it expands outward to accommodate wider annular

spaces between the rim and the shell. In some cases operating companies have been fined for failing to meet the 24-in requirement by perhaps as little as $\frac{1}{2}$ in. The solution to this potential problem is to ensure that there is sufficient height above the liquid to accommodate not only the drop of the show when the mounting mechanism expands but also slight variations in specific gravity which will cause the liquid to vary in relative elevation to the top of the shoe. Some seal manufacturers use a design which has no shoe drop as the annular gap between the roof and the shell changes.

Another important component of the mechanical shoe seal is the primary-seal fabric. Its job is to block emissions from the liquid underneath it. The fabric varies from one manufacturer to another and with service requirements. It is attached to the top of the shoe and the floating-roof rim. This is usually done with baton strips or channel. Federal regulations prohibit any holes, tears, or openings in the primary-seal fabric. It must therefore be seamed radially by the manufacturer where radial joints in the fabric occur.

The mechanism that provides the radial force of the shoe against the tank wall varies with the manufacturer. However, there are several basic classes of mechanisms. See Fig. 8.2.4. One of the earliest designs was a pantograph hanger, shown in Fig. 8.2.1. Gravity acts through the mechanism to provide a constant force on the shoe even at different rim space distances. However, a severe drawback with this design is that the mechanism cannot be serviced or replaced while the tank is in service. Sometimes the weights corrode off, the seal can no longer meet gap specifications, and the tank must be taken out of service. Both coil and plate springs have been used to supply the radial force required. Since spring force is proportional to extension, the force on the shoe varies with the rim space. Mounting details and hardware selection are important in these designs. Early designs used carbon steel springs and bolts and poor mounting details, which allowed the springs to pop out with corrosion or shoe hangups. The plate springs called *pusher bars* exert radial force on the shoe at intervals around the circumference of the tank. The manufacturers have optimized the spacing requirements for these.

The tank owner or operator should pay particular attention to the following items in selecting a mechanical shoe seal:

1. Note the detail mounting the radial force mechanism to the pontoon. If a welded clip is used and the roof hangs up, the clip could tear out a section of the pontoon, allowing it to fill with liquid. Superior designs allow the attachment mechanism to tear out before the integrity of the pontoon is affected. They also allow installation or replacement of the various seal components without removing the tank from service and without hot work.

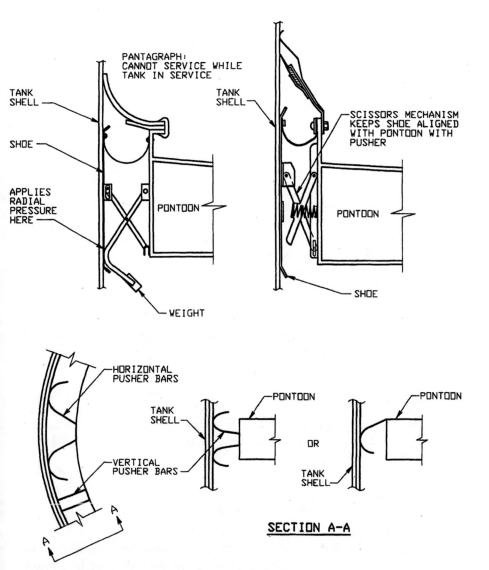

Figure 8.2.4 Mechanical shoe seal mounting methods.

2. The selection of the seal fabric requires compatibility with the stored product. Since the primary-seal fabric is in the vapor, it must withstand the vapor exposure but not necessarily immersion exposure. Typical materials are urethane, buna vinyl, and Teflon. It is important to consider possible changes in service and select a fabric that will meet all anticipated services.

3. Note the optional stainless steel trim and shoe. In many cases the manufacturers supply only stainless steel trim, fasteners, springs,

and other small components because the premium for this is small, but the life of the seal is extended greatly and any future repairs or maintenance is facilitated. The stainless steel shoe is always a premium and should be considered depending on the corrosivity of the vapors, the stored liquid, and the wear characteristics as well as the design life. The selection of the shoe mechanical and material properties is also important. Stainless steel material retains is shape better than softer galvanized sheet steel. The industry standard sheet is 18 gauge thickness for both stainless steel and galvanized shoes. Consideration for thicker 16 gauge shoes should be given where longer life is needed.

4. Either a rim-mounted or shoe-mounted secondary seal can be used with a shoe seal. The shoe-mounted seal is not approved by any regulatory agency since it does not cover the annular space and therefore does not reduce emissions very much.

5. Make certain that all components are replaceable while the tank is in service and without the requirement for hot work.

6. The secondary shoe seal should be able to be removed without affecting the mounting of the primary seal. This allows for inspectors to perform gap measurements on tanks with the secondary seal not mounted, which is much easier than springing the secondary seal out of the way to perform primary gap measurements.

Mechanical shoe seals can tolerate a local variation of tank radius of ± 5 in in a normal 8-in-wide rim space. When the variances of clearance are greater than this, the manufacturers of seals will provide special sections that accommodate the out-of-round condition of the shell.

8.2.2.3 Resilient toroid seals

A resilient toroid seal (see Fig. 8.2.5) is a circular tube similar to a bicycle inner tube that fits in the annular space between the rim and the tank shell. Sometimes these seals are referred to as *log seals*. The tube is made of an elastomeric coated fabric and filled with a liquid, resilient foam or gas. It is attached to the rim. If the bottom contacts the liquid, it is a liquid-mounted seal. If it is mounted up higher where it cannot contact the liquid surface, it is a vapor-mounted seal. A secondary seal or weather shield usually is located above the toroid to protect it from the elements.

Because it is flexible and expansive, it fills all irregularities and out-of-rounds that occur in the tank shell. They are designed to accommodate ± 4 in in a normal 8-in-wide rim space. They are easy to replace, and this can be done while the tank is in service. Because there is no metal-to-metal contact, they are advantageous to use for

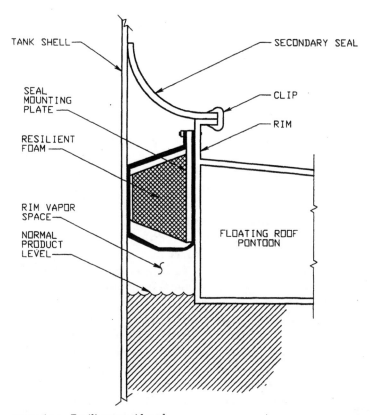

TANK SHELL

SECONDARY SEAL

SEAL
MOUNTING
PLATE

CLIP

RIM

RESILIENT
FOAM

RIM VAPOR
SPACE

NORMAL
PRODUCT
LEVEL

FLOATING ROOF
PONTOON

Figure 8.2.5   Resilient toroid seal.

tanks that have been coated, as the integrity of the coating will not be sacrificed.

Resilient toroids are relatively delicate compared to a shoe seal. They can be torn or punctured by rivet heads, weld imperfections, surface roughness, and so on. Simple abrasion by the action of the roof moving up and down in the tank often wears away the seal envelope. For riveted tanks, the life of these seals is greatly jeopardized, and other seals may be more appropriate. If the seal is filled with a liquid or gas, the seal is immediately lost because of escape fill material. It may be advantageous to avoid these fill media except for unusual reasons.

Resilient toroids are the best fit for vapor-mounted service because if a small tear or puncture occurs, the seal can continue to function. If the seal is liquid-mounted, the envelope quickly fills with product and either destroys the inner material or causes the seal to lose its shape and expansivity. However, they are used in both vapor- and liquid-mounted applications. Resilient toroids have a typical life of only 5 to

7 years compared to at least 20 to 30 years for mechanical shoe seals, but the installed cost is much less. Resilient toroid seals have very few mechanical hardware components to corrode or wear away. However, the toroid materials tend to wear fast and puncture easily as the roof moves up and down.

8.2.2.4 Flexible wiper seals

The flexible wiper seal is often used as a secondary seal. It can be configured in several ways. Most typically, it bridges the annular gap between the top of the rim to the tank shell, using flexible stainless steel plates or shields that are rectangular and overlap one another around the circumference of the floating roof (see Figs. 8.2.6 and 8.2.7). The plates are equipped with a flexible wiper tip that acts to seal the gap even better. There are several different details available for the tip depending on the manufacturer and the air district jurisdiction. The seal can accommodate ± 4 in in a normal 8-in-wide rim space. Some wiper seal tips reverse when the roof direction changes from upward to downward or vice versa, as shown in Fig 8.2.6. In some jurisdictions the flexible wiper tip reversal is considered to be a violation of the gap rules, and for this reason seals designed in a way that is subject to this may not be allowed or used in various air district jurisdictions. Under the shields or seal compressor plates is a continuous vapor barrier to further reduce emissions. The vapor barrier is an elastomeric-coated fabric similar to the primary-seal fabric of a shoe seal. Most manufacturers install the fabric under the shields to protect it from the UV sunlight that tends to degrade these fabrics from the elements. However, some manufacturers have installed the fabric on top of the shields, stating that it makes inspection of the fabric easier. Installation where the vapor barrier is located on top of the compressor plates is probably more subject to damage caused by the elements and may have a shorter life-span. Other manufacturers do not use a fabric seal but rely on a gasket between the seal plates that is either bolted or sealed with a sliding gasket.

Some manufacturers make a double flexible wiper seal that is both a primary and a secondary seal. This design, however, uses two vapor-mounted seals and therefore is rare in domestic applications, because it does not comply with environmental regulations for regulated liquids. The hardness of the flexible wiper tip has a lot to do with the actual gap clearances that a seal will have. The softer it is, the smaller the gap. However, it will also wear faster.

A recent phenomenon is the requirement for a "zero-gap" secondary seal. This resulted from the California air regulatory districts that specified gap tolerances which essentially result in no gaps between the wiper tip and the shell.

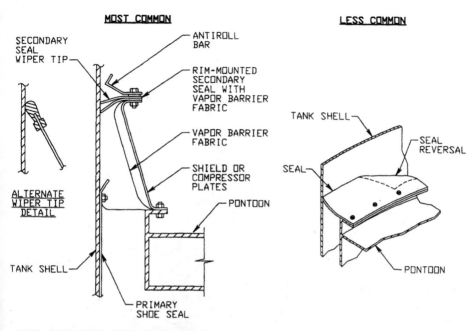

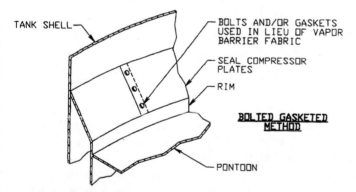

Figure 8.2.6   Secondary seals.

8.2.2.5   Weather shield

This device is essentially a secondary seal without the wiper tip or the seal vapor fabric. It consists of the sheet plates only and functions to keep dust, grit, and large amounts of snow or water out of the product.

8.2.2.6   Other seal hardware

All secondary seal systems are required to have lighting shunts to ground the roof to the shell. Some secondary seals have anti roll tabs,

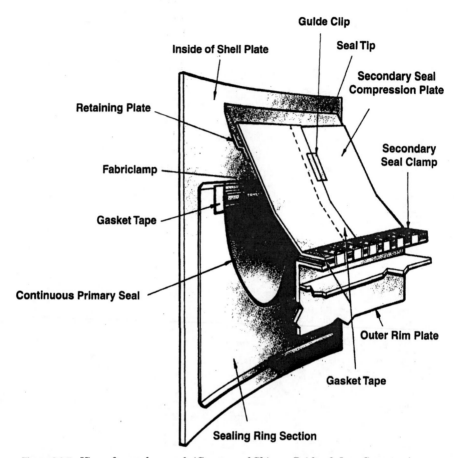

Figure 8.2.7 View of secondary seal. (*Courtesy of Chicago Bridge & Iron Company.*)

as shown in the Fig. 8.2.6. The function of these tabs is to prevent the sheet steel plates from reversing when the roof movement is upward. This would damage the seal plates.

Wax scrapers are shown in Fig. 8.8.2. These act to scrape the waxy buildup off the tank shell as the roof descends. Some designs are devices mounted off the pontoons, and they tend to be more rugged. The shoe-mounted wax scrapers tend to be rather delicate.

8.2.2.7 Fire implications

When an explosion occurs in the rim space under secondary seals, the seals will open to relieve the pressure. However, they also spring back, closing off the oxygen supply and thus starving the fire and explosion. If the seals are permanently deformed, the secondary seal may form a foam dam that allows foam to be applied directly from above and allows it to run into the rim space, extinguishing the fire.

8.2.2.8 How seals become damaged

Seals can be damaged in a catastrophic mode from the following causes:

- When a floating roof is run up high enough, the seal may pop out at the top of the shell and get caught when the roof descends, destroying it.

- Earthquakes cause sloshing waves that have caused seals on tank roofs with high liquid levels to pop out as above, causing extensive damage.

- Some seals have been damaged when the shell has a waxy buildup and the seal is installed on the roof at high level. When the roof descends, the frictional force preventing the seal from moving freely tears it up. The same may happen with newly installed wax scrapers. The key to preventing damage is to ensure that the shell is cleaned prior to installation of seals and wax scrapers.

- Damage is frequently caused to seals when the roof is landed in its low leg configuration because the clearances are inadequate. Frequently, mixer props or swing lines run into the seals when the roof is a low leg position.

- Mixer props or swing lines run into the seals when the roof is in a low leg position.

- There have been cases reported in very cold weather conditions where the moisture between the seal and the shell has frozen. Upon lowering the liquid level, the roof hung up, finally dropping and damaging the roof and seal. The roof movement must be stilled for long enough durations of extended cold for this to occur.

Long-term deterioration of seals may occur because

- Seal fabrics tend to have short life spans. When product incompatibility exists, this exacerbates the deterioration.

- The secondary seal compression plates tend to lose their springiness and configuration. Use of spring steel reduces this possibility.

- Galvanized accessories such as springs, bolts, clips, and fasteners tend to corrode with time. Stainless components are generally superior for this service.

- The wiper seal tips tend to wear as they serve in a rubbing or friction service.

- Shell out-of-roundness causes damage and deformation to compression plates and accessories.

8.2.2.9 Seal inspection

Seal inspectors usually have a device that inserts between the secondary seal and the tank shell and may be rolled around the circumference as it moves the secondary away from the shell. This allows the inspector to examine the primary seal. When gap measurements are required on a primary seal, the inspector can reach through the spread secondary seal and insert the gap-measuring tools.

8.2.3 Internal Floating-Roof Seals

Internal floating-roof seals are generally different from the seals used on external floating roofs because the roof construction differs. Because internal floating-roof tanks are not designed to handle water, rain, and snow loads, as external floating-roof tanks are, IFRTs can be built of lighter construction and are often most economically made of aluminum. Therefore the seals that attach to the internal floating roofs must be of a lighter design, so that they do not exert undue forces on the internal floating roof. Shoe seals are often made smaller and shallower than shoe seals for external floating roofs. More often a wiper seal or foam-filled seal is used. In addition to the mechanical limitations, the regulations include very few requirements for internal floating-roof seals, so that compliance may be achieved with low-cost wiper-type seals or log seals. However, the exception occurs when an external floating-roof is converted to an internal floating-roof tank. Because of the regulatory effect that tends to promote covered internal floating-roof tanks, many conversions are occurring. In many of these cases, the old floating roof can be used. If this happens, then the seals that were in place can be used without modifications.

The seal requirements for internal floating roofs are much less stringent than those for external floating-roof applications. If a liquid-mounted seal is used, then only one seal is required. The liquid-mount seal may be a shoe seal or a resilient toroid. If a vapor-mounted primary seal is used, then two seals are required. There are gap criteria in the federal regulations for internal floating roof tanks; however, local air districts often adopt specific criteria.

SECTION 8.3 FLEXIBLE PIPING SYSTEM FOR ROOFS

8.3.1 General

Because floating roofs can move, the need for flexible piping systems that can connect a fixed shell nozzle to the roof has arisen. The primary applications for flexible piping systems for aboveground storage tanks are

1. Roof drains for external floating roofs

2. Skimming lines for floating-roof tanks

3. Fire protection systems for floating-roof tanks

8.3.2 Types of Flexible Piping Systems

There are currently several types of piping systems that can accommodate the need for flexibility for use in the above applications, and they can be categorized as follows:

8.3.2.1 Rigid piping with flexible joints or articulated joints

The earliest systems developed used rigid piping along with swivel joints. There are a number of variations on this basic system that mostly involve the articulating-joint design. While some designs depend on a swivel joint that can rotate, other designs use a flexible hose that couples the rigid piping.

8.3.2.2 Hoselike flexible piping

These systems are simply hoses that have attaching flanges on each end and no intermediate joints. They are constructed with either an interior or exterior flexible stainless steel jacket. The construction details of these systems are somewhat complex and are covered in Sec. 8.3.5.2.

8.3.3 Applications for Flexible Piping Systems

8.3.3.1 Roof drains

External floating roofs must have a means of draining rainwater from the roof. The accepted and most prevalent way to do this is to route the rainwater to a sump located near the center of the roof and to run the sump drainage through flexible piping to a flange at the base of the tank. Either the rigid-type or the hose-type system may be used for floating-roof drain applications. See Fig. 8.3.1.

8.3.3.2 Skimmers

Most people think of tanks as simply devices to store product or raw materials. However, tanks are particularly useful for some other notable unit operations such as settling and phase separation because of the unusually large residence time associated with storage tanks. In phase separation operations, the lighter phase rests on top, and the bottom phase is withdrawn from a nozzle located low on the shell. Often sludge accumulation will act as a third phase. See Fig. 8.3.2.

A typical application in a refinery might be the control of hydrogen sulfide odors or emissions in wastewater tanks. A layer of light oil, called *sponge oil,* is often added to the wastewater tank. This layer forms a light phase blanket that floats just under the roof. Its function is to absorb hydrogen sulfide gas and to prevent most of it from escaping the roof seals, causing hydrogen sulfide odors and emissions. To optimize this system, you must be able to feed sponge oil into and

Figure 8.3.1 View of floating-roof drain sump and piping.

Figure 8.3.2 Floating-roof skimmer.

out of the tank and to maintain a minimum thickness layer. While a thick layer of sponge oil will not contribute to additional sealing of the hydrogen sulfide in the tank, the extra oil reduces the working volume of the tank and consumes inventory of sponge oil. It is therefore desirable to maintain a minimum, but effective, layer of sponge oil.

Either the rigid-type or hose-type system may be used for skimmer applications, but the hose type system has been by far the most commonly used in this application in recent times.

8.3.3.3 Fire protection

In the area of fire protection application, flexible pipe systems have been used to deliver foam to the rim seal.

8.3.4 Installation Considerations

Where the tank cannot have any hot work performed, then the choice of flexible piping systems is limited to the hose systems since these can be installed with no welding or hot work. Another factor is if the entry time by personnel has to be minimized for either operating reasons or health reasons, the installation time for a hose system is less than that for a rigid system. In a rigidly piped system the components must usually be purchased, prepared, and assembled by the owner or her or his contractor.

8.3.5   Design Considerations and Key Problem Areas

Since in all the applications above the flexible piping is connected from a fixed flange low on the tank shell to a fixed flange just under the roof, it is useful to know the variables that can affect the selection of the type of flexible system to use.

Differences in the equipment can result in different types of problems. Each of the basic types has its own peculiar problems, and within each basic type each supplier of a subtype design has advantages and disadvantages. Some of the major areas of concern in the industry are addressed here.

8.3.5.1   Rigid pipe systems

Rigid pipe systems depend on at least four separate areas of coupling to provide the flexibility necessary to accommodate the full travel of the roof.

In systems that use swivel joints to couple the rigid pipe sections, a number of problems can develop. The swivel joints are probably the largest source of failures in these types of system. The swivel joint depends on the delicate mechanisms of bushings, O rings, snap rings, and the like to provide rotating joints. Because of this, any dirt or grit that gets into the joint can quickly destroy the sealing surfaces and cause leakage. Worse, the joints are subject to freeze-up where binding occurs. There have been reported cases of swivel joint separation, which is extremely serious in that the loose ends of the piping can damage other tank internals or conceivably puncture the shell of the tank as the roof level changes.

If the system depends on flexible hose couplers between the rigid piping, it is at these points that the system is vulnerable. Because of the need for at least four flexible couplings, there are at least eight flanged connections inside the tank. These are subject to leakage due to improper bolt-up and gasket failure. The hose itself must be connected to the flanges, and this is another area where leakage has been a problem. Manufacturers have ameliorated this problem by improving the type of hose used, improving the hose-to-flange joining techniques, and limiting the amount of bending that each joint can take, and pressure testing the hoses prior to installation. These systems are not likely to come apart and break between rigid piped sections because the sections are coupled not only by the hose but also by the angle limiter that removes most of the external forces from the hose.

In applications where leakage is not critical, the above considerations may not be so important to weigh in system selection or specification.

8.3.5.2 Hose system designs

Within the hose system class of flexible piping systems, there are three basic designs:

External jacket, single-coil flexible hose. See Fig. 8.3.3. This construction uses a single coil that can be installed by connecting directly to the sump and the shell nozzles without additional piping. This feature is valuable when hot work is not possible at the time of installation.

Internal jacket, single-coil semiflexible design. See Fig. 8.3.4. This design uses rigid piping to align the ends of the hose connections in the exact pattern required. The internal jacket single-coil designs require more clear floor space (18- to 22-ft diameter) than the multiple-coil designs (14- to 16-ft diameter). The stated diameters include a 3-ft buffer zone of clearance around the coils.

External jacket, multiple-coil semiflexible hose. See Fig. 8.3.5.

Some problems associated with hose designs. The most common failure resulting from these flexible pipes is the damage to the flexible pipe caused by one of the roof legs landing on the flexible pipe when the roof comes to rest on the tank bottom. It should be emphasized

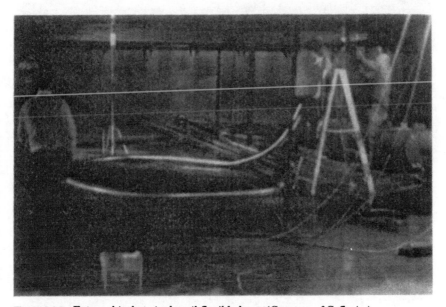

Figure 8.3.3 External jacket single-coil flexible hose. (*Courtesy of Coflexip.*)

Figure 8.3.4 External jacket single-coil semiflexible hose. (*Courtesy of Mesa Rubber Company, Inc.*)

Figure 8.3.5 External jacket multiple-coil semiflexible hose. (*Courtesy of Mesa Rubber Company, Inc.*)

that these problems are avoidable if the recommendations listed below are followed. The reason that these failures have occurred in the past is as follows: The hose systems all depend on what is termed a *repeatable lay pattern*. The basic idea is that the hose must lay down on the tank bottom the same way each time the roof comes down. If it wanders a lot from the original laydown pattern, then roof legs can land on the hose and crush, puncture, or sever it.

Installation problems. These usually occur because of nonconformance to the manufacturer's installation details. The critical dimensions are the actual locations of the fixed flanges to which the flexible pipe is attached. These dimensions and orientations are specified by the vendor, and any variance from them can lead to this type of failure. If possible, the final dimensions should be checked by the vendor's representative prior to returning the tank to service. When this is done it is not unreasonable for the tank owner to ask for guarantees.

Other laydown pattern considerations. The design of the flexible pipe system is such that the coil pattern formed when the roof comes to rest on the tank bottom is repeatable within certain dimensional limits. It is because the pattern is repeatable that the laydown pattern can be determined such that the roof legs never impinge upon the flexible pipe. However, sometimes there is a situation in which the proximity of the roof legs to the flexible pipe warrants additional design measures to ensure that the roof leg does not pinch the hose.

Kinking. Another consideration for flexible pipes is the possibility of kinking in the flexible pipe near the roof connection. To prevent this under specific conditions that may cause excessive loading on the flexible piping, the manufacturer will include a cable that tethers the flexible pipe at some small distance from the end connection with a girth collar around the flexible pipe to the roof. These are only required when the length of the flexible pipe would be such that in high roof levels the weight of the line would damage the flexible pipe. These are often called *bending restrictors*.

Buoyancy. All the hose designs depend on reducing the loads at the interconnecting flanges through the buoyancy of the hose. However, the hose should not be so buoyant that it floats. If it floats, then pockets can develop in the hose, forming traps that could prevent the free flow of water off the roof in roof drain applications. Since the hose material is usually a low-density one, the weight must be controlled by the stainless steel jacket (internal or external).

Flow rate. The hose designs with the smooth inner bore have less resistance to water flow than designs with inner stainless steel jacketing. Sometimes this can result in a smaller size hose's being used on a given tank.

8.3.5.3 Problems common to both rigid sectional and hose systems

Welding. Whenever maintenance work is going on and welding is being done inside the tank, the flexible portions of the flexible piping systems should be covered. Several failures of hoses have occurred because hot weld slag has landed on the hose or plastic nonmetallic portions and burned through. The exterior jacketed hose systems are less subject to this type of failure.

Sandblasting. Flexible hoses must be covered during sandblasting of the tank interior, or else severe damage to the flexible piping will result.

Hose-to-flange joint. Because a hose-to-flange joint is probably the most likely area to fail through leakage or kinking, considerable design effort has been made by the vendors.

Of the three systems, the flexible pipe system and the jointed rigid piping system with flexible connectors have the greatest advantages. They are leakproof. The flexible piping system has no moving mechanical parts to wear out. (Of course, the flexible pipe does move, which might lead to abrasion, but this has been found to be minimal where the manufacturer's recommendations for installation have been followed.)

Another problem with all designs is operating with the tank temperature above the manufacturer's recommendation. Doing so greatly reduces the longevity of the nonmetallic hose components.

8.3.6 Considerations

1. When rigidly piped systems are used, be sure to consider the possible failure modes that are inherent in the design of these systems, as discussed above. Flexible connectors of the jointed system require rigid control of in-line lay of the pipe; otherwise, binding and buckling of the system can occur. The sections of straight pipe must be the right length to prevent locking and damage to piping, tank shell, and sump.

2. When flexible hose systems are used, be sure to deal only with reputable manufacturers that have supplied the drain or skimmer

applications on large numbers of tanks over many years. These manufacturers know the problems associated with storage tanks and design the proper precautions into their product. There are a number of hose manufacturers that are not knowledgeable about problems specific to tanks and have a product that may not perform as expected.

3. Be sure to specify the specific-gravity range of all products expected to be handled as well as the specific gravity of the hydrotest fluid. There have been incidents of hose damage caused by hydrotesting using seawater because the specific gravity of seawater was not considered.

4. After installation, make sure the supplier of the flexible pipe inspects the tank for correct installation. Ask for guarantees or warrantees.

5. Never exceed the design temperature of the hose, as its life will be drastically reduced.

SECTION 8.4   ALUMINUM DOME ROOFS

8.4.1   Introduction

There is a trend in various industries to install aluminum dome roofs on storage tanks of all kinds. Although most dome roofs have been installed on floating-roof tanks, there is a trend to install them on fixed-roof tanks as well, substituting the familiar shallow fixed-cone roof with a geodesic dome. In part, this trend has been caused by EPA requirements causing a greater number of closed tanks to be vented to vapor recovery or vapor destruction systems. Both the aluminum roof manufacturing community and the user have acquired a whole new set of problems associated with the change in dome roof applications from atmospheric to those requiring internal pressure. New problems are just now being dealt with and solved because cost factors tend to make the aluminum dome an economical solution for many cases where sealed tank systems must be used. Because of the increased numbers of geodesic domes as an alternative to a fixed-cone roof tank or as a way to convert an external floating-roof tank to an internal-floating roof tank and as their potential to serve as tools in the environmental arena, this chapter examines all the current issues related to aluminum geodesic domes.

8.4.1.1   Fundamentals

By the late 1970s the aluminum geodesic dome roof gained widespread use. It began to be used to cover both retrofitted and new tanks alike. After various phases of the Clean Air Act were promulgated, the aluminum dome roof enjoyed a resurgence of its application as a means of reducing air emissions.

In the early 1970s there were no aluminum domes, but today they number in the thousands. Most have been used in the petroleum and petrochemical sectors of industry, but they are not limited to these industries. They range in diameters from 6 m (20 ft) to over 60 m (200 ft), although there is no reason that they cannot be applied to much larger-diameter tanks. In fact, some clear-span structures (such as the dome that encapsulated the Spruce Goose airplane in the Los Angeles area) are over 400 ft in diameter. Because of the significant demand for dome roofs on tanks, the American Petroleum Institute issued a new Appendix G (Structurally Supported Aluminum Dome Roofs) to API 650 which not only allows these types of roofs on tanks but also establishes minimum design criteria for them.

Figure 8.4.1 shows a typical aluminum dome roof. The dome is an efficient structural network of aluminum I beams and light = gauge sheets or panels. The beams are linked to one another, forming triangular spaces which are closed by the panels. The overall shape is spherical. The entire structure is assembled from precut beams and panels and is bolted together in the field. Domes may be bolted together on the ground and hoisted to the top of the tank (see Fig. 8.4.2) or fabricated on the floating roof itself. By using the floating roof as a jack, it is lifted to the top of the tank by filling the tank. This makes it possible to install a geodesic dome with no hot work, which is important for tanks storing flammable or combustible liquids.

Figure 8.4.1 An aluminum geodesic dome roof. (*Courtesy of Temcor, Carson, CA.*)

Figure 8.4.2 Hoisting dome into place. (*Courtesy of Tank and Environment, Woodlands, TX.*)

8.4.1.2 Competitive advantage

Aluminum geodesic dome roofs or covers for storage tanks offer two unique advantages to the user that give them a definite niche in the storage tank business. First, they are clear-span structures, meaning that there are no columns within the tank to support the roof (see Fig. 8.4.3). This feature is important for reducing emissions, as discussed later. Clear-span domes that can cover even the largest tanks of up to 300 ft or more are possible. The other advantage of their use is that they are economically competitive and in many cases the lowest-cost option for covering a tank. To make cost comparisons, it is important to establish the appropriate base of comparison. If consideration of maintenance costs (such as painting) is included in the cost feasibility studies for a new tank roof, then a dome roof often is the most economical choice.

8.4.2 Applications

8.4.2.1 Original uses of aluminum domes as weather covers

The original applications for aluminum geodesic dome roofs were to convert external floating-roof tanks to internal floating-roof tanks to

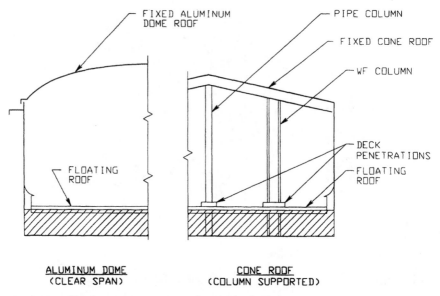

Figure 8.4.3 Aluminum versus cone-roof support columns.

minimize the effects of weather on the maintenance and operation of the tanks in high rainfall areas such as the Gulf Coast or in high snow load areas (see Fig. 8.4.4). The aluminum dome roof (as well as any other fixed tank roof) reduces many of the weather-related problems associated with external floating-roof tanks. External floating-roof tanks must be periodically drained to eliminate the bottom water layers that form because of rainwater that runs down the inside wall of the tank shell and past the roof seals. The bottom water layer must usually be treated to remove environmentally unacceptable materials before it is discharged. Many marketing facilities are simply not equipped to treat this water, and it is sometimes economical to reduce the water infiltration volume as much as possible. External roof drains are subject to freezing and plugging with debris, and they require frequent inspection to ensure that they are working. The use of the dome roof or steel roof eliminates all these problems as well as the maintenance and operational costs associated with them.

8.4.2.2 Product purity

Many diesel fuels are often used in covered tank applications to reduce the infiltration of water. Too much water in the tank can lead to turbidity and off-specification product. To correct this, the tank must be allowed to settle, and water-coalescing units or some other type of water removal system must be used. It is easier to maintain

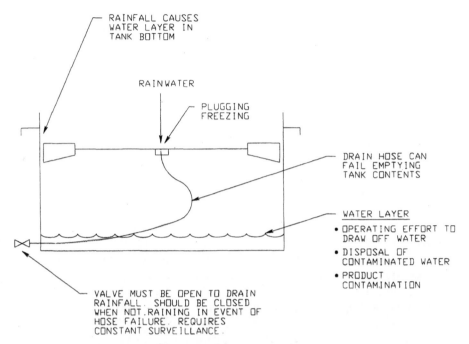

RAINFALL CAUSES
WATER LAYER IN
TANK BOTTOM

RAINWATER

PLUGGING
FREEZING

DRAIN HOSE CAN
FAIL EMPTYING
TANK CONTENTS

WATER LAYER
• OPERATING EFFORT TO
  DRAW OFF WATER
• DISPOSAL OF
  CONTAMINATED WATER
• PRODUCT
  CONTAMINATION

VALVE MUST BE OPEN TO DRAIN
RAINFALL. SHOULD BE CLOSED
WHEN NOT.RAINING IN EVENT OF
HOSE FAILURE. REQUIRES
CONSTANT SURVEILLANCE.

Figure 8.4.4 Weather-related problems of external floating-roof tanks.

product purity by keeping out the water in the first place than to
remove it. Keeping the water out is particularly important for materials which do not easily phase-separate in tanks. Examples are alcohols such as motor fuel oxygenates. Products such as MTBE (methyl
tertiary butyl ether, a gasoline oxygenate additive) are more easily
handled in covered tanks because water may cause the product to go
out of specification. Additives or other chemicals may be selectively
leached out of some products, making them off-specification.

For fixed-roof tanks that have been internally coated to eliminate
product contamination problems resulting from iron, iron salts, or
rust as a result of corrosion of the shell, the use of aluminum tanks
and domes should be considered, if compatible with the product.

8.4.2.3   Applicability to emission reduction

An internal floating-roof tank can have lower evaporative emission
losses than a comparable external floating-roof tank. Emissions from
roof seals are heavily influenced by wind speed. Since the wind speed
above the seals in an internal floating roof is nil, the emission of air
pollutants is minimized. Either a conventional steel roof or a dome
roof will have the same effect on emissions. However, the geodesic

dome has an advantage that is not available to conventional roofs. Large steel roofs on internal floating-roof applications must be supported by columns (Fig. 8.4.3). These support columns must penetrate the roof. At each penetration there is a pathway for emissions. Since the geodesic dome roof is supported only at the perimeter, there are no penetrations in the internal roof for columns.

An important change is occurring for tank roof fitting and seal manufacturing that will increase the use of internal floating-roof tanks and consequently the use of aluminum domes. In the next few years, it is expected that API will produce a testing protocol whereby manufacturers can have their roof seals, fittings, and deck seam evaporative loss factors certified. Because the internal floating-roof manufacturers have perfected much better deck seam techniques and rim seals, it is expected that the internal floating roofs will be recognized as having lower emissions than the external floating-roof tank for a wide variety of tank sizes, stocks, and locations. When this occurs, especially in EPA-designated nonattainment areas (which are areas that do not meet the EPA ambient air quality standards), the use of clear-span roofs as a tool to reduce emissions for both retrofits and new tankage may become widespread. The aluminum dome is expected to play a significant role in this arena.

In addition to the internal floating-roof applications just covered, sealed or closed tank system applications for both internal floating-roof and fixed-roof tanks are increasing. This results from the increased use of vapor recovery systems. The geodesic dome plays a useful role in this area, especially for retrofit applications where open-top tanks must be converted to closed systems. These applications are more fully described in the next section.

8.4.2.4 Applications for small internal pressures

One of the challenges of the aluminum dome application is in its use as a fixed roof designed to contain small internal pressures, which requires a closed-tank system. Although API 650, Appendix G allows pressures under the dome of up to 9 in WC, pressures this high have not been used in practice. However, sealed domes have been used with internal pressures to approximately 2 in WC. Because the geodesic dome is an array of panels and beams, the critical element for pressure containment is the sealing of joints. Sealing the joint between the tank shell and the dome becomes a significant problem as well. These applications are discussed more thoroughly below.

8.4.3 Basic Design Considerations

8.4.3.1 Standards

API 650, Appendix G is the only aluminum dome standard that specifically sets out design criteria for all structurally supported aluminum dome roofs. However, the domes, being part of a larger structure, are often regulated as building structures and are therefore subject to local building permit requirements and fire department requirements. Live and dead loadings as well as maximum height requirements are often regulated by the building permit authorities. API 650, Appendix G recognizes the following applications of aluminum domes on tanks:

- For new tanks
- For existing tanks
- For atmospheric pressure
- For internal pressure up to 9 in WC

However, there are numerous practical design considerations that are not covered by the standard, some of which are covered below.

8.4.3.2 Aluminum dome materials, corrosion and corrosion prevention

Materials. The structural members are typically 6061-T6. The panels are series 3000 or 5000 aluminum with a required minimum thickness of 0.05 in. Fasteners are aluminum or stainless steel. All aluminum must be electrically isolated from carbon steel by an austenitic stainless steel spacer or elastomeric isolator bearing pad. The aluminum dome must be electrically bonded to the tank shell using $\frac{1}{8}$-in stainless steel cable conductors at every third support.

External corrosion resistance. Since aluminum dome roofs are corrosion-resistant to most atmospheres, they can be used in any geographic area. For marine atmospheres, the correct use of the proper aluminum alloys is important for corrosion resistance.

Internal corrosion resistance. Due to corrosion resistance to a number of compounds, the aluminum dome has the advantage over fixed steel roofs in many cases. Hydrogen sulfide or vapors from sour product service, sour crudes, sour wastewater, and many other corrosive environments normally do not attack the aluminum domes. Typical steel roof designs in which underside corrosion is a problem and corrosive vapors condense into the crevices of the lap-welded joints, initiating corrosion, make a much more costly version of tank roof necessary. For sour services, a reverse shingle layout of steel roof plate is often used so that the tendency for condensing liquid on the underside of the plates to fill the crevices is minimized. Seal-welding the laps on the bottomside is another alternative. Some roofs use external rafters so that there are no crevices on the underside of the roof. For these cases, the use of an aluminum dome roof should be considered potentially more economical alternative.

Selection of elastomeric seals. For internal floating-roof applications, the requirement for the dome to act as a sealing membrane is not nearly as important as it is for closed applications. The dome sealing gaskets are also not nearly as critical because internal floating-roof tanks are designed to vent the internal space sufficiently to keep it well below the flammability limit for most volatile organic compounds. The only function of the dome in this case is to keep rain and wind out of the tank. While this means that the elastomer sees only low concentrations of organic compounds and is in relatively undemanding service, it must still withstand the external elements including degradation by UV from sunlight.

In closed systems, the dome must function as a vaportight membrane; in addition, all components must be designed for immersion service because condensate can and does form on the internal surfaces of the dome with diurnal thermal changes. There have been several applications where the extruded gasket seals have worked in some light hydrocarbon services but have failed in closed service after only a few months. The failure mechanism is excessive swelling of the elastomer. Due to absorption of hydrocarbon and swelling, the gasket seals have failed.

Although API 650, Appendix G lists several standards for specifying sealants for dome roofs,* the selection and use of domes for pressurized service should be examined carefully by investigating past experiences in similar services. In addition to careful materials selection and specifications, the user should carefully evaluate the gasket materials and any combinations of chemicals to which they may be

---

*Federal Specification TT-S-00230C establishes performance requirements for single-component sealing compounds and caulk. ASTM C 509, *Standard Specification for Elastomeric Cellular Performed Gasket and Sealing Material,* provides requirements for cellular material used as gaskets in compression joints. Federal Specification ZZ-R-765C applies to silicone only.

exposed. Tests specific to the service should be performed to screen out the unsuitable materials. It must also be borne in mind that the selection of materials available for use is limited by what the dome manufacturers have tested and worked with in the past. While there are numerous resources that rate various elastomers for varied services, the key factor for determining acceptability is that the immersion exposure must produce less than 20 percent swell. Gasoline, kerosene, JP-3, JP-4, JP-5, fuel oil, and diesel acting on nitrile have less than 10 percent swell in immersion service. For immersion service, urethane experiences slightly higher swell, and silicone swells over 20 percent.

As mentioned, developmental work in the materials compatibility area as well as testing can ensure that the proper types of sealing strip materials and caulks are available. These materials must be capable of immersion service, but also the confining joints for the extruded seal strips must be designed to accommodate a certain amount of swelling that will inevitably result from these services. In addition, means of preventing degradation from weathering and UV exposure on the atmospheric side must be maintained by either proper materials selection or, better still, encapsulating the gasket in the bottom in a manner which shields it from UV exposure. Sealing systems that utilize a weather-side surface of one material and a stock-side surface of another material could be considered. Reengineering the sealing strip joint, which would involve redesign of the struts, panels, and compression strips, may prove to be a major contribution to effective pressure service with immersion service requirements.

Another problem to consider is the long-term service requirements. Because diurnal thermal changes cause considerable expansion and contraction, there is the possibility of loosening of joints or caulk. Similar applications in the sewage business where ponds or containment structures are covered with pressurized geodesic structures indicate that domes can be suitable for long-term service without leaking joints. However, the combination of the immersion service for pressurized applications and hydrocarbons, petrochemicals, and oxygenates remains unknown. Dome manufacturers should document, test, and, if necessary, lead industry efforts to confirm the long-term suitability for various stocks that are commonly used and needed in the various industries.

One approach to the problem of creating a sealed membrane is to spray either an internal or an external polyurethane coating of sufficient thickness to form a continuous, durable film over the dome. This would have the advantage of reducing the meticulous and careful effort that is now required to seal the dome roofs for pressurized service. It would have the added benefit for external coatings of allowing the selection of any color. However, although this is now being considered, this has not yet been commercialized in the industry.

Coatings. It is very rare for domes to be coated or painted internally or externally because the cost is quite high. The surfaces must be degreased and etched for good adhesion. Sometimes military installations will use painted aluminum dome roofs on tanks.

8.4.3.3 Basic design limits

The following limits apply to the design of aluminum geodesic domes:

Temperature	200°F
Minimum design temperature range	±120°F
Maximum internal pressure	9 in WC
Minimum design live load	25 psf (per API)
Spherical radius × tank diameter	0.7–1.2
Design wind speed (default design)	100 mi/h
Minimum panel thickness	0.05 in
Finish	Mill

8.4.3.4 Design loadings

The minimum dead loading is the weight of the roof itself and of all accessories attached to it. Typically, aluminum dome roofs average 2.5 psf (0.5 in WC), which is about one-third the weight of conventional $\frac{3}{16}$-in-thick carbon steel roofs. The live load is assigned to be 25 psf or greater if required by the regulatory agencies or building codes. API 650, Appendix G includes requirements for unbalanced loads, panel loading, and concentrated loading. It also gives requirements for the load combinations such as dead load plus seismic. The suppliers are required to run through a series of load combinations to ensure that the roof is structurally adequate for the application. If there is any internal pressure, this must be included in the load combinations.

One of the design loading conditions that requires good communication between the purchaser and the supplier is the means of transferring the roof loads to the tank shell. The tank and foundation must be checked to ensure that they are adequate to assume the increased loading from the added roof. Since the top of an existing tank is rarely round, the dome must be constructed to accommodate this tolerance problem. This is done by the allowance for large tolerances made at the support points. It must also accommodate thermal expansion of the roof within a temperature range of ± 120°F. For existing tanks the easiest way to handle some of these problems is to design the roof-to-shell junction with a sliding surface so that only vertical loads are transferred to the tank shell. For new tanks the tank rim is often strengthened sufficiently that the roof is rigidly attached to the shell, which is designed to take all the roof loading. When tanks have internal pressure, the preferred design is to rigidly affix the roof to the shell. If a sliding joint is used, a sealing fabric to contain the internal pressure must be installed, which is more subject to failure than the other design.

8.4.3.5 Wind loading

Unless specified by the tank owner or operator, the default wind loading condition is 100 mi/h. However, the actual pressure which is assumed to project both horizontally and vertically varies based on the codes used the assumptions made, and the manufacturer. The user should thoroughly understood how wind loads are being applied in the design of the roof.

8.4.3.6 Seismic loading

The seismic loading is presumed to act uniformly over the dome, and the design basis for the dome is

$$F = 0.24ZIW_r \tag{8.4.1}$$

where F = horizontal force
   Z = zone coefficient (refer to API 650, Appendix E)
   I = essential facilities factor = 1.0 for most cases
   W_r = weight of tank roof, lb

8.4.3.7 Shell buckling

Local and general shell buckling requires a minimum factor of safety of 1.65. General shell buckling can be determined from

$$W = 2259 \times 10^6 \ \frac{\sqrt{I_x A}}{(SF)LR^2} \tag{8.4.2}$$

where W = allowable live load, psf
   I_x = moment of inertia of beam about strong axis, in^2
   A = cross-sectional area of beam, in^2
   R = spherical radius of dome, in
   L = average dome beam length, in
   SF = safety factor = 1.65

8.4.3.8 Tension ring area

The minimum tension ring area is determined from

$$A = \frac{D^2}{n \tan \alpha \sin(180/n) \, F_t} \qquad (8.4.3)$$

where A = net area of tension beam, in^2
 D = tank diameter, ft
 n = number of dome supports
  = 0.5 times the central angle of dome or roof slope at tank shell
 F_t = allowable stress of tension ring, psi

8.4.3.9 Temperature limits

API establishes a maximum operating temperature for aluminum dome roofs of 200°F.

8.4.3.10 Design pressure considerations

While the API 650, Appendix G limit for internal pressure applied to aluminum geodesic domes is 9 in WC, the actual limit is usually lower and is controlled by other requirements of the standards. API 650 prohibits net upward forces on the roof or tank shell to prevent tank rupture and subsequent loss of product. Since the light weight of the aluminum dome structure averages out to only about 0.5 in (12 mm) WC, this becomes the practical limit for internal pressure. However, if the user complies with provisions in API 650 paragraphs F.2 through F.4 of Appendix F, then the inclusion of the weight of the tank shell may be credited in the resistance to uplift, which effectively boosts the internal pressure from 150 mm (6 in WC) on a 9-m (30-ft) diameter tank to about 50 mm (2 in) on a 61-m (200-ft) diameter tank.

When the tank is anchored in accordance with the provisions of paragraph F.7 in API 650, then tanks may use internal pressures up to 2.5 psig (69 in WC). However, API limits aluminum dome roof design pressures to 9 in WC.

8.4.3.11 Attachment to shell

Flashing. For atmospheric applications a flashing is used between the dome and the tank to keep rainfall out. Venting required by API 650 for atmospheric internal floating-roof applications is handled by allowing the air to circulate under the flashing or through vent slots in the flashing and with a center vent. For pressurized service the dome must be sealed to the tank. This is usually done by a membrane attached inside the tank between the dome and the top of the shell. All designs should have provisions to drain any condensate or rain from entrapped spaces in this region.

Two basic dome designs. The weight of the aluminum dome structure generates an outward thrust. The flatter the structure, the greater the outward radial thrust. To handle this force, dome manufacturers use a tension ring. The dome is attached to the tank shell by means of support points. The specific details of attachment vary from one manufacturer to another. The details for the fixed- and sliding-base designs are similar; but in the sliding-base design (where the support points must be free to move radially), a sheet of Teflon is used as the bearing surface, and a slotted bolthole allows the radial movement.

8.4.3.12 Appurtenances

Hatches. Hatches are optional. However, most tank applications use only one hatch. If there is a rolling ladder left in a tank, a hatch is often supplied for it.

Nozzles. Nozzles are used for high-level alarms or for thief hatch purposes. Many applications do not have any roof nozzles.

Skylights. Skylights are optional. However, they provide natural lighting for the interior and a means to do visual inspection of roof seals that are required to be performed annually by the EPA. They are recommended, and they should be provided at a rate of 1 percent of the projected area of the dome.

Since the venting of dome roofs is accomplished at the gap between the roof and the shell, peripheral shell vents are not needed and the area requirements of API 650, Appendix H are easily met. One center vent at the top is required per Appendix H and is usually an 8-in vent. Typically, there is no special access provided for this hatch.

Internal rolling ladders. When an existing tank is retrofitted with a dome, the existing rolling ladder can be left in place. Because the dome usually interferes with the operation at the top of the ladder, dome manufacturers often reattach the ladder to the structural members of the dome. This requires that the bottom of the rolling ladder be extended to suit the modifications. Note that the API is considering revising API 650 to make the use of a ladder optional. The reason is that entry into the confined space below the roof should be discouraged, and the user should not have to pay for this option if he or she does not want ladders on internal floating-roof tanks.

Often the tank owner or operator does not wish to make the modifications or there are no modifications that can be made to accommodate the new dome and the full travel range of the floating roof. In these cases the ladder is removed. The majority of tank owners or operators do one of two things for access to the internal roof:

1. They use a rope ladder for access when needed.

2. They wait until the roof is at its high level in the tank and simply access the top of the roof by stepping onto it.

Access to the internal roof is required periodically for seal, appurtenance, and roof condition inspections.

Platforms and walkways. In existing tanks retrofitted with domes, some problems related to the tank gauger's platform often arise. In these cases modification must be made to raise or relocate the platform to clear the dome.

8.4.4 Other Considerations

8.4.4.1 Physical appearance

The spherical radius of the dome must be within 0.7 to 1.2 times the diameter of the tank. While a shallow dome requires heavier members, it may have a better appearance and be more practical for access by ladders, stairways, and platforms. Skylights are sometimes used on tank domes. A typical usage is at a ratio of 1 percent of the projected area of the dome. They can be used to let in light and to do visual inspections required by the EPA. The use of skylights is optional; but when used, they must be constructed of 0.25-in minimum thickness clear acrylic or polycarbonate plastics.

The appearance of new aluminum dome roofs is mirrorlike and bright. There have been cases at airports where the high reflectivity caused problems, requiring that the owner paint the dome or cover it with camouflage netting to reduce glare. With time the color changes to a matte gray which approaches the color of weathered aluminum paint. Some manufacturers have provided the user with an option to specify lightly blasted panels, which produce less glare. It is very rare for domes to be coated or painted internally or externally. This is because the cost is quite high and it is unnecessary for corrosion resistance. The surfaces must be degreased and etched for good adhesion. Sometimes an anodized finish is specified.

8.4.4.2 Tank capacity reduction

When an aluminum dome is retrofitted onto an existing tank, a reduction in capacity may result because the roof legs and other appurtenances may interfere with the edge of the dome when the roof is floating at its highest possible position. Dome manufacturers have provided elevated supports that allow the user to maximize capacity, sealing the space between the dome and the roof with siding. Since this increases costs, the owner or user must make an economically optimized decision about how high to mount the roof.

8.4.4.3 Fire and explosion considerations

A dome roof is never considered to be frangible. Because the external floating-roof tank is particularly subject to rim seal fires caused by lightning strikes, the covered tank virtually eliminates the possibility of fires. API 650, Appendix H requires circulation venting which prevents the buildup of flammable concentrations of vapors.

Note that no fires that have significantly heated the interior vapor space of an internal floating-roof tank with an aluminum dome have occurred. It is possible to speculate about the effects of a tank fire under these circumstances. Since aluminum loses its strength at relatively low temperatures compared to carbon steels, a temperature rise in the vapor space could cause the roof to collapse onto the floating roof. To date, this has not occurred.

In the few cases of explosions that have occurred in domed roof tanks, there is a good indication that the steel tank shell and bottom will remain intact and that the roof will be the point of failure. In one case, in a 52-ft-diameter tank in 1992, an internal explosion in a welded wastewater tank in Indiana caused the point of attachment to the dome and the tank shell to separate in several places (in effect, a frangible joint) and to blow out a 30-in flashing panel as well as to bend the tension ring slightly. While this is not meant to be generalized, it is an indication that the dome manufacturing industry can consider and review for development of possible methods of ensuring pressure relief.

Because the probability of a fire in a floating-roof tank covered with a dome is so low, fixed fire-fighting equipment may not be necessary or useful on these tanks. Fires may be fought, if they do occur, through the hatches or light panels in the roof. Some fire chiefs as well as owners require that fixed foam systems be installed on domed tanks with flammable materials, despite the roof.

8.4.5 Conclusion

The use of the aluminum dome roofs for tanks is well established as an economical and viable alternative to steel roofs. However, new trends indicate that the use of aluminum domes will increase. In part this is driven by the Clean Air Act requirements which tend to increase the use of internal as opposed to external floating-roof tanks. This is a result of an increased use of closed-tank systems for vapor recovery. In addition, current API work will allow for the establishment of floating-roof fitting loss factors which should also tend to drive tanks to internal floating-roof designs. The aluminum dome is expected to be a significant component in both retrofitted and new tank installations. But because its role has changed from simply a roof to exclude the elements to a pressure-containing membrane, more options and new designs for aluminum domes are needed so that the user can select the aluminum dome while ensuring that fire safety, reliability, and long-term use remain intact.

References

1. G. Morovich, "The Use of Aluminum Dome Tank Roofs," *Proceedings of the 2d International Symposium on Aboveground Storage Tanks,* Materials Technology Institute, Houston, TX, January 14–16, 1992,
2. H. Barnes, "New Tank Roofs Capture Evaporating Vapors," *Louisiana Contractor,* 1992.
3. R. Barrett, "Geodesic-Dome Tank Roof Cuts Water Contamination, Vapor Losses," *Oil and Gas Journal,* 1989.
4. R. Kissell and R. Ferry, "Bolted Aluminum Roofs for Pressure Vessels," *Proceedings of the 1994 ASTM Piping and Pressure Vessel Conference.*
5. P. E. Myers, "A Users Perspective on Aluminum Dome Roofs for Aboveground Tanks," *Proceedings of the 1995 ASTM Piping and Pressure Vessel Conference,* Oahu, Hawaii, 1995.

Fire Protection of Tanks

SECTION 9.1   GENERAL PRINCIPLES

9.1.1   Introduction

There are many causes and types of tank fires. Although tank fires are relatively rare, they carry significant potential risk to life and property. In almost all cases, the risk factor is substantial because of the relatively large quantities of fuels or unstable liquids that are stored in one location.

For this reason fire protection principles and designs have been incorporated into national codes and standards and adopted by legislative bodies throughout the country. Many industries have generated additional practices that are more conservative than those required by the codes. Some of the practices found to be effective for specific applications are included in this section.

9.1.1.1   Types of tank fires based on tank type

The type of tank impacts the nature, type, and severity of the fire and/or explosion associated with the storage tank. The greatest impact on the specific hazards associated with tank fires is due to the type of roof system involved, as discussed below.

Fixed-roof tank fires.   Fires that are associated with fixed-roof tanks may be ground fires, vent fires, and fires caused by leakage in the external tank piping. However, fires and explosions do occur in fixed-roof tank vapor spaces, and the losses associated with these fires can be heavy. Fortunately, the occurrence of these types of fires is relatively rare. Because fixed-roof tanks are closed, there is little chance for ignition sources to light flammable vapors. In addition, most storage tanks containing flammable liquids have vapor spaces that are either too rich or too lean to burn.

However, when a fixed-roof tank fire does occur and the roof has collapsed or opened, then this is called a *fully involved fire*. This fire involves the total liquid surface of the tank rather than just a rim seal. It can occur in external roof tanks when the floating roof capsizes or sinks. Then the entire liquid surface can burn, and control is very difficult. In fixed-roof tanks, explosions or sudden vapor releases often tear the roof away from the shell over a substantial percentage of the circumference, resulting in a fully involved fire. These fires are probably the most difficult to extinguish in that hot gases and vapors are emanating from the gap in the roof-to-shell joint and make the addition of foam or other agents difficult.

External floating-roof tank fires.   Although the floating-roof design on storage tanks was in large part implemented to reduce the hazards of fire, a large number of tank fires do involve external floating-roof tanks. Because the roof rests on the liquid surface, a fully involved fire is almost impossible. However, if the roof capsizes or sinks, then the fire can spread over the entire surface. Fires typically associated with floating-roof tanks are rim seal fires. Usually ignited by lightning, these fires burn around the rim area. These tanks are vulnerable to ignition sources such as lightning or static electricity because the rim seal on the floating roof is not a perfect seal and provides a source of flammable vapor near the seal. Fortunately these rim fires are relatively easy to control and are not considered very destructive. Usually, the only loss resulting from these fires is a portion of the floating-roof seals. Injuries are almost unheard of in these types of fires.

Internal floating-roof tank fires.   Because of the cover, these fires are extremely rare. There is little experience with a major fire in an internal floating-roof tank. If the internal roof does not sink or capsize, then the fire in principle should be similar to that in an external floating roof—limited to a rim seal fire. Fighting this type of fire is more difficult because of restricted access and the buildup of smoke inside the fixed roof. Aluminum dome roof tanks may pose a special hazard to internal floating-roof tanks in that the aluminum may melt at relatively low vapor space temperatures during a fire and collapse onto the roof. Depending on the type of internal roof, the collapsed structure could sink the internal roof or possibly damage the shell. To date, there is no experience with a severe internal floating-roof fire that has posed a risk to the integrity of the dome roof.

9.1.1.2 Causes of tank fires

Tank fires may be started at various vents, rims seals, or anyplace there is exposed flammable liquid or vapor due to static electricity or lightning. If flammable vapors of the proper concentration are present in the vapor space of a fixed-roof tank, a deflagration or explosion may result.

Other fires have been initiated by exothermic processes occurring in tanks. Storage tanks with high concentrations of H_2S can produce pyrophoric iron on the walls. When air is allowed to contact this highly reactive substance, it can heat up sufficiently to cause ignition of any flammable vapors present in the tank.

Vapor releases and spills are a common source of fires. Often fires are initiated by vapor trails that emanate from a tank spill or a vent line. Once the vapor trail finds an ignition source, it flashes back to the source, resulting in a fire.

Tanks with flammable vapor levels vented through piping to flares often result in explosions that are caused by a flashback to the vapor space in the tank. This has occurred in spite of the use of flame or detonation arrestors.

Hot spots caused by the heat of reaction taking place in activated carbon beds that are part of a vapor recovery system often result in fires or explosions in the vapor space of a tank. Failure to use flame or detonation arrestors or their improper use and selection have usually been the cause.

Pyrophoric ignition sources, heated tanks, and other special cases of storage of flammable liquids all deserve individual attention to minimize the possibility of fire.

Another related source of fires occurs during the preparation, cleaning, repair, and maintenance of tanks. Mixtures of hydrocarbon vapor and air can be ignited if the concentration of vapors falls within the flammable limits. When tank sludge is disturbed, flammable vapors are given off, resulting in increased concentrations of flammable vapors. Vacuum trucks are a common ignition source. They should be equipped with a spark arrestor muffler or exhaust. The vacuum hose must be electrically conductive and bonded to ground and the equipment being cleaned. Static electricity is a common ignition source in petroleum tank fires and explosions.

Special cases that warrant separate discussion are covered at the end of this section.

9.1.2 Control of Ignition Sources

9.1.2.1 Human factor

Dangerous fire potential situations can be inadvertently created by people providing a possible ignition source, such as open flames, smoking, operating motor vehicles, welding and cutting, operation of power tools, hammering on materials that can generate sparks, rubbing or friction, and ovens, furnaces, and heating equipment. In all these instances, the fire can result only if a flammable vapor is present. Since most flammable vapors are heavier than air, they tend to pool in low depressions and travel as rivers, following a gravity-driven flow path to lower areas until the wind and atmosphere dissipate the concentrations to levels which are not dangerous. Secondary containment areas are often filled with flammable vapors because of releases or overflows. Ignition started some distance away can flashback upstream through the vapor trail caused by overflow or releases, allowing the flame to get back into the tank.

In storage tanks, the generation of static charges is often created by the flow of liquids into the tanks. Under this condition the tank may be full of an initially lean mixture or air. As the liquid flows into the tank, the volatile components of the entering liquid quickly produce large amounts of flammable or rich mixtures in the vapor space of the tank. The flowing liquids provide the source of static charge generation.

Splashing and turbulence as well as high liquid velocities increase the potential for static charges. The presence of water droplets or entrained air or other particles of iron scale or sediment also tends to increase the rate of static charge generation and accumulation potential. Since a floating-roof tank with a landed roof is similar to a fixed-roof tank, it is vulnerable to static-electricity-related fires on first filling. Splash filling is prohibited by NFPA 30, which requires that fill pipes terminate within 6 in of the bottom of the tank.

9.1.2.2 Minimizing spark generation in storage tanks

1. Minimize splashing due to free-fall dropping of the liquid while filling the tank.

2. Limit the velocity of the incoming liquid stream to below 3 ft/s until the fill pipe is submerged at least 2 ft or 2 pipe diameters, whichever is less.

3. Check for ungrounded objects such as empty cans, pieces of wood, and loose gauge floats. These ungrounded objects can build up a static charge and discharge sparks when they come into the proximity of a metal support or the tank shell.

4. Minimize the presence of two-phase flow (e.g., water or air in the incoming product stream).

5. Spark ignitions inside tanks cannot be controlled by external grounding connections.

6. Gaging and sampling through the roof with conductive objects should not be done until the filling is complete and the turbulence has subsided. Wait at least 30 min after these conditions to perform these functions. These functions can be performed inside of slotted guide poles and stilling wells constructed of metal at any time, since they are grounded to the tank bottom and cannot build up sufficient charge to cause a spark.

7. Steam ejected through nozzles represents a source of static electricity buildup. The nozzle of steam lines for steaming out tank sludges or use inside a tank should be grounded to the tank.

8. Sandblasting operations represent static electricity hazards. Proper bonding and grounding of sandblasting equipment should be required and checked.

9.1.2.3 Lightning

Lightning strikes are the primary cause of fires in external floating roofs. The small amount of vapors escaping from the roof rim seals can result in rim seal fires. Fortunately these fires are often easy to deal with and are not catastrophic. Bonding the floating roof to the tank shell allows a path for the lightning charge to pass through to ground without arcing. It has been found that a spacing of 10 ft is adequate if a flexible 303 stainless steel, 28 gage by 2-in-wide strip or shunts with electrical properties equivalent to this are used.

Grounding. Tanks must be grounded to conduct away the lightning current and to reduce the potential for charge buildup.

- The tank may be grounded by connecting it without insulated joints to a grounded metallic piping system.

- Flat-bottom tanks greater than 20 ft in diameter and resting on earth or concrete are grounded by virtue of contact with the earth ground.

- The tank may be bonded to the earth ground through the use of ground rods. There should be at least two ground rods, spaced not more than 100 ft apart around the circumference of the tank.

9.1.3 Fire protection through design and planning

9.1.3.1 Risk assessment

The amount of fixed fire-fighting equipment that should be in place at a facility is a complex subject based on risk assessment. Some of the factors that would be considered in making decisions to install these facilities include tank sizes, tank volumes, NFPA class of liquid stored, type of tank and roof, financial impact, public relations impact, environmental impact, exposure of other operations or facilities, and availability of external fire protection resources.

9.1.3.2 Plant layout and tank spacing

A large effect on the reduction of fire incidents and the control of fires that do occur can be created in the area of layout planning. The NFPA code provides guiding distances which have been optimized through experience. Roads, fire water mains, hydrants, dikes, stairs, access routes, and other aspects of plant layout all contribute to the fire prevention and fire-fighting capability. There should be at least two access roads available to each tank field area, entering at different directions, which would allow equipment access in the event of a major fire. There should be a road on at least one side of all low-flash-point stocks. The road's dimensions should reflect the type of equipment and maneuvers that may be necessary during a fire, with thought given to turnouts, overhead obstructions such as pipeways, and the like.

Fire hydrants should be located at strategic points along the roadways and on the outside of diked areas. The dikes should be constructed to provide easy access for firefighters and equipment. This usually means limiting the dikes to 5 or 6 ft in height and providing stairs or styles.

Tank stairs should be located close to the operator access areas in the tank field or diked area. Another consideration is the prevailing wind. It is advantageous to locate the tops of tank stairways upwind so that rim fires may be fought more easily.

9.1.3.3 Secondary containment

For storage tanks, the key to preventing major fires that cannot be controlled is to ensure that the liquid contents do not escape. Experience shows that containment and impoundment areas are effective means of controlling not only environmentally undesirable releases but also fires. Escape of flammable liquids and vapors can occur due to

- Rupture of the tank from explosions
- Failure of the tank walls due to overheating
- Puncturing of the tank walls due to collapse of the roof structure or appurtenances
- Piping that becomes separated from the tank without provisions to block off the liquid flow with valving that can be operated

- Fires occurring in the secondary containment area
- Operational errors such as overfilling
- Mechanical damage as a result of vandalism, bombings, or war-related incidents

To address this, all the applicable fire codes have adopted the concept of secondary containment.

Secondary containment is an additional volume in the form of either a tank or a diked area that can hold the full contents of the largest tank. The Clean Water Act and NFPA 30 and state and local regulations all have some provision that essentially requires secondary containment for storage tanks storing flammable liquids. In some jurisdictions the requirement is to size the secondary containment to include the volume of the largest tank with an additional freeboard for rainfall (usually 6 in).

9.1.3.4   Local or remote impoundment

Remote impounding is the directing of spillage to a remote containment or impound area so that the collected liquid will not be held against the base of the tank. Since the potential of flammable liquids burning adjacent to tank shells is minimized, it is the preferred method of containing spills from tank fields. However, this type of system requires the largest amount of real estate and is frequently impractical. Frequently partial remote impounding is used where as much liquid as can be removed from the tank field as possible is directed to a remote impoundment area that collects as much of the liquid as possible. The balance of the required volume is made up by diking the tank field area. This allows small spills and overflows that have been ignited to be directed away from the tank area.

9.1.3.5   Diking

If neither partial nor remote impounding is possible, then the next option is to dike or contain the tank area by using dikes or berms. However, the area within the secondary containment area should be sloped to direct liquid away from piping, valves, and tanks with large amounts of flammables or with lower flash points. Specific requirements for slope and drainage, as well as dike requirements, can be found in the NFPA 30 code.

9.1.3.6   Fire water systems

Minimum design conditions.   The water supply pressure should be at least 100 to 150 psig. It should also be segregated from process water or utility water systems. For refineries, pipelines, and major marketing terminals, the fire water system should deliver 1000 to 2000 gpm. Fire water pumps should kick in when the fire water pressure drops below 100 psig.

Water supply sizing for tank fire water system requirements should be designed for the larger of the following two conditions:

- *Foam plus tank cooling*—the maximum foam required to fight a rim seal fire plus enough water to cool the upper half of the tank shell over 50 percent of the circumference

- *Tank cooling*—a fully involved tank fire which would require cooling for 25 percent of the circumference of three adjacent tanks which would be downwind and within 70 ft of the involved tank

For cooling, the water rate that has been found to be adequate is 1.0 gpm per 10 ft^2 as applied in the above surface cooling requirement.

For foam generation, the water flow sizing is based on a single rim seal fire in a floating-roof tank:

1. The rim area is assumed to be 2 ft wide.
2. The foam is applied at 3.0 gpm per 10 ft^2 of rim area.

The drainage system should allow rapid runoff of fire water at the design rates. This will reduce the chance of the fire's spreading or damaging piping and equipment.

Using these guidelines will allow you to fight a worst-case scenario fire. Note that fully involved fires in tanks over 170 ft have not been successfully controlled.

Fire mains. A fire main loop is often used so that if the main is damaged, it can be valved off and water supplied in the other direction as required.

Fire hydrants

1. All hydrants should be on the access road side of fences, dikes, retaining walls, drainage ditches, or other obstructions.
2. All parts of the tank shell and roof can be accessed with a hose not exceeding 500 ft. Hoses that exceed this may experience handling and flow rate problems.
3. Cooling water can be applied to any tank from two directions. This allows access to the tank regardless of wind direction.
4. The hydrant is located more than 50 ft from the tank.
5. Spaced at intervals not exceeding 300 ft in the tank field.

6. Located within 100 ft of any foam lateral run to the road for connection to mobile foam generation trucks.

9.1.3.7 Fixed-roof tanks

Fires in fixed roofs are infrequent; however, the consequences of such a fire can be significant. Because of the large amounts of foam, equipment, and personnel required to fight a fixed-roof tank fire, it may be better to rely on local agencies and private firms for those occasions when a fire does occur. Consideration should be given to installing fixed foam systems on fixed-roof tanks containing class 1 liquids where the quantities stored are large and the liabilities considerable.

Fixed foam equipment installed on fixed-roof tanks may not work as planned because fixed-roof tank fires often result in a separation of the roof-to-shell junction. Part of the roof may fall into the liquid, requiring large amounts of foam applied by special techniques to cover the entire surface. Fighting a fire in this case involves attempting to cover the liquid surface with foam, which can be very difficult because the collapsed roof and supporting structures interfere with the free flow of foam over the liquid surface. The local jurisdiction will often govern the specific requirements for fixed equipment on fixed-roof tanks.

In order to keep the vapor space outside of the flammable range, the practice that can virtually eliminate the chance of a fixed-roof fire is not to store liquids within 20°F of their flash points. This is the best insurance against having a vapor space fire or explosion. Other provisions to reduce the likelihood of fires and explosions in the vapor space include the use of pressure vacuum valves, emergency vents, and compliance with API Standard 2000.

9.1.3.8 Floating-roof tanks

A sunken roof is necessary for a fully involved fire to occur in a floating-roof tank. Since a pan-type roof is subject to sinking due to a hole in the roof caused by corrosion or tipping as the result of foam or fire water application, it should not be used if the risk of a fully involved fire is to be minimized. A more appropriate type of roof to use for tanks containing flammable liquids is the pontoon type or the double-deck roof. Pontoon roofs should provide sufficient stability against tipping and sinking. Double-deck roofs are more resistant to sinking and are usually the economical as well as structural roof of choice in tank sizes over 170 to 200 ft in diameter.

Some external floating-roof tanks use foam dams at least 6 in above the top of the secondary seal. See Fig. 9.1.1. The foam dams allow effective application of the foam against the seal during a rim seal

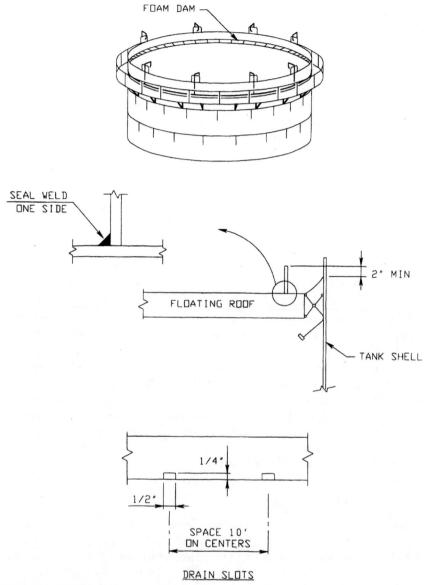

FOAM DAM

SEAL WELD
ONE SIDE

2" MIN

FLOATING ROOF

TANK SHELL

1/4"

1/2"

SPACE 10'
ON CENTERS

DRAIN SLOTS

Figure 9.1.1 Foam dam.

fire by keeping the foam against the seal where the fire can be extinguished. Drainage slots for the foam dam should be designed in accordance with NFPA 11, Appendix A, to cover the maximum rainfall conditions.

Sometimes flammable gases or liquids accumulate in pontoon spaces. This can be extremely dangerous if they are ignited, as the pontoon could explode. Pontoons should be checked regularly to ensure that leaks are not occurring and the vapor is less than 60 percent of the lower flammable limit. Pontoon spaces should be kept tightly closed with bolted-shut manway covers and gooseneck vents at least 18 in above the top of the pontoon.

9.1.3.9 Piping

Carbon steel piping, valves, and fittings should be used between the tank and the first valve for any tanks storing class 1 or 2 liquids. Cast iron, ductile iron, brass, and aluminum will fail when a fire occurs. For remotely located tanks, such as producing operations, with capacities less than 300 bbl this recommendation does not have to be followed.

Water draw, filling, and withdrawing tank nozzles should not be located under stairs. A fire that emanates from these could block access where fire water and equipment are needed.

Tank withdrawal lines will allow the tank to be pumped out in the event of a fire. This reduces the amount of fuel and the burn time for fires that cannot be controlled. For large lines with over 12-in nominal size, the valves should be fireproofed and fail to close on loss of power to the valve actuator. If the pump is located close to the tank, it may not be accessible in the event of a fire and could release additional flammables in the event of a failure. Pumps should not be located in the local impoundment basin at a tank storing class 1, 2, or 3A liquids.

The water draw manifold is an area of higher risk than other tank piping and should be located at least 15 ft from the fill and suction nozzles, which could become involved in a fire as a result of a fire at the manifold.

By connecting external heater circulation piping near the tank fill and withdrawal piping, all the valves in one area are located together for quicker isolation in the event of a fire.

Failure of piping because of settlement can be reduced by designing sufficient flexibility into the piping. This is preferred to providing flexibility with expansion joints and couplings, as these devices can fail unexpectedly in a fire because they may have elastomeric seals and components that will be destroyed by fire.

Because sampling piping has a small diameter and is subject to relatively easy breakage or failure in a fire, the sample piping should include root valves which are in locations that are easily accessible. In the event of a fire involving external sampling lines, the root valve will allow for blocking the tank away from the sample manifold piping.

9.1.3.10 Closed vapor recovery systems

The use of activated carbon on tank vapor spaces is becoming more common as the environmental effort to clean up the air escalates. Any tank vent lines that feed into activated carbon units should have a detonation arrestor just upstream of the activated carbon unit. It should be located less than 10 ft from the activated carbon unit. All systems that have activated carbon units should be evaluated for the need for modifications to prevent the possibility of an ignition source's occurring in the activated carbon units.

Several types of activated carbon are used for emission control. Basically the applications can be categorized as vapor or liquid service. For tank venting lines we consider only the vapor service units. The activated carbon used in the vapor service units is made from coconut shells and is categorized by whether chemicals are impregnated into the carbon for better control of specific compounds. The activated carbon without the addition of chemical reagents is called *nonchemically activated* carbon. The other kind uses a chemically impregnated activated carbon to remove various specific compounds such as H_2S that cannot be removed by the nonchemically activated carbon units. The selection of the carbon type is usually recommended by the carbon vendor.

Generally, the risk of generating high temperatures in chemically impregnated activated carbon units is much higher than with the nonchemically impregnated units. However, it is still possible to generate high temperatures in the straight activated carbon units where easily oxidized compounds such as ketones or chlorine or ozone gases are present. The temperatures are typically generated during a reverse flow condition where outside air comes in contact with the activated carbon laden with trapped vapors. Since the exact composition of the vapor stream is unknown or variable and the rate of oxidation that can occur on the activated carbon is often unpredictable, we have concluded that all vapor lines with flammable gases should be separated from the activated carbon unit by a detonation arrestor. It is important that a detonation arrestor and not a flame arrestor be used, because of the configuration of the system.

9.1.3.11 Area classification

NFPA 70, the National Electric Code, classifies areas according to the potential for flammable vapors to be present and uses this classification to control the electric ignition source potentials. These requirements should be reviewed when one is designing or evaluating a tank for fire prevention.

9.1.4   Fire-Fighting Principles and Practices

9.1.4.1   Fundamentals

The basic principles of fire protection are as follows:

1. Facilities are designed, constructed, maintained, and operated in a manner consistent with industry codes, consensus standards, and good practices.
2. Fire protection programs are kept up to date, maintained, and tested through periodic audits to ensure effectiveness.
3. Fire protection is an integral part of the design, maintenance, and operation of a facility

9.1.4.2   Shell cooling water

Cooling of the shell is effective for maintaining the temperatures within limits that will not cause the steel to collapse. However, water should not be wasted by spraying excess water on the shell because the dikes can fill up, carrying and spreading flames, and it is not effective. A good indicator for shell cooling by water is to cover only those areas that are blistering or discoloring the paint. Generated steam indicates effective cooling. Any water running down the shell is relatively ineffective. It is important not to deluge pan-type roofs with water, or else these will sink, resulting in a fully involved fire.

It should be cautioned that when fires have been burning for long periods and large sections of the tank shell are hot, this poses a special problem. When large quantities of fire water deluge these areas it is possible for the quenched section to buckle and collapse through thermal contraction. The solution is to start shell cooling early; in the event it is not possible to keep the rate of cooling as low as possible, slowly increase it as the temperature of the shell comes down.

9.1.4.3   Water and foam hoses

Foam is generally applied with a 1½-in hand nozzle. A good procedure is to lay a 2½-in hose from the nearest hydrant to the base of the tank and split into two hoses by a Y fitted with two 1½-in hose couplers. One of the hoses can provide shell cooling, and the other can be used for foam making.

9.1.4.4   Rim seal fires

Lightning is the most common source of ignitions causing rim seal fires. Sometimes the vapor space between the liquid surface and the seal explodes, exposing a portion of the annulus between the rim and the shell to the atmosphere, where combustion may continue indefinitely. Secondary seals make these types of fires much less probable. The first secondary seals were installed on tanks in the 1970s, and there is no experience with fires on tanks with dual seals.

Oil should generally not be removed from a tank that has a rim seal fire unless other attempts to extinguish the fire have failed. The lowering of the roof exposes more of the shell to the heat and makes access to the fire more difficult.

9.1.4.5 Special cases

Vent fires. Flammable vapors that are discharging from tank vents during filling operations, during rising ambient temperatures in hot weather, or when subject to heat from a local fire can be extinguished by shutting off the flow to the tank, withdrawing liquid from the tank, covering the vent, or cooling the tank with water.

Ground fires. Fires that occur near tanks due to spills require, as a first step, that the flow to the area be stopped. If the fire is being fed from piping, the valves should be closed to limit this flow. Water can be used for personnel protection. If it is not possible to close off the valve, flow can sometimes be stopped by other means. For example, raising a swing line above the liquid level can stop flow from a nozzle connected to the swing line. Adding enough water to cause the water level to be above the leaking valve or nozzle is often effective. Tanks, equipment, and piping should be water-cooled by hose streams for severe fires where the intensity of radiation is sufficient to rupture or cause release of product.

Refined oil fires. These products do not have boilover potential. Foam can be applied to the liquid surface by water cannons or crane-mounted nozzles.

For oils with high flash points such as kerosene and heavier oils, ignition of the liquid surface is difficult. Foam applied to the surface will easily quench such a fire. For viscous oils, water sprayed in a mist will cool the surface by frothing slightly and covering the surface with cool bubbles. Another method for extinguishing heavy refined oil fires is to agitate the tank with injected air. As soon as cooler stock comes to the surface, the temperature at the liquid surface can drop to below the flash point and the flames self extinguish.

For light oils such as gasolines, fire can usually only be suppressed by covering the liquid surface completely with foam. Air agitation has no effect because the flash point is well below any possible stock temperature. The addition of air would actually increase the intensity of the flames.

Black crude, asphalt, and heavy fuel oil fires. Water or foam applied to these fires, where the product is stored above about 250°F, is likely to froth over and cause a spread of fire outside the tank. When the product storage temperature is below 250°F, any foam that is applied within about a $\frac{1}{2}$ h or before the liquid surface has heated up significantly will not cause a frothing. The more time that passes between the start of the fire and the application of foam or water increases the difficulty of fire fighting. These tanks should be pumped out as rapidly as possible, and application of foam or water delayed until the level is at least 5 ft below the top of the tank shell. If the tank contains crude oil, the foam or water should be applied right away.

Pyrophoric fires. Asphalt tanks containing asphalt cements, cutbacks, emulsions, and no. 6 fuel oil may contain pyrophoric iron sulfide. When these tanks are emptied for repair or cleaning, the heating coils should be covered at least several feet and the contents heated to 250 to 300°F as well as circulated. The resulting material can then be pumped out as much as possible, and the tank cooled for cold removal of asphalt. The interior should be checked for flammable vapors, H_2S, and oxygen. If the H_2S level is high and the oxygen level low, assume that pyrophoric iron sulfide exists in the tank and follow the necessary precautions and procedures. If flammable vapor levels are above 10 percent of the lower explosive limit (LFL), then the tank should be steamed out. Steaming the asphalt tanks also keeps the pyrophoric iron sulfide wet so that it remains cool. Steam hoses must be bonded to ground or the tank because static electricity is generated as the steam ejects from the nozzle. Asphalt tanks also contain sludge which tends to release flammable vapors on being disturbed. Special precautions must be followed.

Sour crude oil tanks or tanks with high levels of dissolved hydrogen sulfide or sulfur compounds are subject to pyrophoric iron sulfide buildup and should be treated accordingly.

Hot tanks. Storage of materials above the boiling point of water poses the risk of large vapor releases. Hot tanks are subject to fires and explosions for the following reasons:

- Water entering a hot tank can evaporate quickly, causing a frothover of flammables or causing large quantities of vapor to be released. If the release rate is great enough, the tank can be damaged, allowing the escape of flammable vapors.

- Water on the tank bottom, when exposed to hot entering stock, can suddenly boil and cause the generation of flammable vapors or froth.

Heated tanks using steam or hot water have an increased risk of incidents due to the possibility of water entering the stock during a leak.

Froth-overs. Froth-overs or slop-overs occur when water enters a hot tank. This can result in minor or major releases. Some problems have arisen because water unexpectedly entered a tank. Where hot hydrocarbon streams enter a tank after being cooled by shell-and-tube heat exchangers using cooling water, it has sometimes happened that a leak or tube failure resulted in water getting into the hot stream and resulting in a froth-over once in the tank. This type of incident can be prevented by the design, by ensuring that cooling water pressures are lower than hot process streams and the hot process leaks into the cooling water, rather than the other way round.

Boilovers. See Fig. 9.1.2. Boilovers are probably the most frightening concept involving the destructive potential of fully involved tank fires. Boilovers start as relatively contained surface fires but suddenly start releasing large quantities of burning fuel over the tank walls. Boilovers can only happen in the specific combination of storing crude oil in a fixed-roof tank of significant size. Today it is rare to see anyone storing crude oil in other than a fixed-roof tank except for smaller tanks.

The conditions that allow a boilover are limited. The interesting process that leads to the event known as a boilover occurs when there is a fully involved fire in an open-top tank storing crude oil. An open-top tank in this context might be the result of a sunken roof or the loss of a roof on a fixed-roof tank due to an internal fire or explosion. As the liquid surface heats up, a hot layer of crude oil is created. The more volatile components of the crude feed the fire, the less volatile components remain behind, and the density of the remaining liquid near the surface increases as its temperature increases. The surface layer, usually several feet thick, may reach temperatures as high as 600 to 800°F. When the density is sufficiently high, it slowly starts to sink at a rate of several feet per hour as a hot layer, often called a *heat wave.* Once the hot layer approaches the bottom, any water in the form of a bottom heel or emulsion becomes superheated and soon starts to boil. This boiling generates a steam-hydrocarbon foam that causes the liquid level to rise, resulting in loss of burning material outside of the tank. This in turn reduces the pressure head on the superheated water, and the process becomes progressively more rapid

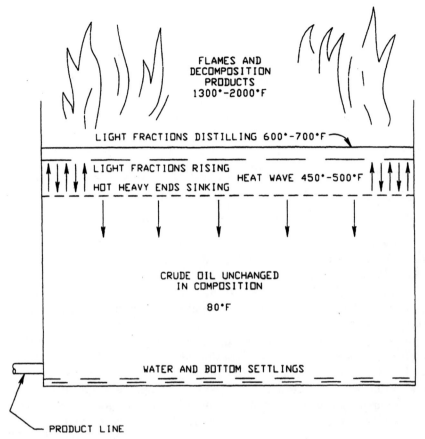

Figure 9.1.2   Boilover process in fuel oil tank fire.

and violent, resulting in the sudden release of large quantities of burning crude. The total boilover volume exceeds by many times the volume of the tank and could even exceed the capacity of the secondary containment areas and flow out over them.

If there is a possibility of a boilover during a fire, an attempt should be made early on to determine the location and rate of movement of the heat wave by use of an optical pyrometer. After this determination, an estimate of the time available for local fire-fighting efforts can be made before an evacuation must be ordered. Boilovers can be anticipated within a few minutes if there is a sudden brightening and increase of flames.

Storage of oxidizers.   Oxidizing gases should be stored at least 25 ft from petroleum gases and liquids. An impermeable dike wall at least 5 ft high may be used for the separation of fuel and oxidizers.

SECTION 9.2   VENTING

9.2.1   Causes of Tank Venting

This section covers the venting requirements for both normal and emergency conditions for nonrefrigerated fixed-roof tanks. In designing, operating, or maintaining storage tanks, consideration must be given to the concept of venting to relieve excessive buildup of internal pressures or vacuum. Either condition can lead to serious damage to the tank. The basic concepts for venting a fixed-roof storage tank include

- Liquid level changes that cause inbreathing and outbreathing of vapors
- Thermal effects that cause inbreathing and outbreathing of vapors
- Outbreathing of vapors as a result of a fire in or near the tank

It should be borne in mind that venting requirements apply equally to flammable and nonflammable liquid storage. For flammable liquids, the various industrial and federal codes generally require compliance with API Standard 2000. (See Sec. 9.2.5.) However, the same principles should be applied to the venting of materials that are nonflammable in order to prevent damage to the tank. Table 9.2.1 summarizes the computation for liquid displacement venting requirements from API 2000. Even if the liquid level does not change, there will be the need to vent a storage tank for both thermal inbreathing and outbreathing. Thermal inbreathing and outbreathing are caused by a change in the temperature of the vapor in the tank. The temperature of the vapor in the tank can be affected by several factors:

- Heat gain by direct radiation of the sun during the day
- Radiation losses during the night

TABLE 9.2.1   Venting Required for Liquid Displacement

Determine the capacity needed due to the maximum pump rate
by the following table:

If the pump-out rate is in:	Multiply by:	To obtain
U.S. gpm	8.0210	scfh air required
U.S. gph	0.1337	scfh air required
bbl/h	5.6150	scfh air required
bbl/day	0.2340	scfh air required

The values computed in the table represent the venting required that is equivalent to the volume displacement caused by liquid flow into or out of a tank. These values may be used for both pump-in and pump-out rates for any liquid with a flash point above 100°F.

Important: For liquids with flash points above 100°F, the venting requirements for exhaling (the venting caused by pumping into a tank) should be doubled. The inbreathing requirements for these liquids are per the table.

- Convective heat gain by an ambient air temperature that is greater than the tank vapor temperature

- Convective heat loss by an ambient air temperature that is less than the tank vapor temperature

- Convective heat gain or loss due to the temperature difference between the tank vapor and the stored liquid temperature

- Other effects such as the quenching of internal temperature caused by rainfall

It is not necessary to attempt to calculate the thermal breathing requirements from first principles since these requirements have been standardized by API 2000. Table 9.2.2 provides the required thermal venting capacity in cubic feet per hour versus tank size. For example,

TABLE 9.2.2 Requirements for Thermal Venting Capacity

		Thermal venting capacity, ft³/h of free air*		
			Outbreathing (pressure)	
Tank capacity		Inbreathing	Flash point ≥100°F	Flash point <100°F
bbl	gal	(vacuum)	(37.78°C)	(37.78°C)
60	2,500	60	40	60
100	4,200	100	60	100
500	21,000	500	300	500
1,000	42,000	1,000	600	1,000
2,000	84,000	2,000	1,200	2,000
3,000	126,000	3,000	1,800	3,000
4,000	168,000	4,000	2,400	4,000
5,000	210,000	5,000	3,000	5,000
10,000	420,000	10,000	6,000	10,000
15,000	630,000	15,000	9,000	15,000
20,000	840,000	20,000	12,000	20,000
25,000	1,050,000	24,000	15,000	24,000
30,000	1,260,000	28,000	17,000	28,000
35,000	1,470,000	31,000	19,000	31,000
40,000	1,680,000	34,000	21,000	34,000
45,000	1,890,000	37,000	23,000	37,000
50,000	2,100,000	40,000	24,000	40,000
60,000	2,520,000	44,000	27,000	44,000
70,000	2,940,000	48,000	29,000	48,000
80,000	3,360,000	52,000	31,000	52,000
90,000	3,780,000	56,000	34,000	56,000
100,000	4,200,000	60,000	36,000	60,000
120,000	5,040,000	68,000	41,000	68,000
140,000	5,880,000	75,000	45,000	75,000
160,000	6,720,000	82,000	50,000	82,000
180,000	7,560,000	90,000	54,000	90,000

*At 14.7 psia (1.014 bar) and 60°F (15.56°C).

for a 100,000-bbl tank (4.2 million gal) the required inbreathing rate is 60,000 ft³/h, and the outbreathing rate for liquids with a flash point greater than 100°F is 36,000 ft³/h. The inbreathing rates are higher than the outbreathing rates because cooling can occur very quickly in tanks with warm stock and vapor spaces that are quenched by a sudden cool rainfall. For volatile stocks with flash points less than 100°F, the outbreathing rate is set equal to the inbreathing rate because of the possibility of stock evaporation, as discussed earlier.

9.2.1.1 Emergency venting

When tanks are exposed to fire, the vapor that must be released from the tank can exceed the flow rate of vented vapor that is required for normal conditions. Consideration must be given to vent this vapor, and it is termed *emergency venting.* The typical means of venting for fire conditions are the use of a frangible roof construction and the use of additional or larger emergency venting valves than required for normal venting.

Table 9.2.3 provides the required emergency venting rate. It is based upon the area of the tank exposed to fire and a heat input rate that is empirically based. The required venting may be reduced by 50 percent when the drainage is away from the tank. This is because pools of burning liquids will not form at the base of the tank, causing the amount of heat input to be reduced. Other reduction factors are included for insulated tanks because the heat flux into the tank is reduced by the insulation.

9.2.2 Venting Floating-Roof Tanks

Since operating floating-roof tanks have no vapor space below the roof, there is normally no need for venting thermal breathing, filling and emptying losses. Venting does occur on initial filling until the roof floats, however. The space between the floating roof and the shell is called the *rim space,* and this volume is relatively small. However, rim space vents are installed to allow this volume to breathe.

Another problem that can develop with floating-roof tanks is boiling or vaporization under the roof for stocks with highly volatile components. If the roof is large and flexible, it will bulge into a spherical shape, trapping the vapors under the tank. This is not desirable because fatigue of the welds is possible, roof drainage is impaired, and there is always the possibility of damaging the seals. Pan roofs and single-deck roofs are subject to this type of problem while the more rigid double-deck roof is less likely to distort. The vapor bleeds out from the periphery of the tank and through the seals. The solution to these problems is to select the proper type of tank or roof. In

TABLE 9.2.3   Emergency Venting Requirements

Wetted area,* ft^2	Venting requirement, ft^3/h free air†	Wetted area,* ft^2	Venting requirement, ft^3/h free air†
20	21,100	350	288,000
30	31,600	400	312,000
40	42,100	500	354,000
50	52,700	600	392,000
60	63,200	700	428,000
70	73,700	800	462,000
80	84,200	900	493,000
90	94,800	1000	524,000
100	105,000	1200	557,000
120	126,000	1400	587,000
140	147,000	1600	614,000
160	168,000	1800	639,000
180	190,000	2000	662,000
200	211,000	2400	704,000
250	239,000	2800	742,000
300	265,000	>2800	—

Note: Interpolate for intermediate values. The total surface area does not include the area of ground plates but does include roof areas less than 30 ft above grade.
*The wetted area of the tank or storage vessel shall be calculated as follows: For spheres and spheroids, the wetted area is equal to 55 percent of the total surface area or the surface area to a height of 30 ft (9.14 m), whichever is greater. For horizontal tanks, the wetted area is equal to 75 percent of the total surface area. For vertical tanks, the wetted area is equal to the total surface area of the shell within a maximum height of 30 ft (9.14 m) above grade.
†At 14.7 psia (1.014 bar) and 60°F (15.56°C).

Datum	F factor
Bare-metal vessel	1.0
Insulation thickness	
6 in (152 mm)	0.05
8 in (203 mm)	0.037
10 in (254 mm)	0.03
12 in (305 mm) or more	0.025
Concrete thickness	—
Water application facilities	1.0
Depressurizing and emptying facilities	1.0
Underground storage	0
Earth-covered storage above grade	0.03

some circumstances a PV valve can be used to bleed off the gassing that occurs under a floating roof. See Sec. 9.9.4 below for this application. EPA regulations prohibit storage of liquids with a true vapor pressure exceeding 11 psia. This should provide a margin of 3.7 psia against boiling. However, the rate of vapor pressure increases exponentially with temperature, and a few degrees' increase in stock temperature or the warming of the upper strata of the storage vessel by the roof can cause the vapor pressure to exceed 14.7 psia—that required for atmospheric boiling.

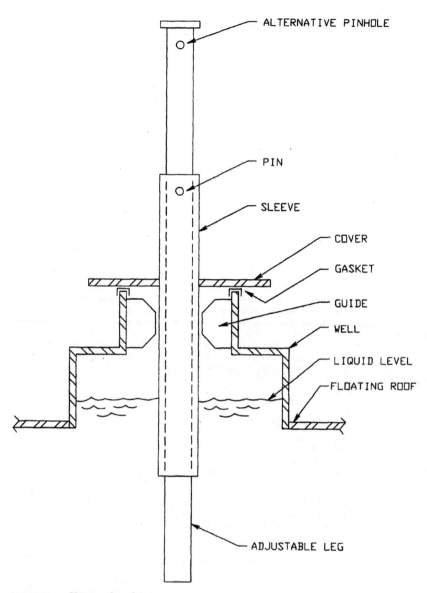

Figure 9.2.1 Vacuum breakers.

The conventional manner of handling filling losses until the roof is able to float is to use landing actuated vent valves. Figure 9.2.1 shows the details. These devices operate when the roof leg touches the bottom. Some local regulations restrict open vents on floating-roof tanks that have landed roofs. This rules out the mechanical landing-actuated vent valves discussed above. Instead, one solution has been to use pressure vacuum (PV) valves on floating roofs.

9.2.3 Design Guidance

1. The provisions of API 2000 for determining venting requirements for fixed-roof tanks should be applied to floating-roof tanks using the landed roof condition. These are the same as the NFPA 30 requirements and OSHA requirements and will therefore satisfy legal requirements. Apply the principles of venting to both flammable and nonflammable liquids.

2. Determine the true vapor pressure of liquid to be stored and tank size to assess whether the local, state, or federal regulations limit the design to a particular tank configuration. Refer to Table 9.2.4.

3. Store liquids in floating-roof tanks only if the true vapor pressure is 11 psia or less. For liquids above this vapor pressure, there are special engineering considerations such as storage in pressure vessels or the use of fixed roof tanks with vapor recovery systems.

4. Double-deck roofs reduce boiling losses and handle vapor buildup under the roof better than single-deck roofs. However, they usually cost more.

9.2.4 Hardware

To relieve excess pressure from a tank or to allow inflow of air or some other gas into a tank to maintain a relatively constant pressure near atmospheric pressure, many methods have been used. The oldest method simply uses an open-top tank. The problem with an open-top tank is that rain, snow, and atmospheric contamination such as dust get into the tank. Open-top tanks are also very rare because the fluid usually evaporates and results in air pollution and product loss.

9.2.4.1 Open vents

The next-simplest method to vent a fixed-roof tank is to use an open vent. The "gooseneck" design is simply a pipe installed on the roof of the tank that is turned down to keep out rain and snow. A screen can be welded to the end so that birds and insects do not enter.

9.2.4.2 PV valves

General. Pressure vacuum vent valves are often referred to as *breather valves* or *conservation valves*. The latter term came into use when it was realized that evaporation losses were substantially reduced by using PV valves in lieu of open vents. See Fig. 9.2.2. PV valves are the workhorse of the industry. They have a number of very useful characteristics that have made them standard apparatus on storage tanks:

TABLE 9.2.4 Tank Capacity and Vapor Pressure Cutoffs

Regulation	Source category	Existing tanks				New tanks			
		0< <10,000 gal	10,000≤ <20,000 gal	20,000≤ <40,000 gal	40,000 gal ≤	0< <10,000 gal	10,000≤ <20,000 gal	20,000≤ <40,000 gal	40,000 gal ≤
40CFR60 Kb (NSPS)*†	VOL storage	—	—	—	—	—	—	4.0 psia	0.75 psia
40CFR61 Y†	Benzene storage	—	No cutoff	No cutoff	No cutoff	—	No cutoff	No cutoff	No cutoff
HONeshaps*†	Socmi	—	—	1.9 psia	0.75 psia	—	1.9 psia	1.9 psia	0.1 psia
Baaqmd*§									
Reg. 8, Rule 5	VOL storage	—	—	1.5 psia	0.5 psia	—	—	1.5 psia	0.5 psia

*Closed vent system and control device (capture system) are required if the true vapor pressure is greater than or equal to 11.1 psia.
†This regulation does not apply to pressure vessels designed to operate in excess of 29.7 psia and without emissions to the atmosphere.
§Tanks smaller than 20,000 gal must either meet the control requirements set forth for larger tanks or have a P/V valve.
SOURCE: Myers and Ferry, *Oil & Gas Journal*, June 7, 1993.

SERIES NO. 8540H

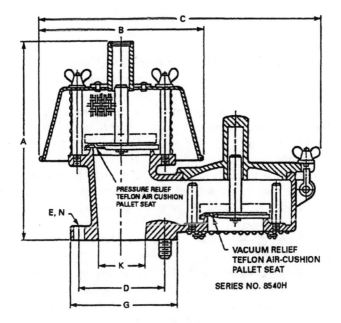

Side View

PRESSURE RELIEF
TEFLON AIR CUSHION
PALLET SEAT

VACUUM RELIEF
TEFLON AIR-CUSHION
PALLET SEAT

SERIES NO. 8540H

MATERIALS OF CONSTRUCTION

Series	Housing	Pallets*	Pallet Diaphragm
8540H	Aluminum 356	Aluminum	Teflon*
C8540H	Ductile Iron A536316	Stainless Steel	Teflon*
F8540H	316 Stainless Steel	316 Stainless Steel	Teflon*
RE8540H	Aluminum 356	316 Stainless Steel	Teflon*

*On Ductile Iron and Aluminum vents, weights are Steel or Lead. On Stainless Steel vents, weights are Stainless Steel or Lead.
Aluminum flanged to mate with 125 # FF A.N.S.I. Ductile Iron and Stainless Steel flanged to mate with 150 # RF A.N.S.I.

Figure 9.2.2 Typical PV valves. (*Courtesy of the Protectoseal Company.*)

1. They protect tanks against over- and underpressure.

2. They reduce evaporation losses compared to open vents.

3. They can double as flame arrestors, and they eliminate the need for flame arrestors in some cases.

4. Because the atmospheric oxygen concentration is apt to be lower in a tank using a PV valve than in a tank with open venting, the internal corrosion in the vapor space will often be reduced.

5. PV valves are generally required by EPA, OSHA, NFPA, etc.

API Bulleting 2521 spells out the basic operations as well as design and requirements for PV valves. This bulletin addresses all types of PV valves including

1. Solid pallet (hard pallet to hard seat)

2. Diaphragm pallet

3. Liquid seal valve

How the PV valve works.   By far, the most common type of PV valve is the solid pallet type. Figure 9.2.3 shows a schematic diagram of this valve. It can be seen that the PV valve is really two valves in one. One is for pressure, and the other is for vacuum. The principle of operation is the same, however.

As the pressure on the pressure side of a PV valve rises, the force due to pressure reduces the seating force of the pallet, and it starts to leak. At least one manufacturer specifies that the leakage rate is certified to be less than 1.0 scfh at 90 percent of the set point pressure. Leakage, however, is relatively insignificant until the set point is reached at which point the flow increases dramatically and follows the flow curves, given by the manufacturer. Once the set point pressure is reached, defined as the pressure at which the valve first cracks open, the pallets will flutter or chatter on the seat in an unstable state. Only when the pressure has increased beyond the unstable operating regime will the PV valve operate in stable equilibrium. Operation in the unstable regime is not advisable because the rapid flutter of the pallet will tend to fatigue the seat materials or may cause premature wear.

Beyond the set point, PV valves do not "pop" open, but slowly lift as the overpressure (the actual upstream pressure above the value of the set point) increases. Between the set point and the maximum open position, the PV valve is essentially a variable-area orifice as the curtain area (cylindrical area between the seat and pallet) increases. When it is fully open, the flow may still increase with increasing pres-

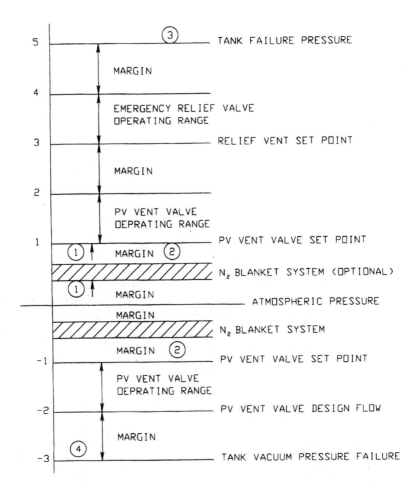

① THE HIGHER THIS VALUE IS THE LESS WILL BE THE EVAPORATIVE EMISSION LOSSES.

② IF THE MARGIN IS TOO SMALL SIGNIFICANT LOSS OF BLANKETING GAS MAY RESULT.

③ SEE API 650 APPENDIX F FOR GUIDANCE.

④ THIS MAY BE ASSUMED TO BE 1051 WITHOUT CALCULATIONS. FOR HIGHER VACUUMS, USE DETAILED ENGINEERING ANALYSIS TO DETERMINE WHAT VACUUM MAY BE USED.

Figure 9.2.3 Operating characteristics of PV valving.

sure; but since the curtain area is now fixed, the flow follows the pressure drop and flow characteristics of a fixed orifice.

Overpressure is an inherent factor in the use of PV valves. The amount of overpressure depends on the manufacturer and the specific conditions. Depending on the PV valve manufacturer, an overpressure may vary from 10 to 120 percent for fully open PV valves and provide the design flow for this condition. Some manufacturers advise a minimum overpressure to prevent operating in the unstable regime. Some manufacturers request that the overpressure be on the order of 80 to 100 percent. A high overpressure is an obvious disadvantage because it broadens the band of pressure that must be allocated to normal venting. Since a tank does not have a lot of bandwidth for pressure, as shown by Fig. 9.2.3, a device with a lower overpressure is more desirable. A narrow operating pressure range becomes particularly more important for systems that have inert gas blanketing or large tanks with shallow roof angles that have a very low failure pressure. Manufacturers' flow curves and recommendations should be consulted for sizing these valves.

The problems with sufficient margin to allow vents to operate within the design pressure of the tank become more acute for

- Large tank diameters. Smaller tanks can frequently take the higher pressures without the need for special design consideration whereas large tanks will be damaged if the internal pressure exceeds the design pressure. Large tanks, however, will often be damaged by overpressure as soon as the internal pressure exceeds the weight of the roof plates (for example, $1\frac{1}{2}$ in WC for $\frac{3}{16}$-in steel plate).

- Tanks with manifolded vent systems

- Tanks that have inert gas blankets

Emergency vent valves. Emergency vent valves are simply large PV valves capable of venting the greater-than-normal venting loads caused by emergency conditions. The excessive pressure caused by a fire near the tank creates a heat flux into the liquid space in the tank. The walls of the tank cause the liquid to boil and generate vapor. Emergency venting may be accomplished by

1. Larger or additional open vents

2. Larger or additional PV vents or pressure relief valves

3. A gauge hatch which permits the cover to lift under abnormal internal pressure

4. A manhole cover which permits the cover to lift under abnormal internal pressure

5. A connection between the roof and shell which is frangible

Design guidance for hardware

1. Do not use flame arrestors unless required by regulations or for special situations. Many tank failures have been caused by plugged

flame arrestors. Instead use PV valves, which in most cases act as flame arrestors. Your insurer or fire protection engineer should review each situation and should have the final say on this matter.

2. Use the standard gravity-operated PV valve for most cases. Spring-loaded pallets are often required where there are higher set point pressures (for example, 1 psi and above). Size the PV valve for 80 to 100 percent overpressure. Make sure the soft components of the valve, such as the seats, are compatible with stored fluids and vapors. Be sure to specify operating temperatures, as these valves may require special provisions for below-freezing temperatures.

3. Do not use pilot-operated vent valves unless it is necessary, as these can malfunction owing to the more intricate parts and mechanical functions of these devices. These valves often require pressurized air for operation and are therefore subject to air failure problems. Select a stand-alone pilot-operated vent that does not require plant utilities. However, as pressures increase above a few inches of water column, usually a pilot-operated valve is needed.

9.2.5   Codes and Standards Governing Tank Venting

9.2.5.1   API 2000, Venting Atmospheric and Low Pressure Storage Tanks

The most widely used industry standard for emergency venting is API Standard 2000. It covers tanks designed from $\frac{1}{2}$ ounce per square inch (osi) to 15 psig. The standard provides formulas and tables that standardize the computation of required normal and emergency venting flow rates, so that venting valves can be sized. Although the standard does not apply to floating-roof tanks, it is advisable to use the principles for the floating-roof tank condition where the roof is landed. This code does not address detonations and deflagrations that occur in the vapor space of the tank.

9.2.5.2   NFPA 30, Flammable and Combustible Liquids Code

NFPA 30 is the code upon which other legal codes such as OSHA and the Uniform Fire Code are based. Like API 2000, NFPA 30 requires that a tank not be damaged as a result of pressure or vacuum for normal or emergency venting. It allows sizing of vents in accordance with API 2000 or some other accepted standard. It also allows the venting size to be uncalculated if it is at least as large as the filling or withdrawal connection, whichever is larger, but in no case less than $1\frac{1}{4}$-in nominal inside diameter. The allowance of computing by "some other accepted standard" was included to cover special cases such as off-gassing when crude oil is delivered from the oil field or extreme temperature changes.

9.2.5.3 Uniform Fire Code

The requirements of this code are similar to the NFPA 30 requirements.

SECTION 9.3   FRANGIBLE ROOFS

9.3.1   What Is a Frangible Roof?

Frangible is a word that means easily broken. However, in the context of tanks, the word has a very specific meaning and is defined in API Standard 650. The concept of a frangible roof only applies to flat-bottom, cone-roof tanks with a limited roof apex angle. A frangible roof is a roof-to-shell joint or junction that is weaker than the rest of the tank and will preferentially fail if the tank is overpressurized. Since this junction will fail before any other part of the tank, such as the shell, the bottom, or the shell-to-bottom joint, the bottom and shell can be relied upon to be intact. Since failure at the roof is the least damaging mode of failure for a liquid storage tank, liquid chemicals or stored product will probably not be released.

9.3.2   Why Are Frangible Roof Joints Desirable?

Figure 9.3.1 shows a tank that was overpressurized and did not have a frangible roof. Frangible roofs have been preferred by many companies as the means of choice for emergency venting for several reasons. Most obvious is that with a frangible roof, there is no cost associated with emergency venting devices, since they are obviated by the frangible roof. However, when emergency venting occurs, the tank will suffer damage in the roof-to-shell region. Typically, this area buckles into a wave-shaped pattern, and the roof tears away from the top angle rim, allowing gases to escape. It could be argued that the damage to the tank is more severe than would occur if release of the gases occurred through an emergency venting device. However, when an emergency venting is required, it might be because the entire tank is engulfed in flames, which would also have the potential to damage the tank material.

Another reason that frangible roofs are preferred is that they have the capability to relieve overpressure due to explosions, detonations, and deflagrations. Once the roof-to-shell junction starts to tear, it can

Figure 9.3.1 Explosion in tank without frangible roof. (*Courtesy of the National Fire Protection Association.*)

open as required to relieve any quantity of gas that can be generated by such events. Experience shows that frangible roofs have maintained the integrity of the tank shell and bottom during the following types of events:

- Emergency venting
- Explosions and deflagrations

9.3.3   What Is the Criterion for Frangibility?

Section 3.10.2.5 of API Standard 650 defines the criteria for frangibility of the roof:

1. The roof must be a cone roof with a slope not exceeding 2-in rise in a 12-in run.

2. The weld joining the roof plates to the top angle cannot exceed $\frac{3}{16}$ in.

3. The cross-sectional area of the compression region at the roof-to-shell junction cannot exceed the area defined by

$$A = 0.153 \frac{W}{30,800 \tan \theta} \qquad (9.3.1)$$

where A = participating area of roof-shell junction (defined by Fig. F-1 of API 650, Appendix F)

W = weight of tank shell and roof framing supported by shell, lb

θ = roof angle, deg

4. The details of the frangible roof must be in accordance with API.

9.3.4 Frangibility Limitations

Although it may be desirable to have frangible roofs in all aboveground storage tanks, this is not possible. It becomes progressively more difficult to meet the criterion stated by the inequality in Eq. (9.3.1) as the tank diameter becomes smaller. Experience also shows that the frangibility of the roof is not reliable at smaller diameters.

To understand the effect of diameter on frangibility, it is useful to review Fig. 9.3.2. The curve is included for explanatory purposes only. In Fig. 9.3.2 note that the height required for frangibility increases dramatically at the smaller tank diameters. Even with the roof slope at a minimum, for tank diameters less than about 30 ft the shell height is impractically high and, therefore, it is not possible to have a

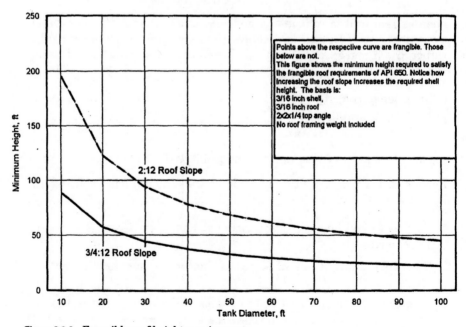

Figure 9.3.2 Frangible roof height requirements.

frangible roof. This figure also shows the effect of increasing the roof slope to the maximum allowed of 2:12. The 2:12 slope is the maximum slope for frangible roof designs and the minimum slope for self-supported roofs. It can be concluded, therefore, that frangible roof designs for self-supported cone-roof tanks are not feasible.

9.3.5 Summary

- Frangible roofs are the means of choice to satisfy emergency venting requirements and are the safest designs where explosions, detonations, and deflagrations are a possibility.

- To achieve a frangible roof design requires meeting the criteria spelled out in API 650.

- Frangible roof slopes cannot exceed 2:12. It is easiest to obtain frangibility with minimally sloped roofs (for example, $3/4$:12).

- Frangibility is not reliable for small tank diameters of less than 30 to 40 ft.

- The API criteria for frangibility are hard to achieve in small tank diameters.

- It is impractical to have self-supported cone-roof tanks that have a frangible roof.

SECTION 9.4   FLAME AND DETONATION ARRESTORS

9.4.1   Introduction

This section covers the principles of flame- and detonation-arresting devices as applied to aboveground storage tanks. The function of a flame or detonation arrestor is to prevent a flame from passing through it, that is, from the unprotected side, where an ignition source may exist, to the protected side. A distinction must be made between a flame arrestor and a detonation arrestor. A flame arrestor is useful only for end-of-line applications or applications where the amount of piping connected to the unprotected side of the device is minimal. Detonation arrestors are made for in-line applications where much higher flame velocities and pressures can occur. There is a tendency to believe that because an agency such as Underwriters' Laboratories, Factory Mutual Research, or the U.S. Coast Guard has tested or listed these devices that they are suitable for any application. This is specifically not the case, and inappropriate application of any of these devices may render them useless for their intended purpose.

9.4.2   Application to Tanks

The use of flame arrestors should, in general, be discouraged if there are alternative means of fire protection at the disposal of the designer. An example of this principle would be to store a class 1 liquid in a floating-roof tank rather than in a fixed-roof tank vented to the atmosphere. This principle is recommended because there is often uncertainty when attempting to match the actual conditions which these devices are expected to operate to the conditions that really occur in the field. If all process conditions were within specified conditions, then the uncertainty of the device might be able to be removed. However, these devices are also sensitive to installation configurations. For example, a flame arrestor meant for use on end-of-line applications that have been installed with even a single elbow when straight pipe is the basis for listing the device by UL or FM might prevent it from performing its function.

The vapor spaces of tanks can be protected from ignition sources in several ways:

- Use of a floating-roof tank where possible for class 1 liquids
- Inerting and blanketing gases that keep the vapor space too rich or too lean to burn
- Active flame-arresting devices that isolate segments of the tank from the process
- Passive devices or flame or detonation arrestor

This section is limited to the use of the flame and detonation arrestor.

On tanks where the venting is to atmosphere, a PV valve is often used with a flame arrestor because the PV valve reduces evaporative emissions and the flame arrestor reduces the chance of ignition sources accessing the vapor space in the tank. In these situations the sizing requirements and the use of a flame arrestor are relatively straightforward. Part of the objection to the use of a flame arrestor is that it must be unbolted and removed for inspection. Newer types that surround the PV valve have been developed which make the inspection job easier.

- Although NFPA 30 discourages manifolded venting of tanks, the implementation of the Clean Air Act has seen an increasing trend of manifolding tank vapor spaces together and routing the vapor to vapor recovery or vapor destruction units. Because vapors are confined within piping and various tanks are connected, there are a number of hazards and precautions to note:
- The most significant of these that is not encountered with a tank that is vented to the atmosphere is the possibility of misapplying a

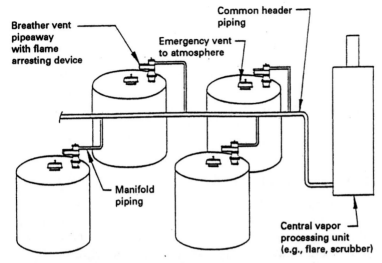

Breather vent pipeaway with flame arresting device

Common header piping

Emergency vent to atmosphere

Manifold piping

Central vapor processing unit (e.g., flare, scrubber)

Figure 9.4.1   Flame arrestor on manifolded tanks.

flame arrestor. In these circumstances a detonation arrestor may be used, but even a detonation arrestor has limitations.

- Manifolding tank vents together creates increased explosion hazards because there is the possibility that the entire manifold system may be within the combustible range and provide an ignition source to any of the connected tanks. To reduce this possibility, class 1 tanks should not be manifolded to tanks with class 2 or 3 liquids unless a low flow of inert gas is vented into each tank with class 2 or 3 liquids.

- Since passive flame arrestor devices are effective only under specific conditions (temperature, pressure, flow, composition, piping configuration), all the variability to which a process subjects the device must fall within its ability to operate as intended.

9.4.3   PV Valves as a Substitute for a Flame Arrestor

There is controversy about whether a PV valve is an effective flame arrestor under all conditions. NFPA 30 (paragraph 2-2.4.6) recognizes either a flame arrestor or a PV valve as an acceptable alternative for fire protection for storing class 1 liquids. API 2210, *Flame Arrestors for Vents of Tanks Storing Petroleum Products,* discourages the use of flame arrestors because they frequently plug and cause excess tank vacuum or pressure to damage the tank. API 2000 states that flame arrestors are not considered necessary if a PV valve is used for class 1 liquids. While PV valves are probably effective in most circumstances in acting as flame arrestors, there is uncertainty as to whether they can arrest a flame under all conditions (e.g., simultaneous occurrence of ignition source just as valve starts to open, stuck-open valve, confined deflagration, and associated pressure or velocity wave). Fortunately, the conditions necessary for a PV valve to malfunction as a flame arrestor occur rarely, and industry experience seems to support this. The use of a flame arrestor implies the need for periodic inspection and maintenance. Because flame arrestors can clog due to dust or process residues, they can ice up under cold-weather conditions and therefore require regular maintenance and inspection.

9.4.4 Regulatory Requirements

NFPA 30, *Flammable and Combustible Liquids Code* (paragraphs 2-3.4.6 and 2-3.4.7), and OSHA 1910.106 are the primary regulatory drivers that control the use of flame arrestors. Since the OSHA flammable liquids standard was based upon NFPA 30 the requirements in the area of flame arrestors and venting are virtually the same. Table 9.4.1 specifies the requirements.

Class 1A liquids that boil below 100°F could give off flammable vapors, and therefore control of these emissions through the use of PV valves is required. Tanks with class 1A liquids have vapor spaces that are too rich to burn. In class 1A or 1B, storage tanks may have vapors in the flammable range so that the flame arrestor is allowed as an alternative to a PV valve.

Small crude tanks are exempt because they are usually located in sparsely populated areas and the vapor space is generally too rich to burn. Tanks smaller than 1000 gal are exempt due to the insignificance of the rate of vapor release, and the vent location requirements reduce the probability of a continuous vapor trail in the flammable regime that could flash back to the tank. The exemption does not apply to class 1A liquids.

It is interesting to note that these standards allow an exemption for the use of flame arrestors or PV valves for class 1B and 1C liquids if the service is such that corrosivity, plugging, fouling, polymerization, etc., could occur. In these services where engineering judgment calls for the use of these devices, frequent inspection and periodic maintenance should be instituted to ensure that the devices are operating properly.

The regulatory standards refer to *listed equipment* and to *authority having jurisdiction* in reference to the acceptability of this type of equipment. More specific information is given in the next sections.

9.4.5 Flame Arrestor Testing

There is a need for comprehensive standards that guide the use of flame-arresting devices for any application within piping because of

TABLE 9.4.1   Regulatory Requirements

Combustible liquid	Requirement
1A	PV valve
1B	PV valve or flame arrestor
1C	PV valve or flame arrestor

Open vents are permitted for

1. Crude tank in producing areas <3000 bbl

2. Tanks <1000 gal (excluding class 1A)

NFPA 30-2-2.4.62   Tanks and pressure vessels storing class 1A liquids shall be equipped with venting devices that shall be normally closed except when venting under pressure or vacuum conditions. Tanks and pressure vessels storing class 1B and 1C liquids shall be equipped with venting devices that shall be normally closed except when venting under pressure or vacuum conditions or with listed flame arrestors. Tanks of 3000-bbl capacity or less containing crude petroleum in crude-producing areas and outside aboveground atmospheric tanks under 23.8-bbl capacity containing other than class 1A liquids may have open vents.

NFPA 30-2-2.4.7   Flame arrestors or venting devices required in 2-2.4.6 may be omitted for class 1B and 1C liquids where conditions are such that their use may, in case of obstruction, result in tank damage. Liquid properties justifying the omission of such devices include, but are not limited to, condensation, crystallization, polymerization, freezing, and plugging. When any of these conditions exists, consideration may be given to heating, use of devices employing special materials of construction, the use of liquid seals, or inerting.

NFPA 30   General Provisions

Listed—equipment or materials included in a list published by an organization acceptable to the "authority having jurisdiction" and concerned with product evaluation, which maintains periodic inspection of production of listed equipment or materials and whose listing states that the equipment or material either meets the appropriate standards or has been tested and found suitable for use in a specified manner.

OSHA 1910.106(b)(2)(iv)(f)1   Tanks and pressure vessels storing class 1A liquids shall be equipped with venting devices which shall be normally closed except when venting to pressure or vacuum conditions, or with approved flame arrestors. Exemption—tanks of 3000-bbl capacity or less containing crude petroleum in crude-producing areas; and outside aboveground atmospheric tanks under 1000-gal capacity containing other than class 1A flammable liquids may have open vents. (g) Flame arrestors or venting devices required in subdivision (f) may be omitted for class 1B and 1C liquids where conditions are such that their use may, in case of obstruction, result in tank damage.

the increasing use of manifolded vapor recovery systems on storage that results from efforts to minimize pollution. Although current testing or listing is performed by nationally recognized testing laboratories such as Underwriters' Laboratories (UL) or Factory Mutual Research (FM), they are specific, very limited conditions for which these devices may be applied.

UL and FM both test and list flame arrestors. The test conditions are limited to flame arrestors with various lengths of piping using group D vapors (see below). UL has testing and listing services for detonation arrestors. The Coast Guard also has listing services for detonation arrestors in marine vapor recovery applications. ASTM is considering the development of national standards. Testing criteria for detonation arrestors require that the devices be subject to multiple, stable, and overdriven detonations.

There have been numerous incidents in which the device did not prevent flame propagation because of its misapplication. Perhaps the most common misapplication is a flame arrestor that is installed with too much piping connected to it or with piping that has elbows or reducers in it. Clearly, the tests validate the device only for the conditions at which it was tested. In the field, many different process and configuration variables may affect the suitability of the device for service.

9.4.6 Fundamental Principles

9.4.6.1 Flame propagation principles

There are three modes by which a flame front in a flammable gas mixture may propagate:

Unconfined deflagration. This term applies to a flame in unrestricted space. For example, an unconfined deflagration may occur if there are flammable vapors outside a tank where a spill has occurred. Unconfined deflagrations are characterized by a relatively slow-moving flame front (15 to 20 ft/s for hydrocarbons). Because the expansion generated by the flame is unrestricted, there is no significant pressure buildup or shock wave.

Confined deflagration. This term applies to flame propagation in a confined space. Confined deflagrations may occur in the vapor space of tanks, piping, equipment, buildings, or structures. This mode of flame propagation is characterized by the buildup of pressure and increasing flame propagation velocities away from the initial ignition source. Turbulence in the unburned mixture just ahead of the flame front causes the flame velocity to increase dramatically. The confined deflagration is defined as a flame front traveling at less than sonic speeds. Confined deflagrations can produce velocities up to several hundred feet per second. Because of confinement, one of three things happens.

- The flame goes out if it encounters a flame or detonation arrestor.

- A buildup of internal pressure results in failure of the confining structure. Tanks or buildings simply burst at a weak seam and relieve the pressure this way.

■ The confined deflagration becomes a detonation. In piping and pressure vessels which can withstand substantial pressure, the possibility of a detonation is high.

Detonations. Detonations occur in confined spaces where the propagation velocity of the flame front exceeds sonic velocities. One of the most common locations of detonations in industrial loss incidents is in piping. The propagation of flames through piping systems is highly variable and is subject to the specific nature of the piping. When a flame front is confined by piping, the flame front velocities can increase dramatically due to the induced turbulence in the piping and the internal pressure buildup. In detonations where the flame front travels above sonic speeds, the velocities might reach several thousand feet per second. For a stoichiometric propane-air mixture (4.3 percent propane), measurements show the velocity can approach 5800 ft/s accompanied by a pressure buildup to 500 psi.

Refer to API Publication 2028, *Flame Arrestors in Piping Systems,* for further guidance.

9.4.6.2   Flame-arresting principles

There are essentially three primary concepts to prevent flame propagation through confined conduits that connect other piping or equipment.

1. Mechanical isolation closes off the flame propagation path by using, for example, a quick-acting valve that closes before the flame front reaches it. These systems are usually complex because they depend on sensors that can detect flame fronts and related instrumentation systems that can respond quickly and reliably. These systems are rarely used on aboveground tanks.

2. Velocity methods depend on maintaining a sufficient velocity of flammable gases passing through the conduit that even if it ignited downstream, the flames could back up through the conduit, because the velocities are greater than the flame front velocities.

In some companies a PV valve is considered comparable to a flame arrestor because it depends on the velocity concept to prevent a flashback of an ignition source into the tank. Even when a flame is burning on flammable mixtures being expelled from PV valves, a flame will not pass back through the PV valve provided that the flammable gas velocity exceeds approximately 10 ft/s. This is easily accomplished in practice by having the set point pressure for the relief valve exceed approximately ¾ in WC. This discussion does not apply to PV valves connected to piping because the flame front velocities may be much higher and go right through the PV valve.

3. Quenching is the cooling of the flame front. This may be accomplished by passing the flammable mixture through water seals, packed beds, or passages constructed of metal. For tanks, the off-the-shell flame arrestor or detonation arrestor constructed with quenching passages is typically used.

9.4.6.3 Flame-arresting devices

A flame arrestor is a device that is intended to stop the propagation of a flame, deflagration, or detonation through equipment and piping. The detonation arrestor may be broadly categorized by two types:

- Active devices
- Passive devices

Active means that there are some moving parts that in some way interfere with the propagation of the flame front. Passive devices contain no moving parts. Generally, passive devices are simpler and more dependable, but the specific conditions of the application need to be taken into account for proper selection of the type of device to be used. Only quench-type passive devices are considered here. These devices depend on quenching a burning mixture to below the temperature that will support the burning. Since each of these types has very specific conditions under which it will operate, it is essential to apply the device to expected conditions that are within its capabilities.

Some specific flame arrestor types are:

9.4.6.4 Passive flame arrestor

For tanks, the flame-arresting method that has been used to satisfy fire protection requirements is the passive, quenching flame arrestor design. Passive flame-arresting designs consist of the following types:

Group 1	Wire gauze
	Wire gauze in packs
	Perforated plates
	Sintered
	Wire pack
	Metal foam
Group 2	Crimped metal
	Parallel plates

Group 1 is considered unreliable and unrepeatable and should be used only in very specialized situations. Examples of designs from group 2 are shown in Fig. 9.4.2. For standard industrial applications

**Series 25000 Bidirectional
Detonation Arrestor**

**Series 27000 Crimped Metal,
End-of-Line Arrestor**

**Series 3000 Crimped Metal,
Vent-Line Flame Arrestor**

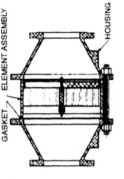

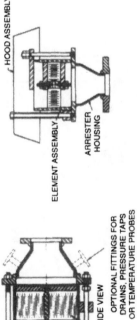

GASKET   ELEMENT ASSEMBLY

HOUSING

HOOD ASSEMBLY

ELEMENT ASSEMBLY

ARRESTER
HOUSING

OPTIONAL FITTINGS FOR
DRAINS, PRESSURE TAPS
OR TEMPERATURE PROBES

SIDE VIEW

CRIMPED METAL
ARRESTER ELEMENT

Figure 9.4.2 Typical flame and deteriorating arrestors. (*Courtesy of The Protectoseal
Company.*)

such as for use on tanks, the crimped-metal and parallel-plate designs are recommended as they have been extensively tested.

Crimped-metal flame arrestor. This design has the cellular construction shown in Fig. 9.4.3. In this design, the combustible gas mixture

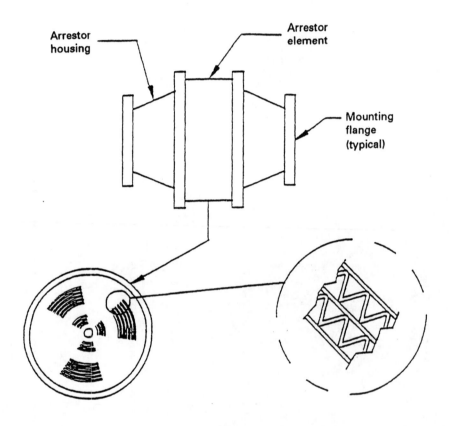

Typical crimped metal
element construction

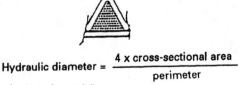

$$\text{Hydraulic diameter} = \frac{4 \times \text{cross-sectional area}}{\text{perimeter}}$$

Figure 9.4.3 Typical crimped-metal flame arrestor.

is directed through numerous small passages or apertures which allow vapors to flow, but which are too small for a flame to pass through. The size of the opening is based on the quenching diameter. The length of the quench passages is based upon tests. The flame arrestor may have to withstand high pressures and velocity flame fronts that can develop in the attached piping.

Parallel-plate flame arrestor.   Another popular design is the parallel-plate design shown in Fig. 9.4.4. In this design, the combustible gas mixture passes between numerous parallel plates which allow vapors to flow, but through which a flame cannot pass. The spacing between the plates is based on the quenching diameter. As with the first design, the length of the quench passages is based upon tests. Other considerations with regard to pressures and flame velocities apply as well.

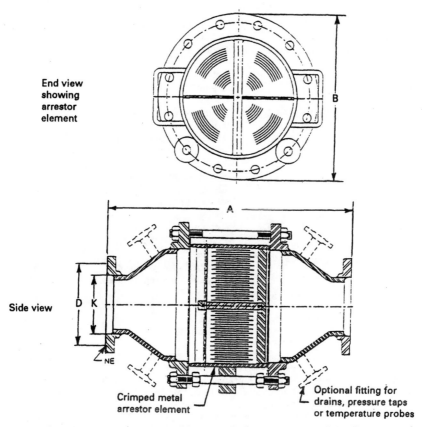

Figure 9.4.4   Parallel-plate flame arrestor detail. (*Courtesy of the Protectoseal Company.*)

Chemistry. The National Electric Code groups flammable chemicals by their burning and explosion characteristics into groups A, B, C, and D. See Table 9.4.2 and 9.4.3. Flame arrestors are typically tested for one member of a flammable group. The manufacturers usually consider that a flame arrestor will be suitable for another member of

TABLE 9.4.2   Chemical Atmosphere Hazards by NFPA Groupings

Group A atmospheres	Group D atmospheres
Acetylene	Acetone
	Acrylonitrile
Group B atmospheres	Ammonia
	Benzene
Butadiene	Butane
Ethylene oxide	1-Butanol (butyl alcohol)
Hydrogen	2-Butanol (secondary butyl alcohol)
Manufactured gases containing more than 30% hydrogen (by volume)	n-Butyl acetate
	Isobutyl acetate
Propylene oxide	Ethane
	Ethanol (ethyl alcohol)
Group C atmospheres	Ethyl acetate
	Ethylene dichloride
Acetaldehyde	Gasoline
Cyclopropane	Heptanes
Diethyl ether	Hexanes
Ethylene	Isoprene
Unsymmetric dimethyl hydrazine	Methane (natural gas)
	Methanol (methyl alcohol)
	3-Methyl-1-butanol (isoamyl alcohol)
	Methyl ethyl ketone
	Methyl isobutyl ketone
	2-Methyl-1-propanol (isobutyl alcohol)
	2-Methyl-2-propanol (tertiary butyl alcohol)
	Petroleum naphtha
	Octanes
	Pentanes
	1-Pentanol (amyl alcohol)
	Propane
	1-Propanol (propyl alcohol)
	2-Propanol (isopropyl alcohol)
	Propylene
	Styrene
	Toluene
	Vinyl acetate
	Vinyl chloride
	Xylenes

Group D, C, B, and A atmospheres represent progressively more difficult flame-arresting situations. The stable detonation velocities of the chemicals increase progressively for group D to group A atmospheres. Arresting devices tested with one member of a group are generally considered to be suitable for use with the other members of that group as long as initial temperature and pressure conditions are comparable. Arrestors tested with a higher-level group are considered to be suitable for use with chemicals from lower-level groups. A group C arrestor would be suitable for use in a group D atmosphere but not in a group B atmosphere.

TABLE 9.4.3   Conventions Relating to Chemical Groups

1. Arrestors tested for one member of a group are generally considered to be suitable for other members of the group under similar conditions.
2. Groups D, C, B, and A represent progressively more difficult flame-arresting situations:
    a. Smaller quenching diameters
    b. Higher detonation pressures and velocities

Gas	Group	Quenching diameter, in	Detonation velocity, ft/s
Methane	D	0.145	
Propane	D	0.105	5800
Butane	D	0.110	
Hexane	D	0.120	
Ethylene	C	0.075	
Hydrogen	B	0.034	7000

Note: Underwriters' Laboratories and Factory Mutual testing of vent line arrestors is performed with group D materials (gasoline, propane).

the same group. Groups D, C, B, and A represent a progressively more difficult flame-arresting function. This, however, may not be a good assumption. Testing is also based upon stoichiometric mixtures which are assumed to be the worst case. It has been suggested that these mixtures may not represent the worst case.

Strength of the passages and housing. Because of the shock wave and high pressures that occur in detonations, the devices must be able to quench the flame front while withstanding the pressure without damage. For this reason, detonation arrestors are built with heavier plates or passages as well as housing. Notice the housing shown in Fig. 9.4.2.

9.4.6.5 Guidance

- In general, the use of a flame or detonation arrestor should be minimized where alternatives are reasonably possible.

- When used on tanks, flame arrestors have been known to plug and to cause subsequent damage because of resulting excess pressure or vacuum. Therefore, scheduled inspections should be included for any installation that has these devices.

- The performance of flame arrestors is strongly affected by the connected piping and configuration. It is up to the system designer to ensure that the application is within the parameters specified by the manufacturer for proper operation, because the Underwriters' Laboratories and Factory Mutual approved flame arrestors are strictly limited to specific conditions.

- Be wary of configuration problems. Flame or detonation arrestors should not be installed where the piping is reduced or expanded in diameter, as there have been reports of flame fronts passing through as a result of the necking down.

- If flame arrestors are installed, then a support program to inspect and clean or maintain them is essential.

References

Section 9.1

1. Consultation with Dave Bloomquist, Bloomquist Fire Protection Company.
2. API Recommended Practice 2003, *Protection against Ignitions Arising out of Static, Lightning, and Stray Currents,* 4th ed., March 1982.
3. NFPA 77, *Static Electricity,* National Fire Protection Association, Quincy, MA, 988 ed.
4. API Bulletin 1003, *Precaution against Electrostatic Ignition during Loading of Tank Truck Motor Vehicles,* American Petroleum Institute, New York.
5. AAR Circular 17-D, *Recommended Practice for the Prevention of Electric Sparks that May Cause Fire during the Transfer of Flammable Liquids or Flammable Compressed Gases to or from Rail Equipment and Storage Tanks.*
6. NFPA 78, *Lightning Protection Code,* National Fire Protection Association, Quincy, MA.
7. NFP 30, *Flammable and Combustible Liquids Code,* National Fire Protection Association, Quincy, MA.
8. Arthur E. Cote (ed.), *NFPA Fire Protection Handbook,* 16th ed., National Fire Protection Association, Quincy, MA, 1986.
9. API Publication 2021, *Fighting Fires in and around Flammable and Combustible Liquid Atmospheric Storage Tanks,* American Petroleum Institute, New York, January 1991.
10. Charles H. Vervalin (ed.), *Fire Prevention Manual,* Gulf Publishing Company, Houston, TX.
11. Wilbur L. Walls (ed.), *Liquefied Petroleum Gases Handbook,* NFPA 58-1986, National Fire Protection Association, Quincy, MA, 1986.

Section 9.2

1. API Publication 2210, *Flame Arrestor for Vents of Tanks Storing Petroleum Products,* American Petroleum Institute, New York.
2. API Publication 2028, *Flame Arrestors in Piping Systems,* American Petroleum Institute, New York.
3. *Flame Arrestor for Use on Vents of Storage Tanks for Petroleum, Oil and Gasoline,* Underwriters Laboratories.
4. API Publication 2028, *Flame Arrestor in Piping Systems,* American Petroleum Institute, New York.
5. Thomas C. Piotrowski, "Specification of Flame Arresting Devices for Manifolded Low Pressure Storage Tanks," presented at the 1990 Summer National Meeting, American Institute of Chemical Engineers, San Diego, CA, August 19–22, 1990.
6. John Hutter, *Detonation Flame Arresters' Function, Inspection and Maintenance.*
7. Thomas C. Piotrowski, *Review of Proposed Testing and Approval Standards for Detonation Arrestor.*
8. Thomas C. Piotrowski, *Recommended Approval Testing Procedures for Detonation Arresters in Vapor Recovery Systems.*
9. *Flame Arrestor for Use on Vents of Storage Tanks for Petroleum Oil and Gasoline,* UL Standard 525, Underwriters' Laboratories.
10. actory Mutual Research, *Class 6061—Flame Arresters for Vent Pipes of Storage Tanks.*

10

Tank Emissions

SECTION 10.1 OVERVIEW OF TANK EMISSION CONCEPTS

10.1.1 Tank Type and Emissions

Figure 10.1.1 shows the development of tank emission control technology as well as the relative effectiveness of the different types of storage tanks concerning emissions. All storage tanks can be broadly classified as follows:

10.1.1.1 Fixed-roof tanks

These tanks expose the entire liquid surface to the vapor space above the liquid, which results in a high potential for evaporation and subsequent emissions. As a result of both liquid-level changes and diurnal temperature changes, the emission rate for these tanks may be substantial for liquids with a relatively high vapor pressure. The first effective improvement that was made was to fit these tanks with pressure vacuum vent valves (PV valves). A sealed tank with a PV valve represents a significant improvement in the ability to reduce evaporative losses. In fact, the primary driving force that caused the petroleum industry to upgrade from wood roof tanks (now obsolete), open-top tanks (also obsolete), and fixed-roof tanks with open vents and/or lots of holes caused by corrosion was economic savings (both evaporation losses and fires). The savings more than paid for the costs of using the sealed roofs with a PV valve. Note that the evaporative losses increase with vapor pressure and that for very low-vapor-pressure material the open-vent, fixed-roof tank is still appropriate and in common use.

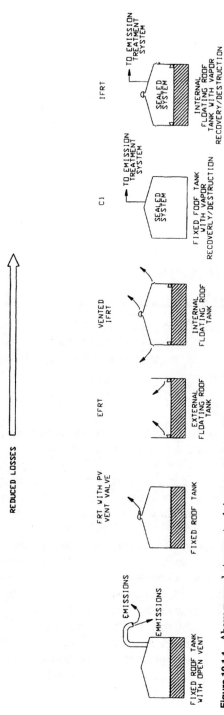

Figure 10.1.1 Aboveground storage tank improvements. (*Courtesy CB&I*)

10.1.1.2   External floating-roof tanks

Figures 10.1.2 and 10.1.3 show external floating-roof tanks fitted with the pontoon and double-deck type of roofs. The floating-roof design is an order-of-magnitude increase in improvement of the reduction of evaporative losses. The reason is that the roof covers over 95 percent of the surface area. In addition, floating-roof tanks are fitted with rim seals that further reduce emissions from the remaining exposed surface area. The emissions are limited to the amount that can escape past the rim seals and roof fittings and that which clings to the walls of the tank as the liquid level descends.

10.1.1.3   Internal floating-roof tanks

Figure 10.1.4 shows a typical internal floating roof (IFR). This is a floating roof inside a fixed-roof tank. These types of installations are common in several instances:

- When a fixed-roof tank changes service to a more volatile product or when regulations require a tank owner to convert an existing fixed-roof tank to a covered or floating-roof tank, IFRs are common. For existing tanks it is usually more cost-effective to simply install

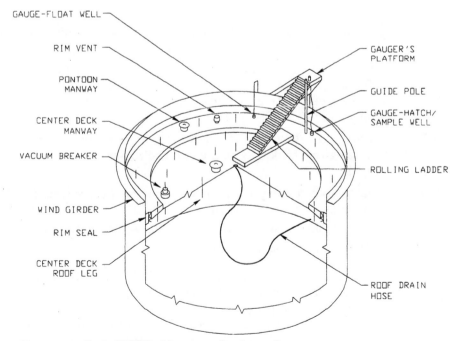

Figure 10.1.2 Typical EFRT with pontoon floating roof.

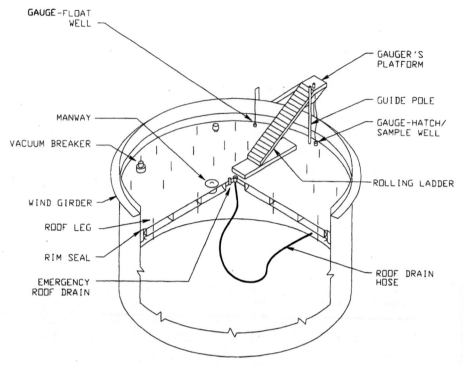

GAUGE-FLOAT WELL

GAUGER'S PLATFORM

GUIDE POLE

MANWAY

GAUGE-HATCH/ SAMPLE WELL

VACUUM BREAKER

ROLLING LADDER

WIND GIRDER

ROOF LEG

RIM SEAL

ROOF DRAIN HOSE

EMERGENCY ROOF DRAIN

Figure 10.1.3   Typical EFRT with double-deck floating roof.

or build a roof inside a fixed-roof tank than to convert a fixed-roof tank to an external floating-roof tank (EFR).

- Because the emissions are often lower than those for a comparable external floating-roof tank in areas where the average wind speed is high, the internal floating-roof installation may help reduce overall emissions.

- When highly hazardous toxins are stored, a properly designed internal floating-roof tank may be the best choice (for example, benzene storage).

- When it is desired to keep the rainwater out of the product (for example, MTBE), IFRs may be used.

- When it is desirable to minimize the possibility of rim seal fires resulting from lightning strikes, IFRs may be used.

IFRs are more effective at reducing emissions than EFRs at high wind speeds due to the wind-induced loss mechanism described later.

Most IFRs are vented to prevent the accumulation of hazardous or flammable mixtures in the space between the internal roof and the tank roof. However, some designs require a sealed space above the

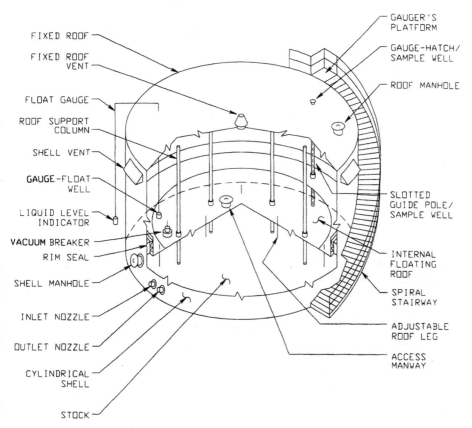

Figure 10.1.4 Typical IFRT with a welded floating roof.

floating roof. In these instances the space above the internal floating roof is sealed and operates under either slight pressure or vacuum. The space must also be connected to some form of vapor recovery or destruction system.

10.1.2 Trend and Outlook for Emission Controls

Although economics drove the initial improvements in evaporation loss methods (PV valves, floating-roof tanks, etc.), this is no longer the case. The EPA's regulations will essentially drive the technology to better methods of emission control. For example, the EPA has recently promulgated rules for both hazardous organic NESHAPs (National Emission Standards for Hazardous Air Pollutants), often referred to as HON, and maximum-achievable control technology (MACT) for petroleum refineries. These rules are mandated by the Clean Air Act

Amendments of 1990 which are based upon the top 12 percent of the best-performing facilities in the emission source category.

There is also the element of increased liability and accountability for pollution. Tank owners who handle toxic chemicals must submit Toxic Chemical Release Inventory (TRI) Reports (called Section 313 reports) to comply with the Emergency Planning and Community Right-to-Know Act (Superfund, Title 3 and SARA).

Tank owners must have an understanding of the causes of emission and the methods of controlling these emissions, a knowledge of the options available and the methods of estimating emissions, and a way of obtaining permitting and compliance with the regulatory jurisdictions by obtaining credit for reduced emissions.

10.1.3 Various Options for Controlling Tank Vapors

Some basic options that relate to the sealed-tank systems shown in Fig. 10.1.1 are

1. Conversion to internal floating-roof tank

2. Incineration

3. Catalytic processes

4. Vapor compression

5. Refrigeration

6. Activated carbon adsorption

Table 10.1.1 compares these alternatives.

For existing installations where a fixed roof is under consideration, an internal floating roof in the existing tanks is usually the recommended solution. There may be instances in which it is not feasible to retrofit an internal floating roof, because of structural reasons, because it is not possible to take the tank out of service, or because there are existing vapor recovery systems usable on site. When emission reduction is not possible by converting a fixed-roof tank to a floating-roof tank, then other types of pollution and odor control can be considered.

Some of the alternatives to reduce emissions from fixed-roof tanks are discussed in greater detail.

10.1.3.1 Conversion of a fixed-roof tank to an internal floating-roof tank

Conversion of fixed-roof tanks to internal floating-roof tanks is a relatively simple operation. The easiest method is by retrofitting aluminum

TABLE 10.1.1   Comparison of Vapor Recovery Systems for Fixed-Roof Tanks

	Capital investment, $M	Operating cost, $M/yr	Operating problems	Comments
Internal floating roof	−	−	−	Simple, requires little maintenance.
Incineration	−	0	+	Has rotating equipment, requires surveillance.
Catalytic	0	0	0	
Compression/ reinjection	0	+	+	Puts the problem of emissions downstream.
Refrigeration	+	+	−	Not as effective as other methods. Uses rotating machinery and has operability and maintenance problems.
Activated carbon	0	+	−	Carbon disposal may be problem.

+ = high, 0 = average, and − = low

roofs into these tanks. A small door sheet or roof cutout is made through which the roof components are inserted and fastened together. When a fixed-roof tank is retrofitted with a steel pontoon or pan-type roof, a substantial amount of welding must be done inside the tank and the tank must remain out of service for a long time. Either type of roof requires a complete shutdown and tank cleaning to make the installation.

Installation of an internal floating roof in a fixed-roof tank can reduce the vent gas emitted by over 95 percent.

Advantages

1. Low capital cost

2. Vapor emissions reduced by more than 95 percent

3. Easy to operate

4. Reduces chance of tank fire

Disadvantage

1. Tank must be taken out of service to install internal floating roof.

10.1.3.2 Incineration (thermal oxidizer)

This option is probably the least expensive alternative after the internal floating roof. It is easy to operate, but the stack gas may contain unacceptably high amounts of SO_x or other acid compounds. Also, there may be community concern about incineration.

Advantages

1. It is the least expensive non-floating-roof alternative.
2. It can be installed without taking the tank out of service.
3. It is easy to operate.

Disadvantages

1. Products of combustion may exceed SO_x limits.
2. It requires a source of fuel gas.
3. There may be community bias against incineration.
4. Future liabilities may be associated with trace toxic combustion by-products.

10.1.3.3 Catalytic processes

These processes are usually aimed at specific compounds such as H_2S or mercaptans, where in addition to volatile organic compound (VOC) losses odor is a problem. The catalytic processes must be used in addition to other hydrocarbon-removing processes such as refrigeration or activated carbon.

Advantages

1. It can be installed without taking the tank out of service.
2. It is easy to operate.
3. It removes both H_2S and mercaptans—product gas has an acceptable odor.

Disadvantages

1. The process does not remove hydrocarbons. The vapor may additionally require incineration to meet local VOC specifications.
2. The vapor may have to be water-saturated, requiring additional equipment.
3. Large-diameter vessels are required to minimize pressure drop and to allow vessels to be constructed as tanks rather than under the pressure vessel code.
4. Laboratory data may be needed to develop a design basis.

10.1.3.4 Vapor compression

One process was developed in Houston for vapor recovery from oil production tanks in west Texas. It is based on compression of the tank vapor to a pressure high enough to allow reinjection into oil and gas piping systems. The system is fully computer-controlled and skid-mounted. The attractiveness of the process depends on the pressure to which the vapor must be compressed and the availability of pipelines with sufficiently low gas pressure to accommodate the compressor discharge flows.

Advantages

1. It can be installed without taking the tank out of service.
2. It allows recovery of gas and condensed liquids, providing some economic return on investment.
3. It is fully computer controlled, thus easy to operate.
4. Essentially there are no emissions.

Disadvantages

1. It may require a large compressor if the pipeline pressure is high.
2. Maintenance problems are possible if the system malfunctions.
3. It transfers the problems in a downstream direction or to the "end of pipe."

10.1.3.5 Refrigeration

This process requires a refrigeration unit to cool the tank vapor to −30°C to condense as much of the hydrocarbons and/or sulfur compounds as is economically possible. It is relatively costly and more difficult to operate than the other alternatives.

Advantages

1. It can be installed without taking the tank out of storage.
2. It allows recovery of condensed liquids, providing some economic return on investment.

Disadvantages

1. It is the highest investment of all options.
2. It is more difficult to operate.
3. There is the potential fouling of the condenser due to ice and hydrate deposits.
4. Product with odor-laden vapors may still require deodorization (possibly with activated carbon) when vented to the atmosphere.

10.1.3.6   Activated carbon adsorption

Adsorption of mercaptan vapors from air on activated carbon is a well-demonstrated technology. The operating cost of this type of system, particularly for large tanks storing high-vapor-pressure substances, is likely to be prohibitive, since the carbon will quickly be expended and require regeneration.

Advantages

1. It is applicable to a wide variety of specialty chemical and hydrocarbon tanks.
2. It can be installed without taking the tank out of service.
3. It is easy to operate.
4. It removes specific compounds such as H_2S and mercaptans that cause odor problems.

Disadvantages

1. Usage of activated carbon will normally be high.
2. Disposal of spent carbon may be a problem.
3. Activated carbon often supplies enough heat to be a fire and explosion safety hazard.

10.1.4   Mechanisms of Evaporation Losses in Storage Tanks

Emissions from organic liquids in storage occur because of evaporative losses of stored liquid. The emission sources vary with tank design, as do the relative contributions of each emission source. Emissions from fixed-roof tanks resulting from evaporative losses due to storage of volatile fluids are known as *breathing losses* (or *standing storage losses*). The evaporative losses that occur in fixed-roof tanks during filling and emptying operations are known as *working losses*.

External and internal floating-roof tanks are emission sources because of evaporative losses that occur during standing storage and withdrawal of liquid from the tank. Standing storage losses result from evaporative losses through rim seals, deck fittings, and/or deck seams. The loss mechanisms for fixed-roof and external and internal floating-roof tanks are described in greater detail in the following sections. Variable vapor space tanks are also emission sources because of evaporative losses that result during filling operations. The loss mechanism for variable vapor space tanks is also described in this section. Emissions occur from pressure tanks as well. However, loss mechanisms from these sources are not described in this chapter.

10.1.4.1 Fixed-roof tanks

The two significant types of emissions from fixed-roof tanks are storage and working losses. Storage loss is the expulsion of vapor from a tank through vapor expansion and contraction, which are the results of changes in temperature and barometric pressure. This loss occurs without any liquid-level change in the tank.

The combined loss from filling and emptying is called the *working loss*. Evaporation during filling operations is a result of an increase in the liquid level in the tank. As the liquid level increases, the pressure inside the tank exceeds the relief pressure and vapors are expelled from the tank. Evaporative loss during emptying occurs when air drawn into the tank during liquid removal becomes saturated with organic vapor and expands, thus exceeding the capacity of the vapor space.

Fixed-roof tank emissions vary as a function of vessel capacity, vapor pressure of the stored liquid, utilization rate of the tank, and atmospheric conditions at the tank location. Several methods are used to control emissions from fixed-roof tanks. Emissions from fixed-roof tanks can be controlled by installing an internal floating roof and seals to minimize evaporation of the product being stored. The control efficiency of this method ranges from 60 to 99 percent, depending on the type of roof and seals installed and on the type of organic liquid stored.

Vapor balancing is another means of emission control. Vapor balancing is probably most common in the filling of tanks at gasoline stations. As the storage tank is filled, the vapors expelled from the storage tank are directed to the emptying gasoline tanker truck. The truck then transports the vapors to a centralized station where a vapor recovery or control system is used to control emissions. Vapor balancing can have control efficiencies as high as 90 to 98 percent if the vapors are subjected to vapor recovery or control. If the truck vents the vapor to the atmosphere instead of to a recovery or control system, no control is achieved.

Vapor recovery systems collect emissions from storage vessels and convert them to liquid product. Several vapor recovery procedures may be used, including vapor/liquid absorption, vapor compression, vapor cooling, vapor/solid adsorption, or a combination of vapors recovered and the mechanical condition of the system. In a typical thermal oxidation system, the air/vapor mixture is injected through a burner manifold into the combustion area of an incinerator. Control efficiencies for this system can range from 96 to 99 percent.

10.1.4.2 External floating-roof tanks

Total emissions from external floating-roof tanks are the sum of withdrawal losses and standing storage losses. Withdrawal losses occur as the liquid level, and thus the floating roof, is lowered. Some liquid remains attached to the tank surface and is exposed to the atmosphere. Evaporative losses will occur until the tank is filled and the exposed surface (with the liquid) is again covered. Standing storage losses from external floating-roof tanks include rim seal and roof fitting losses. Rim seal losses can occur through many complex mechanisms, but the majority of rim seal vapor losses have been found to be wind-induced. Other potential standing storage loss mechanisms include breathing losses as a result of temperature and pressure changes. Also, standing storage losses can occur through permeation of the seal material with vapor or via a wicking effect of the liquid. Testing has indicated that breathing, solubility, and wicking loss mechanisms are small in comparison to the wind-induced loss. Also, permeation of the seal material generally does not occur if the correct seal fabric is used. The rim seal loss factors incorporate all types of losses. The roof fitting losses can be explained by the same mechanisms as the rim seal loss mechanisms. However, the relative contribution of each is not known. The roof fitting losses identified in this section account for the combined effect of all the mechanisms.

A rim seal system is used to allow the floating roof to travel within the tank as the liquid level changes. The seal system also helps to fill the annular space between the rim and the tank shell and therefore minimizes evaporative losses from this area. A rim seal system may consist of just a primary seal or a primary seal and a secondary seal, which are mounted above the primary seal. Examples of primary and secondary seal configurations are shown in Figs. 10.1.5 through 101.7. Three basic types of primary seals are used on external floating roofs: mechanical (metallic) shoe, resilient toroid (nonmetallic), and flexible wiper. The resilient seal can be mounted to eliminate the vapor space between the seal and liquid surface (liquid-mounted) or to allow a vapor space between the seal and liquid surface (vapor-mounted). A primary seal serves as a vapor conservation device by closing the annular space between the edge of the floating roof and the tank wall. Some primary seals are protected by a metallic weather shield. Additional evaporative loss may be controlled by a secondary seal. Secondary seals can be either flexible wiper seals or resilient toroid seals. Two configurations of secondary seals are currently available: shoe-mounted and rim-mounted. Although there are other seal systems, the systems described here include the majority in use today.

Roof fitting loss emissions from external floating-roof tanks result from penetrations in the roof by deck fittings, the most common of which are described below. Some of the fittings are typical of both external and internal floating-roof tanks.

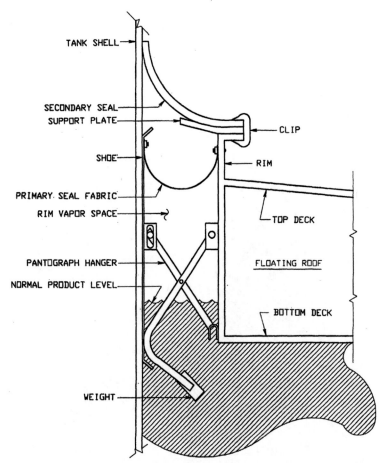

Figure 10.1.5 Mechanical-shoe primary seal with rim-mounted secondary seal.

Access hatch (Fig. 10.1.8). An access hatch is an opening with a peripheral vertical well that is large enough to provide passage for workers and materials through the deck for construction or servicing. Attached to the opening is a removable cover that may be bolted and/or gasketed to reduce evaporative loss. On internal floating-roof tanks with noncontact decks, the well should extend down into the liquid to seal off the vapor space below the noncontact deck. A typical access hatch is shown in Fig. 10.1.8.

Gauge float well (Fig. 10.1.9). A gauge float is used to indicate the level of liquid within the tank. The float rests on the liquid surface and is housed inside a well that is closed by a cover. The cover may be bolted and/or gasketed to reduce evaporative loss. As with other similar deck

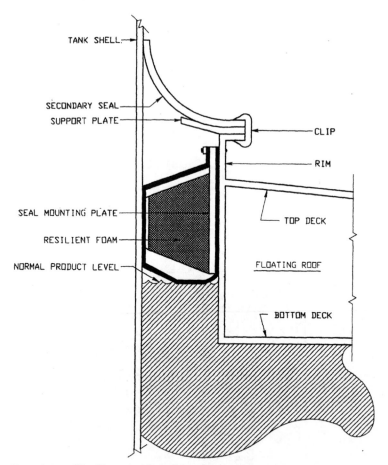

Figure 10.1.6 Liquid-mounted resilient filled primary seal.

penetrations, the well extends down into the liquid on noncontact decks in internal floating-roof tanks. A typical gauge float well is shown in Fig. 10.1.9.

Gauge hatch/sample well (Fig. 10.1.10). A gauge hatch sample well consists of a pipe sleeve equipped with a self-closing gasketed cover (to reduce evaporative losses) and allows hand gauging or sampling of the stored liquid. The gauge hatch/sample well is usually located beneath the gauger's platform, which is mounted on top of the tank shell. A cord may be attached to the self-closing gasketed cover so that the cover can be opened from the platform. A typical gauge hatch/sample well is shown in Fig. 10.1.10.

Rim vents (Fig. 10.1.11). Rim vents are usually used only on tanks equipped with a mechanical-shoe primary seal. A typical rim vent is

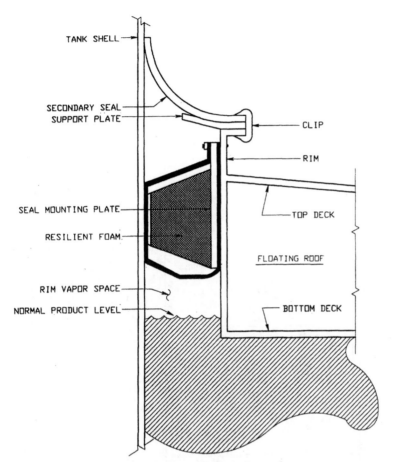

Figure 10.1.7 Vapor-mounted resilient filled primary seal.

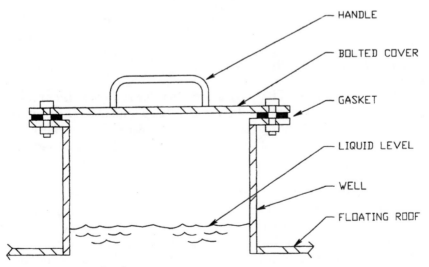

Figure 10.1.8 Access hatch.

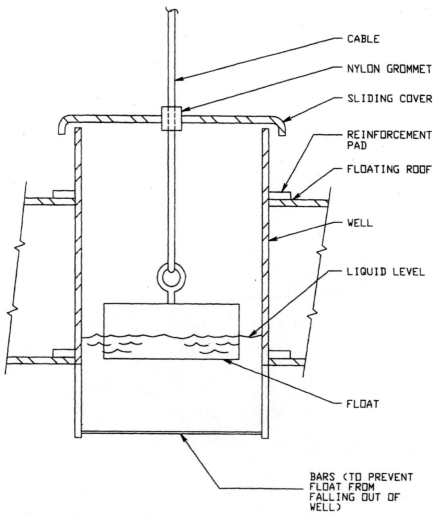

Figure 10.1.9 Gauge-float well.

shown in Fig. 10.1.11. The vent is used to release any excess pressure or vacuum that is present in the vapor space bounded by the primary-seal shoe and the floating-roof rim, and the primary seal fabric and the liquid level. Rim vents usually consist of weighted pallets that rest on a gasketed opening.

Roof drains (Fig. 10.1.12).   Currently, two types of roof drains are in use (closed and open) to remove rainwater from the floating-roof surface. Closed-roof drains carry rainwater from the surface of the roof through a flexible hose or an articulated piping system that runs through the stored liquid to a shell-mounted nozzle exiting the tank.

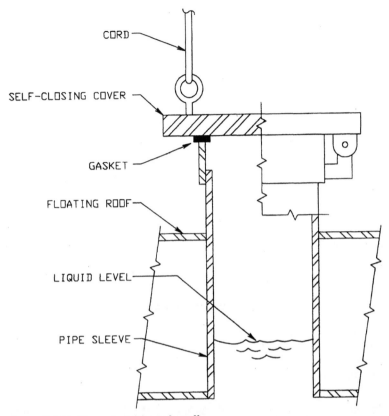

Figure 10.1.10 Gauge hatch/sample well.

The rainwater does not come in contact with the liquid, so no evaporative losses result.

Open-roof drains can be either flush or overflow drains, and they are used only on double-deck external floating roofs. Both types consist of a pipe that extends below the roof to allow the rainwater to drain into the stored liquid. The liquid from the tank enters the pipe, so evaporative losses can result from the tank opening. Flush drains are flush with the roof surface. Overflow drains are elevated above the roof surface. A typical overflow roof drain is shown in Fig. 10.1.12. Overflow drains are used to limit the maximum amount of rainwater that can accumulate on the floating roof, providing emergency drainage of rainwater, if necessary. Overflow drains are usually used in conjunction with a closed drain system to carry rainwater outside the tank.

Roof leg (Fig. 10.1.13). To prevent damage to fittings underneath the deck and to allow for tank cleaning or repair, supports are provided to

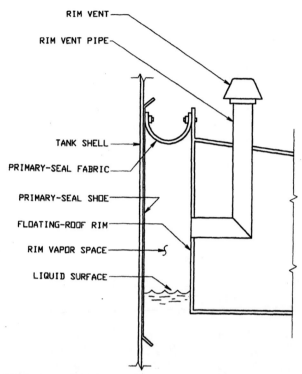

RIM VENT

RIM VENT PIPE

TANK SHELL

PRIMARY-SEAL FABRIC

PRIMARY-SEAL SHOE

FLOATING-ROOF RIM

RIM VAPOR SPACE

LIQUID SURFACE

Figure 10.1.11 Rim vent.

hold the deck at a predetermined distance off the tank bottom. These supports consist of adjustable or fixed legs attached to the floating deck or hangers suspended from the fixed roof. For adjustable legs or hangers, the load-carrying element passes through a well or sleeve into the deck. With noncontact decks, the well should extend into the liquid. Evaporative losses may occur in the annulus between the roof leg and its sleeve. A typical roof leg is shown in Fig. 10.1.13.

Unslotted guide pole wells (Fig. 10.1.14). A guide pole well is an antirotational device that is fixed to the top and bottom of the tank, passing through the floating roof. The guide pole is used to prevent adverse rotational movement of the roof and thus damage to roof fittings and the rim seal system. It may also be used for sampling and tank gauging but is much less efficient for this purpose than the slotted guide pole.

Slotted guide pole/sample wells (Fig. 10.1.15). The function of the slotted guide pole/sample well is similar to the unslotted guide pole well but also has additional features. A typical slotted guide pole well is shown in Fig. 10.1.15. As shown in this figure, the guide pole is slotted to allow stored liquid to enter. The liquid entering the guide pole

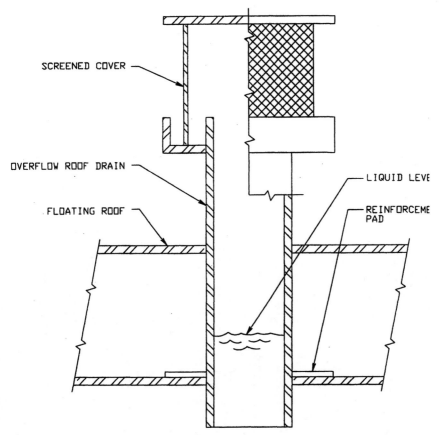

SCREENED COVER

OVERFLOW ROOF DRAIN

FLOATING ROOF

LIQUID LEVE

REINFORCEME
PAD

Figure 10.1.12 Overflow roof drain.

is well mixed, having the same composition as the remainder of the stored liquid, and is at the same liquid level as the liquid in the tank. Representative samples can therefore be collected from the slotted guide pole. The opening at the top of the guide pole and along the exposed sides is typically the emission source. However, evaporative loss from the top of the guide pole can be reduced in several ways.

Vacuum breaker (Fig. 10.1.16). A vacuum breaker equalizes the pressure of the vapor space across the deck as the deck is either being landed on or floated off its legs. A typical vacuum breaker is shown in Fig. 10.1.16. As depicted in this figure, the vacuum breaker consists of a well with a cover. Attached to the underside of the cover is a guided leg long enough to contact the tank bottom as the floating deck approaches. When in contact with the tank bottom, the guided leg mechanically opens the breaker by lifting the cover off the well; otherwise, the cover closes the well. The closure may be gasketed or ungasketed. Because

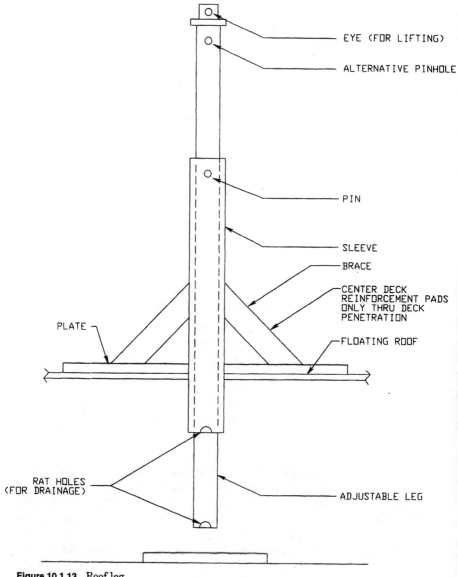

Figure 10.1.13 Roof leg.

the purpose of the vacuum breaker is to allow the free exchange of air and/or vapor, the well does not extend appreciably below the deck.

10.1.4.3 Internal floating-roof tanks

Total emissions from internal floating-roof tanks are the sum of withdrawal losses and standing storage losses. Withdrawal losses occur in the same manner as in external floating-roof tanks: As the floating roof

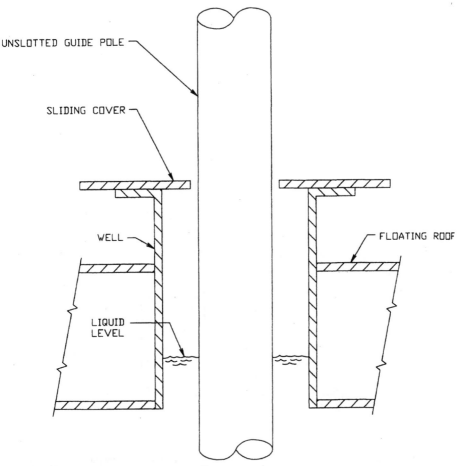

UNSLOTTED GUIDE POLE

SLIDING COVER

WELL

FLOATING ROOF

LIQUID
LEVEL

Figure 10.1.14 Unslotted guide pole well.

lowers, some liquid remains attached to the tank surface and evaporates. Also, in internal floating-roof tanks that have a column-supported fixed roof, some liquid clings to the columns. Standing storage losses from internal floating-roof tanks include rim seal, deck fitting, and deck seam losses. The loss mechanisms described for external floating-roof rim seal and roof fitting losses also apply to internal floating roofs. However, unlike external floating-roof tanks in which wind is the predominant factor affecting rim seal loss, no dominant wind loss mechanism has been identified for internal floating-roof tank rim seal losses. Deck seams in internal floating-roof tanks are a source of emissions to the extent that these seams may not be completely vaportight. The loss mechanisms described for external floating-roof tank rim seals and roof fittings can describe internal floating-roof deck seam losses. As with internal floating-roof rim seal and roof fittings, the relative importance

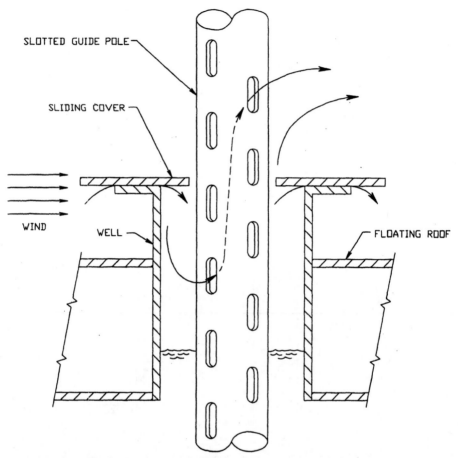

Figure 10.1.15 Wind-related convection loss mechanism, slotted guide pole.

of each of the loss mechanisms is not known. Note that welded steel internal floating roofs do not have deck seam losses.

Internal floating roofs typically incorporate one of two types of flexible, product-resistant seals: resilient foam-filled seals or wiper seals. Similar to those used on external floating roofs, each of these seals closes the annular vapor space between the edge of the floating roof and the tank shell to reduce evaporative losses. They are designed to compensate for small irregularities in the tank shell and allow the roof to move freely up and down in the tank without binding.

A resilient foam-filled seal used on an internal floating roof is similar in design to that described for external floating roofs. Two types of resilient foam-filled seals for internal floating roofs are shown in Fig. 10.1.17. These seals can be mounted either in contact with the liquid surface (liquid-mounted) or several centimeters above the liquid surface (vapor-mounted).

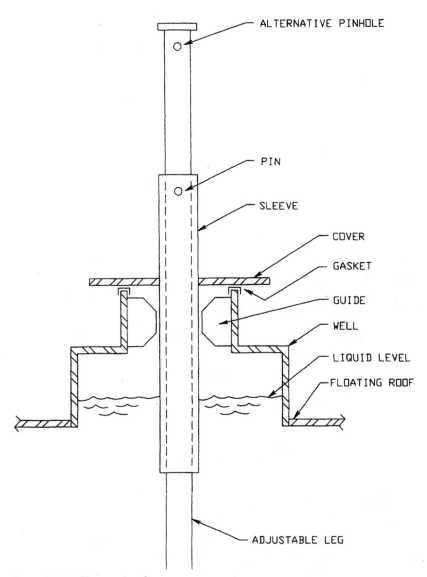

ALTERNATIVE PINHOLE

PIN

SLEEVE

COVER

GASKET

GUIDE

WELL

LIQUID LEVEL

FLOATING ROOF

ADJUSTABLE LEG

Figure 10.1.16　Vacuum breaker.

Resilient foam-filled seals work because of the expansion and contraction of a resilient material to maintain contact with the tank shell while accommodating varying annular rim space widths. These seals consist of a core of open-cell foam encapsulated in a coated fabric. The elasticity of the foam core pushes the fabric into contact with the tank shell. The seals are mounted on the deck perimeter and are continuous around the roof circumference. Polyurethane-coated nylon fabric and polyurethane foam are commonly used materials. For emission control,

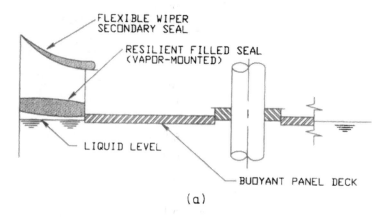

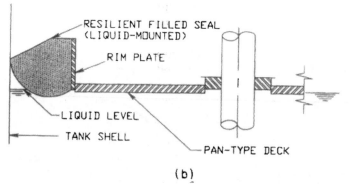

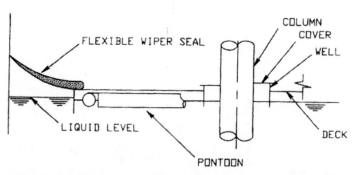

Figure 10.1.17 Seals for internal floating roofs for storage of organic liquids. (*a*) Resilient foam-filled seal (vapor mounted); (*b*) resilient foam-filled seal (liquid mounted); (*c*) elastomeric wiper seal.

it is important that the mounting and radial seal joints be vaportight and that the seal be in substantial contact with the tank shell.

Wiper seals are commonly used as primary seals for internal floating-roof tanks. This type of seal is depicted in Fig. 10.1.17. New tanks with wiper seals may have dual wipers, one mounted above the other. Wiper seals generally consist of a continuous annular blade of flexible material fastened to a mounting bracket on the deck perimeter that spans the annular rim space and contacts the tank shell. The mounting is such that the blade is flexed, and its elasticity provides a sealing pressure against the tank shell. Such seals are vapor-mounted; a vapor space exists between the liquid stock and the bottom of the seal. For emission control, it is important that the seal be continuous around the circumference of the roof and that the blade be in substantial contact with the tank shell.

Two types of material are commonly used to make the wipers. One type consists of a cellular, elastomeric material tapered in cross section with the thicker portion at the mounting. Buna-N rubber is a commonly used material. All radial joints in the blade are joined.

A second type of wiper seal construction uses a foam core wrapped with a coated fabric. Polyurethane on nylon fabric and polyurethane foam are common materials. The core provides the flexibility and support, while the fabric provides the vapor barrier and wear surface.

Secondary seals may be used to provide some additional evaporative loss control over that achieved by the primary seal. The secondary seal is mounted to an extended vertical rim plate, above the primary seal, as shown in Fig. 10.1.18. Secondary seals can be either

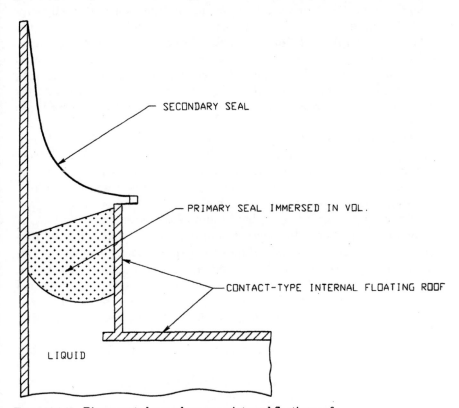

Figure 10.1.18 Rim-mounted secondary on an internal floating roof.

a resilient foam-filled seal or an elastomeric wiper seal, as previously described. For a given roof design, using a secondary seal further limits the operating capacity of a tank due to the need to keep the seal from interfering with the fixed-roof rafters when the floating roof is at its highest level.

Numerous deck fittings penetrate or are attached to an internal floating roof. These fittings accommodate structural support members or allow for operational functions. The fittings can be a source of evaporative loss in that they require penetrations in the deck. Other accessories are used that do not penetrate the deck and are not, therefore, sources of evaporative loss. The most common fittings relevant to controlling vapor losses are described in the following paragraphs.

The access hatches, guide pole wells, roof legs, vacuum breakers, and automatic gauge float wells for internal floating roofs are similar fittings to those already described for external floating roofs. Other fittings used on internal floating-roof tanks include column wells, ladder wells, and stub drains.

Column wells.   The most common fixed-roof designs are normally supported from inside the tank by means of vertical columns, which must penetrate an internal floating deck. (Some fixed roofs are entirely self-supporting and, therefore, have no support columns.) Columns are made of pipe with circular cross sections or of structural shapes with irregular cross sections (built-up). The number of columns varies with tank diameter from a minimum of 1 to over 50 for very large tanks.

The columns pass through deck openings through wells. With noncontact decks, the well should extend down into the liquid stock. Generally, a closure device exists between the top of the well and the column. Several proprietary designs exist for this closure, including sliding covers and fabric sleeves, which must accommodate the movements of the deck relative to the column as the liquid level changes. A sliding cover rests on the upper rim of the column well (which is normally fixed to the roof) and bridges the gap or space between the column well and the column. The cover, which has a cutout, or opening, around the column slides vertically relative to the column as the roof raises and lowers. At the same time, the cover slides horizontally relative to the rim of the well, which is fixed to the roof. A gasket around the rim of the well reduces emissions from this fitting. A flexible fabric sleeve seal between the rim of the well and the column (with a cutout or opening, to allow vertical motion of the seal relative to the columns) similarly accommodates limited horizontal motion of the roof relative to the column. A third design combines the advantages of the flexible fabric sleeve seal with a well that excludes all but a small portion of the liquid surface from direct exchange with the vapor space above the floating roof.

Ladder wells. Some tanks are equipped with internal ladders that extend from a manhole in the fixed roof to the tank bottom. The deck opening through which the ladder passes is constructed with similar design details and considerations to deck openings for column wells, as previously discussed.

Stub drain. Bolted internal floating-roof decks are typically equipped with stub drains to allow any stored product that may be on the deck surface to drain back to the underside of the deck. The drains are attached so that they are flush with the upper deck. Stub drains are approximately 1 in in diameter and extend down into the product on noncontact decks.

10.1.5 Resources For Emission Determination

The following API documents have been the primary tools for computing emission losses from tanks. They have become the basis for EPA's rules for computing emissions.

Tank type	API Publication
EFRT	API Publication 2517, 3rd ed., February 1989
FRT	API Publication 2518, 2d ed., October 1991
IFRT	API Publication 2519, 3rd ed., June 1983

These publications are the result of API-sponsored laboratory and field testing programs that included analytical developments to model storage tank evaporative loss processes.

Another useful document is API Publication 2557, "Vapor Collection and Control Options for Storage and Transfer Operations in the Petroleum Industry." This publication should be consulted by those designing or purchasing vapor recovery systems for storage tanks. It describes the various control technologies and relative costs for different options.

API will soon issue a publication that covers the speciation of evaporative loss sources. This is useful for determining the emissions where multicomponent liquids are involved (which is quite common) or when a specific hazardous compound is present in the stored liquid.

The EPA Office of Air Quality Planning and Standards maintains a technology transfer network. One of the features of this system is an electronic bulletin board called *The Chief Bulletin Board System.* This system includes various databases, some of which are useful for computing tank emissions. One of these is a program called TANKS (currently version 2.0) that can be downloaded at no charge to the user. This program computes the emissions for tanks in accordance with the procedures outlined in the following sections. It is based upon the EPA document AP-42. A user's manual may also be downloaded. The EPA office that maintains this information is listed in the References below.

SECTION 10.2   COMPUTING EMISSIONS FROM INTERNAL AND EXTERNAL FLOATING ROOFS

10.2.1   Loss Components

API 2517 and 2519 as well as EPA's document AP-42 have complete graphs and tables that allow a determination of each of the required variables necessary to perform the emission-estimating calculations for tanks. The data allowing solution to most of the common problems are presented here:

The total evaporative losses from an internal or external floating-roof tank can be expressed by

$$L_T = L_S + L_W$$
$$L_T = \text{total tank loss, lb}\cdot\text{mol/yr}$$
$$L_S = \text{standing losses, lb}\cdot\text{mol/yr} \qquad (10.2.1)$$
$$L_W = \text{withdrawal losses, lb}\cdot\text{mol/yr}$$

As indicated by the equation, there are two sources of loss:

1. *Standing storage losses.* These losses result from the fact that the tank is not a completely closed container, but has openings and gaps in the seals and fittings that allow the escape of volatile components.

2. *Withdrawal losses.* These losses result from clingage of product that remains on the walls of the tank as the roof is being lowered. For IFRTs the clingage loss also results from clingage that occurs on the roof support columns that may penetrate the internal roof.

The withdrawal losses are usually insignificant in comparison with the standing storage losses.

10.2.1.1   Standing storage losses

The losses may be calculated from

$$L_S = (F_T)(P^*M_VK_C) \qquad (10.2.2)$$

where F_T = total roof fitting loss factor, lb•mol/yr

M_V = average molecular weight of stock vapor, lb/(lb•mol). Use 64 for gasoline, 50 for crude oil, and the actual molecular weight for other single-component substances. Table 10.2.1 lists physical properties for selected hydrocarbons and chemicals.

K_C = Product factor (1.0 for refined stocks or single-component liquids, 0.4 for crude oils)

P^* = vapor pressure function

$$P^* = \frac{P/P_a}{[1-(1-P/P_a)^5]^2}$$

and P is the true stock vapor pressure at the bulk liquid temperature in psia and P_a is atmospheric pressure in psia.

Figure 10.2.1 is a plot of the vapor pressure function. It relates the true vapor pressure in pounds per square inch to the vapor pressure function used in the loss equations. It is based on empirical data that are applicable between 1.5 psia to a vapor pressure just under atmospheric pressure (i.e., below the boiling point). For refined stocks and crude oils, Figs. 10.2.2 and 10.2.3 can be used to determine the true vapor pressure. For single-component pure chemical liquids, the vapor pressures may be extracted from Tables 10.2.1 and 10.2.2 as a function of the bulk liquid storage temperature.

Since true vapor pressure is related to the stock temperature, it is necessary to determine the temperature of the stock. The average annual ambient temperature is selected based upon the location given in Table 10.2.3. Then an adjustment for tank color can be made by using Table 10.2.4.

In Eq. (10.2.2) there are two factors which are independent: The term F_T is a factor that depends only on the physical characteristics of the tank such as size, type of seal, and numbers and types of fittings. The terms of the second factor $(P^*M_VK_C)$ are all dependent on physical characteristics of the stored liquid.

The total roof fitting loss factor F_T results from two components, the rim seal losses and the fitting losses.

$$F_T = F_rD + F_f + F_d \qquad (10.2.3)$$

where F_r = rim seal loss factor, lb•mol/yr

F_d = deck seam loss factor, lb•mol/yr. This factor applies only to an IFRT and is 0 for an EFRT

D = tank diameter, ft

T_f = total roof-fitting loss factor, lb•mol/yr

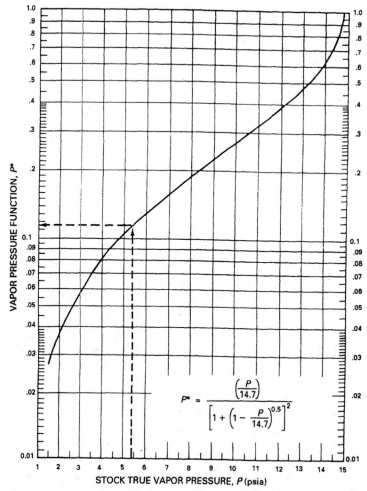

Figure 10.2.1 Vapor pressure function. (Note: Dashed line illustrates sample problem for $P = 5.4$ lb/in^2 absolute. Atmospheric pressure = 14.7 lb/in^2 absolute.) (*Courtesy of the American Petroleum Institute.*)

Rim seal loss factor F_r. Since the losses for seals are a function of the length of the seal, the rim seal loss factor F_r is multiplied by the diameter of the tank. This factor depends on the specific type and number of seals used and on the ambient wind speed for external floating roof tanks.

Internal floating roof. Table 10.2.5 shows the rim seal loss factor for average- and tight-fitting seals.

External floating roof. For an EFRT seal, the rim seal loss factor F_r is a function of the wind speed and is given by Eq. (10.2.4) and by

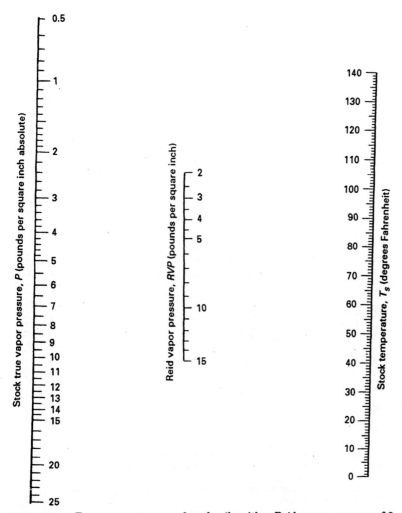

Figure 10.2.2 True vapor pressure of crude oils with a Reid vapor pressure of 2 to 15 psi. (*Courtesy of the American Petroleum Institute.*) .

Table 10.2.6. Table 10.2.6 shows the values for K_r and exponent n that may be used to determine the rim seal loss factor at any wind speed for average-fitting seals. API 2517 has additional factors for tight-fitting seals. Equation (10.2.4) is applicable only to wind speeds within 2 to 15 mph. In addition, the values of these factors are computed for 5, 10, and 15 mph. Table 10.2.3 gives annual average wind speeds for selected U.S. locations. Note that the wind speeds range from a low of about 5 mph to a high of about 18 mph.

$$F_r = K_r V^n \tag{10.2.4}$$

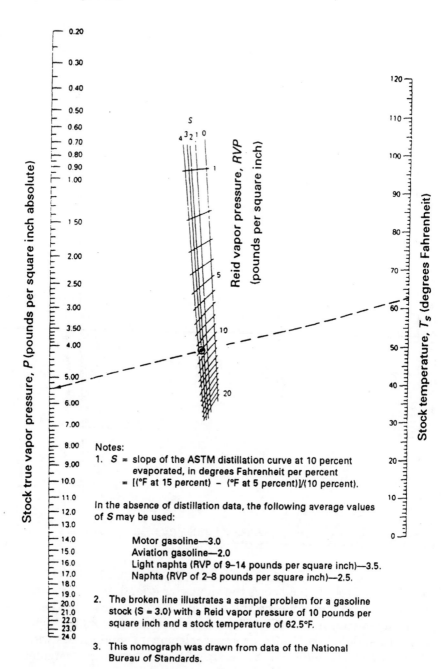

Notes:

1. S = slope of the ASTM distillation curve at 10 percent evaporated, in degrees Fahrenheit per percent
   = [(°F at 15 percent) − (°F at 5 percent)]/(10 percent).

In the absence of distillation data, the following average values of S may be used:

   Motor gasoline—3.0
   Aviation gasoline—2.0
   Light naphta (RVP of 9–14 pounds per square inch)—3.5.
   Naphta (RVP of 2–8 pounds per square inch)—2.5.

2. The broken line illustrates a sample problem for a gasoline stock (S = 3.0) with a Reid vapor pressure of 10 pounds per square inch and a stock temperature of 62.5°F.

3. This nomograph was drawn from data of the National Bureau of Standards.

Figure 10.2.3 True vapor pressure of refined petroleum stocks with a Reid vapor pressure of 1 to 20 psi. (*Courtesy of the American Petroleum Institute.*)

TABLE 10.2.1 Physical Properties (M_V, W_V, P_V, A, B) of Selected Petrochemicals

Petroleum liquid	Vapor molecular weight M_V, lb·mol	Boiling temperature (at 14.7 psia), °F	Condensed vapor density (at 60°F) W_V, lb/gal	Vapor pressure P_V, psia at: 40°F	50°F	60°F	70°F	80°F	90°F	100°F	Vapor pressure equation constants* A (dimensionless)	B, °R	Temperature range for constants A and B Minimum	Maximum
Alcohols														
Methyl alcohol	32.04	148.5	6.641	0.735	1.006	1.412	1.953	2.610	3.461	4.525	15.9482	8131.3	−47.2	435.2
Ethyl alcohol	46.07	172.9	6.643	0.193	0.406	0.619	0.870	1.218	1.682	2.320	16.3801	8760.7	−24.3	467.6
Allyl alcohol	58.08	205.9	7.127	0.135	0.193	0.261	0.387	0.522	0.716	1.006	17.1073	9579.2	−4.0	205.9
Isopropyl alcohol	60.10	180.1	6.607	0.213	0.329	0.483	0.677	0.928	1.296	1.779	16.7687	9113.6	−15.0	449.6
Tertiary butyl alcohol	74.12	180.4	6.586	0.174	0.290	0.425	0.638	0.909	1.238	1.702	17.2230	9430.3	−4.7	432.5
Amines														
Diethyl amine	73.14	131.9	5.942	1.644	1.992	2.862	3.867	4.892	6.130	7.541	13.7881	6617.7	−27.4	410.0
Aromatic hydrocarbons														
Benzene	78.11	176.2	7.361	0.638	0.870	1.160	1.508	1.972	2.610	3.287	14.0920	7377.5	−34.1	554.5
Toluene	92.14	231.1	7.289	0.174	0.213	0.309	0.425	0.580	0.773	1.006	13.8288	7770.6	−16.1	606.2
Chlorine compounds														
Methylene chloride	84.93	103.6	11.150	3.094	4.254	5.434	6.787	8.702	10.329	13.342	14.8970	6857.5	−94.0	105.3
Chloroform	119.38	142.1	12.503	1.528	1.934	2.475	3.191	4.061	5.163	6.342	13.8649	6792.5	−72.4	489.2
Carbon tetrachloride	153.82	170.0	13.357	0.793	1.064	1.412	1.798	2.301	2.997	3.771	13.5218	6908.7	−58.0	528.8
Vinylidene chloride	96.94	98.6	10.164	4.990	6.344	7.930	9.806	11.799	15.280	23.210	14.6756	6531.1	−107.0	89.1
1,2-Dichloroethane	98.96	182.2	10.509	0.561	0.773	1.025	1.431	1.740	2.243	2.804	13.8035	7200.2	−48.1	545.0
Trichloroethylene	131.40	189.0	12.234	0.503	0.677	0.889	1.180	1.508	2.030	2.610	14.3744	7529.8	−46.8	188.1
1,1,1-Trichloroethane	133.40	165.4	11.229	0.909	1.218	1.586	2.030	2.610	3.307	4.199	14.3734	7256.4	−61.6	165.4
Allyl chloride	76.53	112.3	7.828	2.998	3.772	4.797	6.015	7.447	9.110	11.025	14.4564	6689.5	−94.0	112.3

TABLE 10.2.1 Physical Properties (M_V, W_V, P_V, A, B) of Selected Petrochemicals (Continued)

Petroleum liquid	Vapor molecular weight M_V, lb•mol	Boiling temperature (at 14.7 psia), °F	Condensed vapor density (at 60°F) W_V, lb/gal	Vapor pressure P_V, psia at:							Vapor pressure equation constants*		Temperature range for constants A and B,	
				40°F	50°F	60°F	70°F	80°F	90°F	100°F	A (dimensionless)	B, °R	Minimum	Maximum
Acrylonitrile	53.06	171.5	6.726	0.812	0.967	1.373	1.779	2.378	3.133	4.022	14.1319	7191.8	−59.8	173.3
Cyano														
Cyclopentane	70.13	120.7	6.338	2.514	3.287	4.177	5.240	6.517	8.063	9.668	14.3384	6711.5	−90.4	120.7
Cyclohexane	84.16	177.3	6.532	0.677	0.928	1.218	1.605	2.069	2.610	3.249	13.6969	7091.7	−49.5	495.5
Cycloparaffins														
Methyl acetate	74.08	134.5	7.847	1.489	2.011	2.746	3.693	4.699	5.762	6.961	14.3340	7002.9	−71.0	437.0
Vinyl acetate	86.09	162.5	7.827	0.735	0.986	1.296	1.721	2.262	3.113	4.022	15.0324	7670.9	−54.4	162.5
Methyl acrylate	86.09	176.9	7.957	0.599	0.773	1.025	1.354	1.798	2.398	3.055	14.9971	7786.4	−46.7	176.4
Ethyl acetate	88.11	170.7	7.548	0.580	0.831	1.102	1.489	1.934	2.514	3.191	14.4776	7517.5	−46.1	455.0
Methyl methacrylate	100.11	212.5	7.928	0.116	0.213	0.348	0.541	0.773	1.064	1.373	14.7995	8127.7	−22.9	213.8
Esters														
Diethyl ether	74.12	94.0	6.002	4.215	5.666	7.019	8.702	10.443	13.342	>14.7	13.9146	6290.5	−101.7	361.9
Ethers														
Acetone	58.08	133.3	6.647	1.682	2.185	2.862	3.713	4.699	5.917	7.251	14.2539	6920.2	−74.9	418.1
Methyl ethyl ketone	72.11	175.4	6.753	0.715	0.928	1.199	1.489	2.069	2.668	3.345	14.3812	7380.2	−54.9	175.3
Ketones														
n-Pentane	72.15	96.9	5.262	4.293	5.454	6.828	8.433	10.445	12.959	15.474	13.2999	5972.6	−105.9	376.3
n-Hexane	86.18	155.7	5.534	1.102	1.450	1.876	2.436	3.055	3.906	4.892	13.8236	6907.2	−65.0	408.9
n-Heptane	100.20	209.2	5.738	0.290	0.406	0.541	0.735	0.967	1.238	1.586	13.9835	7615.8	−29.2	477.5
Paraffins														

*The vapor pressure equation is $P_V = \exp(A - B/T_l)$, where P_V is the vapor pressure in psia, T_l is the liquid surface temperature in °R, and exp is the exponential function.

TABLE 10.2.2 Properties (M_V, W_V, P_{VA}, W_l) of Selected Petroleum Liquids

Petroleum liquid	Vapor molecular weight (at 60°F) M_V, lb/(lb·mol)	Condensed vapor density (at 60°F) W_V, lb/gal	Liquid density W_l, lb/gal, at 60°F	True vapor pressure P_{VA}, psi, at:						
				40°C	50°F	60°F	70°F	80°F	90°F	100°F
Gasoline RVP 13	62	4.9	4.9	4.7	5.7	6.9	8.3	9.9	11.7	13.8
Gasoline RVP 10	66	5.1	5.1	3.4	5.7	5.2	6.2	7.4	8.8	10.5
Gasoline RVP 7	68	5.2	5.2	2.3	2.9	3.5	4.3	5.2	6.2	7.4
Crude oil RVP 5	50	4.5	4.5	1.8	2.3	2.8	3.4	4.0	4.8	5.7
Jet naphtha (JP-4)	80	5.4	5.4	0.8	1.0	1.3	1.6	1.9	2.4	2.7
Jet kerosene	130	6.1	6.1	0.0041	0.0060	0.0085	0.011	0.015	0.021	0.029
Distillate fuel oil no. 2	130	6.1	6.1	0.0031	0.0045	0.0074	0.0090	0.012	0.016	0.022
Residual oil no. 6	190	6.4	6.4	0.00002	0.00003	0.00004	0.00006	0.00009	0.00013	0.00019

TABLE 10.2.3   Average Annual Ambient Temperature T_a and Annual Wind Speed V for Selected U.S. Locations

Location	Ambient temp., °F	Wind speed, mph	Location	Ambient temp., °F	Wind speed, mph
Alabama			Colorado		
Birmingham	62.0	7.3	Colorado Springs	48.9	10.1
Huntsville	60.6	8.1	Denver	50.3	8.8
Mobile	67.5	9.0	Grand Junction	52.7	8.1
Montgomery	64.9	6.7	Pueblo	52.8	8.7
Alaska			Connecticut		
Anchorage	35.3	6.8	Bridgeport	51.8	12.0
Annette	45.4	10.6	Hartford	49.8	8.5
Barrow	9.1	11.8	Delaware		
Barter Island	9.6	13.2	Wilmington	54.0	9.2
Bethel	28.4	12.8	District of Columbia		
Bettles	21.2	6.7	Dulles Airport	53.9	7.5
Big Delta	27.4	8.2	National Airport	57.5	9.3
Cold Bay	37.9	16.9	Florida		
Fairbanks	25.9	5.4	Apalachicola	68.2	7.9
Gulkana	26.5	6.8	Daytona Beach	70.3	8.8
Homer	36.6	7.2	Fort Myers	73.9	8.2
Juneau	40.0	8.4	Jacksonville	68.0	8.2
King Salmon	32.8	10.7	Key West	77.7	11.2
Kodiak	40.7	10.6	Miami	75.7	9.2
Kotzebue	20.9	13.0	Orlando	72.4	8.6
McGrath	25.0	5.1	Pensacola	68.0	8.4
Nome	25.5	10.7	Tallahassee	67.2	6.5
St. Paul Island	34.3	18.3	Tampa	72.0	8.6
Talkeetna	32.6	4.5	West Palm Beach	74.6	9.5
Valdez	26.4	6.0	Georgia		
Yakutat	38.3	7.4	Athens	61.4	7.4
Arizona			Atlanta	61.2	9.1
Flagstaff	45.4	7.3	Augusta	63.2	6.5
Phoenix	71.2	6.3	Columbus	64.3	6.7
Tucson	68.0	8.2	Macon	64.7	7.7
Winslow	54.9	8.9	Savannah	65.9	7.9
Yuma	73.8	7.8	Hawaii		
Arkansas			Hilo	73.6	7.1
Fort Smith	60.8	7.6	Honolulu	77.0	11.6
Little Rock	61.9	8.0	Kahului	75.5	12.8
California			Lihue	75.2	11.9
Bakersfield	65.5	6.4	Idaho		
Blue Canyon	50.4	7.7	Boise	51.1	8.9
Eureka	52.0	6.8	Pocatello	46.6	10.2
Fresno	62.6	6.4	Illinois		
Long Beach	63.9	6.4	Cairo	59.1	8.5
Los Angeles (city)	65.3	6.2	Moline	49.5	10.0
Los Angeles			Peoria	50.4	10.1
International Airport	62.6	7.5	Rockford	47.8	9.9
Mount Shasta	49.5	5.1	Springfield	52.6	11.3
Red Bluff	62.9	8.6	Indiana		
Sacramento	60.6	8.1	Evansville	55.7	8.2
San Diego	63.8	6.8	Fort Wayne	49.7	10.2
San Francisco (city)	56.8	8.7	Indianapolis	52.1	9.6
San Francisco (airport)	56.6	10.5	South Bend	49.4	10.4
Santa Maria	56.8	7.0			
Stockton	61.6	7.5			

TABLE 10.2.3   Average Annual Ambient Temperature T_a and Annual Wind Speed V for Selected U.S. Locations (Continued)

Location	Ambient temp., °F	Wind speed, mph	Location	Ambient temp., °F	Wind speed, mph
Iowa			**Missouri (cont.)**		
Des Moines	49.7	10.9	Springfield	55.9	10.9
Sioux City	48.4	11.0	**Montana**		
Waterloo	46.1	10.7	Billings	46.7	11.3
Kansas			Glascow	41.6	10.8
Concordia	53.2	12.3	Great Falls	44.7	12.8
Dodge City	55.1	13.9	Havre	42.3	9.9
Goodland	50.7	12.6	Helena	43.3	7.8
Topeka	54.1	10.2	Kalispell	42.5	6.6
Wichita	56.4	12.4	Miles City	45.4	10.2
Kentucky			Missoula	44.1	6.1
Cincinnati airport	53.4	9.1	**Nebraska**		
Jackson	52.6	7.0	Grand Island	49.9	12.0
Lexington	54.9	9.5	Lincoln	50.5	10.4
Louisville	56.2	8.3	Norfolk	58.3	11.8
Louisiana			North Platte	48.1	10.3
Baton Rouge	67.5	7.7	Omaha (city)	49.5	10.6
Lake Charles	68.0	8.7	Scotts Bluff	48.5	10.6
New Orleans	68.2	8.2	Valetine	46.8	10.0
Shreveport	68.4	8.6	**Nevada**		
Maine			Elko	46.2	6.0
Caribou	38.9	11.2	Ely	44.4	10.4
Portland	45.0	8.7	Las Vegas	66.2	9.2
Maryland			Reno	49.4	6.5
Baltimore	55.1	9.2	Winnemucca	48.8	7.9
Massachusetts			**New Hampshire**		
Blue Hills Observatory	48.6	15.4	Concord	45.3	6.7
Boston	51.5	12.4	Mt. Washington	26.6	35.1
Worcester	46.8	10.2	**New Jersey**		
Michigan			Atlantic City	54.1	10.2
Alpena	42.2	7.9	Newark	54.2	10.2
Detroit	48.6	10.2	**New Mexico**		
Flint	46.8	10.3	Albuquerque	56.2	9.1
Grand Rapids	47.5	9.8	Roswell	61.4	8.7
Houghton Lake	42.9	8.9	**New York**		
Lansing	47.2	10.1	Albany	47.2	8.9
Muskegon	47.2	10.7	Binghamton	45.7	10.3
Sault Ste. Marie	39.7	9.4	Buffalo	47.6	12.1
Minnesota			New York (Central Park)	54.6	9.4
Duluth	38.2	11.2	New York (JFK Airport)	53.2	12.2
International Falls	36.4	9.0	New York	54.3	12.3
Minneapolis-St. Paul	44.7	10.5	(La Guardia Airport)		
Rochester	43.5	12.9	Rochester	48.0	9.8
St. Cloud	41.4	8.0	Syracuse	47.7	9.7
Mississippi			**North Carolina**		
Jackson	64.6	7.4	Asheville	55.5	7.6
Meridian	64.1	6.0	Cape Hatteras	61.9	11.4
Missouri			Charlotte	60.0	7.5
Columbia	34.1	9.8	Greensboro	57.8	7.6
Kansas City (city)	59.1	9.9	Raleigh	59.0	7.8
Kansas City (airport)	56.3	10.7	Wilmington	63.4	8.9
St. Louis	55.4	9.7			

TABLE 10.2.3   Average Annual Ambient Temperature T_a and Annual Wind Speed V for Selected U.S. Locations (Continued)

Location	Ambient temp., °F	Wind speed, mph	Location	Ambient temp., °F	Wind speed, mph
North Dakota			Texas (cont.)		
Bismarck	41.3	10.3	Austin	68.1	9.3
Fargo	40.5	12.5	Brownsville	73.6	11.6
Williston	40.8	10.1	Corpus Christi	72.1	12.0
Ohio			Dallas-Ft. Worth	66.0	10.8
Akron	49.5	9.8	Del Rio	69.8	9.9
Cleveland	49.6	10.7	El Paso	63.4	9.2
Columbus	51.7	8.7	Galveston	69.6	11.0
Dayton	51.9	10.1	Houston	68.3	7.8
Mansfield	49.5	11.0	Lubbock	59.9	12.4
Toledo	48.6	9.4	Midland-Odessa	63.5	11.1
Youngstown	48.2	10.0	Port Arthur	68.7	9.9
Oklahoma			San Angelo	65.7	10.4
Oklahoma City	59.9	12.5	San Antonio	68.7	9.4
Tulsa	60.3	10.4	Victoria	70.1	10.0
Oregon			Waco	67.0	11.3
Astoria	50.6	8.5	Wichita Falls	63.5	11.7
Eugene	52.5	7.6	Utah		
Medford	53.6	4.8	Salt Lake City	51.7	8.8
Pendleton	52.5	9.0	Vermont		
Portland	53.0	7.9	Burlington	44.1	8.8
Salem	52.0	7.0	Virginia		
Sexton Summit	47.7	11.8	Lynchburg	56.0	7.8
Pennsylvania			Norfolk	59.5	10.5
Allentown	51.0	9.2	Richmond	57.7	7.5
Avoca	49.1	8.4	Roanoke	56.1	8.3
Erie	47.5	11.2	Washington		
Harrisburg	53.0	7.7	Olympia	49.6	6.7
Philadelphia	54.3	9.5	Quillayute	48.7	6.1
Williamsport	50.1	7.9	Seattle International	51.4	9.1
Rhode Island			Airport		
Providence	50.3	10.6	Spokane	47.2	8.7
South Carolina			Walla Walla	54.1	5.3
Charleston	66.1	8.7	Yakima	49.7	7.1
Columbia	63.3	6.9	West Virginia		
Greenville-Spartanburg	60.1	6.7	Beckley	50.9	9.3
South Dakota			Charleston	54.8	6.4
Aberdeen	43.0	11.2	Elkins	49.3	6.2
Huron	44.7	11.7	Huntington	55.2	6.5
Rapid City	46.7	11.2	Wisconsin		
Sioux Falls	45.3	11.1	Green Bay	43.6	10.1
Tennessee			La Crosse	46.1	8.8
Bristol-Johnson City	55.9	5.6	Madison	45.2	9.8
Chattanooga	59.4	6.2	Milwaukee	46.1	11.6
Knoxville	48.9	7.1	Wyoming		
Memphis	61.8	9.0	Casper	45.2	12.9
Nashville	59.2	8.0	Cheyenne	45.7	12.9
Oak Ridge	57.5	4.4	Lander	44.4	6.9
Texas			Sheridan	44.6	8.1
Abilene	64.5	12.2			
Amarillo	57.3	13.7			

TABLE 10.2.4 **Average Annual Stock Storage Temperature** T_S
as a Function of Tank Paint Color

Tank color	Average annual stock storage temperature T_S, °F
White	T_a
Aluminum	$T_a + 2.5$
Gray	$T_a + 3.5$
Black	$T_a + 5.0$

Note: T_a = average annual ambient temperature in degrees Fahrenheit.

TABLE 10.2.5 **IFRT Rim Seal Loss Coefficient** K_R *for* **Average- and Tight-Fitting Seals**

	F_R, lb•mol/(ft•yr)	
Tank type and seal type	Average fit	Tight fit
Welded tank		
Mechanical-shoe seal		
Primary only	3.0	2.6
Rim-mounted secondary	1.6	1.2
Liquid-mounted resilient filled seal		
Primary only	3.0	2.6
Rim-mounted secondary	1.6	1.2
Vapor-mounted resilient filled seal		
Primary only	6.7	5.6
Rim-mounted secondary	2.5	2.3
Flexible-wiper seal		
Primary only	6.7	5.6
Rim mounted secondary	2.5	2.3

TABLE 10.2.6 **EFRT Rim Seal Loss Coefficient** K_R **for Average-Fitting Seals**

Tank construction and rim seal system	Average-fitting seals		F_r, lb•mol/(ft diam•yr)		
	K_r, lb•mol/ [(mi/h)^n•ft•yr]	n (dimensionless)	5	10	15
Welded tanks					
Mechanical-shoe seal					
Primary only	1.2	1.5	13.4	37.9	69.7
Shoe-mounted secondary	0.8	1.2	5.52	12.7	20.6
Rim-mounted secondary	0.2	1.0	1.00	2.00	3.00
Liquid-mounted resilient filled seal					
Primary only	1.1	1.0	5.50	11.0	16.5
Weather shield	0.8	0.9	3.41	6.35	9.15
Rim-mounted secondary	0.7	0.4	1.33	1.76	2.07
Vapor-mounted resilient filled seal					
Primary only	1.2	2.3	48.6	239.0	608.0
Weather shield	0.9	2.2	31.0	143.0	348.0
Rim-mounted secondary	0.2	2.6	13.1	79.6	228.0
Riveted tanks					
Mechanical-shoe seal					
Primary only	1.3	1.5	14.5	41.1	75.5
Shoe-mounted secondary	1.4	1.2	9.66	22.2	36.1
Rim-mounted secondary	0.2	1.6	2.63	7.96	15.2

Note: The rim seal loss factors K_r and n may only be used for wind speeds from 2 to 15 mph.

where K_r = rim seal loss coefficient, lb•mol/yr
V = average wind speed, mph
n = rim-seal-related wind speed exponent, dimensionless

Roof fitting loss factor F_f.   The total roof fitting loss factor in Eq. (10.2.3) F_f can be computed from

$$F_f = (N_{f1}K_{f1} + N_{f2}K_{f2} + \cdots + N_{fk}K_{fk}) \qquad (10.2.5)$$

where F_f = total roof fitting loss factor, lb•mol/yr
N_{fi} = number of roof fittings of particular type (dimensionless)
K_{fi} = loss factor for particular type of roof fitting, lb•mol/yr
i = 1, 2,...,k (dimensionless)
k = total number of different types of roof fittings (dimensionless)

For internal floating roof.   The roof fitting loss factor may be computed by knowing the fittings used and with the help of Tables 10.2.5, 10.2.7, 10.2.8, and 10.2.9. Alternatively, a representative value that does not require a tally of fitting number and type can be found from generic equations. These equations will give good estimates for a typical roof configuration.

Column-supported roofs.   For bolted decks,

$$F_f = 0.0481D^2 + 1.392D + 134.2 \quad \text{lb•mol/yr}$$

For welded decks,

$$F_f = 0.0385D^2 + 1.392D + 134.2 \quad \text{lb•mol/yr}$$

Self-supported roofs.   For bolted decks,

$$F_f = 0.0228D^2 + 0.79D + 105.2 \quad \text{lb•mol/yr}$$

For welded decks,

$$F_f = 0.0132D^2 + 0.79D + 105.2 \quad \text{lb•mol/yr}$$

These equations are plotted in Figs. 10.2.4 and 10.2.5.

For EFRTs.   For EFRTs the loss factors for fittings are not constants but depend on the wind speed (as does the rim seal loss factor for EFRTs). For EFRTs the loss factor for a particular type of roof fitting K_{fi} can be estimated as follows:

$$K_{fi} = K_{fai} + K_{fbi}V^{mi} \qquad (10.2.6)$$

TABLE 10.2.7 IFRT Roof Fitting Loss Factor F_F for Typical Construction Details

API fitting no.	Roof fitting type; construction details	Typical no. of fittings N_F	F_F lb·mol/yr	Notes
8	Vacuum breaker; weighted actuation, gasketed	1	0.7	
7	Stub drain	See Note	1.2	$N_{F7} = 0$ for welded floating decks. $N_{F7} = D^2/125$ for bolted floating decks where D = tank diameter, ft
1	Access hatch; bolted cover, gasketed	1	1.6	$N_{F3} = 5 + D/10 + D_2/600$, where D = tank diameter, ft.
3	Roof leg; adjustable	See Note	7.9	
4	Gauge-float well; unbolted cover, ungasketed	15	28.0	Valve from Table 10.2.8
2	Column well; built-up column, sliding cover, ungasketed	See Note	47.0	
6	Unslotted guide pole well; sliding cover, ungasketed	1	32.0	
6	Slotted guide pole sample well; sliding cover, ungasketed	1	57.0	
5	Ladder well, 36-in-diameter; sliding cover, ungasketed	1	76.0	

TABLE 10.2.8   Typical Number of Columns N_c for Tanks with Column-Supported Fixed Roofs

Tank diameter range D, ft	Typical number of columns N_c
$0 < D \leq 85$	1
$85 < D \leq 100$	6
$100 < D \leq 120$	7
$120 < D \leq 135$	8
$135 < D \leq 150$	9
$150 < D \leq 170$	16
$170 < D \leq 190$	19
$190 < D \leq 220$	22
$220 < D \leq 235$	31
$235 < D \leq 270$	37
$270 < D \leq 275$	43
$275 < D \leq 290$	49
$290 < D \leq 330$	61
$330 < D \leq 360$	71
$360 < D \leq 400$	81

TABLE 10.2.9   Deck Seam Length Factors S_D for Typical Deck Constructions for Internal Floating-Roof Tanks

Deck	Typical deck seam length factor S_D, ft/ft²
Continuous-Sheet Construction*	
5 ft wide	0.20†
6 ft wide	0.17
7 ft wide	0.14
Panel Construction‡	
5 × 7.5 ft rectangular	0.33
5 × 12 ft rectangular	0.28

Deck seam loss applies to bolted decks only.
*$S_D = 1/W$, where W = sheet width, ft.
†If no specific information is available, this factor can be assumed to represent the most common bolted decks currently in use.
‡$S_D = (L + W)/(LW)$, where W = panel width and L = panel length.

where K_{fai} = loss factor for a particular type of roof fitting, lb•mol/yr

K_{fbi} = loss factor for a particular type of roof fitting, lb•mol/[mi/h]^m•yr

m_i = loss factor exponent for a particular type of roof fitting (dimensionless)

$i = 1, 2,..., k$ (dimensionless)

V = Average wind speed, mph

Tables 10.2.10 to 10.2.12 show roof fitting loss factors for some typical fittings. For convenience, the factors have been used to determine the loss factors at 5, 10, and 15 mph in the last three columns of Tables 10.2.10 to provide an idea of the value of the factors and how

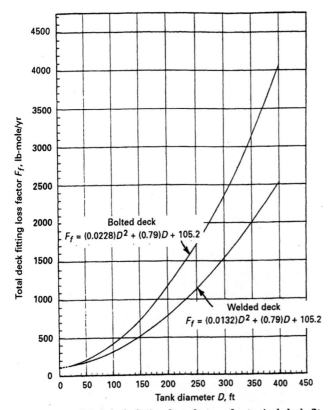

Figure 10.2.4 Total deck fitting loss factors for typical deck fittings in tanks with self-supporting fixed roofs. (*Courtesy of the American Petroleum Institute.*)

they vary with wind speed. Where multiple fittings are involved such as roof legs or rim vents, the specific number can be estimated from Table 10.2.11 or 10.2.12. A shortcut to computing all the roof fitting losses is to use Figs. 10.2.6 and 10.2.7 which give typical total roof fitting loss factors for single- and double-deck roofs.

Deck seam loss factor F_d (applies to IFRs only). It should be understood that all the API factors the apply to deck seams are generic and do not take into account the true value of loss factors. To ensure that manufacturers of internal floating roofs are credited with better designs with reduced evaporative losses through deck seams and fittings, API has undertaken a program of certification and development of individualized roof loss factors which should be available for use by manufacturers and users in the late 1990s.

Losses from decks can occur as the result of the seams which may be used in the construction to connect sections of roof deck. The equation below is used to determine these losses:

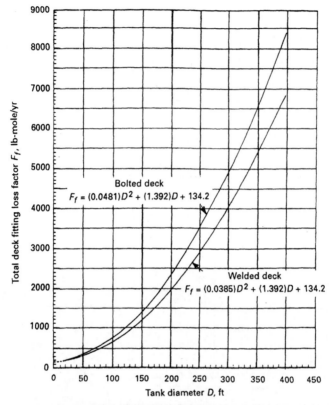

Figure 10.2.5 Total deck fitting loss factors for typical deck fittings in tanks with column-supported fixed roofs. (*Courtesy of the American Petroleum Institute.*)

$$F_d = K_d S_d D^2 \qquad (10.2.7)$$

where F_d = deck seam loss factor, lb•mol/yr
  K_d = deck seam loss per unit seam length factor, lb•mol/(ft•yr)
  S_d = deck seam length factor, ft/ft²
  D = tank diameter, ft

The factor K_d represents the losses per unit length of deck and is assumed to be constant. The following may be used to determine K_d for a great number of cases; however, there are numerous proprietary deck seam designs, and the manufacturer should be contacted to determine a better value for this factor.

$$K_d = \begin{cases} 0 & \text{for welded steel decks} \\ 0.34 & \text{for bolted decks} \end{cases}$$

TABLE 10.2.10 EFRT Roof-Fitting Loss Factors K_{fa}, K_{fb}, and m and Typical Number of Roof Fittings N_f

| Roof fitting type; construction details | Loss factors | | | Typical number of fittings N_f | F_F lb·mol/yr | | |
| | K_{fa}, lb·mol/yr | K_{fb}, lb·mol/[(mi/h)^m·yr] | m (dimensionless) | | Wind speed, mi/h | | |
					5	10	15
Access hatch; bolted cover, gasketed	0.	0	0	1	0.	0.	0.
Rim vent; weighted actuation, gasketed	0.71	0.10	1.0	1	1.21	1.71	2.21
Gauge-hatch/sample well; weighted actuation, gasketed	0.95	0.14	1.0	1	1.65	2.35	3.05
Vacuum breaker; weighted actuation, gasketed	1.2	0.17	1.0	Table 10.2.11	2.05	2.90	3.75
Roof leg; adjustable pontoon area	1.5	0.20	1.0	Table 10.2.12	2.50	3.50	4.50
Gauge-float well; unbolted cover, ungasketed	2.3	5.7	1.0	1	31.8	61.3	90.8
Overflow roof drain; open	0	7.0	1.4	Table 10.2.11	66.6	176	310
Unslotted guide pole well; sliding cover, ungasketed (see Sec. 10.4)							
Slotted guide pole/sample well; cover, ungasketed (See Sec. 10.4)							

TABLE 10.2.11   Typical Number of Vacuum Breakers N_{fb} and Roof Drains N_{ft}

| Tank diameter, D, ft | Number of vacuum breakers N_{fb} | | Number of roof drains N_{ft} (double-deck roof) |
	Pontoon roof	Double-deck roof	
50	1	1	1
100	1	1	1
150	2	2	2
200	3	2	3
250	4	3	5
300	5	3	7
350	6	4	—
400	7	4	—

Contact the manufacturer for other designs such as adhesively joined or caulked seams.

The deck seam length factor simply relates the ratio of seams to the area of the deck. If the deck is assembled with numerous narrow panels, the deck seam length factor will be larger than that for a deck with fewer seams. The deck seam length factor may be expressed as

$$S_d = \frac{L_{seam}}{A_{deck}} \qquad (10.2.8)$$

where S_d = deck seam length factor, ft/ft²
L_{seam} = total length of deck seams, ft
A_{deck} = total deck area, ft²

Table 10.2.9 may be used in lieu of the equation.

10.2.1.2 Withdrawal loss

The withdrawal loss from both internal and external floating-roof tanks can be established by

$$L_w = \frac{0.943QCW_l}{D}\left(1 + \frac{N_c F_c}{D}\right) \qquad (10.2.9)$$

where L_W = withdrawal loss, lb/yr
Q = annual net throughput (associated with lowering liquid stock level in tank), bbl/yr
C = clingage factor, bbl/1000 ft²
W_l = average liquid stock density at average storage temperature, lb/gal
N_c = number of columns, dimensionless

TABLE 10.2.12 Typical Number of Roof Legs N_{rs}

| Tank diameter D, ft | Pontoon roof | | Number of legs on double-deck roof |
	Number of pontoon legs	Number of center legs	
30	4	2	6
40	4	4	7
50	6	6	8
60	9	7	10
70	13	9	13
80	15	10	16
90	16	12	20
100	17	16	25
110	18	20	29
120	19	24	34
130	20	28	40
140	21	33	46
150	23	38	52
160	26	42	58
170	27	49	66
180	28	56	74
190	29	62	82
200	30	69	90
210	31	77	98
220	32	83	107
230	33	92	115
240	34	101	127
250	35	109	138
260	36	118	149
270	36	128	162
280	37	138	173
290	38	148	186
300	38	156	200
310	39	168	213
320	39	179	226
330	40	190	240
340	41	202	255
350	42	213	270
360	44	226	285
370	45	238	300
380	46	252	315
390	47	266	330
400	48	281	345

F_c = effective column diameter, ft
D = tank diameter, ft

For an EFRT, the term in parentheses on the right side of Eq. (10.2.9) is equal to 1.0. The constant, 0.943, has dimensions of $1000 \text{ ft}^3 \times \text{gal/bbl}^2$.

The withdrawal loss is converted from pounds per year to barrels per year as follows:

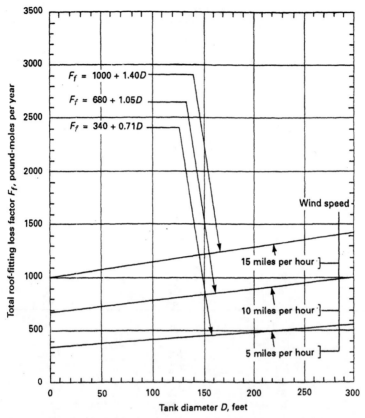

Figure 10.2.6 Total roof fitting loss factors for typical fittings on pontoon floating roofs. (*Courtesy of the American Petroleum Institute.*)

$$L_w \text{ (bbl/yr)} = \frac{\text{lb/yr}}{42W_l} \qquad (10.2.10)$$

where W_l = average liquid stock density at 60°F, in pounds per gallon.
Clingage factors may be determined as follows:

| | | Average clingage factor C (bbl/1000 ft²) Shell and column condition | |
Product	Light rust (see note)	Dense rust	Gunite-lined
Gasoline	0.0015	0.0075	0.15
Single-component stocks	0.0015	0.0075	0.15
Crude oil	0.0060	0.030	0.60

Note: If no specific information is available, these values can be assumed to represent the most common or typical condition of tanks currently in use.

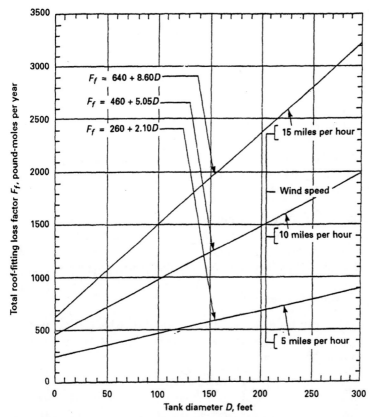

Figure 10.2.7 Total roof fitting loss factors for typical fittings on double-deck floating roofs. (*Courtesy of the American Petroleum Institute.*)

Average liquid density. For gasoline, 6.1 lb/gal may be used. For other stocks, the actual liquid density should be used.

SECTION 10.3 EMISSION ESTIMATION PROCEDURES FOR FIXED-ROOF TANKS

10.3.1 Losses from Fixed-Roof Tanks

Total losses from fixed-roof tanks are equal to the sum of the standing storage loss and working loss:

$$L_T = L_S + L_W \tag{10.3.1}$$

where L_T = total losses, lb/yr
L_S = standing storage losses, lb/yr
L_W = working losses, lb/yr

10.3.1.1 Standing storage loss

Fixed-roof tank breathing or standing storage losses can be estimated from

$$L_S = 365 V_V W_V K_E K_S \qquad (10.3.2)$$

where L_S = standing storage loss, lb/yr
V_V = vapor space volume, ft^3
W_V = vapor density, lb/ft^3
K_E = vapor space expansion factor, dimensionless
K_S = vented vapor saturation factor, dimensionless
365 = constant, days/yr

Computing the variables in a strategic manner simplifies the computations. This order is presented below. Meteorological data are the starting point for determining other data. They are used to determine the liquid temperatures which are then used to determine vapor pressures.

Temperature-related calculations

Daily average ambient temperature. First calculate the *daily average ambient temperature* T_{AA} and *daily ambient temperature range* ΔT_A, using the following equations:

$$T_{AA} = \frac{T_{AX} + T_{AN}}{2} \qquad (10.3.3)$$

$$\Delta T_A = T_{AX} - T_{AN} \qquad (10.3.4)$$

where T_{AA} = daily average ambient temperature, °R
T_{AX} = daily maximum ambient temperature, °R
T_{AN} = daily minimum ambient temperature, °R
ΔT_A = daily ambient temperature range, °R

Table 10.3.1 gives values of T_{AX} and T_{AN} for selected locations. A more complete listing can be found in API Publications 2517, 2518, and 2519.

Liquid bulk temperature. These values may then be used to determine the *liquid bulk temperature* T_B by using the following equation:

$$T_B = T_{AA} + 6\alpha - 1 \qquad (10.3.5)$$

where T_B = liquid bulk temperature, °R
T_{AA} = daily average ambient temperature, °R
α = tank paint solar absorptance, dimensionless; see Table 10.3.2

TABLE 10.3.1   Meteorological Data (T_{AX}, T_{AN}, I) for Selected U.S. Locations

Location	Symbol*	Units	Jan.	Feb.	Mar.	Apr.	May	June	July	Aug.	Sep.	Oct.	Nov.	Dec.	Annual average
Birmingham	T_{AX}	°F	52.7	57.3	65.2	75.2	81.6	87.9	90.3	89.7	84.6	74.8	63.7	55.9	73.2
airport, AL	T_{AN}	°F	33.0	35.2	42.1	50.4	58.3	65.9	69.8	69.1	63.6	50.4	40.5	35.2	51.1
	I	Btu/ (ft²• day)	707	967	1296	1674	1857	191	1810	1724	1455	1211	858	661	1345
Montgomery, AL	T_{AX}	°F	57.0	60.9	68.1	77.0	83.6	89.8	91.5	91.2	86.9	77.5	67.0	59.8	75.9
	T_{AN}	°F	36.4	38.8	45.5	53.3	61.1	68.4	71.8	71.1	66.4	53.1	43.0	37.9	53.9
	I	Btu/ (ft²• day)	752	1013	1341	1729	1897	1972	1841	1746	1468	1262	915	719	1388
Homer, AK	T_{AX}	°F	27.0	31.2	34.4	42.1	49.8	56.3	60.5	60.3	54.8	44.0	34.9	27.7	43.6
	T_{AN}	°F	14.4	17.4	19.3	28.1	34.6	41.2	45.1	45.2	39.7	30.6	22.8	15.8	29.5
	I	Btu/ (ft²• day)	122	334	759	1248	1583	1751	1598	1189	791	437	175	64	838
Phoenix, AZ	T_{AX}	°F	65.2	69.7	74.5	83.1	92.4	102.3	105.0	102.3	98.2	87.7	74.3	66.4	85.1
	T_{AN}	°F	39.4	42.5	46.7	53.0	61.5	70.6	79.5	77.5	70.9	59.1	46.9	40.2	57.3
	I	Btu/ (ft²• day)	1021	1374	1814	2355	2677	2739	2487	2293	2015	1577	1151	932	1869
Tucson, AZ	T_{AX}	°F	64.1	67.4	71.8	80.1	88.8	98.5	98.5	95.9	93.5	84.1	72.2	65.0	81.7
	T_{AN}	°F	38.1	40.0	43.8	49.7	57.5	67.4	73.8	72.0	67.3	56.7	45.2	39.0	54.2
	I	Btu/ (ft²• day)	1099	1432	1864	2363	2671	2730	2341	2183	1979	1602	1208	996	1872
Forth Smith, AR	T_{AX}	°F	48.4	53.8	62.5	73.7	81.0	88.5	93.6	92.9	85.7	75.9	61.9	52.1	72.5
	T_{AN}	°F	26.6	30.9	38.5	49.1	38.2	66.3	70.5	68.9	62.1	49.0	37.7	30.2	49.0
	I	Btu/ (ft²• day)	744	999	1312	1616	1912	2089	2065	1877	1502	1201	851	682	1404
Little Rock, AR	T_{AX}	°F	49.8	54.5	63.2	73.8	81.7	89.5	92.7	92.3	85.6	75.8	62.4	53.2	72.9
	T_{AN}	°F	29.9	33.6	41.2	50.9	59.2	67.5	71.4	69.6	63.0	50.4	40.0	33.2	50.8
	I	Btu/ (ft²• day)	731	1003	1313	1611	1929	2107	2032	1861	1518	1228	847	674	1404
Bakersfield, CA	T_{AX}	°F	57.4	63.7	68.6	75.1	83.9	92.2	98.8	96.4	90.8	81.0	67.4	57.6	77.7
	T_{AN}	°F	38.9	42.6	45.5	50.1	57.2	64.3	70.1	68.5	63.8	54.9	44.9	38.7	53.3
	I	Btu/ (ft²• day)	731	1003	1313	1611	1929	2107	2032	1861	1518	1228	847	674	1404
Long Beach, CA	T_{AX}	°F	66.0	67.3	68.0	70.9	73.4	77.4	83.0	83.8	82.5	78.4	72.7	67.4	74.2
	T_{AN}	°F	44.3	45.9	47.7	50.8	55.2	58.9	62.6	64.0	61.6	56.6	49.6	44.7	53.5
	I	Btu/ (ft²• day)	928	1215	1610	1938	2065	2140	2300	2100	1701	1326	1004	849	1594
Los Angeles	T_{AX}	°F	64.6	65.5	65.1	66.7	69.1	72.0	75.3	76.5	76.4	74.0	70.3	66.1	70.1
airport, CA	T_{AN}	°F	47.3	48.6	49.7	52.2	55.7	59.1	62.6	64.0	62.5	58.5	52.1	47.8	55.0
	I	Btu/ (ft²• day)	926	1214	1619	1951	2060	2119	2308	2080	1681	1317	1004	849	1594

TABLE 10.3.1   Meteorological Data (T_{AX}, T_{AN}, I) for Selected U.S. Locations (Continued)

Location	Symbol*	Units	Jan.	Feb.	Mar.	Apr.	May	June	July	Aug.	Sep.	Oct.	Nov.	Dec.	Annual average
Sacramento, CA	T_{AX}	°F	52.6	59.4	64.1	71.0	79.7	87.4	93.3	91.7	87.6	77.7	63.2	53.2	73.4
	T_{AN}	°F	37.9	41.2	42.4	45.3	50.1	55.1	57.9	57.6	55.8	50.0	42.8	37.9	47.8
	I	Btu/ (ft²• day)	597	939	1458	2004	2435	2684	2688	2368	1907	1315	782	538	1643
San Francisco airport, CA	T_{AX}	°F	55.5	59.0	60.6	63.0	66.3	69.6	71.0	71.8	73.4	70.0	62.7	56.3	64.9
	T_{AN}	°F	41.5	44.1	44.9	46.6	49.3	52.0	53.3	54.2	54.3	51.2	46.3	42.2	48.3
	I	Btu/ (ft²• day)	708	1009	1455	1920	2226	2377	2392	2117	1742	1226	821	642	1553
Santa Maria, CA	T_{AX}	°F	62.8	64.2	63.9	65.6	67.3	69.9	72.1	72.8	74.2	73.3	68.9	64.6	68.3
	T_{AN}	°F	38.8	40.3	40.9	42.7	46.2	49.6	52.4	53.2	51.8	47.6	42.1	38.3	45.3
	I	Btu/ (ft²• day)	854	1141	1582	1921	2141	2349	2341	2106	1730	1353	974	804	1608
Denver, CO	T_{AX}	°F	43.1	46.9	51.2	61.0	70.7	81.6	88.0	85.8	77.5	66.8	52.4	46.1	64.3
	T_{AN}	°F	15.9	20.2	24.7	33.7	43.6	52.4	58.7	57.0	47.7	36.9	25.1	18.9	36.2
	I	Btu/ (ft²• day)	840	1127	1530	1879	2135	2351	2273	2044	1727	1301	884	732	1568
Grand Junction, CO	T_{AX}	°F	35.7	44.5	54.1	65.2	76.2	87.9	94.0	90.3	81.9	68.7	51.0	38.7	65.7
	T_{AN}	°F	15.2	22.4	29.7	38.2	48.0	56.6	63.8	61.5	52.2	41.1	28.2	17.9	39.6
	I	Btu/ (ft²• day)	791	1119	1554	1986	2380	2599	2465	2182	1834	1345	918	731	1659
Wilmington, DE	T_{AX}	°F	39.2	41.8	50.9	63.0	72.7	81.2	85.6	84.1	77.8	66.7	54.8	43.6	63.5
	T_{AN}	°F	23.2	24.6	32.6	41.8	51.7	61.2	66.3	65.4	58.0	45.9	36.4	27.3	44.5
	I	Btu/ (ft²• day)	571	827	1149	1480	1710	1883	1823	1615	1318	984	645	489	1208
Atlanta, GA	T_{AX}	°F	51.2	55.3	63.2	73.2	79.8	85.6	87.9	87.6	82.3	72.9	62.6	54.1	71.3
	T_{AN}	°F	32.6	34.5	47.1	50.4	58.7	65.9	69.2	68.7	63.6	51.4	41.3	34.8	51.1
	I	Btu/ (ft²• day)	718	969	1304	1686	1854	1914	1812	1709	1422	1200	883	674	1345
Savannah, GA	T_{AX}	°F	60.3	63.1	69.9	77.8	84.2	88.6	90.8	90.1	85.6	77.8	69.5	62.5	76.7
	T_{AN}	°F	37.9	40.0	46.8	54.1	62.3	68.5	71.5	71.4	67.6	55.9	45.5	39.4	55.1
	I	Btu/ (ft²• day)	795	1044	1399	1761	1852	1844	1784	1621	1364	1217	941	754	1365
Honolulu, HI	T_{AX}	°F	79.9	80.4	81.4	82.7	84.8	86.2	87.1	88.3	88.2	86.7	83.9	81.4	84.2
	T_{AN}	°F	65.3	65.3	67.3	68.7	70.2	71.9	73.1	73.6	72.9	72.2	69.2	66.5	69.7
	I	Btu/ (ft²• day)	1180	1396	1622	1796	1949	2004	2002	1967	1810	1540	1266	1133	1639
Chicago, IL	T_{AX}	°F	29.2	33.9	44.3	58.8	70.0	79.4	83.3	82.1	75.5	64.1	48.2	35.0	58.7
	T_{AN}	°F	13.6	18.1	27.6	38.8	48.1	57.7	62.7	61.7	53.9	42.9	31.4	20.3	39.7
	I	Btu/ (ft²• day)	507	760	1107	1459	1789	2007	1944	1719	1354	969	566	402	1215

TABLE 10.3.1   Meteorological Data (T_{AX}, T_{AN}, I) for Selected U.S. Locations (Continued)

Location	Symbol*	Units	Jan.	Feb.	Mar.	Apr.	May	June	July	Aug.	Sep.	Oct.	Nov.	Dec.	Annual average
Springfield, IL	T_{AX}	°F	32.8	38.0	48.9	64.0	74.6	84.1	87.1	84.7	79.3	67.5	51.2	38.4	62.6
	T_{AN}	°F	16.3	20.9	30.3	42.6	52.5	62.0	65.9	63.7	55.8	44.4	32.9	23.0	42.5
	I	Btu/ (ft²• day)	585	861	1143	1515	1866	2097	2058	1806	1454	1068	677	490	1302
Indianapolis, IN	T_{AX}	°F	34.2	38.5	49.3	63.1	73.4	82.3	85.2	83.7	77.9	66.1	50.8	39.2	62.0
	T_{AN}	°F	17.8	21.1	30.7	41.7	51.5	60.9	64.9	62.7	55.3	43.4	32.8	23.7	42.2
	I	Btu/ (ft²• day)	496	747	1037	1398	1638	1868	1806	1644	1324	977	579	417	1165
Wichita, KS	T_{AX}	°F	39.8	46.1	55.8	68.1	77.1	87.4	92.9	91.5	82.0	71.2	55.1	44.6	67.6
	T_{AN}	°F	19.4	24.1	32.4	44.5	54.6	64.7	69.8	67.9	59.2	46.9	33.5	24.2	45.1
	I	Btu/ (ft²• day)	784	1058	1406	1783	2036	2264	2239	2032	1616	1250	871	690	1502
Louisville, KY	T_{AX}	°F	40.8	45.0	54.9	67.5	76.2	84.0	87.6	86.7	80.6	69.2	55.5	45.4	66.1
	T_{AN}	°F	24.1	26.8	35.2	45.6	54.6	63.3	67.5	66.1	59.1	46.2	36.6	28.9	46.2
	I	Btu/ (ft²• day)	546	789	1102	1467	1720	1904	1838	1680	1361	1042	653	488	1216
Baton Rouge, LA	T_{AX}	°F	61.1	64.5	71.6	79.2	85.2	90.6	91.4	90.8	87.4	80.1	70.1	63.8	78.0
	T_{AN}	°F	40.5	42.7	49.4	57.5	64.3	70.0	72.8	72.0	68.3	56.3	47.2	42.3	57.0
	I	Btu/ (ft²• day)	785	1054	1379	1681	1871	1926	1746	1677	1464	1301	920	737	1379
Lake Charles, LA	T_{AX}	°F	60.8	64.0	70.5	77.8	84.1	89.4	91.0	90.8	87.5	80.8	70.5	64.0	77.6
	T_{AN}	°F	42.2	44.5	50.8	58.9	65.6	71.4	73.5	72.8	68.9	57.7	48.9	43.8	58.3
	I	Btu/ (ft²• day)	728	1010	1313	1750	1849	1970	1788	1657	1485	1381	917	706	1365
New Orleans, LA	T_{AX}	°F	61.8	64.6	71.2	78.6	84.5	89.5	90.7	90.2	86.8	79.4	70.1	64.4	77.7
	T_{AN}	°F	43.0	44.8	51.6	58.8	65.3	70.9	73.5	73.1	70.1	59.0	49.9	44.8	58.7
	I	Btu/ (ft²• day)	835	1112	1415	1780	1968	2004	1814	1717	1514	1335	973	779	1437
Detroit, MI	T_{AX}	°F	30.6	33.5	43.4	57.7	69.4	79.0	83.1	81.5	74.4	62.5	47.6	35.4	58.2
	T_{AN}	°F	16.1	18.0	26.5	36.9	46.7	56.3	60.7	59.4	52.2	41.2	31.4	21.6	38.9
	I	Btu/ (ft²• day)	417	680	1000	1399	1716	1866	1835	1576	1253	876	478	344	1120
Grand Rapids, MI	T_{AX}	°F	29.0	31.7	41.6	56.9	69.4	78.9	83.0	81.1	73.4	61.4	46.0	33.8	57.2
	T_{AN}	°F	14.9	15.6	24.5	35.6	45.5	55.3	59.8	58.1	50.8	40.4	30.9	20.7	37.7
	I	Btu/ (ft²• day)	370	648	1014	1412	1755	1957	1914	1676	1262	858	446	311	1135
Minneapolis- St. Paul, MN	T_{AX}	°F	19.9	26.4	37.5	56.0	69.4	78.5	83.4	80.9	71.0	59.7	41.1	26.7	54.2
	T_{AN}	°F	2.4	8.5	20.8	36.0	47.6	57.7	62.7	60.3	50.2	39.4	25.3	11.7	35.2
	I	Btu/ (ft²• day)	464	764	1104	1442	1737	1928	1970	1687	1255	860	480	353	1170

TABLE 10.3.1   Meteorological Data (TAX, TAN, I) for Selected U.S. Locations (Continued)

Location	Symbol*	Units	Jan.	Feb.	Mar.	Apr.	May	June	July	Aug.	Sep.	Oct.	Nov.	Dec.	Annual average
Jackson, MS	T_{AX}	°F	56.5	60.9	68.4	77.3	84.1	90.5	92.5	92.1	87.6	78.6	67.5	60.0	76.3
	T_{AN}	°F	34.9	37.2	44.2	52.9	60.8	67.9	71.3	70.2	65.1	51.4	42.3	37.1	52.9
	I	Btu/ (ft²• day)	754	1026	1369	1708	1941	2024	1909	1781	1509	1271	902	709	1409
Billings, MT	T_{AX}	°F	29.9	37.9	44.0	55.9	66.4	76.3	86.6	84.3	72.3	61.0	44.4	36.0	57.9
	T_{AN}	°F	11.8	18.8	23.6	33.2	43.3	51.6	58.0	56.2	46.5	37.5	25.5	18.2	35.4
	I	Btu/ (ft²• day)	486	763	1190	1526	1913	2174	2384	2022	1470	987	561	421	1325
Las Vegas, NV	T_{AX}	°F	56.0	62.4	68.3	77.2	87.4	98.6	104.5	101.9	94.7	81.5	66.0	57.1	79.6
	T_{AN}	°F	33.0	37.7	42.3	49.8	59.0	68.6	75.9	73.9	65.6	53.5	41.2	33.6	52.8
	I	Btu/ (ft²• day)	978	1340	1824	2319	2646	2778	2588	2355	2037	1540	1086	881	1864
Newark, NJ	T_{AX}	°F	38.2	40.3	49.1	61.3	71.6	80.6	85.6	84.0	76.9	66.0	54.0	42.3	62.5
	T_{AN}	°F	24.2	25.3	33.3	42.9	53.0	62.4	67.9	67.0	59.4	48.3	39.0	28.6	45.9
	I	Btu/ (ft²• day)	552	793	1009	1449	1687	1795	1760	1565	1273	951	595	454	1165
Roswell, NM	T_{AX}	°F	55.4	60.4	67.7	76.9	85.0	93.1	93.7	91.3	84.9	75.8	63.1	56.7	75.3
	T_{AN}	°F	27.4	31.4	37.9	46.8	55.6	64.8	69.0	67.0	59.6	47.5	35.0	28.2	47.5
	I	Btu/ (ft²• day)	1047	1373	1807	2218	2459	2610	2441	2242	1913	1527	1131	952	1810
Buffalo, NY	T_{AX}	°F	30.0	31.4	40.4	54.4	65.9	75.6	80.2	78.2	71.4	60.2	47.0	35.0	55.8
	T_{AN}	°F	17.0	17.5	25.6	36.3	46.3	56.4	61.2	59.6	52.7	42.7	33.6	22.5	39.3
	I	Btu/ (ft²• day)	349	546	889	1315	1597	1804	1776	1513	1152	784	403	283	1034
New York, NY (LaGuardia Airport)	T_{AX}	°F	37.4	39.2	47.3	59.6	69.7	78.7	83.9	82.3	75.2	64.5	52.9	41.5	61.0
	T_{AN}	°F	26.1	27.3	34.6	44.2	53.7	63.2	68.9	68.2	61.2	50.5	41.2	30.8	47.5
	I	Btu/ (ft²• day)	548	795	1118	1457	1690	1802	1784	1583	1280	951	593	457	1171
Cleveland, OH	T_{AX}	°F	32.5	34.8	44.8	57.9	69.7	78.7	83.9	82.3	75.2	64.5	52.9	41.5	61.0
	T_{AN}	°F	18.5	19.9	28.4	38.3	47.9	57.2	61.4	60.5	54.0	43.6	34.3	24.6	40.7
	I	Btu/ (ft²• day)	388	601	922	1350	1681	1843	1828	1583	1240	867	466	318	1091
Columbus, OH	T_{AX}	°F	34.7	38.1	49.3	62.3	72.6	81.3	84.4	83.0	76.9	65.0	50.7	39.4	61.5
	T_{AN}	°F	19.4	21.5	30.6	40.5	50.2	59.0	63.2	61.7	54.6	42.8	33.5	24.7	41.8
	I	Btu/ (ft²• day)	459	677	980	1353	1647	1813	1755	1641	1282	945	538	387	1123
Toledo, OH	T_{AX}	°F	30.7	34.0	44.6	59.1	70.5	79.9	83.4	81.8	75.1	63.3	47.9	35.5	58.8
	T_{AN}	°F	15.5	17.5	26.1	36.5	46.6	56.0	60.2	58.4	51.2	40.1	30.6	20.6	38.3
	I	Btu/ (ft²• day)	435	680	997	1384	1717	1878	1849	1616	1276	911	498	355	1133

TABLE 10.3.1   Meteorological Data (TAX, TAN, I) for Selected U.S. Locations *(Continued)*

Location	Symbol*	Units	Jan.	Feb.	Mar.	Apr.	May	June	July	Aug.	Sep.	Oct.	Nov.	Dec.	Annual average
Oklahoma City, OK	T_{AX}	°F	46.6	52.2	61.0	71.7	79.0	87.6	93.5	92.8	84.7	74.3	59.9	50.7	71.2
	T_{AN}	°F	25.2	29.4	37.1	48.6	57.7	66.3	70.6	69.4	61.9	50.2	37.6	29.1	48.6
	I	Btu/ (ft²• day)	801	1055	1400	1725	1918	2144	2128	1950	1554	1233	901	725	1461
Tulsa, OK	T_{AX}	°F	45.6	51.9	60.8	72.4	79.7	87.9	93.9	93.0	85.0	74.9	60.2	50.3	71.3
	T_{AN}	°F	24.8	29.5	37.7	49.5	58.5	67.5	72.4	70.3	62.5	50.3	38.1	29.3	49.2
	I	Btu/ (ft²• day)	732	978	1306	1603	1822	2021	2031	1865	1473	1164	827	659	1373
Astoria, OR	T_{AX}	°F	46.8	50.6	51.9	55.5	60.2	63.9	67.9	68.6	67.8	61.4	53.5	48.8	58.1
	T_{AN}	°F	35.4	37.1	36.9	39.7	44.1	49.2	52.2	52.6	49.2	44.3	39.7	37.3	43.1
	I	Btu/ (ft²• day)	315	545	866	1253	1608	1626	1746	1499	1183	713	387	261	1000
Portland, OR	T_{AX}	°F	44.3	50.4	54.5	60.2	66.9	72.7	79.5	78.6	74.2	63.9	52.3	46.4	62.0
	T_{AN}	°F	33.5	36.0	37.4	40.6	46.4	52.2	55.8	55.8	51.1	44.6	38.6	35.4	44.0
	I	Btu/ (ft²• day)	310	554	895	1308	1663	1773	2037	1674	1217	724	388	260	1067
Philadelphia, PA	T_{AX}	°F	38.6	41.1	50.5	63.2	73.0	81.7	86.1	84.6	77.8	66.5	54.5	43.0	63.4
	T_{AN}	°F	23.8	25.0	33.1	42.6	52.5	61.5	66.8	66.0	58.6	46.5	37.1	28.0	45.1
	I	Btu/ (ft²• day)	555	795	1108	1434	1660	1811	1758	1575	1281	959	619	470	1169
Pittsburgh, PA	T_{AX}	°F	34.1	36.8	47.6	60.7	70.8	79.1	82.7	81.1	74.8	62.9	49.8	38.4	59.9
	T_{AN}	°F	19.2	20.7	29.4	39.4	48.5	57.1	61.3	60.1	53.3	42.1	33.3	24.3	40.7
	I	Btu/ (ft²• day)	424	625	943	1317	1602	1762	1689	1510	1209	895	505	347	1069
Providence, RI	T_{AX}	°F	36.4	37.7	45.5	57.5	67.6	76.6	81.7	80.3	73.1	63.2	51.9	40.5	59.3
	T_{AN}	°F	20.0	20.9	29.2	38.3	47.6	57.0	63.3	61.9	53.8	43.1	34.8	24.1	41.2
	I	Btu/ (ft²• day)	506	739	1032	1374	1655	1776	1695	1499	1209	907	538	419	1112
Columbia, SC	T_{AX}	°F	56.2	59.5	67.1	77.0	83.8	89.2	91.9	91.0	85.5	76.5	67.1	58.8	75.3
	T_{AN}	°F	33.2	34.6	41.9	50.5	59.1	66.1	70.1	69.4	63.9	50.3	40.6	34.7	51.2
	I	Btu/ (ft²• day)	762	1021	1355	1747	1895	1947	1842	1703	1439	1211	921	722	1380
Sioux Falls, SD	T_{AX}	°F	22.9	29.3	40.1	58.1	70.5	80.3	86.2	83.9	73.5	62.1	43.7	29.3	56.7
	T_{AN}	°F	1.9	8.9	20.6	34.6	45.7	56.3	61.8	59.7	48.5	39.7	22.3	10.1	33.9
	I	Btu/ (ft²• day)	533	802	1152	1543	1894	2100	2150	1845	1410	1005	608	441	1290
Memphis, TN	T_{AX}	°F	48.3	53.0	61.4	72.9	81.0	88.4	91.5	90.3	84.3	74.5	61.4	52.3	71.6
	T_{AN}	°F	30.9	34.1	41.9	52.2	60.9	68.9	72.6	70.8	64.1	51.3	41.1	34.3	51.9
	I	Btu/ (ft²• day)	683	945	1278	1639	1885	2045	1972	1824	1471	1205	817	629	1366

TABLE 10.3.1   Meteorological Data (TAX, TAN, I) for Selected U.S. Locations *(Continued)*

Location	Symbol*	Units	Jan.	Feb.	Mar.	Apr.	May	June	July	Aug.	Sep.	Oct.	Nov.	Dec.	Annual average
Amarillo, TX	T_{AX}	°F	49.1	53.1	60.8	71.0	79.1	88.2	91.4	89.6	82.4	72.7	58.7	51.8	70.7
	T_{AN}	°F	21.7	26.1	32.0	42.0	51.9	61.5	66.2	64.5	56.9	45.5	32.1	24.8	43.8
	I	Btu/ (ft²•day)	960	1244	1631	2019	2212	2393	2281	2103	1761	1404	1033	872	1659
Corpus Christi, TX	T_{AX}	°F	66.5	69.9	76.1	82.1	86.7	91.2	94.2	94.1	90.1	83.9	75.1	69.3	81.6
	T_{AN}	°F	46.1	48.7	55.7	63.9	69.5	74.1	75.6	75.8	72.8	64.1	54.9	48.8	62.5
	I	Btu/ (ft²•day)	898	1147	1430	1642	1866	2094	2186	1991	1687	1416	1043	845	1521
Dallas, TX	T_{AX}	°F	54.0	59.1	67.2	76.8	84.4	93.2	97.8	97.3	89.7	79.5	66.2	58.1	76.9
	T_{AN}	°F	33.9	37.8	44.9	55.0	62.9	70.8	74.7	73.7	67.5	56.3	44.9	37.4	55.0
	I	Btu/ (ft²•day)	822	1071	1422	1627	1889	2135	2122	1950	1587	1276	936	780	1468
Houston, TX	T_{AX}	°F	61.9	65.7	72.1	79.0	85.1	90.9	93.6	93.1	88.7	81.9	71.6	65.2	79.1
	T_{AN}	°F	40.8	43.2	49.8	58.3	64.7	70.2	72.5	72.1	68.1	57.5	48.6	42.7	57.4
	I	Btu/ (ft²•day)	772	1034	1297	1522	1775	1898	1828	1686	1471	1276	924	730	1351
Midland-Odessa, TX	T_{AX}	°F	57.6	62.1	69.8	78.8	86.0	93.0	94.2	93.1	86.4	77.7	65.5	59.7	77.0
	T_{AN}	°F	29.7	33.3	40.2	49.4	58.2	66.6	69.2	68.0	61.9	51.1	39.0	32.2	49.9
	I	Btu/ (ft²•day)	1081	1383	1839	2192	2430	2562	2389	2210	1844	1522	1176	1000	1802
Salt Lake City, UT	T_{AX}	°F	37.4	43.7	51.5	61.1	72.4	83.3	93.2	90.0	80.0	66.7	50.2	38.9	64.0
	T_{AN}	°F	19.7	24.4	29.9	37.2	45.2	53.3	61.8	59.7	50.0	39.3	29.2	21.6	39.3
	I	Btu/ (ft²•day)	639	989	1454	1894	2362	2561	2590	2254	1843	1293	788	570	1603
Richmond, VA	T_{AX}	°F	46.7	49.6	58.5	70.6	77.9	84.8	88.4	87.1	81.0	70.5	60.5	50.2	68.8
	T_{AN}	°F	26.5	28.1	35.8	45.1	54.2	62.2	67.2	66.4	59.3	46.7	37.3	29.6	46.5
	I	Btu/ (ft²•day)	632	877	1210	1566	1762	1872	1774	1601	1348	1033	733	567	1248
Seattle, WA (Sea-Tac Airport)	T_{AX}	°F	43.9	48.8	51.1	56.8	64.0	69.2	75.2	73.9	68.7	59.5	50.3	45.6	58.9
	T_{AN}	°F	34.3	36.8	37.2	40.5	46.0	51.1	54.3	54.3	51.2	45.3	39.3	36.3	43.9
	I	Btu/ (ft²•day)	262	495	849	1294	1714	1802	2248	1616	1148	656	337	211	1053
Charleston, WV	T_{AX}	°F	41.8	45.4	55.4	67.3	76.0	82.5	85.2	84.2	78.7	67.7	55.6	45.9	65.5
	T_{AN}	°F	23.9	25.8	34.1	43.3	51.8	59.4	63.8	63.1	56.4	44.0	35.0	27.8	44.0
	I	Btu/ (ft²•day)	498	707	1010	1356	1639	1776	1683	1514	1272	972	613	440	1123

TABLE 10.3.1   Meteorological Data (TAX, TAN, I) for Selected U.S. Locations _(Continued)_

Location	Symbol*	Units	Jan.	Feb.	Mar.	Apr.	May	June	July	Aug.	Sep.	Oct.	Nov.	Dec.	Annual average
		Property						Monthly averages							
Huntington, WV	T_{AX}	°F	41.1	45.0	55.2	67.2	75.7	82.6	85.6	84.4	78.7	67.6	55.2	45.2	65.3
	T_{AN}	°F	24.5	26.6	35.0	44.4	52.8	60.7	65.1	64.0	57.2	44.9	35.9	28.5	45.0
	I	Btu/ (ft²• day)	526	757	1067	1448	1710	1844	1769	1580	1306	1004	638	467	1176
Cheyenne, WY	T_{AX}	°F	37.3	40.7	43.6	54.0	64.6	75.4	83.1	80.8	72.1	61.0	46.5	40.4	58.3
	T_{AN}	°F	14.8	17.9	20.6	29.6	39.7	48.5	54.6	52.8	43.7	34.0	23.1	18.2	33.1
		Btu/ (ft²• day)	766	1068	1433	1771	1995	2258	2230	1966	1667	1242	823	671	1491

*Where T_{AA} = daily average ambient temperature, °R
T_{AX} = daily maximum ambient temperature, °R
T_{AN} = daily minimum ambient temperature, °R
ΔT_A = daily ambient temperature range, °R

This equation may be used to determine the tank paint solar absorptance α.

$$\alpha = \frac{\alpha_R + \alpha_S}{2}$$

where α = tank paint solar absorptance (dimensionless)
α_R = tank roof paint solar absorptance (dimensionless)
α_S = tank shell paint solar absorptance (dimensionless)

TABLE 10.3.2   Solar Absorptance (α) for Selected Tank Paints

Paint color	Paint shade or type	Solar absorptance α (dimensionless) with paint condition:	
		Good	Poor
Aluminum	Specular	0.39	0.49
Aluminum	Diffuse	0.60	0.68
Gray	Light	0.54	0.63
Gray	Medium	0.68	0.74
Red	Primer	0.89	0.91
White	—	0.17	0.34

Note: If specific information is not available, a white shell and roof, with the paint in good condition, can be assumed to represent the most common or typical tank paint in use.

Daily average liquid surface temperature. Next the values of T_{AA} and T_B are used to determine the _daily average liquid surface temperature_ T_{LA} according to

$$T_{LA} = 0.44T_{AA} + 0.56T_B + 0.0079\alpha I \qquad (10.3.6)$$

where T_{LA} = daily average liquid surface temperature, °R
$\quad\quad T_{AA}$ = Daily average ambient temperature, °R
$\quad\quad T_B$ = liquid bulk temperature, °R
$\quad\quad\alpha$ = tank paint solar absorptance, dimensionless; see Table 10.3.2
$\quad\quad I$ = daily total solar insolation factor, Btu/(ft² • day); see Table 10.3.1

If T_{LA} is used to calculate P_{VA} from Fig. 10.2.2 or 10.2.3, then T_{LA} must be converted from degrees Rankine to degrees Fahrenheit (°F = °R−460). Equation (10.3.6) should not be used to estimate emissions from insulated tanks. In the case of insulated tanks, the average liquid surface temperature should be based on liquid surface temperature measurements from the tank.

Daily vapor temperature range. The *daily vapor temperature range* Δ_{TV} may be calculated by using the following equation:

$$\Delta T_V = 0.72 \, \Delta T_A + 0.028\alpha I \qquad (10.3.7)$$

where ΔT_V = daily vapor temperature range, °R
$\quad\quad \Delta T_A$ = daily ambient temperature range, °R
$\quad\quad \alpha$ = tank paint solar absorptance, dimensionless; see Table 10.3.2
$\quad\quad I$ = daily total solar insolation factor, Btu/(ft² • day); see Table 10.3.1

Daily maximum and minimum liquid surface temperatures. The *daily maximum and minimum liquid surface temperatures* are the last temperatures that need to be computed.

$$T_{LX} = T_{LA} + 0.25 \, \Delta T_V \qquad (10.3.8)$$

$$T_{LN} = T_{LA} - 0.25 \, \Delta T_V$$

where T_{LX} is the daily maximum liquid surface temperature, T_{LA} is as defined in Eq. (10.3.6), ΔT_V is as defined in Eq. (10.3.7), and T_{LN} is the daily minimum liquid surface temperature. Once the temperature variables are established, the other variables in the stock vapor density computation may be determined.

Vapor-pressure-related calculations. The required temperatures having been established, the vapor pressure may now be found.

Method 1. True vapor pressures for organic liquids can be determined from Table 10.2.1. True vapor pressure can also be determined for crude oils by using Fig. 10.2.2. For refined stocks (gasolines and naphthas), Table 10.2.3 can be used. In order to use Fig. 10.2.2 or 10.2.3, the stored liquid surface temperature T_{LA} must be determined in degrees Fahrenheit.

Method 2. True vapor pressure for selected petroleum liquid stocks, at the stored liquid surface temperature, can be determined from

$$P = \exp\left(\frac{A - B}{T}\right) \tag{10.3.9}$$

where exp = exponential function e
$\quad A$ = constant in vapor pressure equation, dimensionless
$\quad B$ = constant in vapor pressure equation, °R
$\quad T$ = temperature of surface of liquid, which may be any of the following three:
$\quad T_{LA}$ = daily average liquid surface temperature, °R
$\quad T_{LX}$ = daily maximum liquid surface temperature, °R
$\quad T_{LN}$ = daily minimum liquid surface temperature, °R
$\quad P$ = true vapor pressure at T_{LA}, T_{LX}, or T_{LN}, psia

Another form of this equation for organic liquids at the stored liquid temperature can be estimated by

$$\log P = A - \frac{B}{T_{LA} + C} \tag{10.3.10}$$

where A = constant in vapor pressure equation
$\quad B$ = constant in vapor pressure equation
$\quad C$ = constant in vapor pressure equation
$\quad T_{LA}$ = average liquid surface temperature, °C
$\quad P$ = vapor pressure at average liquid surface temperature, mmHg

For organic liquids, the values for the constants A, B, and C are listed in Table 10.3.3. Note that in Eq. (10.3.9), T_{LA} is determined in degrees Celsius instead of degrees Rankine. Also, in Eq. (10.3.9), P is determined in millimeters of Hg rather than pounds per square inch absolute (760 mmHg = 14.696 psia).

Tank vapor space volume. The tank vapor space volume is calculated by using the following equation:

$$V_V = \frac{\pi}{4} D^2 H_{VO} \tag{10.3.11}$$

where V_V = vapor space volume, ft³
$\quad D$ = tank diameter, ft; see Fig. 10.3.1
$\quad H_{VO}$ = vapor space outage, ft

TABLE 10.3.3   Vapor Pressure Equation Constants for Organic Liquids

Liquid	Vapor pressure equation constants		
	A (dimensionless)	B, °C	C, °C
Acetaldehyde	8.005	1600.017	291.809
Acetic acid	7.387	1533.313	222.309
Acetic anhydride	7.149	1444.718	199.817
Acetone	7.117	1210.595	229.664
Acetonitrile	7.119	1314.4	230
Acrylamide	11.2932	3939.877	273.16
Acrylic acid	5.652	648.629	154.683
Acrylonitrile	7.038	1232.53	222.47
Aniline	7.32	1731.515	206.049
Benzene	6.905	1211.033	220.79
Butanol (iso)	7.4743	1314.19	186.55
Butanol-(1)	7.4768	1362.39	178.77
Carbon disulfide	6.042	1169.11	211.59
Carbon tetrachloride	6.934	1242.43	230
Chlorobenzene	6.978	1431.05	217.55
Chloroform	6.493	929.44	196.03
Chloroprene	6.161	783.45	179.7
Cresol (-M)	7.508	1856.36	199.07
Cresol (-O)	6.911	1435.5	165.16
Cresol (-P)	7.035	1511.08	161.85
Cumene (isopropylbenzene)	6.963	1460.793	207.78
Cyclohexane	6.841	1201.53	222.65
Cyclohexanol	6.255	912.87	109.13
Cyclohexanone	7.8492	2137.192	273.16
Dichloroethane (1,2)	7.025	1272.3	222.9
Dichlorethylene (1,2)	6.965	1141.9	231.9
Diethyl (N,N) anilin	7.466	1993.57	218.5
Dimethyl formamide	6.928	1400.87	196.43
Dimethyl hydrazine (1,1)	7.408	1305.91	225.53
Dimethyl phthalate	4.522	700.31	51.42
Dinitrobenzene	4.337	229.2	−137
Dioxane (1,4)	7.431	1554.68	240.34
Epichlorohydrin	8.2294	2086.816	273.16
Ethanol	8.321	1718.21	237.52
Ethanolamine (mono-)	7.456	1577.67	173.37
Ethyl acrylate	7.9645	1897.011	273.16
Ethyl chloride	6.986	1030.01	238.61
Ethylacetate	7.101	1244.95	217.88
Ethylbenzene	6.975	1424.255	213.21
Ethylether	6.92	1064.07	228.8
Formic acid	7.581	1699.2	260.7
Furan	6.975	1060.87	227.74
Furfural	6.575	1198.7	162.8
Heptane (iso)	6.8994	1331.53	212.41
Hexane (-N)	6.876	1171.17	224.41
Hexanol (-1)	7.86	1761.26	196.66
Hydrocyanic acid	7.528	1329.5	260.4
Methanol	7.897	1474.08	229.13
Methyl acetate	7.065	1157.63	219.73
Methyl ethyl ketone	6.9742	1209.6	216
Methyl isobutyl ketone	6.672	1168.4	191.9

TABLE 10.3.3   Vapor Pressure Equation Constants for Organic Liquids (*Continued*)

Liquid	Vapor pressure equation constants		
	A (dimensionless)	B, °C	C, °C
Methyl methacrylate	8.409	2050.5	274.4
Methyl styrene (alpha)	6.923	1486.88	202.4
Methylene chloride	7.409	1325.9	252.6
Morpholine	7.7181	1745.8	235
Naphthalene	7.01	1733.71	201.86
Nitrobenzene	7.115	1746.6	201.8
Pentachloroethane	6.74	1378	197
Phenol	7.133	1516.79	174.95
Picoline (-2)	7.032	1415.73	211.63
Propanol (iso)	8.117	1580.92	219.61
Propylene glycol	8.2082	2085.9	203.5396
Propylene oxide	8.2768	1656.884	273.16
Pyridine	7.041	1373.8	214.98
Resorcinol	6.9243	1884.547	186.0596
Styrene	7.14	1574.51	224.09
Tetrachloroethane (1,1,1,2)	6.898	1365.88	209.74
Tetrachloroethane (1,1,2,2)	6.631	1228.1	179.9
Tetrachloroethylene	6.98	1386.92	217.53
Tetrahydrofuran	6.995	1202.29	226.25
Toluene	6.954	1344.8	219.48
Trichloro(1,1,2)trifluoroethane	6.88	1099.9	227.5
Trichloroethane (1,1,1)	8.643	2136.6	302.8
Trichlorethane (1,1,2)	6.951	1314.41	209.2
Trichloroethylene	6.518	1018.6	192.7
Trichlorofluoromethane	6.884	1043.004	236.88
Trichloropropane (1,2,3)	6.903	788.2	243.23
Vinyl acetate	7.21	1296.13	226.66
Vinylidene chloride	6.972	1099.4	237.2
Xylene (-M)	7.009	1426.266	215.11
Xylene (-O)	6.998	1474.679	213.69

The vapor space outage H_{VO} is the height of a cylinder of tank diameter D whose volume is equivalent to the vapor space volume of a fixed-roof tank, including the volume under the cone or dome roof. The vapor space outage H_{VO} is estimated from

$$H_{VO} = H_S - H_L + H_{RO} \qquad (10.3.12)$$

where H_{VO} = vapor space outage, ft
H_S = tank shell height, ft
H_L = liquid height, ft
H_{RO} = roof outage, ft

Use Fig. 10.3.1 to estimate H_{RO} for a cone roof, a dome roof, or a horizontal tank.

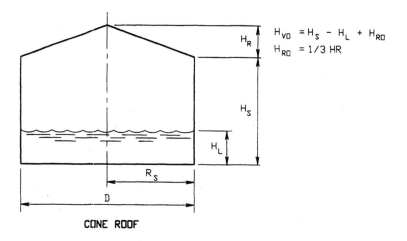

$$H_{VD} = H_S - H_L + H_{RO}$$
$$H_{RO} = 1/3 \ HR$$

CONE ROOF

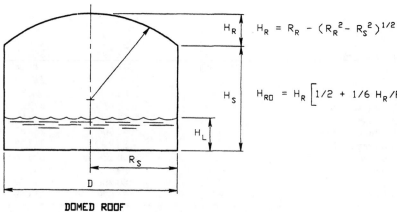

$$H_R = R_R - (R_R^2 - R_S^2)^{1/2}$$

$$H_{RO} = H_R \left[1/2 + 1/6 \ H_R/R_S \right]^{1/2}$$

DOMED ROOF

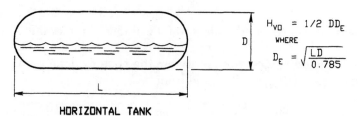

$$H_{VD} = 1/2 \ DD_E$$
WHERE
$$D_E = \sqrt{\frac{LD}{0.785}}$$

HORIZONTAL TANK

Figure 10.3.1 Fixed-roof tank geometry.

Vapor density W_V. The density of vapor is

$$W_V = \frac{M_V P_{VA}}{RT_{LA}} \tag{10.3.13}$$

where W_V = vapor density, lb/ft^3

M_V = vapor molecular weight, lb/(lb•mol)

R = ideal gas constant, 10.731a psia•ft^3/(lb•mol•°R)

P_{VA} = vapor pressure at daily average liquid surface temperature, psia

T_{LA} = daily average liquid surface temperature, °R

Molecular weight of vapor M_V. Determine M_V from Tables 10.2.1 and 10.2.2 for selected petroleum liquids and volatile organic liquids or by analyzing vapor samples. Where mixtures of organic liquids are stored in a tank, M_V can be calculated from the liquid composition. The molecular weight of the vapor M_V is equal to the sum of the component molecular weights M_i multiplied by the *vapor* mole fractions y_i for each component. The *vapor* mole fraction y_i is equal to the partial pressure P_i of the component, divided by the total vapor pressure P_{NP}. The partial pressure of component i is equal to the true vapor pressure of component P_{VPi}, multiplied by the *liquid* mole fraction x_i. Therefore,

$$M_V = \Sigma M_i y_i = \Sigma \frac{M_i P_{VPi} X_i}{P_{TVP}} \qquad (10.3.14)$$

where P_{TVP}, the total vapor pressure of the stored liquid, by Raoult's law is

$$P_{VA} = \Sigma P x_i \qquad (10.3.15)$$

In the absence of more precise data, use a molecular weight of 64 for refined petroleum products. For crude oil a value of 50 should be used. For pure substances of a single component, the vapor molecular weight equals the liquid molecular weight.

Vapor space expansion factor K_E. The vapor space expansion factor K_E is computed from the temperatures ΔT_V and T_{LA} which were computed in Eq. (10.3.7).

$$K_E = \frac{\Delta T_V}{T_{LA}} + \frac{\Delta P_V - \Delta P_B}{P_A - P_{VA}} \qquad (10.3.16)$$

where ΔT_V = daily vapor temperature range, °R

ΔP_V = daily vapor pressure range, psi

ΔP_B = breather vent pressure setting range, psi

P_A = atmospheric pressure, 14.7 psia

P_{VA} = vapor pressure at daily average liquid surface temperature, psia

T_{LA} = daily average liquid surface temperature, °R

Daily vapor pressure range ΔP_V. Calculate from the following equation:

$$\Delta P_V = P_{VX} - P_{VN} \qquad (10.3.17)$$

where ΔP_V = daily vapor pressure range, psia

P_{VX} = vapor pressure at daily maximum liquid surface temperature, psia; calculate from Eqs. (10.3.8) and (10.3.9) and T_{LX}

P_{VN} = vapor pressure at daily minimum liquid surface temperature, psia; calculate from Eqs. (10.3.8) and (10.3.9) and T_{AN}

The following method can be used as an alternate means of calculating ΔP_V for petroleum liquids:

$$\Delta P_V = \frac{0.50 B P_{VA} \Delta T_V}{T_{LA}^2} \qquad (10.3.18)$$

where ΔP_V = daily vapor pressure range, psia

B = constant in vapor pressure equation of Table 10.3.1, °R

P_{VA} = vapor pressure at daily average liquid surface temperature T_{LA}, psia

T_{LA} = daily average liquid surface temperature, °R

ΔT_V = daily vapor temperature range, °R

Breather vent pressure setting range ΔP_B. Calculate from the following equation:

$$\Delta P_B = P_{BP} - P_{BV} \qquad (10.3.19)$$

where ΔP_B = breather vent pressure setting range, psig

P_{BP} = breather vent pressure setting, psig

P_{BV} = breather vent vacuum setting, psig

If specific information on the breather vent pressure setting and vacuum setting is not available, assume 0.03 psig for P_{BP} and −0.03 psig for P_{BV} as typical values. If the fixed-roof tank is of bolted or riveted construction in which the roof or shell plates are not vaportight, assume that $\Delta P_B = 0$, even if a breather vent is used. The estimating equations for fixed-roof tanks do not apply to either low- or high-pressure tanks. The pressure setting of the breathers should normally be less than 1 psig.

Vapor pressures associated with daily maximum and minimum liquid surface temperatures P_{VX} and P_{VN}, respectively. Calculate by substituting the corresponding temperatures, T_{LX} and T_{LN} into the pressure function discussed in Eq. (10.3.8).

Vented vapor saturation factor K_S. The vented vapor saturation factor K_S is calculated from

$$K_S = \frac{1}{1 + 0.053 P_{VA} H_{VO}}$$ (10.3.20)

where K_S = vented vapor saturation factor, dimensionless
P_{VA} = vapor pressure at daily average liquid surface temperature, psia
H_{VO} = vapor space outage, ft, as calculated in Eq. (10.3.4)

The constant 0.053 in Eq. (10.3.20) has units of $(\text{psi} \cdot \text{ft})^{-1}$. This equation is plotted in Fig. 10.3.2.

10.3.1.2 Working loss

The working loss L_W can be estimated from

$$L_W = 0.0010 M_V P_{VA} Q K_N K_P$$ (10.3.21)

where L_W = working losses, lb/yr
M_V = vapor molecular weight, lb/(lb•mol)
P_{VA} = vapor pressure at daily average liquid surface temperature [Eq. (10.3.8) with T_{LA}], psia

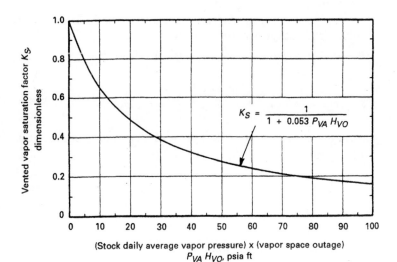

Figure 10.3.2 Vented vapor saturation factor K_S. (*Courtesy of the American Petroleum Institute.*)

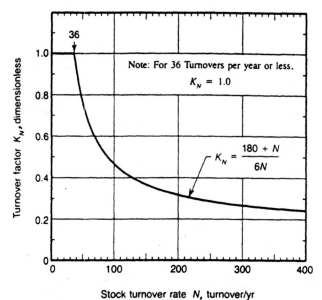

Figure 10.3.3 Working loss turnover factor K_N. (*Courtesy of the American Petroleum Institute.*)

Q = annual net throughput, bbl/yr
K_N = turnover factor, dimensionless; see Fig. 10.3.3
      for turnovers >36, $K_N = (180 + N)/(6N)$
      for turnovers ≤36, $K_N = 1$
K_P = working loss product factor, dimensionless, 0.75 for crude
      oils; for all other organic liquids, $K_P = 1$
N = number of turnovers per year, dimensionless

$$N = \frac{5.614Q}{V_{LX}} \qquad (10.3.22)$$

where Q = annual net throughput, bbl/yr, and V_{LX} = tank maximum liquid volume, ft³, and

$$V_{LX} = \frac{\pi}{4} D^2 H_{LX} \qquad (10.3.23)$$

where D = diameter, ft
H_{LX} = maximum liquid height, ft

SECTION 10.4   EMISSIONS FROM SLOTTED AND UNSLOTTED GUIDE POLES

10.4.1   Introductory Remarks

Although evaporative losses, or emissions, of volatile organic compounds can be easily controlled and significantly reduced by using simple technologies and techniques, they have been the subject of regulatory scrutiny and a source of irritation to those who own or maintain them. This chapter addresses the evaporative loss, or emissions, of petroleum products to the atmosphere from aboveground storage tanks through the guide pole fitting of external floating-roof tanks.

This source of loss had first been identified and quantified in a research program sponsored by the American Petroleum Institute in 1984. The evaporative loss factors that were developed from that research were included in API Publication 2517, *Evaporative Loss from External Floating-Roof Tanks,* and the U.S. Environmental Protection Agency emissions estimation document entitled *USEPA Report AP-42.* These factors have been relied upon by both industry and regulatory agencies to estimate the amount of product that is lost each year to the atmosphere through guide pole fittings.

API's investigation of the sources of evaporative loss from petroleum storage tanks demonstrated that the guide pole usually represents a relatively high portion of the total emissions from external floating-roof tanks. This, in turn, has resulted in the guide pole fitting being the subject of both additional industry research and increased regulatory scrutiny. While some regulatory agencies are feeling compelled to impose new restrictions on the use of guide poles, new data provide the real measure of emissions and should prove to be sound guidance in the implementation of regulations.

10.4.1.1   Guide pole loss factors

The factors for estimating evaporation loss from external floating-roof tank (EFRT) fittings in the current third edition of API Publication 2517[1] are based on a research program that was conducted in 1984. Figure 10.4.1 illustrates the relative contribution to emissions from the various features of an EFRT, based on these factors. It is readily evident from this graph that a significant portion of the total emissions from an EFRT complying with EPA's New Source Performance Standards (NSPs) are attributable to slotted guide poles.

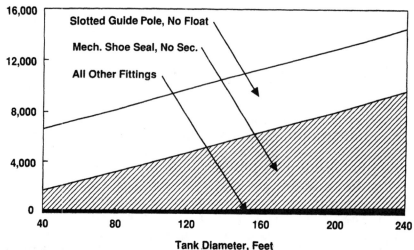

Figure 10.4.1 Typical uncontrolled EFRT components of emission loss.

This discovery prompted the industry to develop improved controls for this fitting. API undertook a CELM testing program in 1993 to establish loss factors for these improvements. The results of the latest research effort have recently been published as an Addendum to API Publication 2517.

10.4.1.2 Guide pole description

The primary mechanical function of a guide pole is to prevent the floating roof from rotating as it rests on the stored liquid or when it is landed. The guide pole is mounted from the tank at points above and below the range of travel of the floating roof, and it passes through an opening in the floating-roof deck (see Fig. 10.4.2).

The guide pole is typically located adjacent to a platform at the top of the tank and has an opening through which personnel may access the liquid below the floating roof. This access may be for the purpose of measuring the depth of the stored product or for obtaining samples of it. When used as a sampling or gauging port, the guide pole usually has a series of slots along its length to allow product to flow freely through it. The sectional view shown in Fig. 10.4.3 depicts a typical slotted guide pole with a sliding cover.

Evaporation that occurs at the liquid surface fills the vapor space of the well with product vapors. These vapors are then flushed to the atmosphere by the action of wind, as shown in Fig. 10.4.4, thus creating emissions.

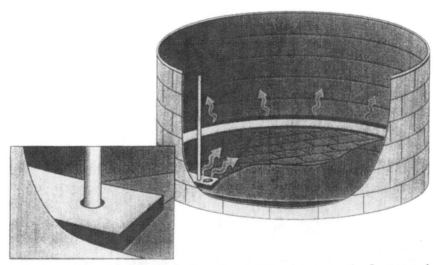

Figure 10.4.2 The guide pole passes through a well that penetrates the floating-roof pontoon.

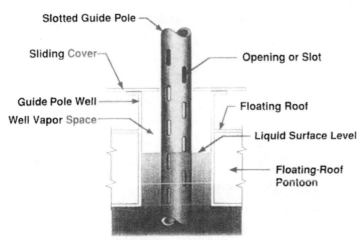

Figure 10.4.3 Guide pole section.

The two emission control techniques identified at the time of the 1984 test program were the addition of a gasket under the sliding cover and the introduction of a float into the guide pole itself (Fig. 10.4.5). The addition of a *well gasket* and a *guide pole float* was shown through testing to roughly halve the emissions from a slotted guide pole, but the remaining emissions were still significantly greater than those from any other EFRT fitting and the rim seal, when a double seal is used.

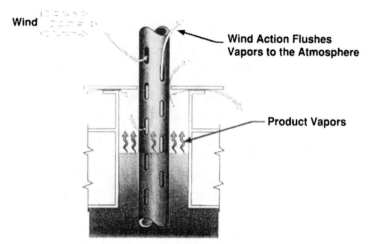

Figure 10.4.4 Emission paths from a slotted guide pole. [Loss factors from API 2517 Addendum, wind speed = 10 mph (no wind speed correction factor).]

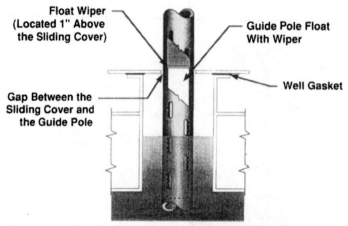

Figure 10.4.5 Slotted guide pole with well gasket and float.

The new emission controls tested in the latest research are a wiper-type gasket between the lid of the well and the guide pole, called a *pole wiper*, and a *pole sleeve* that attaches to the underside of the lid and extends downward into the liquid (Fig. 10.4.6). The pole sleeve may be a length of pipe, the inside diameter of which is nominally larger than the outside diameter of the guide pole. By extending downward from the lid into the product, this sleeve surrounds the guide pole, isolating it from the vapor space of the well.

In the process of developing these new loss factors, it was found that while the loss factor of those fittings actually tested in 1984 compared well with the loss factors measured in the 1993 test results, some loss factors that had been developed by extrapolation did not compare well with the new test results. The most significant deviation observed was for the unslotted guide pole, which was found to have much higher loss factors than had been previously predicted.

10.4.1.3 Test results

Figure 10.4.7 illustrates some of the results of the 1993 testing program.

Affixing a well gasket and a guide pole float to an uncontrolled slotted guide pole is still shown to reduce emissions by about 50 percent. The addition of a pole wiper brings the loss control effectiveness up to about 90 percent, and 98 to 99 percent can be achieved by the further addition of a pole sleeve. These results were achieved with the wiper on the float positioned 1 inch above the sliding cover. It was determined that positioning the float wiper at the same elevation as the sliding cover did not reduce emissions.

Emissions from unslotted guide poles are reduced about 50 percent when controlled with existing technology. This reduction is achieved simply by adding a well gasket because a float would serve no purpose in an unslotted guide pole. Similar to results of the for slotted guide poles, the combination of a well gasket with either a pole wiper or a pole sleeve achieves reductions of 98 to 99 percent, as compared to the uncontrolled case.

Tables 10.4.1 and 10.4.2 illustrate the results of the recent research for each configuration of guide pole that was tested. Table 10.4.1 contains the emission factors, while Table 10.4.2 shows the emission factors at 5-, 10-, and 15-mph wind speeds. In response to industry concern that some regulatory jurisdictions may require the removal of slotted guide poles from existing EFRTs, API asked CELM to expedite the publication of the test data in an Addendum to API Publication 2517.

10.4.2 Regulatory Situation

The New Source Performance Standards[2] for aboveground storage tanks, published as subpart Kb to 40CFR, Part 60, have been a model for regulatory actions affecting tanks. Subpart Kb is generally clear and explicit about the tank fitting and hardware requirements needed to reduce emissions.

Confusion has arisen, however, in one area. There has been considerable debate concerning what is required for controlling emissions from external floating-roof guide poles, particularly when the pole is slotted.

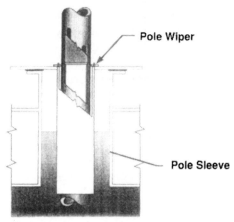

Figure 10.4.6 Slotted guide pole with well gasket, float, float wiper, pole sleeve, and pole wiper.

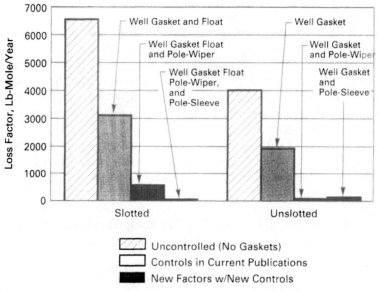

Figure 10.4.7 Comparison of guide pole loss factors.

10.4.2.1   EPA requirements

EPA published Report AP-42[3] for the estimation of air emissions from various sources, including storage tanks with floating roofs. The estimation of emissions attributable to an individual floating-roof fitting is a function of that fitting's loss factor. These fitting loss factors, in turn, are dependent on the design of the fitting. Modifications, such as installing gaskets under the lids of fittings, or bolting them closed,

TABLE 10.4.1 Comparison of Guide Pole Emission Loss Factors from API Publication 2517, Third Edition, 1989, with API Publication 2517, Addendum 1994

| Guide pole type | API fitting device | Guide pole description | | | | | K_{fa} | | K_{fb} | | m | |
		Well gasket	Float	Float wiper	Pole sleeve	Pole wiper	Old	New	Old	New	Old	New
Unslotted	18	N	N	N	N	N	0	31.1	67	372	0.98	1.03
	27	Y	N	N	N	N	0	25.0	3.0	1.05	1.4	3.26
	28	N	N	N	Y	N	N/A	25.0	N/A	0.0267	N/A	4.02
	19	Y	N	N	Y	N	N/A	8.63	N/A	13.8	N/A	0.755
	17	Y	N	N	N	Y‡	N/A	13.7	N/A	5.78	N/A	0.587
Slotted	1	N	N	N	N	N	0	45.4	310	698	1.2	0.974
	25	Y	N	N	N	N	0	40.7	260	311	1.2	1.29
	3	N	Y	Y*	N	N	0	35.7	29	102	2.0	1.71
	26	Y	Y	Y*	N	N	0	25.8	8.5	9.08	2.4	2.54
	20	Y	N	N	N	Y‡	N/A	41.2	N/A	130	N/A	1.23
	2	Y	N	N	Y	N	N/A	16.3	N/A	132	N/A	1.19
	30	Y	Y¶	N	N	Y‡	N/A	13.8	N/A	13.7	N/A	1.94
	23	Y	Y	Y†	N	Y‡	N/A	17.9	N/A	54.2	N/A	1.10
	4	Y	Y	Y*	N	Y‡	N/A	24.2	N/A	6.14	N/A	1.95
	24	Y	Y	Y†	Y	Y‡	N/A	19.2	N/A	6.19	N/A	1.25
	29	Y	Y	Y*	Y	Y§	N/A	9.09	N/A	13.4	N/A	0.512

*Float wiper is 1 in above the sliding cover.
†Float wiper is at the same elevation as the sliding cover.
‡Pole wiper is at the same elevation as the sliding cover.
§Pole wiper is 6 in above the sliding cover.
¶Float with no wiper and ⅛-in clearance.

445

TABLE 10.4.2 Comparison of Guide Pole Emission Loss Factors at Wind Speeds of 5, 10, and 15 mph

Guide pole type	Guide pole description						K_p, lb·mol/yr^f					
	API fitting device	Well gasket	Float	Float wiper	Pole sleeve	Pole wiper	5 mph		10 mph		15 mph	
							Old^g	New	Old^g	New	Old^g	New
Unslotted	18	N	N	N	N	N	380	1980	750	4020	1120	6,080
	27	Y	N	N	N	N	35.9	224	94.6	1940	167	7,190
	28	N	N	Y	Y	N	N/A	42.2	N/A	305	N/A	1,450
	19	Y	N	Y	Y	N	N/A	55.1	N/A	87.1	N/A	115
	17	Y	N	N	N	Y^c	N/A	28.6	N/A	36.0	N/A	42.0
Slotted	1	N	N	N	N	N	2600	3390	5970	6620	9710	9,800
	25	Y	N	N	N	N	2180	2520	5010	6100	8150	10,300
	3	N	Y	Y^a	N	N	1000	1630	4010	5270	9030	10,500
	26	Y	Y	Y^a	N	N	598	567	3150	3170	8350	8,840
	20	Y	N	N	N	Y^c	N/A	982	N/A	2250	N/A	3,680
	2	Y	N	N	Y	N	N/A	912	N/A	2060	N/A	3,330
	30	Y	Y^e	N	N	Y^c	N/A	325	N/A	1210	N/A	2,630
	23	Y	Y	Y^b	N	Y^c	N/A	336	N/A	700	N/A	1,080
	4	Y	Y	Y^a	N	Y^c	N/A	166	N/A	571	N/A	1,230
	24	Y	Y	Y^b	Y	Y^c	N/A	65.5	N/A	129	N/A	202
	29	Y	Y	Y^a	Y	Y^d	N/A	39.6	N/A	52.7	N/A	62.7

^{a}Float wiper is 1 in above the sliding cover.
^{b}Float wiper is at the same elevation as the sliding cover.
^{c}Pole wiper is 6 in above the sliding cover.
^{d}Float with no wiper and ⅛-in clearance.
^{e}Float with no wiper and ⅛-in clearance.
$^f K_f = K_{fa} + K_{fb} V^m$
^{g}The old values were calculated using the loss factors from API Publication 2517, where the wind speed V was corrected to $V/0.85$ in accordance with the Addendum to API Publication 2517.

may be employed to reduce evaporation losses. The storage tank loss factors and formulas in AP-42 are reprinted from API's evaporation loss publications 2517,[1] 2518,[4] and 2519.[5]

Several fittings share the feature of being located in a well in the deck of the floating roof. These wells are covered with lids which may be sealed to their curbs with gaskets. The 1984 API test program developed loss factors for each of these fittings for both the gasketed and ungasketed conditions. Some of these fittings, such as the vacuum breaker, the gauge float, and the guide pole, also have a penetration through their lid. The 1984 testing was done with an intentional gap between the lid and any penetration through it, because in actual practice a gap was necessary to prevent the lid from hanging up on the device that penetrated it.

The other loss control feature was tested for slotted guide poles. The float that was tested in 1984 had a wiper gasket that sealed between the float and the inside of the guide pole pipe. This fitting was tested with the float wiper positioned 1 in above the level of the sliding cover on the well, as shown in Fig. 10.4.5.

10.4.2.2 No visible gap

Subpart Kb requires that fittings with covers or lids be equipped with gaskets and stipulates that these gasketed covers, seals, or lids must "be maintained in a closed position at all times (i.e., no visible gap) except when the device is in actual use." The reference in this statement is to the item that is capable of being opened or closed, which is the lid that covers the fitting well. The *visible gap* refers to the fit of this lid when it is resting on its curb.

The requirement of a gasket pertains to the more stringent of the tested conditions, which is the gasketed case. This gasket is for the purpose of sealing between the lid and the curb when the lid is closed, so as to prevent the prohibited visible gap. Subpart Kb does not explicitly require that slotted guide poles have floats or any other loss control beyond a gasketed cover over the well.

EPA is developing maximum-achievable control technology (MACT) which, like previous regulations, is based on available controls that have already been demonstrated to be effective in the regulated industry. It is not EPA's policy to require new or unproven technology.

The first MACT rule containing storage tank provisions is the Hazardous Organic National Emission Standards for Hazardous Air Pollutants for the Synthetic Organic Chemical Manufacturing Industry (the HON).[6] The controls stipulated in this rule are very similar to subpart Kb, with the notable addition of a requirement for gasketed floats in slotted guide poles.

The no-visible-gap provision in subpart Kb is clarified in the HON by describing "a gasketed cover, seal or lid which is to be maintained in a closed position (i.e., no visible gap) at all times except when the cover or lid must be open for access." It is the closure of the cover or lid on its curb which is to occur with no visible gap. The additional requirement of a float in the slotted guide pole does not prohibit the gaps between the cover and the slotted guide pole that are required to prevent the lid from hanging up on the guide pole. Unfortunately, it is this very point about which some local regulators have misunderstood the rule and required absolutely no gaps, which is essentially impossible and unnecessary with current and normal practices.

10.4.2.3 New loss controls

To establish the effectiveness of new and additional controls, the 1993 API testing program investigated the loss-control effectiveness of a pole wiper and pole sleeve in reducing emissions from guide pole fittings. While these new loss controls do not meet the test of having been demonstrated in practice and, therefore, are not eligible as mandatory MACT requirements, they do offer tank operators a potential option for dramatically reducing guide pole emissions.

Table 10.4.3 lists the data that were shown graphically in Fig. 10.4.7. Adding a pole wiper, in conjunction with the existing technology of the well gasket and guide pole float, attains emission reductions of about 90 percent, compared to the uncontrolled case. The resulting emission level is considerably below that of an unslotted guide pole controlled with existing technology (i.e., a well gasket). Should both of the new loss controls be added to those of the existing technology, then the emission reductions are on the order of 98 to 99 percent.

Clearly, the new loss control technologies have the potential to adequately control the emissions from slotted guide poles.

TABLE 10.4.3 Emission Loss Factor Reduction of Guide Pole Loss Controls

Guide pole type	Loss controls	K_p, lb•mol/yr*	Emission reduction, %
Unslotted	Uncontrolled	4020	—
	Well gasket	1940	51.7
	Well gasket, pole wiper, or pole sleeve	36.0–87.1†	99.1–97.8
Slotted	Uncontrolled	6620	—
	Well gasket, float, float wiper, and pole wiper	571	91.4
	Well gasket, float, float wiper, pole wiper, and pole sleeve	52.7	99.2

*For a 10-mph wind speed.
†The range of values is due to the difference in effectiveness between a pole wiper and a pole sleeve.

10.4.3   Impact on the User

The requirements as delineated in the NSPS and NESHAPS for implementation of emission control devices on most tank fittings are explicit. However, for guide poles, the rules and requirements are unclear. State and local agencies are left with little practical guidance for applying reasonable-cost and emission-effective rules. It is the intent of the following section to show both users and regulators how effective and practical emission control options can be applied and at the same time meet the intent of the Clean Air Act.

10.4.3.1   Purpose and function of guide poles

Guide poles are devices used on external floating-roof tanks to fulfill the following purposes:

- Prevent the roof from rotating within the tank
- Provide a stilling well from which product samples can be drawn
- Provide a means of accurately measuring the quantity of liquid stored in the tank
- Provide inventory reconciliation
- Provide monitoring for stock loss control
- Provide leak detection and overfill prevention

The slotted guide pole was an evolutionary advance over the unslotted guide pole because it allowed better tank sampling and gauging. The slots in the guide pole allow the tank liquid contents to freely enter and leave the guide pole so that a representative sample is available and so that the liquid level is representative of the bulk liquid level in the tank. In contrast, unslotted guide poles trap liquid that is not representative of the tank contents. Figure 10.4.8 shows the errors in sampling and level measurement that can occur when an unslotted guide pole is used. Because of its success in meeting the purposes outlined above, the slotted guide pole enjoys widespread use today.

10.4.3.2   Disadvantages of unslotted guide poles

The use of the unslotted guide pole has serious disadvantages:

- The unslotted guide pole gives erroneous results for both sampling and level-gauging tanks. For custody transfers, problems can arise from inaccuracies, and additional metering facilities and proving

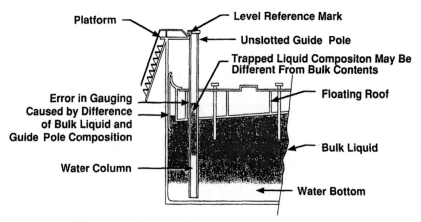

Figure 10.4.8   Effect of "solid" (unslotted guide pole on level measurement).

facilities are required that reduce the cost-effectiveness of this approach. In extreme cases, its use may lead to spills as a result of the inaccuracies of the unslotted guide pole. See Fig. 10.4.8.

- The likelihood of unnoticed groundwater contamination from bottom leaks is increased. Some facilities rely on the measurements taken through the slotted guide poles as their primary leak detection system, and the use of the unslotted guide pole can result in significant erroneous volumetric measurements.

- Replacing slotted guide poles with unslotted guide poles while the tank is in service creates the need to clean the tank and make it gas-free so that the replacement guide pole can be installed. The emissions created by this operation can be significant compared to the benefit derived from the new guide pole configuration. The costs can also be extremely high due to the tank cleaning and degassing operations.

10.4.3.3   Guide pole emissions

Comparing the loss factors from the third edition of API Publication 2517 for various configurations of guide poles, as shown in Tables 10.4.4 and 10.4.5, has led some regulatory agencies to believe that all unslotted guide poles were significantly better for the environment than the slotted guide poles. However, as shown in these tables, the same comparison made using the data based upon the new API work shows that there is no longer the orders-of-magnitude difference between the two types of guide poles. In addition, it is clear that various configurations of controlled slotted guide poles have loss factors

TABLE 10.4.4   Comparison of Old and New Loss Factors for Unslotted Guide Poles and Regulatory Compliance

Guide pole configuration	K_f, lb•mol/yr*		Complies with regulation
	API 2517 (old)†	Addendum (new)	
Uncontrolled (worst case)	750.	4020	Pre-NSPS
With gasketed cover	94.6	1940	NSPS and HON MACT
Addition of new controls:			
With gasketed cover and pole sleeve	NA‡	87.1	§
With gasketed cover and pole wiper	NA‡	36.0	§

*For a 10-mph wind speed.
†The old values were calculated using the loss factors from API Publication 2517, where the wind speed V was corrected to $V/0.85$ in accordance with the Addendum to API Publication 2517.
‡The emission loss factor for this configuration of unslotted guide pole was not included in the third edition of API 2517 and cannot, therefore, be compared against the new Addendum.
§The new controls do not meet the test of existing technology in the MACT rules and, therefore, are not eligible as mandatory MACT requirements.

TABLE 10.4.5   Comparison of Old and New Loss Factors for Slotted Guide Poles and Regulatory Compliance

Guide pole configuration	K_f, lb•mol/yr*		Complies with regulation
	API 2517 (old)†	Addendum (new)	
Uncontrolled (worst case)	5970	6620	Pre-NSPS
With gasketed cover	5010	6100	NSPS
With gasketed cover, float, and float wiper	3150	3170	HON MACT
Addition of new controls			
With gasketed cover, float, float wiper, and pole wiper	NA‡	571	§
With gasketed cover, float, float wiper, pole wiper, and pole sleeve	NA‡	52.7–129¶	§

*For a 10-mph wind speed.
†The old values were calculated using the loss factors from API Publication 2517, where the wind speed V was corrected to $V/0.85$ in accordance with the Addendum to API Publication 2517.
‡The emission loss factor for this configuration of slotted guide pole was not included in the third edition of API 2517 and cannot, therefore, be compared against the new Addendum.
§The new controls do not meet the test of existing technology in the MACT rules and, therefore, are not necessarily eligible as mandatory MACT requirements.
¶The fully controlled case for the slotted guide pole has a range which results from the pole wiper and float wiper being at different elevations. The higher value applies to the condition where the pole wiper and float wiper are at the elevation of the sliding cover.

as low as those for unslotted guide poles for emissions reduction purposes.

An examination of Tables 10.4.4 and 10.4.5 leads to these significant implications:

- The emissions from uncontrolled slotted and unslotted guide poles are reduced by approximately 100 times with the best new loss controls.

- Table 10.4.5 shows that the controlled slotted guide pole has loss control technology that exceeds the NSPS, as well as the MACT requirements of NESHAPS, by a wide margin—571 versus 3170 lb•mol/yr loss factor.

- Using loss controls on guide poles effectively reduces emissions.

- Table 10.4.4 shows that for unslotted guide poles, the use of a well gasket and pole wiper is extremely effective for reducing emissions.

- Table 10.4.5 shows that for slotted guide poles, the use of a well gasket, float, float wiper, and pole wiper is extremely effective for reducing emissions.

Although the pole sleeve for a slotted guide pole has been tested, this equipment has not been commonly used in practice and represents new technology. The fully controlled case without pole sleeve is currently available technology, and it produces an emission reduction of over 90 percent. The use of a pole sleeve may have the potential to cause roof binding. It may also require that existing tanks be shut down, degassed, and cleaned in order to install the pole sleeve.

In an attempt to comply with various interpretations of the no-visible-gap and float-level requirements, some companies have resorted to alternative schemes which seem to have been more acceptable to regulators. Several exotic examples of alternatives considered to meet some of the local regulatory interpretations that would have required converting existing guide poles have been reported.[7] The problem with these alternatives is that while they may satisfy a local regulator,

- They may not reduce emissions as effectively as the new API-tested guide pole loss controls and may, in fact, increase the emissions.

- They may have a short service life.

- They may not be as cost-effective as the new API-tested guide pole controls.

- They may introduce other potential hazards and unknowns.

10.4.3.4 New developments

The API CELM has expedited the testing and publication of the data available in the Addendum to API Publication 2517. This Addendum provides many options for loss control of guide poles, and a wide range of reasonable available control technologies have been tested.

In an effort to solve the loss control problem, some vendors have developed retrofit devices which fulfill the basic requirements of the new loss control options. Some of these devices can be installed while the tank is in service, and they may offer excellent solutions to the problems of emissions control from slotted guide poles. At this time, however, emission factors based upon wind tunnel testing are not available for these other options. It would be an obvious environmental benefit for innovative companies to be able to apply their technologies to these loss control problems. Thus, there should be a method for regulators to more quickly recognize the loss control effectiveness of equipment improvements.

10.4.3.5 Conclusion

An analysis of emissions produced by slotted and unslotted guide poles using the old data published in API Publication 2517 indicated that the slotted guide pole produced significantly greater emissions than the unslotted guide pole. It is natural that interpretive regulatory control would discourage the use of the slotted guide pole, as has been the case in some jurisdictions. However, the new data show that various configurations using the controls make the emissions from slotted guide poles comparable to those of unslotted guide poles.

Remember that, in addition to consideration for emissions, many other factors impact the need for slotted guide poles, including tank operability, maintenance, and safety, that are addressed on a case-by-case basis, taking into consideration many specific variables. This provides the compelling reason why slotted guide poles should be preserved as an option for the tank guide pole user.

Because the new information presented in API Publication 2517 provides additional options for control of emissions from guide poles, changes in the control of guide poles are certain to come. However, the momentum of past practice and thinking that governs current trends may be more quickly and efficiently aligned with this new information by

- Renegotiating the option to use slotted guide poles where they are currently prohibited.

- Drafting new language that can be effectively used by the EPA and local air agencies.

- Encouraging the development and testing of equipment that can reduce emissions from slotted and unslotted guide poles without impairing operability, decreasing safety, or increasing the potential for emissions to other media.

References

Sections 10.1 and 10.3

1. American Petroleum Institute, *Evaporative Loss from External Floating-Roof Tanks,* Publication 2517, 1st ed., February 1989.

2. American Petroleum Institute, *Manual of Petroleum Measurement Standards,* chap. 19, "Evaporation Loss Measurement," section 1, "Evaporative Loss from Fixed-Rig Tanks," Publication 2517, 2d ed., Washington, October 1991.
3. American Petroleum Institute, *Evaporative Loss from Internal Floating-Roof Tanks,* Publication 2519, 3rd ed., Washington, June 1983, Reaffirmed, March 1980.
4. U.S. Environmental Protection Agency, *Compilation of Air Pollutant Emission Factors,* USEPA Report AP-42, 3rd ed., Section 4.3, "Storage of Organic Liquids," September 1985.
5. CBI Industries, Inc., "Testing Program to Measure Hydrocarbon Evaporation Loss from External Floating-Roof Fittings," CBI Contract 41851, Final Report, prepared for the Committee on Evaporation Loss Measurement, American Petroleum Institute, Washington, September 13, 1985.
6. R. J. Laverman, "Evaporative Loss from External Floating Roof Ranks," presented at the 1989 Pipeline Conference of the American Petroleum Institute, Dallas, TX, April 17, 1989 (CBI Technical Paper CBT-5536).
7. American Petroleum Institute, *Welded Steel Tanks for Oil Storage,* Standard 650, 8th ed., Washington, November 1988.
8. American Petroleum Institute, *Venting Atmospheric and Low-Pressure Storage Tanks (Nonrefrigerated and Refrigerated),* Standard 2000, 3rd ed., January 1982, reaffirmed December 1987.
9. R. J. Laverman, "Evaporative Loss from Storage Tanks," presented at the Fourth Oil Loss Control Conference on Real and Apparent Losses in Refining and Storage, The Institute of Petroleum, London, October 30–31, 1991 (CBI Technical Paper CBT-5562).
10. FluiDyne Engineering Corporation, "Wind Tunnel Tests of a 0.017 Scale Model to Determine Pressure, Flow, and Venting Characteristics of a Liquid Storage Tank," FluiDyne Report 1114, prepared for Chicago Bridge & Iron Co., CBI Research Contract R-1050, June 1977.
11. Chicago Bridge & Iron Co., "Testing Program to Measure Hydrocarbon Emissions from a Controlled Internal Floating Roof Tank," CBI Contract 05000, Final Report, prepared for the Committee on Evaporative Loss Measurement, American Petroleum Institute, Washington, March 1982.
12. Radian Corp., "Field Testing Program to Determine Hydrocarbon Emissions from Floating-Roof Tanks," Final Report, prepared for the Committee on Evaporative Loss Measurement, American Petroleum Institute, Washington, May 1979.
13. R. J. Laverman, "Emission Measurements on a Floating Roof Pilot Test Tank," Paper 31-37, presented at the 1979 Midyear Refining Meeting of the American Petroleum Institute, San Francisco, May 16, 1979. Published in the 1979 *Proceedings of the API Refining Department,* vol. 58, pp. 301–322, 1979 (CBI Technical Paper CBT-5364).
14. R. J. Laverman, "Emission Loss Control Effectiveness for Storage Tanks," presented at the 1984 Midyear Refining Meeting of the American Petroleum Institute, New Orleans, LA, May 14–17, 1984 (CBI Technical Paper CBT-5543).
15. R. J. Laverman, "Emission Reduction Options for Floating Roof Tanks," presented at the Second International Symposium on Aboveground Storage Tanks, sponsored by the Material Technology Institute of Chemical Process Industries, Inc., and the National Association of Corrosion Engineers, Houston, TX, January 14–16, 1992 (CBI Technical Paper CBT-5565).
16. P. J. Winters, and R. J. Laverman, "Emission Reductions from Aboveground Storage Tanks," presented at the 12th Annual Operation Conference of the Independent Liquid Terminals Association, Houston, TX, June 23, 1992 (CBI Technical Paper CBT-5571).
17. U.S. Environmental Protection Agency, Emission Inventory Ranch (MD-14), Research Triangle Park, NC 27711, (919) 541-5285

Section 10.4

1. API, API Publication 2517, *Evaporative Loss from External Floating Roof Tanks,* 3rd ed., Washington, February 1989.

2. U.S. Environmental Protection Agency, "Standards of Performance for New Stationary Sources: Volatile Organic Liquid Storage Vessels (Including Petroleum Liquid Storage Vessels)," 40 CFR Part 60 Subpart Kb; promulgated April 8, 1987.
3. U.S. Environmental Protection Agency, *Compilation of Air Pollutant Emission Factors,* USEPA Report AP-42, 4th ed., supplement F, chap. 12, "Storage of Organic Liquids," Washington, 1993.
4. American Petroleum Institute, API Publication 2518, *Manual of Petroleum Measurement Standards,* chap. 19, "Evaporative Loss Measurement," section 1, "Evaporative Loss from Fixed-Roof Tanks," 2d ed., Washington, October 1991.
5. American Petroleum Institute, API Publication 2519, *Evaporation Loss from Internal Floating Roof Tanks,* 3rd ed., Washington, June 1983, reaffirmed March 1990.
6. U.S. Environmental Protection Agency, National Emission Standards for Organic Hazardous Air Pollutants (NESHAPS) for Source Categories, from the Synthetic Organic Chemical Manufacturing Industry (SOCMI) for Process Vents, Storage Vessels, Transfer Operations, and Wastewater, 40 CFR Part 63, Subpart G. Published in the *Federal Register,* vol. 59, no. 78, April 22, 1994.
7. P. E. Myers, *Alternatives Considered to Comply with BAAQMD's Guidepoles Requirements.* (To obtain a copy, write to P. E. Myers, 34 Martha Road, Orinda, CA 94563.)
8. Letter from L. Slagel (API) to M. Morris (EPA/CPB), January 28, 1994. Comments on the draft CTG, "Control of Volatile Organic Compound Emissions from Volatile Organic Liquid Storage in Floating and Fixed Roof Tanks."

11

Tanks Constructed of Other Materials

SECTION 11.1 STAINLESS STEEL TANKS

11.1.1 Use and Selection of Stainless Steel Tanks

Stainless steel tanks have been used for many years in the chemical, pulp and paper, and food industries. However, there has been little guidance for design because few standards-writing organizations undertook to produce a standard of construction specifically for stainless steel tanks. There have been several basic approaches to designing and specifying stainless steel tanks. One approach has been to pattern the specification and build an appendix to API 650 with modifications to account for the change from steel to stainless steel. This is a practice widely used by the chemical industry. Another approach has been to rely on the informal standard for stainless steel tanks developed by the American Iron and Steel Institute and the Steel Plate Fabricators Association.[1] Still another approach could be to purchase tanks based upon API 620, Appendix Q. However, this appendix is specifically for tanks storing liquified ethane, ethylene, and methane. It is conservative and is not commonly used. Table 11.1.1 compares some major differences between the API 650 stainless steel appendix and API 620Q.

In 1993 under pressure from the chemical industries, API undertook to revise API Standard 650 to develop a standard for general-purpose stainless steel tanks. API decided to make an appendix that modifies the basic requirements of API 650. In general, the use of stainless steel may be dictated by one of the following reasons:

TABLE 11.1.1 Comparison of API Requirements

Description	API Standard 650	API stainless steel appendix	API 620 Appendix Q
Service	General	General	Liquefied methane, ethane, and ethylene
Material	Carbon steels	304, 316, 317, and L grades	304, 304L, and nickel steels
Refrigerated	No	No	Yes
Geometry	Vertical, cylindrical	Vertical, cylindrical	Single or double curvature
Number of shells	Single	Singlet	Single or double
Minimum thickness, in	0.25 Bottom $\frac{3}{16}$ shell and roof	$\frac{3}{16}$ bottom, shell, and roof	$\frac{3}{16}$ bottom, shell, and roof
Design temperature	Ambient–500°F	Up to 500°F	–270°F to ambient
Impact toughness testing	Sometimes	No	Impact testing of welds
Stress relieving	Sometimes	No	No
Annular bottom plates required	Sometimes	No	Yes
Shell joints	Butt	Butt	Lap or butt
Allowable stress basis	$0.66Y$ or $0.4T$	$0.90Y$ or $0.30T$ 304: 30,000 psi 304L: 25,000 psi	$0.66Y$ or $0.33T$ 304: 22,500 psi 304L: 18,750 psi
Joint efficiency	1.0; Appendix A tanks 1.0–0.7	1.0–0.7	1.0–0.35

- For corrosion resistance
- To maintain strength at higher temperatures
- For maintenance of product purity (discoloration, dissolved iron, imparting of off-tastes in the food and beverage industry, etc.)
- As a lower-cost alternative to coated carbon steel tanks
- When required by regulations such as various food processing applications

11.1.1.1 Types of stainless steels

Although there are dozens of stainless steels, the API stainless steel appendix (not published as of this printing) applies only to the austenitic series stainless steels including grades 304, 304L, 316, 316L, 317, and 317L. API felt that this grouping of materials covered the vast majority of the need for stainless steels and did not want to complicate the appendix by the addition of other materials.

The austenitic stainless steels have about 18 percent chromium and 8 percent nickel as the primary alloying elements. Although other elements are added, the austenite phase is stabilized by the presence of the nickel. This group of stainless steels is characterized by its excellent general corrosion resistance, ductility, ease of fabricability and weldability, and relatively low cost. This group comprises almost all stainless steel tanks. Types 316 and 317 contain molybdenum to provide better aqueous corrosion resistance. The addition of molybdenum also increases resistance to pitting as encountered in the paper and pulp industry as well as a host of other chemical applications.

11.1.1.2 Corrosion

Stress corrosion cracking. The austenitic stainless steels are subject to a phenomenon called *stress corrosion cracking*. Stress corrosion cracking requires the presence of stress and specific chemical exposure. Although the subject of stress corrosion cracking is complex and involved, some general principles apply that make selection of stainless steel for tank applications simpler:

- It is usually possible when there is exposure to hot chloride.
- Below 160°F stress corrosion failures are not likely to occur.
- Above 160°F the time to failure is temperature- and composition-dependent.
- Type 316 tends to be more resistant to chloride stress cracking than type 304.

Effect of cold-forming stainless steel parts. Cold forming of stainless steel does not generally decrease its corrosion resistance. However, for severely cold-worked stainless steel subject to exposure known to cause pitting, there have been reports of increased pitting in the cold-worked zone. Chloride solutions might behave this way as well as cause stress corrosion cracking. Low-temperature stress relieving (500 to 800°F) is sufficient to reduce cracking, but it will not stop it for specific aggressive environments known to cause stress corrosion cracking. Complete elimination of stress is required which would require much higher stress-relieving temperatures.

Crevice corrosion. In stainless steel tanks, the lap welds of tank bottoms and roofs produce crevices that are subject to oxygen deprivation which produces pitting and increased corrosion. Sometimes, it is beneficial to use butt-welded or seal-welded lap-welded roofs which prevent the interior environment from getting into the crevices of the roof. For tank bottoms, a concrete slab or a foundation which is free of mineral salts is the most effective way to ensure that crevice corrosion does not initiate from the underside.

11.1.1.3 When to use the L grades?

The low-carbon stainless steel grades should be used when the tank is susceptible to intergranular attack on the as-welded 0.08 percent standard stainless steel grades. The general corrosion and pitting resistance of the low-carbon grades may not be better than those of the higher-carbon counterparts, and there is no fabricability or weldability advantage. However, the low-carbon grades have a lower design stress than the higher-carbon counterparts.

11.1.2 Design

11.1.2.1 General

The design of stainless steel tanks follows the principles outlined in API Standard 650. In fact the new stainless steel appendix in API 650 was created as an exception document that points out those areas where adjustments need to be made to account for the properties of stainless steel. Otherwise, most of the rules in API 650 apply. These are some of the major exceptions that apply in the API appendix:

- Design temperature at temperature up to 500°F
- Materials
- Allowable design stresses

- Notch toughness testing not required for any stainless steel materials
- Design minimum thickness is $\frac{3}{16}$ instead of $\frac{1}{4}$ in for bottom plates
- Maximum shell thickness of 0.5 in
- Thermal stress relief not required
- Annular bottom plates not required

The API appendix applies only to fixed-roof and open-top tanks. It has no rules for the design of floating roofs. One reason is that the primary demand for stainless steel tanks is for small-diameter, fixed-roof tanks in chemical and paper and pulp applications. Therefore, the floating-roof tanks are a small percentage of the total number of stainless steel tanks, and the effort to devise a standard to cover roof design was not warranted. Because the intent of the standard was to apply mostly to small-diameter tanks and because stainless steel is quite ductile, the use of annular bottom plates was not included as a requirement. Some API design requirements are as follows:

11.1.2.2 Minimum thickness

Prior to the advent of a general-purpose API standard for stainless steel tanks, 10 gauge (0.1325-in, 5.6-psf) stainless steel sheet was often used for tank construction. However, because of the difficulty of maintaining good tolerance and appearance with such thin material, API decided to specify a minimum bottom, shell, and roof thickness of $\frac{3}{16}$ in. The primary motivation to use minimum thickness is, of course, the relatively high cost of stainless steel material compared to carbon steel. In fact, one of the hallmarks of stainless steel tank use is that little or no corrosion allowance is needed, so that lighter and lower-cost materials may be used.

11.1.2.3 Minimum required thickness

The shell minimum thickness is determined by

$$t_d = \frac{2.6D(H-1)G}{S_dE} + CA$$

where t_d = design shell thickness, in
t_t = hydrostatic test shell thickness, in
D = nominal diameter of tank, ft
H = design liquid level, ft
G = specific gravity of liquid to be stored
E = joint efficiency, 1.0, 0.85, or 0.7 based on radiography specified

CA = corrosion allowance
S_d = design allowable stress, psi
S_t = hyrostatic test allowable stress, psi

This differs from the 1-ft method used by the main body of API 650 by introducing a joint efficiency which is assumed to be 1.0. It is the same equation that is used in Appendix A of API 650 for small tanks.

11.1.2.4 Design stresses

Stainless steels have a relatively low yield point when considered as a percentage of the tensile strength compared to carbon steels. The design stresses as a percentage of yield or of tensile stresses have been modified to account for the difference in the shape of the stress-strain curves of stainless steels. This also accounts for differing design stresses for plate materials used on the shell and plate materials used for flanges, which must not be allowed to deflect and thus leak.

For shell material the design stress is the lesser of 0.3 times the tensile strength or 0.9 times the yield strength. For carbon steels the design stress basis is 0.3 times tensile strength or two-thirds the yield strength. By increasing the percentage of yield strength for which stainless steels may be designed, a greater permanent deformation can occur, and for stainless steel this turns out to be a permanent strain of 0.10 percent. Since strain levels are critical for applications which depend on sealing surfaces such as gasketed joints, API decided to use the lesser of 0.3 times the tensile strength or two-thirds of the minimum yield strength for plate ring flanges. This gives a permanent strain of less than 0.02 percent. For temperatures up to 100°F, the values for design allowable stresses are based only on 0.3 times the tensile strength for the low-carbon steel grades because the actual yield strengths are almost always considerably above the specified minimum value. Design allowable stresses are given in API 650.

11.1.2.5 Adjustment of stresses

For roofs designed in accordance with Appendix F, an adjustment was made to account for the difference in yield stress between carbon steels and stainless steel. Since the design of roofs is based upon yield strength, the 30800 used in the Appendix F must be reduced by the ratio of the yields strength of stainless steel at temperature to the value assumed by API 650 Appendix F for carbon steels, or 32,000 psi.

For tanks with design temperatures above 100°F, similar adjustments to increase thicknesses should be made for various components such as flush cleanouts, roof structural components, and other components by using the ratio of the allowable stress at temperature to the allowable stress at 100°F. In addition, the modulus-of-elasticity basis that is used for self-supporting roof thicknesses and for wind girders must be adjusted by the ratio of the modulus of elasticity at temperature to that at 100°F.

11.1.2.6 Joint efficiency and radiography

Joint efficiency is determined according to Table 11.1.2. Since many stainless steel tanks are small-diameter tanks, it may often be more cost-effective for a user or fabricator to omit radiography. For example, assuming no radiography, a 10-ft-diameter by 12-ft-high 317L tank requiring no corrosion allowance and storing a liquid with a specific gravity of 1.0 requires a wall thickness of 0.0182 in according to the equation above. However, it must be a minimum of $^{3}\!/_{16}$ (0.1875) in thick. Performing radiography on this tank will not reduce the required thickness.

11.1.2.7 Material toughness and ductility

Testing has indicated that stainless steels retain ductility down to cryogenic temperatures. Therefore, the stainless steel appendix has no lower-temperature bound for design. This property also eliminates the need to test for toughness for low-temperature applications.

11.1.3 Fabrication Considerations

11.1.3.1 Comparison to carbon steels

To form stainless steel uses essentially the same processes as for carbon steel fabrication, although there are differences in mechanical properties that must be considered. One such property is the work-hardening character of stainless steel. Since stainless steels work harder more rapidly than carbon steels for a given elongation, more power

TABLE 11.1.2 Joint Efficiencies for Stainless Steel Tanks

Joint Efficiency	Radiography required	Comment
1.0	Radiography per API 650, paragraph 6.1.2	Same requirements as main body of API 650
0.85	Radiography per API 650, paragraph A.5.3	Same as API 650, Appendix A requirements Slightly less stringent than main body of API 650
0.7	No radiography	Same as API 650, Appendix A requirements

is required for simple operations such as forming, shearing, or bending. Another result of this characteristic is the noticeably greater spring-back that occurs during bending operations. When dies are used, they must account for the greater spring-back. Stainless steel tanks can be welded by the gas metal arc or submerged arc processes for production welding. Alternatively, shielded metal arc welding is used extensively as well. Typically, type 309 weld electrodes are used. Cutting is usually performed with the plasma arc process or by carbon arc.

11.1.3.2 Openings and reinforcements

The minimum nominal thickness of connections and openings should meet the following criteria:

Nominal diameter, in	Minimum nominal neck thickness
2	Schedule 80
3–4	Schedule 40
Over 4	0.25-in wall thickness

The same reinforcement requirements that apply to the main body of API 650 apply to the stainless steel appendix.

11.1.3.3 Carbon steel attachments

API prohibits the use of carbon steel components unless they are external to the tank such as wind girders, anchor chairs, attachment clips, stiffeners, and the like. Reinforcing pads, however, must be stainless steel. Because of the greater costs of preparing and coating carbon steel, many purchasers require the use of all stainless steel for all components. In addition, the environmental causes of problems such as spills or corrosive atmospheres are always present which could attack the outer surfaces of the tank. Although carbon steel can be welded directly to the tank, a "poison pad" is often used. This is a fillet-welded stainless steel pad which is welded to the surface of the tank. To this pad is welded the carbon steel attachment. This method provides isolation of the shell from the carbon steel and acts as reinforcement for the attachment component.

11.1.3.4 Cold forming and heat treating

The unstabilized austenic stainless steels (for example, 304, 316, and 317) cannot be appreciably hardened by heat treatment, but they do harden as a result of cold working. To remove residual stresses, thermal stress relieving is sometimes applied to stainless steels. However, the corrosion resistance may be impaired by the formation of chromium carbides at the grain boundaries. Stress relieving cannot be done without impairing the corrosion resistance of the unstabilized stainless steel unless

1. The workpiece is held at the carbide-dissolving temperature for a specific time and is rapidly cooled from 1500 to 800°F.
2. The workpiece is heated to some temperature below 800°F and is held at this temperature for a long time.

API allows for cold, warm, or hot forming of stainless steel tank components. This is defined by the following:

Cold forming	Ambient temperature forming
Warm forming	1000–1200°F
Forming not permitted	1200–1650°F
Hot forming	1650–2200°F

When one is warm-forming components, it can be assumed that the stainless steel will be sensitized by intergranular carbide precipitation. The prohibition against forming at 1200 to 1650°F is intended to minimize sensitization of the stainless steel. Unless this is acceptable, the low-carbon grades (304L, 316L, and 317L) should be used or the material should be solution-annealed. If the unstabilized austenitic stainless steels are heated to above 1500°F and are not cooled below 800°F within 3 min, then sensitization occurs, impairing the corrosion resistance of the stainless steel. Since tanks are too large to stress-relieve, the L grades should be used when a sensitized stainless steel cannot be used.

For thinner sections, it is most likely that cold forming would be used. Few problems have resulted from lack of heat treatment after cold forming, and few companies require it. In addition, heat treatment may not effectively solve the stress corrosion-related problems where they exist, and it may be appropriate to consider other materials in these cases. Care should be taken to avoid insulating tanks with chloride-containing materials which could cause stress corrosion problems.

11.1.4   Cleaning and Passivation

API specifies measures to prevent contamination of the stainless steel while it is being shipped, delivered, or stored at the job site. If the surface is clean, then it can self-passivate in the atmosphere. Iron particle contamination or oil and grease on the surfaces are the most common form of debris encountered. The iron tends to stain and streak the stainless steel, damaging its appearance or, worse, creating a service problem where iron contamination or discoloring cannot be tolerated, such as in food-grade services. The primary source of iron contamination is from fabrication in a shop that works with both ordinary carbon steel and stainless steel without proper cleaning operations.

11.1.5 Liquid-Metal Embrittlement

One company reported zinc liquid-metal embrittlement in a stainless steel weld made after a manway flange had been near an inorganic zinc coating setup due to overspray. Cracks developed in the flange, and the theory propounded was that welding caused liquid-metal embrittlement cracking to occur. The lesson to be learned here is to be aware of surrounding equipment when spray-painting with inorganic zinc. Also, welds between galvanized pipe and stainless steel are subject to cracking and have cracked. API prohibits welding of galvanized components to the tank. Awareness of this form of failure increased as a result the major fire at ICI's Flixborough plant in the 1970s that directly resulted from liquid-metal embrittlement of stainless steel by zinc.

11.1.6 Temperature Differences between Shell and Bottom

Since the thermal expansion rate of stainless steel is higher than that of carbon steel, rapidly filling the tank with hot liquids could impose significant stresses on the bottom-to-shell weld. Consideration should be given to this factor by limiting the rate of heat-up or by designing to accommodate it. Even carbon steel tanks that have been heated too quickly by filling with hot liquids have developed waves or ripples in the bottom plates at the lap welds which have a high potential to become cracked.

11.1.7 Cathodic Protection

Although cathodic protection has been applied to stainless steel, a specialized form of cathodic protection called *anodic protection* has been used. It is a process similar to cathodic protection except that the current flow is reversed. The polarity reversal tends to stabilize and promote the oxide film that protects stainless steel. However, the procedures and processes used for anodic protection are too exacting for tank applications. In general, neither anodic nor cathodic protection is applied either internally or externally to aboveground stainless steel storage tanks.

11.1.8   Testing and Inspection

11.1.8.1   Liquid-penetrant examination

The shell-to-bottom inside attachment weld, all welds of opening connections the tank shell which are not completely radiographed, all butt-welded joints in the tank shell and annular plate on which backing strips are to remain, and all external attachment welds to the shell such as stiffeners, compression rings, or clips must be examined by the liquid-penetrant method before the hydrostatic test.

11.1.8.2   Hydrostatic testing

Hydrostatic testing of new tanks is required. The test water must meet the following criteria:

Free chlorine	0.2 ppm maximum
Total chloride	40 ppm maximum
pH range	6–8.3
Temperature	120°F maximum
Maximum test time	21 days for potable water
	7 days for other fresh water

11.1.8.3   Radiography

Radiography shall be conducted in accordance with Table 11.1.2.

SECTION 11.2   ALUMINUM TANKS

11.2.1   Introduction

Aluminum has a number of attributes that ensure it a niche in the structural metals consumption market. The best-known properties are its light weight (approximately one-third the density of carbon steel, 0.1 lb/in^3) and its corrosion resistance. The strength-to-weight ratio is of obvious value; however, its relatively low modulus of elasticity requires attention to control of deflections and buckling. By alloying aluminum with other elements, physical properties comparable to those of carbon steels may be achieved. The reflectivity of aluminum is another attribute that in many cases eliminates the need for surface coatings and other treatments. In nonstructural applications its high thermal and electrical conductivities as well as reflectance are well known. Aluminum may be formed, machined, joined, welded, and fastened by standard methods and equipment. Welding of aluminum requires shielded-gas techniques comparable to stainless steel welding.

11.2.2   Corrosion of Aluminum

One of the most important properties of aluminum is its resistance to corrosion. Aluminum's corrosion resistance is due to a thin aluminum oxide film which quickly forms when aluminum is exposed to the oxygen in air or certain aqueous solutions. The surface-treating process of *anodizing* simply builds a thicker oxide layer on the surface of the metal. Because the tenacious oxide film forms so readily, it renews itself when abraded away or chemically removed. Aluminum is sensitive to crevice corrosion and will often build up voluminous quantities of "white rust" or aluminum oxide. This is common where an aluminum surface is tightly pressed against another surface. Lap-welded tank bottoms or roofs are an area where the potential for crevice corrosion is high.

The corrosion chemistry of aluminum is complex. For example, a level of 0.1 percent water in methanol prevents corrosion, even at high temperatures, whereas a trace of water in liquid SO_2 accelerates corrosion. Aluminum is, however, completely immune to the corrosive effects of many chemicals and is, therefore, a candidate for tank materials of construction. The external conditions of the tank such as the atmosphere and the foundation must also be considered. Relative to stainless steel, the material selection problem is much more difficult.

Aluminum tends to pit with water that has chloride ions in it. At levels as low as 0.1 ppm, copper in water can react with aluminum, depositing metallic copper at local sites, which acts to initiate pitting. Therefore, aluminum is not suitable for any tanks which may have trace heavy metals in the stored liquid.

Some aluminum alloys are subject to stress corrosion cracking. This should be investigated during the materials selection phase of a project.

To meet the demand for increased corrosion resistance of aluminum wrought products, clad aluminum products were developed. Aluminum-clad products are high-strength alloy cores in sheet or tubing form that have clad layers of pure aluminum or aluminum alloys bonded to the core. The cladding is engineered to be anodic or sacrificial to the core and essentially creates a built-in cathodic protection system. The clad material is usually less than 10 percent of the thickness of the total material thickness. These products are usually not heat-treatable. Because of the sacrificial cladding the corrosion progresses through the cladding but stops at the core. It is a highly efficient method of reducing through-wall pitting.

11.2.3   Aluminum Alloys

Numerous alloys are available for industrial applications. Each alloy is available in a broad range of tempers. The Aluminum Association has established a system of numerical designations for all alloy grades in general commercial use. These designations act to standardize the specifications and properties of the material regardless of the source.

The wrought alloys and their temper designations are as follows:

Aluminum (99+ percent pure)	1xxx
Alloying element	
Copper	2xxx
Manganese	3xxx
Silicon	4xxx
Magnesium	5xxx
Magnesium and silicone	6xxx
Zinc	8xxx
Other	9xxx
Temper designation:	
F	As fabricated
O	Annealed
H	Strain-hardened
W	Solution heat-treated
T	Thermally treated to produce stable tempers other than F, O, or H

Aluminum as a pure element has relatively low strength. The strength is enhanced by addition of small amounts of other elements, heat treatment and/or strain hardening, or cold working. Some alloys are *heat-treatable,* meaning that the strength can be enhanced by heat treatment. *Non-heat-treatable* alloys can be cold-worked for strength enhancement.

Table 11.2.1 is a list of the alloys acceptable for use in aluminum tanks.

11.2.4 Applications

Aluminum is commonly used in hoppers and silos for storage of plastics and resins. It is commonly used in the chemical industry for storage of fertilizers and other chemicals.

11.2.4.1 Chemicals

Some liquids that can be stored in aluminum tanks are nitric acid in concentrations above 82% by weight, acetic acid, other organic acids, alcohols, aldehydes, ammonium nitrate, aqua ammonia, esters, glyc-

TABLE 11.2.1   Material Specifications

Plate and sheet (ASTM B 209)		Pipe and Tube (ASTM B 210, ASTM B 241, ASTM B 345)		Rod, bar, and shapes (ASTM B 211, ASTM B 221*, ASTM B 308)		Forgings (ASTM B 247)		Castings [ASTM B 26 (sand), ASTM B 108 (permanent mold)]	
Alloy	Temper	Alloy	Temper	Alloy	Temper	Alloy	Temper	Alloy	Temper
1060	All	1060	All	1060†	All	—	—	—	—
1100	All	1100†	All	1100†	All	1100	H112	—	—
—	—	—	—	2024†	T4	—	—	—	—
3003	All	3033	All	—	—	3003	H112	—	—
Alclad 3003	All	Alclad 3003	All	—	—	—	—	—	—
3004	All	—	—	3004†	All	—	—	—	—
Alclad 3004	All	—	—	—	—ll	—	—	—	—
5050	All	5050†	All	—	—	—	—	—	—
—	—	—	—	5052†	All	—	—	—	—
5052	All	5052†	All	—	—	—	—	—	—
5083	All	5083	All	5083†	All	5083	H111, H112	—	—
5086	All	5086	All	5086†	All	—	—	—	—
5154	All	5154†	All	5154†	All	—	—	—	—
5254	All	5254†	—	—	—	—	—	—	—
5454	All	5454†	All	5454†	All	—	—	—	—
5456	All	5456†	All	5456†	All	—	—	—	—
5652	All	5652†	—	—	—	—	—	—	—
6061	T4, T6 ‡.	6061†	T4, T6	6061	T6	6061	T6	—	—
Alclad 6061	T4, T6 ‡	—	—	—	—	—	—	—	—
		6063	T6	6063†	T6	—	—	—	—
				6262†§	T9	—	—	—	—
								514.0†	F
								443.0	F
								356.0	T6, T71

*Tubular shapes handling fluid pressure not included.
†Not included in all specifications.
‡Thickness 0.006 to 0.249 in; for thickness of 0.250 in and over, corresponding tempers are T451 and T651.
§Nuts only.

erin, highly purified water, hydrogen peroxide, hydrocyanic acid, ketones, nitroparaffins, organic acid anhydrides, and food and paint stuffs. Because aluminum shows no low-temperature embrittlement, it has been used in cryogenic storage. The nonspark characteristics of aluminum alloys make it useful for some applications where flammability is involved. Although cryogenic liquids were once commonly stored in aluminum vessels, this application now relies almost entirely on vessels constructed of austenitic stainless steels. Consideration for the relatively low melting point of aluminum needs to be given for fire safety where the exposure potential is sufficiently high.

11.2.4.2 Maintaining product purity

Aluminum is sometimes used to store polymers or resins that would be tainted with color if stored in steel tanks or bins. Other applications include chemicals and products that are to be maintained in a water-white condition.

11.2.4.3 Water storage

Because aluminum is compatible with high-purity water, distilled water, deionized water, uncontaminated rainwater, and heavy water used in nuclear reactors, aluminum storage tanks offer a good, cost-effective alternative for these applications. Since surface preparations and coatings are not necessary, the aluminum storage tank will often be competitive with coated carbon steel storage systems. In the proper applications, there is virtually no metal contamination of waters. Potable waters can be safely stored in aluminum tanks without the threat of metal contamination.

Freshwater categories by chemistry

1. Waters containing heavy metals such as copper, nickel, and lead. Aluminum is not recommended for these services because the heavy metals may contribute to high pitting rates.
2. Neutral or near neutral waters. For waters in the pH range of 6 to 9 there is little or no concern about corrosion.
3. Alkaline waters. A pH range of 8.5 to 9 is acceptable.
4. Acid waters. A pH of 4 or higher is acceptable.

Treated water. Dissolved gases such as carbon dioxide or oxygen in condensate applications or where amines, chromates, and polyphosphates or other alkaline inhibitors have been used do not adversely affect the use of aluminum.

Recirculated water may become corrosive to aluminum because it picks up copper and iron from the various equipment such as pumps, pipes, and instrumentation. The dissolved metals "plate" out on the aluminum, causing localized pitting. If the water is treated with inhibitors and cathodic protection, the problem may sometimes be controlled. However, aluminum tubes in heat exchangers have a poor history of success in refinery applications. Generally, cathodic protection of process plant components for aluminum does not work.

High-purity water systems can be a candidate for aluminum storage systems. Aluminum is often used for heavy water for nuclear reactors.

Steam condensate. If the water is free from boiler carryover, aluminum is inert to condensate; however, alkaline water-treating compounds may be corrosive.

Seawater. Aluminum is generally not an acceptable material for seawater service, except under special conditions, such as these:

- Seaplanes, boats, and outboard motors which must be flushed with fresh water after use in seawater.

- The more corrosion-resistant grades of aluminum (3003, 5052, etc.) are acceptable in marine environments such as near shore use or above sea level.

Alloys containing over 0.3 percent copper should not be used at all anywhere near seawater. The usual mode of attack on aluminum by seawater is through pitting on free surfaces or crevice corrosion. The copper-containing alloys also have a heavy general surface buildup of "white rust" with accompanying high rates of general corrosion.

11.2.5   ASME B96.1, Aluminum Tank Standard

The recognized standard that covers the details for cylindrical aluminum storage tanks is ASME B96.1, *Welded Aluminum-Alloy Storage Tanks.* This standard covers the fundamental mechanical design principles needed for the design, material, fabrication, erection, inspection, and testing of vertical, aboveground, cylindrical, fixed-roof or floating-roof storage tanks approximating atmospheric temperatures and pressures.

B96.1 was largely patterned from API Standard 650 with modifications to cover differences in material properties and fabrication and erection processes associated with aluminum structures. Many of the detail drawings for nozzles, flush cleanouts, manholes, and roof-shell fittings are nearly identical to the figures shown in API 650.

11.2.5.1   Pressure classification of tanks

The standard covers internal tank pressures from approximately atmospheric to 1.0 psig maximum. It departs from API 650 by making a fundamental classification of tanks based on pressure:

Class I tanks are less than 100 ft in diameter with an internal pressure not exceeding 0.5 psi.

Class II tanks are less than 100 ft in diameter with an internal pressure greater than 0.5 psi, which is great enough to produce uplift of the roof plate load.

Class III tanks are those whose internal pressure produces uplift forces greater than the tank weight of the roof plus the shell plus any framing supported by the shell or roof.

This classification of tanks allows for simplification of the design rules that apply to roof-shell compression area, anchorage, and shell thickness calculations.

SECTION 11.3 FIBERGLASS-REINFORCED PLASTIC TANKS

11.3.1 Resin Fundamentals

FRP materials and chemical resistance.   Table 11.3.1 is a generalization about the chemical resistance of various available resins on the market that may be used for tank construction. Probably the simplest rule of thumb for FRP tank selection is that if a stainless steel tank cannot be used due to inadequate corrosion resistance or because it is noncompetitive with fiberglass-reinforced plastic (FRP), then the FRP should be considered.

11.3.1.1 Basic resin types

Several resins are used in FRP construction including vinyl esters, polyesters, epoxies, furans, and phenolics. The vast majority of FRP tanks are made from polyesters and vinylesters because of their relatively low cost and the wide pH range (from basic to acidic) to which the material is chemically resistant. Although FRP pipe is sometimes fabricated using epoxy resins, tanks generally do not use epoxy resins. Although epoxies have excellent mechanical and thermal properties, they are not compatible with very acidic environments below a pH of 3. The furans have excellent general chemical resistance. In fact for chemical sewer applications where the composition or type of chemical exposure is not known, the furan resins will often be selected. However, furans are more costly and difficult to work with, and they are harder to inspect since the resin is black (as opposed to the vinylester or polyester resins which are honey-colored or clear). For tanks, the use of furans is quite rare.

11.3.1.2 Glass content

The increasing content of glass reinforcement increases the mechanical strength properties. Other variables that can affect the properties are resin selection, type of reinforcement (chopped glass mat, unidirectional roving, woven roving, etc.), orientation of reinforcing fibers,

TABLE 11.3.1   Chemical Compatibility Chart for FRP Tanks

	Aqueous solutions			Organic solvents	Heat resistance
	Acid	Base	Oxidation		
Polyesters					
Structural, examples: Dion FR-6604T Hetron 99-P	F	NR	NR	NR	F
Isophthalic polyesters, examples Dion ISO 6631T Hetron 92 Aropol 7241	G–VG	P	P	F–G	F
Het-acid (chlorinated) polyester, example Hetron 197	VG	P	VG	F–G	G–VG
Bisphenol-A/fumaric acid polyester, examples Atlac 382 Dion Cor-Res 6694 Hetron 700	G–VG	VG	G	F–G	G
Vinyl esters					
Epoxy-based, examples Derakane 411/470 Hetron 922/980	VG	G	G	G–VG	G–VG
Bisphenol-based, example Atlac 580	G	F	G	F	G–VG
Epoxies Matcote Matstick 1-01 (amine adduct) Porter MCR-65 (amine-cured) Napko 5684 (polyamide-cured)	F	VG	P–G	G–VG	G–VG
Furans Examples: Quacorr Hetron 800	G	G	NR	VG	VG

VG = Very good   F = Fair   NR = Not recommended   G = Good   P = Poor

and ratio of fiberglass content to resin. Table 11.3.2 shows typical mechanical properties of FPR.

11.3.1.3   Resin curing and catalysis

Two crucial factors yield a composite structure that meets the required specifications: the extent to which the laminate cures and the system used to initiate the crosslinking or polymerization of the resin.

TABLE 11.3.2   Nominal Properties of FRP Laminates

Property	Contact-molded	Filament-wound
Laminate density, lb/in^3	0.05–0.06	0.06–0.07
Specific gravity	1.5–1.8	1.8–2.1
Tensile strength, psi	8,000–18,000	25,000–50,000 (hoop)
Flexural strength, psi	16,000–22,000	20,000–40,000 (hoop)
Compressive edge strength, psi	18,000–24,000	20,000–24,000
Flexural modulus of elasticity, psi	0.7–1.0×10^6	1.8–3.2×10^6 (hoop)
Tensile modulus of elasticity, psi	0.8–1.1×10^6	2.0–3.5×10^6 (hoop)
		0.9–1.4×10^6 (axial)
Poisson's ratio	0.33	0.33
Impact strength, ft•lb izod	30–40	40–50
Thermal conductivity, Btu•in/(h•ft^2•°F)	1.3–1.8	1.3–1.8
Linear coefficient of expansion, in/(in•°F)	15–20×10^{-6}	12–16×10^{-6}
Heat distortion temperature (resin), °F, at 264 psi	170–300	170–300
Barcol hardness	27–45	27–45

A resin can cure to varying degrees, and the greater the degree of cure, the better the final properties. Oven curing at 125 to 180°F is one way to ensure a complete cure of the resin. However, this is impractical for most tanks because of the large sizes. For tanks the ambient temperature method is used where the heat of reaction of the curing process provides sufficient heat to provide the curing. The most widely used indicator of proper cure is the Barcol hardness test in accordance with ASTM method D 2583. Another way to test for proper cure is the acetone test. In this test a small amount of acetone is smeared onto the surface of the laminate until it evaporates. If the surface becomes softened or tacky, it is an indication of undercure.

Another important aspect of attaining the desired mechanical properties is the initiation of crosslinking. There are several important types of initiators or catalysis. Methyl ethyl ketone peroxide-cobalt naphthanate (MEKP-CoNap), benzoyl peroxide-dimethyl aniline (BPO-DMA), and cumene-cobalt naphthenate (cumene-CoNap) are the most common for polyester resins. MEKP-CoNap is the most common for polyester or vinyl esters. It generates little heat and is the easiest to use, as it requires a single feed into the mixing gun at the time of the application. However, like all catalysts, it is not used in the resin reaction, and therefore cobalt remains in the resin which under certain conditions can initiate chemical attack. An example is tanks initiated with MEKP-CoNap which should not be used with hypochlorite solutions. In these cases, the BPO-DMA may be used. However, it is more costly and difficult to work with. It requires two-component mixing at the time of application. A premium of approximately 5 percent would be applicable to a tank constructed using the BPO-DMA catalyst.

11.3.1.4   Addition of thixotropic agents

These agents are used to control the viscosity of the resin. The addition of such an agent, however, can reduce the clarity of the resin, making visual inspection more difficult. It may also reduce the corrosion resistance of the resin to certain chemicals. API Standard 12P limits the addition of thixotropic agents to 5 percent.

11.3.1.5   Fire retardants

Antimony trioxide or antimony pentoxide is sometimes specified as the fire retardant which should be supplied in accordance with the FRP tank manufacturer's recommendations. The resins must be halogenated with chlorides or bromides and have a class 1 rating per ASTM test method E-84.

11.3.1.6   Conductivity

Metal powder, carbon, or other substances are sometimes added to the resin blend to control static electricity. Such additions may interfere with the clarity of the resin and hence the ability to visually inspect for laminate quality.

11.3.2   Layered Wall Construction

The wall thickness of an FRP tank comprises several functions. Because of this requirement several different kinds of layers are used, as shown in Fig. 11.3.1. Figure 11.3.2 shows the appearance of various reinforcing glasses and fabrics used in FRP construction.

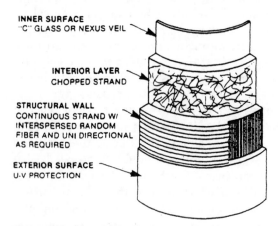

INNER SURFACE
"C" GLASS OR NEXUS VEIL

INTERIOR LAYER
CHOPPED STRAND

STRUCTURAL WALL
CONTINUOUS STRAND W/
INTERSPERSED RANDOM
FIBER AND UNI DIRECTIONAL
AS REQUIRED

EXTERIOR SURFACE
U-V PROTECTION

Figure 11.3.1 Layered wall construction. (*Courtesy of Ershigs, Inc.*)

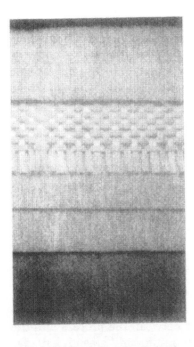

1 1/2 oz
Chopped Strand

24 oz Wozen Roving

2 Layers 1 1/2 oz
Chopped Strand

"C" Glass Veil

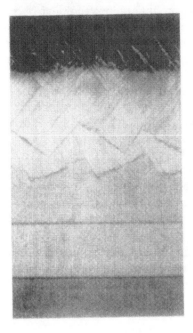

Bi directional
Filament Wound Strand

2 Layers 1 1/2 oz
Chopped Strand

"C" Glass Veil

Figure 11.3.2 Various glass fabrics used in FRP construction.

11.3.2.1 The veil

This is the inner layer that is exposed to the chemicals stored. It is usually about 10 to 20 mils thick and consists of about 10 percent glass or polyester fibers and 90 percent resin. The purpose of this resin-rich layer is to resist chemical attack by the process fluids stored. The reinforcement may be a C glass veil or a synthetic fiber such as polyester. A polyester veil is often specified because it is more corrosion-resistant than the glass fiber materials. However, the polyester veil is usually about $0.12 per square foot versus $0.06 per square foot for glass veil.The polyester resists strong oxidants such as chlorine dioxide and hypochlorites as well as basic solutions that would attack glass. The reinforcement can be chopped-strand mat, chopped strand, surface veil, or synthetic veil matrix. API Standard 12P requires that this layer contain less than 20 percent by weight reinforcing material.

11.3.2.2 The liner

The next layer, called the *liner,* prevents weeping of the inner surface and provides a chemically resistant-liner between the process and the structural layer. By using noncontinuous reinforcing glass fibers, the fibers may corrode longitudinally but not propagate into the next fiber. This corrosion-resistant liner layer is approximately 30 percent reinforcement and 70 percent resin. The liner can be applied in a number of ways, but there are two basic processes. In the first method, resin and noncontinuous glass-fiber strands are applied in a minimum of two plies equivalent to 3 oz/ft². In the second method, the resin and chopped-fiber strands are sprayed up to an equivalent weight of 3 oz/ft². Each ply is rolled prior to the application of the next layer, and the combined thickness of the liner is usually 180 to 250 mils.

When the spray-up method is used, there are some resins which will have inferior mechanical properties compared to the layup method of application. Styrene during the spray process may evaporate and result in inadequate levels, preventing the crosslinking and curing from fully occurring. Spray tends to introduce air as well which must be rolled out by hand. Some companies prohibit the spray method in their specifications, but this should not be a universal requirement and should be based upon individual consideration of the supplier, the resin, the equipment application, and other variables.

11.3.2.3 The structural layer

This layer provides the structural strength of the tank including all loadings and stresses. The glass content of this layer is usually about 30 to 45 percent depending on the amount of woven roving used. This layer is most often made for tanks using continuous helical-wound glass fiber reinforcement, in which case the glass content approaches 55 to 70 percent by weight. Table 11.3.2 is a comparison of typical properties attained by either the contact-molded process or the filament-wound process.

11.3.2.4   External layer

An exterior layer is applied after the structural layer to protect the tank from ultraviolet light degradation and to provide color selection. It may consist of chopped strand, chopped-strand mat, or surfacing mate with sufficient resin to avoid exposure of any surface-reinforcing material. UV inhibitors are added as well as color pigmenting. Another means of coating the exterior is to use a gel coat or paint.

11.3.3   FRP Construction Methods

11.3.3.1   Filament-wound construction

This process applies continuous glass fiber roving in a helical pattern either by rotating the mold or by winding the fiber onto a stationary mold. Variation in the angle of the helix will result in differing strengths in the hoop versus longitudinal direction. By applying chopped-strand mat, woven fabric, or chopped strands in addition to filament-wound material, increased strength in the longitudinal direction may be achieved. The thickness of FRP tanks often varies because the hoop stress is greatest at the base of the tank where the liquid pressure is the greatest. It is wise to specify that at least one ply of 15.7 oz/yd^2 vertically oriented unidirection roving be applied to filament-wound tanks.

11.3.3.2   Contact-molded construction

This method uses multiple layers of fiberglass chopped-strand mat or woven roving saturated with resin and built up to the required thickness. The glass layers are laid by hand onto the mold saturated with resin. Rolling is required to remove air bubbles and to ensure that the glass fibers are saturated with resin. The spray-up method is another form of contact molding, in which a spray gun equipped with a chopper gun applies resin spray and chopped glass at a predetermined rate (see Fig. 11.3.3).

11.3.4   Design

Differences of design properties as compared with metals.   The design of FRP tanks differs substantially from that of metal tanks for several reasons. First, the dimensional tolerances for metals are small and

Figure 11.3.3 Spray-up method of FRP construction. (*Courtesy of Ershigs, Inc.*)

reproducible. FRP construction tolerances depend upon workmanship. Metals are isotropic, meaning that they have the same mechanical properties in all directions. FRP will usually have differing properties in different directions. FRP materials do not yield before failure, and therefore this cannot be a basis for strength. FRP has a modulus of elasticity of 1 to 4 million psi versus about 30 million psi for steel. FRP is much more flexible than steel. FRP is useful only to temperatures of about 200 to 300°F.

11.3.4.1 Design standards

Engineers rely on several standards for FRP tank procurement. By far the most common standard has been ASTM D 3299, *Filament-Wound Glass-Fiber-Reinforced Thermoset Resin Chemical-Resistant Tanks*. Another less-used standard is ASTM D 4096, *Standard Specification for Contact-Molded Glass-Fiber-Reinforced Thermoset Resin Chemical-Resistant Tanks*. As indicated in the title, this standard applies to contact-molded tanks. The petroleum industry relies primarily on API Specification 12P, *Specification for Fiberglass Reinforced Plastic Tanks*. This standard relies heavily on the ASTM standards but includes additional information such as

- Standard tank sizes and designs for both closed- and open-top tanks (Table 11.3.3 and Figs. 11.3.4 and 11.3.5)

TABLE 11.3.3   Typical API Tank Dimensions

1	2	3	4	5	6	7	8	9
							Size of connections, in	
Nominal capacity, bbl	Approximate working capacity, bbl (see note)	Inside diameter, ft-in A	Height, ft-in B	Height of overflow connection, ft-in C	Height of walkway lugs, ft-in D	Location of fill-line connection, in E	C-1, C4	C-3, C-2, C-4
90	74	8-0	10-0	9-6	7-7	14	2	2
110	96	8-0	12-6	12-0	10-1	14	2	2
150	92	8-0	16-6	16-0	14-1	14	2	2
150	122	10-0	10-6	10-0	8-1	14	2	2
210	185	10-0	15-0	14-6	12-7	14	2	4
210	176	10-6	10-6	10-0	8-1	14	2	4
250	217	12-0	12-6	12-0	10-1	14	4	4
300	267	12-0	15-0	14-6	12-7	14	4	4
400	368	12-0	20-0	19-6	17-7	14	4	4
500	459	14-0	18-6	18-0	16-1	14	4	
500	445	15-6	16-0	15-6	13-7	14	4	
500	466	12-0	25-0	24-6	22-7	14	4	
750	705	15-6	24-0	23-6	21-7	14	4	4
1000	955	15-6	30-0	29-6	27-7	14	4	4
Tolerances (all sizes)		± ½	± ½	± ⅛	± ⅛	± ⅛		

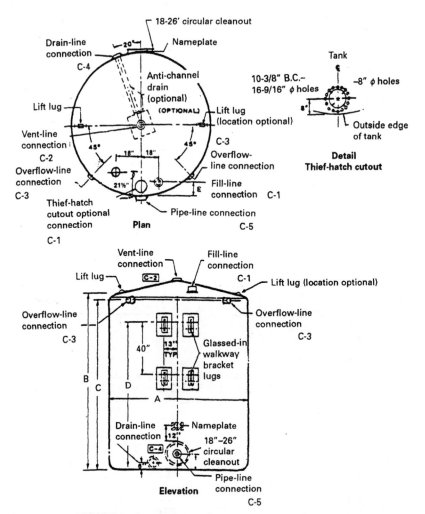

Figure 11.3.4 API FRP tanks. (*Courtesy of the American Petroleum Institute.*)

- Appendices with information on walkways, stairways, and ladders
- Miscellaneous information on handling of FRP tanks, marking, the API monogram, and inspection and testing

The latest standard to be applied to FRP tanks is ASME/ANSI RTP-1 which is modeled after section VIII, division 1 of the ASME Boiler and Pressure Vessel Code. This standard addresses all the stresses to be encountered such as discontinuity stresses and local stresses due to wind, seismic, or thermal load conditions. This standard has requirements for shop certification that when specified weed out the "garage operations." However, specifying tanks to this stan-

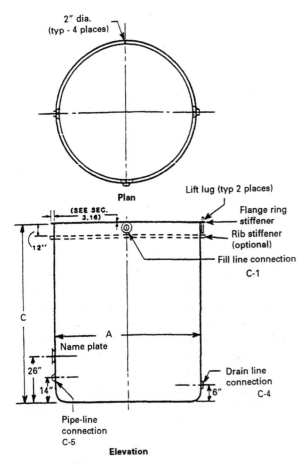

Figure 11.3.5 API open-top tanks. (*Courtesy of the American Petroleum Institute.*)

dard commands a premium in that additional paperwork and materials traceability are required. For most industries, specifying tanks according to API Standard 12P1 should be adequate provided that a reputable manufacturer is involved.

11.3.4.2 Internal pressure

Internal pressure above the liquid produces stresses that require special design considerations. ASTM D 4096, *Standard Specification for Contact-Molded Glass-Fiber-Reinforced Thermoset Resin Chemical-Resistant Tanks,* covers only atmospheric-pressure tanks. ASTM D 3299, *Filament-Wound Glass-Fiber-Reinforced Thermoset Resin Chemical-Resistant Tanks,* covers tanks with internal pressures up to 14 in WC. Internal pressure may produce a condition called *uplift* which tends to lift the shell. Anchoring the tank may be required for conditions of uplift.

11.3.4.3 Allowable stress

The minimum wall thickness for an FRP tank is $\frac{3}{16}$ in overall. However, the hoop stress developed by typical storage tanks requires a thicker wall, as determined by the classical hoop stress formula

$$t = \frac{PD}{2S_a}$$

where t = required thickness at shell elevation considered, in
P = internal pressure including hydrostatic plus any gas pressure above liquid, psi
D = inside diameter of tank, in
S_a = design allowable hoop stress, psi

The design allowable stress values for shell thickness are based upon an allowable strain. Tests have shown that an acceptable strain rate considering fatigue failure is 0.001 in/in. In addition, a factor of safety of 10 is used to account for variations in manufacturing capability, tolerances, not fully cured resins, localized and discontinuity stresses, and strength reduction due to long-term chemical exposure. Under no conditions is the design allowable stress to exceed one-tenth of the ultimate strength. The modulus of elasticity and the ultimate strength are determined from ASTM D 638.

11.3.4.4 Roof

Various components such as the dome or roof of a filament-wound tank as well as nozzles, supports, and joints may be constructed by the contact-molded method. The roof is required to support a single 250-lb load at any 4-in by 4-in square area without deflecting more than 0.5 percent of the tank diameter. This requirement does not provide the ability to compute thickness but is determined by the experience and judgment that a manufacturer has. In addition the minimum thickness, as with the shell, is $\frac{3}{16}$-in.

The required shape is ellipsoidal, flanged, dished, or conical with a 1:12 slope.

11.3.4.5 Bottom design

Minimum thickness. The minimum thicknesses for fully supported flat bottoms are based upon diameter as follows:

	ASTM D 3299	API 12P
2–6 ft	$\frac{3}{16}$ in	$\frac{1}{4}$ in
>6–12 ft	$\frac{1}{4}$ in	$\frac{1}{4}$ in
>12 ft	$\frac{3}{8}$ in	$\frac{3}{8}$ in

Knuckle. The bottom detail must comply as a minimum requirement with the detail shown in Fig. 11.3.6. This detail provides added reinforcement to account for the discontinuity stresses at this point as well as high local stresses caused by potential uneven loading of the bottom due to variations in foundations. API and ASTM have slight differences in the vertical dimensions for reinforcements as well as minimum inside radius. For more conservative designs the API 12P should be used.

11.3.4.6 Joints

Longitudinal joints should not be permitted in FRP tanks, as there is no fabrication advantage and this joint could be a likely source of failure. However, circumferential joints are often used to connect shell sections and to connect top and bottoms to the shell.

These joints are made by the use of a laminate bond. The joint should be made of alternating layers of mat or chopped strand and woven roving and should be made at least as thick as the shell. Extending the reinforcement up the sides a sufficient distance is important to ensure that the joint is as strong as the shell. ASTM D 3299 and D 4097 provide some requirements for these joints. Prior to joining, the surfaces to be joined should be roughened by sanding or abrasive blasting to remove all the resin-rich surface and to expose glass fibers. The roughened area should extend beyond the work area

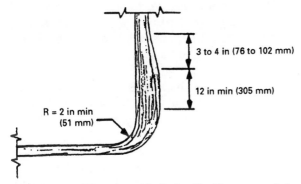

Figure 11.3.6 Bottom-to-shell detail. (*Courtesy of the American Society for Testing and Materials.*)

and subsequently is coated with resin. Care should be taken to limit crevices to less than the workpiece thickness and to limit offset to approximately one-third to one-half of the wall thickness. All exposed fibers must be coated with resin.

11.3.4.7 Open-top tanks

Open-top tanks may also be fabricated. These are required to have a horizontal reinforcing ring which stiffens the shell at the top of the tank.

11.3.4.8 Appurtenances

Opening. Openings in the shell should be reinforced according to the following formula:

$$t_r = \frac{PDK}{2S_a}$$

where $K = 1.0$ for nozzles with 6-in diameter and larger
$K = (d_r - d)$ for nozzles with less than 6-in diameter
d = nozzle outside diameter, in
d_r = reinforcement diameter, in, which must be $2d$ for nozzles 6 in and larger and $d + 6$ for nozzles less than 6 in
P = hydrostatic pressure at point of nozzle installation, psi
D = inside diameter of tank, in
S_a = allowable tensile stress, psi

For $t_r < \frac{1}{8}$ in, no reinforcement is required other than for the overlay for glassed-in nozzles.

Nozzles 4 in and smaller must be supported so that they do not break due to piping loads. Figure 11.3.7 shows the two most common methods of nozzle reinforcement. Glassed-in appurtenances such as nozzles and manways are installed using overlay laminate with thickness and length sufficient to withstand the anticipated load conditions. These should generally be fabricated according to Figs. 11.3.8 and 11.3.9.

Lift lugs. Lift lugs should be designed to carry a minimum load of 5 times the service load, and the service load of each lug should be considered at least equal to the weight of the empty tank. Lugs should not be attached by fasteners that penetrate the shell of the tank. Verification tests for lug strength are often specified by purchasers of FRP tanks. It is important for the purchaser and the manufacturer to agree on the forces and moments to be applied for these tests. Sometimes previous test results for similar tanks or prototype test results can be used.

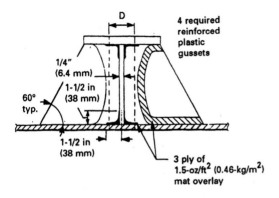

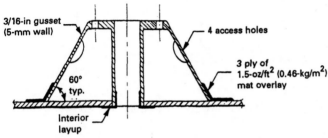

Figure 11.3.7 Reinforcement of nozzles. (*Courtesy of the American Society for Testing and Materials.*)

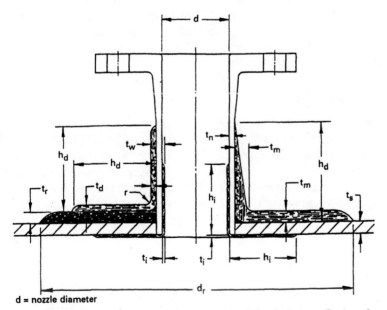

d = nozzle diameter

Figure 11.3.8 Flush nozzle detail. (*Courtesy of the American Society for Testing and Materials.*)

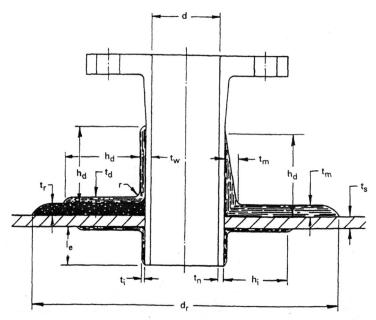

Figure 11.3.9 Penetrating nozzle detail. (*Courtesy of the American Society for Testing and Materials.*)

Hold-down lugs. For internal pressure, seismic loads, or wind loading, the bottom must have lugs sufficient to hold the tank down. The seismic load conditions should be determined in accordance with API 650, Appendix E. The wind load conditions should be determined in accordance with ANSI/ASCE 7-88.

11.3.4.9 Venting requirements

It is not required for tanks storing volatile organic compounds to comply with the emergency venting requirements of API Standard 2000 because FRP tanks will normally fail at very low temperatures (approximately 200 to 300°F) below which significant vaporization of most stocks could take place. For this reason, weak roof-to-shell joints are not specified. However, for pressure tanks special attention should be given to designing and specifying the venting system and components.

11.3.4.10 Ladders

API 12P1 provides a useful function in Appendix B by specifying load conditions for typical ladders, walkways, and platforms. Walkways are to be furnished by the purchaser unless otherwise specified.

Item	Dimensional requirements	Loading criteria
Walkway	26 in wide minimum	Uniform load 50 psf Concentrated load 1000 lb Maximum deflection $\frac{1}{360}$ span
Railings	42 in high	Point load in any direction 200 lb
Stairs	Normal angle 45° Acceptable angle 30 to 50° Rise height and tread width uniform throughout	100 lb/lin ft of tread or 1000-lb point load Deflection maximum $\frac{1}{360}$ span
Ladders	Maximum open ladder 20 ft high, caged ladder required for ladders 20 to 30 ft high 26-in by 30-in minimum rest platforms required when used Rungs three-quarters of diameter, 12-in spacing	Minimum rung load 200 lb

11.3.5 Fabrication and Testing

11.3.5.1 Tolerances

ASTM provides acceptable tolerances for fabrication as shown below and in Fig. 11.3.10.

Description	Tolerance
Diameter	± 1 percent
Shell taper	± ½°
Height	± ½ in or ½%
Nozzles	See Fig. 11.3.10

11.3.5.2 Inspection

The importance of visual inspection of FRP tanks during the fabrication process cannot be overemphasized. Visual inspection can detect chips, cracks, delamination, inclusions, air bubbles, failure to saturate the fibers, wrinkles, crazing, etc. Visual defects are described in ASTM D 2563, *Classifying Visual Defects in Glass Reinforced Plastic Laminate Parts.* Inspection should be made before painting, gel coating, or application of UV inhibitors, all of which interfere with the visual inspection process. The interior surface should be smooth, free of cracks and crazing, and limited to 2 pits per square foot with depths less than $\frac{1}{32}$ in. The outer surface should be free of exposed fibers. Table 11.3.4 provides a sample checklist for inspection and allowable defects.

The Barcol hardness test specified by ASTM D 2583 should be done on surfaces of the flange faces, secondary overlays, and other places

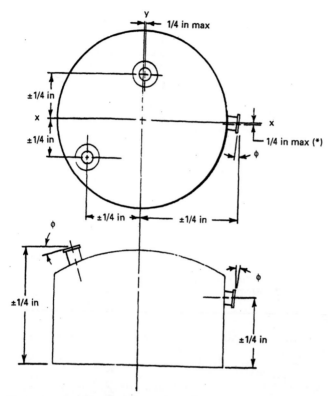

Figure 11.3.10 Nozzle tolerances. (*Courtesy of the American Society for Testing and Materials.*)

to determine the degree of cure. If a wax surface layer is used to enable the proper curing, the wax layer must be removed prior to hardness testing. When an organic surfacing veil is used, hardness values may have to be revised by agreement between purchaser and manufacturer. The acetone sensitivity tests should be used to verify resin cure. Tests should be made on all separately manufactured components and secondary overlays on both internal and external surfaces.

Hydrostatic testing can be performed in the shop or field. The advantage of testing the tank after installation is that any shipping and handling damage will be indicated. The water should contain a wetting agent to ensure that potential leaks will show up. The test should be conducted for at least 24 h. Strain in the shell should be measured. If defects are found, they should be repaired and the tank retested.

TABLE 11.3.4 Allowable Defects of FRP Laminates

Condition defects	Inner surface	Interior mat layers	Structural wall	Exterior surface
Chip	None	None	None	None
Crack	None	None	None	None
Crazing	None	None	None	None
Delamination	None	See air bubble	See air bubble	See air bubble
Dry spot	None	See air bubble	See air bubble	See air bubble
Foreign inclusion	None	Maximum dimension $1/32$ in	See air bubble	See air bubble
Fracture	None	None	None	None
Air bubble (void)	Less than $1/64$ in: unlimited $1/64$ to $1/16$ in: $2/\text{in}^2$ Greater than $1/16$ in: none	Less than $1/32$ in: unlimited $1/32$ to $1/8$ in: $5/\text{in}^2$ Greater than $1/8$ in: none	Maximum size: 0.25 in^2 Maximum $2/\text{ft}^2$	Maximum size: 0.25 in^2 Maximum $2/\text{ft}^2$
Blister	See air bubble	See air bubble	See air bubble	See air bubble
Burned	None	None	None	None
Pit (pinhole)	Greater than $1/16$-in dia.: none $1/32$ to $1/16$-in dia.: $10/\text{ft}^2$ Less than $1/32$-in dia.: $50/\text{ft}^2$ Maximum depth $= 1/32$ in	N/A	N/A	None $>3/16$-in dia. If $<3/16$-in dia., $2/\text{ft}^2$
Resin pocket	None	Maximum 1 in^2 per occurrence	Maximum 1 in^2 per occurrence	Maximum 1 in^2 per occurrence
Wrinkles	Allowable if laminate is glass-reinforced. No sharp edges allowed.	Allowable if laminate is glass-reinforced and full mat layer thickness and total thickness are maintained	Allowable if laminate is glass-reinforced and full mat layer thickness and total thickness are maintained	Allowable if laminate is glass-reinforced; no sharp edges allowed
Scratch	None	N/A	N/A	None
Fiber prominence	None	Maximum 10 fibers visible/ 1 in^2	Maximum 20 fibers visible/ 1 in^2	N/A

Note: Defects shall be classified per ASTM D 2563, *Classifying Visual Defects in Glass Reinforced Laminate Parts.*

SECTION 11.4   TANKS CONSTRUCTED OF MISCELLANEOUS MATERIALS

11.4.1   Factory-Coated Bolted Steel Tanks

11.4.1.1   Introduction

This category of tank is a field-erected, bolted tank using steel panels or plates that have been factory-coated on either the internal or internal and external surfaces. The panels are made from standard-size steel sheets that are sheared, punched and rolled, and coated with linings. The life expectancy of these tanks is at least 40 years. They have been used primarily for water storage tanks and standpipes. Figure 11.4.1 shows potable water supply tanks using a glass-lined bolted steel tank.

ANSI/AWWA D103-87, *Standard for Factory Coated Bolted Steel Tanks for Water Storage,* is the main specification used. Although this standard was created to address the needs of the municipal water storage industry, there are potentially a great number of other applications in other industries. However, since engineers in other industries are unfamiliar with these tanks, the tanks are often not used when they might perform more economically, taking all things into consideration.

11.4.1.2   Why these tanks are used

These tanks have several features that make them a good choice for a number of applications:

- They are bolted and can be easily relocated when used in applications which are temporary or move from site to site.
- They have excellent corrosion resistance and require very little maintenance.
- They can be more quickly erected than welded field-erected tanks.
- They can be competitive with coated, field-erected, welded steel storage tanks.
- The capacity can be increased by adding shell courses.

Glass-lined tanks are commonly used for a variety of applications:

- *Potable water.*   In potable water service, the glass lining meets the National Sanitation Foundation (NSF) requirements. When welded steel tanks are coated, then specific coating systems must be approved by NSF. Another reason why local water municipalities use the glass-lined tank is that future capacity requirements may

20' Dia. x 29' High, 65,000 Gal. Aquastore Standpipe
Located Near the Bottom of Grand Canyon, Provides
Potable Water to Nearby Indian Reservation

1.7 m Gal., 120' x 20' Aquastore Potable Water Reservoir,
With Free-Span Aluminum

Figure 11.4.1   Glass-lined bolted steel potable water storage tank.
(Courtesy of A. O. Smith Harvestore Products Inc.)

increase. These tanks may be increased in capacity by adding shell
courses to either the bottom or top of the tank. Since communities
tend to grow around these supply tanks, space often becomes a limiting factor that pushes the user toward the concept of increased
capacity. Another reason why these tanks are popular is their low

maintenance costs. In fact, the water districts justify the higher initial costs of concrete and glass-lined tanks based upon the reduced overall costs when maintenance is taken into account.

- *Leachate storage.* The more strict regulations regarding leachate from landfills have driven the increasing share of private landfill owners to install storage tank systems to contain the leachate prior to and after treatment. The relatively short installation time coupled with the corrosion resistance of glass-lined tanks has made them popular for this application. A 60-ft-diameter tank can be erected in 4 to 6 weeks.

- *Wastewater.* These tanks have been used for trickling filters, surge basins, sludge storage, and digestion in the wastewater arena.

- *Other applications.* Although these tanks have not caught on yet in other industries, they should be considered for applications which make them a good choice, such as expandability, portability, or corrosion resistance.

11.4.1.3 Coating systems

AWWA D103 allows the following types of coatings on these tanks:

- Fused glass coatings applied by wet spraying, flow coating, dipping, or electrophoretic deposition. The coating is required to be 7 to 11 mils thick. To apply glass coatings requires a temperature of 1450 to 1600°F.

- Galvanized steel hot-dipped in accordance with ASTM A123 and A153.

- Thermoset liquid suspension epoxies.

- Thermoset powder epoxies.

In order to apply the coatings, the manufacturer must adhere to strict surface preparation procedures. The glass-coated tanks are popular in the water industry for both potable water and corrosive waters due to superior performance and long life. Glass-coated panels require a white metal blast and a primer of nickel oxide to prevent "fish scaling" of the steel surface. Visual inspection is used as well as wet-pad resistance tests to ensure the integrity of the coating against defects.

11.4.1.4 Design requirements

The design requirements for glass-lined tanks are covered in AWWA D103. Some of these requirements cover

- Material requirements for the steel, bolting and gaskets, and sealants.

- Design loads including dead, water, snow, wind, seismic, and appurtenance loads.

- Tank shell. The computations are specified to determine bolt spacing and size as well as allowable stresses. Tank plates must be at least 0.094 in thick but no more than $\frac{5}{16}$ in thick.

- Opening reinforcement design.

- Allowable shell stresses. These may be increased by one-third for any combination of seismic or wind loading as with almost all other codes.

- Circumferential wind girders or stiffeners.

- Roof design.

- Accessories such as pipe connections, overflows, ladders, openings, and vents.

- Welding of shop- and field-joined applications.

- Erection, inspection, and testing. These tanks are hydrostatically tested, and seams are vacuum-boxed. The AWWA code requires that the tanks be disinfected by the tank owner prior to use.

11.4.1.5 Foundations

The AWWA code recognizes that specific provisions regarding the foundation must be followed to have a successful and long-lasting installation. The following foundation designs are recognized:

- Type 1: tank on concrete ringwall foundation.

- Type 2: tank on concrete slab.

- Type 3: tank inside a ringwall. In the configuration the inside of the ringwall is no more than $\frac{3}{4}$ in away from the bottom plates.

- Type 4: tank on berm. This is essentially a compacted soil foundation.

- Type 5: tank within a steel retaining band. This is similar to the type 3 foundation except that a steel plate at least 12 in from the shell is used to confine the foundation soil.

- Type 6: tank with sidewall embedded in concrete foundation. In this design the concrete slab foundation is the floor of the tank since the tank has no steel plates on the bottom. Figure 11.4.3 shows this detail.

It is recognized that the potential for damage to the lining or for leakage is greater than for welded steel tanks, should the foundation allow for significant movement of the tank. Before these tanks are considered at a given site, it is essential that adequate soil investigations be performed by a qualified geotechnical engineer, including recommendations for the type of foundation to be used, the depth of foundation, and the design soil bearing pressures. If the tanks will be expanded in the future, this should be considered in the foundation design.

Other considerations for foundations include these:

- The top of the foundation should be at least 6 in and preferably 12 in above local average grade.
- The depth of the foundation is governed by the frost line depth for which AWWA D103 provides a chart.
- The foundation should project at least 3 in beyond the outside of the tank shell.

Seismic design provisions generally follow those of API Standard 650, Appendix E. However, it is recommended that when uplift occurs, glass-lined tanks be anchored or that they not be used. This is because of the damage that is likely to occur throughout the tank to the lining if uplift should occur.

11.4.1.6   Use of aluminum domes

When the tank diameter exceeds about 30 ft, the roof can often be installed most economically be installed using an aluminum geodesic dome. Note the geodesic dome applied to the top of the tank in Fig. 11.4.1.

11.4.2   Concrete Tanks

11.4.2.1   Introduction

Concrete tanks have been used primarily for potable water suppliers and for wastewater treatment operations. Interestingly, the design and construction have become geographically characterized within the eastern and western areas of the United States.

In western states, the use of concrete tanks is mostly limited to underground applications in which the roof becomes usable land such as for parking lots or tennis courts. The roofs for these tanks are, of course, flat. There are two reasons for this. First, land and space are at a premium, and views about appearance favor placing the tank underground. Second, an underground tank is considered safer in a major seismic event. Dome-roof tanks are also used, but tanks with domes tend to be partially or completely buried up to the top of the shell. Figure 11.4.2 shows an example of an underground, flat-roof

Figure 11.4.2   Basketball court on underground tank. (*Courtesy of DYK, Inc.*)

tank where the roof has been designated for use as a basketball court. Figure 11.4.3 is an example of an underground tank with a 10-in-thick prestressed flat concrete roof designed for 2 ft of soil cover. On the east coast, most concrete tanks are aboveground, and people tend to be less concerned with appearance and beautification projects.

Figure 11.4.3   Construction of an underground concrete tank. (*Courtesy of Mesa Rubber Co.*)

Concrete tanks have been built to liquid depths to over 100 ft. The economic size of concrete tanks is usually over 200 ft in diameter. However, in certain areas of the United States such as in Florida, a few contractors specialize in small-diameter, hand gunited or precast tanks from 50,000 to 200,000 gal. The smallest economical prestressed tank would be on the order of 500,000 gal.

A few applications for concrete tanks were for petroleum storage, and they have been in use for over 50 years. Most of these were military applications for buried tanks. The early tanks were normally reinforced. These leaked and were either lined or taken out of service. Later versions consisted of steel walls, floors, and roofs with concrete top and bottom slabs. Most recently, the prestressed tanks with steel membrane liners have been constructed with sizes up to 100,000-bbl capacities for use on the Alaska pipeline.

Generally the use of concrete tanks will not be competitive with that of welded steel tanks, but concrete tanks may be appropriate where high security and integrity is required, such as for military applications, or where resistance to fire and other types of damage and the necessity of keeping the tank in service during emergencies are premiere.

11.4.2.2 Kinds of concrete tanks

Classifying concrete tanks can be difficult. For example, classification can be based upon whether they are above- or belowground, square or round in plan, prestressed or cast-in-place, for water storage or for other purposes, coated or uncoated. Generally, the choice between precast or cast-in-place is a function of the site-specific conditions and economics more than any other factor.

By far the most common tank is the wrapped or prestressed tank which represents approximately 90 percent of the concrete tank population. The remaining concrete tank population is pour-in-place reinforced concrete. Pour-in-place tanks use external strapping for generating the hoop forces required to contain the concrete walls. The vertical joints are usually spaced approximately 50 ft apart or more. Walls are typically 6- to 8-in minimum thickness. In wrapped construction, the walls are sandblasted to remove laitance and for roughening, wrapped with high-strength steel cable, and coated with shotcrete several times.

When used aboveground, prestressed tanks have a shell height of up to 40 ft, but heights to 80 ft are possible. When used underground, the shell height is limited to about 20 ft. For underground reservoirs the bottom is either dished or flat (with a slope). Flat-bottom tanks are much easier to construct so that they are watertight and do not settle as much.

11.4.2.3 Basic prestressing principles

Prestressing is the application of a permanent compressive force to concrete by using steel cables, rods, or tendons to ensure that the working loads do not subject any part of the concrete structure to tensile forces for which concrete is considered to be ineffective. The amount of prestress in the shell should be such that it will maintain a residual final concrete compression of about 200 psi. The final stress levels are not the same as initial tensioning due to elastic shortening of the concrete, frictional losses from the curvature, and slip at anchorage points. There are also long-term losses due to creep of the steel and concrete and shrinkage in the concrete due to moisture content. The prestressed tank is made by wrapping steel wire under tension around the concrete walls. Both vertical prestressing and horizontal prestressing are required.

11.4.2.4 Watertightness

Although some unlined concrete tanks have performed well, they generally leak. When leakproof construction is required, there are several methods of accomplishing this:

- The vertical joints and shell-to-bottom joints in concrete tanks are subject to leakage. Usually a water stop is cast into the joints to prevent leaking. But the concrete itself is subject to leaks through hairline cracks that develop due to high stresses caused by differential drying or settlement.

- A diaphragm which is a corrugated sheet steel cylinder that is placed inside the concrete walls or on the interior surface of the tank walls prevents leakage. This diaphragm also functions as vertical reinforcement, eliminating the need for vertical prestressing. When used for water storage applications, the diaphragm should be hot-dip galvanized to reduce metal corrosion. Interestingly, on the east coast of the United States, most municipal water storage tanks use a diaphragm whereas on the west coast it is not common.

- Sometimes a liner is placed in the interior of the tank which must be coated and bonded to the concrete system. When this is done, the primary containment is the steel liner, and the leak detection system may consist of a monitoring system between the liner and the concrete tank.

- One company has developed the use of a fiberglass lining on the interior of the tank. This increases the choice of liquids which may be stored in these tanks.

11.4.3 Bladder-Lined Tanks

Bladder liners are flexible membranes which can line either part of the tank or the entire internal surface of the tank. These liners are distinguished from other tank liners in that they are not bonded to the tank surfaces. They are really plastic bags inside a steel tank shell. These liners have been used in applications for storage of fertilizer solutions such as phosphates, nitrates, and urea; and acids, bases, and other chemicals in the chemicals, paper and pulp, and plating industries. Although specific petroleum products such as diesels or kerosenes have been stored, the liners have not been used to store oxygenated products due to the deleterious effects of alcohols or ethers blended in the fuel.

Liners can be applied on tanks up to 1 million gal and 40 to 50 ft in height. However, most bladder-lined tanks are less than 40 ft in diameter by 40 ft high. The bladders can be installed on horizontal or vertical, above- or belowground tanks. Some of the open-top tanks in the plating and chemicals business come both round and rectangular.

11.4.3.1 Applications of bladder-lined tanks

Bladders are suitable for various configurations and have been used for the following reasons:

- When stainless steel tanks are inadequate for chemical storage, the bladder liner should be considered. The liner in this case covers all interior surfaces. Attention should be paid to the details of isolating penetrations so that small leaks do not occur. In these systems, vacuum is one of the most effective means for adhering the liner to the tank, and the liner serves as a leak detection system as well.

- Another application is to use the liner to extend the life of an existing tank. This application is a relining of the tank.

- Bladders are used as bottoms for factory-coated bolted tanks. The liner is sealed near the tank bottom with batton strips and covers a concrete slab which functions as the structural tank bottom.

11.4.3.2 Corrosion

There has been concern about corrosion in the internal steel surface between the liner and uncoated carbon steel. While it is possible for moisture to condense, it has been found that uncoated carbon steel will provide an acceptable life. If this is a concern, a coating should be applied to the tank. Since all liners have some degree of permeability, when corrosive substances can permeate the liners, coatings should be considered for the steel surfaces of the tank. Some local authorities require coatings as well.

11.4.3.3 Liner materials

Polyethylene, polyvinyl chloride, polyurethane ethers or esters, and most recently polyurethane-PVC blends have been used in bladders in tanks. The liners should be at least 30 mils thick. Some manufacturers standardized on 40-mil-thick material. Polyetheylene and polypropylene are stiff materials which must be creased. Providing a good seam is a rather difficult field operation. This material is also subject to stress cracking. It is generally not used today for these applications, as it has been replaced by the polyurethanes and PVC which provide more consistently adequate seams. Blends of PVC range from virgin PVC to various blends which include plasticizers. Each application should be carefully reviewed by the manufacturer and the user. The polyurethanes are resistant to a wide range of pH values and chemistries. However, they tend to have high permeability rates. If you rub you finger along the outside surface of a bladder containing gasoline, the odor of gasoline will adhere to your finger.

Due to the potential movement of the liner, the internal surfaces of the steel tank should be prepared by removing all ridges, sharp edges, or gouges that could abrade or puncture the liner. Some liner manufacturers place a geotextile fabric between the liner and the tank or supply a liner that has the fabric.

11.4.3.4 Type of tank suitable for bladder

The bladder liners cannot be used with floating-roof tanks because of the roof legs, which would puncture a liner on the bottom of the tank. The ideal candidate for a bladder liner is a horizontal or vertical tank with no internal support columns or rafters. On vertical tanks the roof should be a self-supported cone roof or a dome roof. If the liner only extends up to the shell and is not a full liner, then a rafter roof tank can be used. The support column must be booted and sealed separately.

11.4.3.5 Mounting systems

On tanks which are fully lined, a good method of ensuring that the liner does not move during operations is to subject the interstitial space between the liner and the tank shell to a vacuum of about 6 in of mercury (Hg). A slow leak should be expected in that it is difficult to have tanks which do not leak some small amount or have no pinholes. The vacuum must be maintained by a small on-site pump, or else periodically maintenance must be performed to maintain the vacuum. The vacuum system is most appropriate for fully lined tanks, although it can be applied to liners which are shell-supported if a good sealing strip between the shell and the liner is used.

Large tanks or tanks without the vacuum system are mounted by draping the liner from a ring mounted near the top of the shell to which the fabric has been attached. The ring functions to facilitate liner installation and to reduce the stresses acting on the tank liner.

11.4.3.6 Primary and secondary containment

Some suppliers have argued that the bladder is the primary containment and the tank shell the secondary containment, thereby eliminating the need for diked secondary containment. While this might be true for isolated cases, most tanks exist in small or large groups, and therefore secondary containment is usually present or required. If the bladder liner is deemed to be primary containment, then the tank bottom should be examined to ensure that it has no leaks, especially in a retrofit situation. In addition, when the bladder is considered the primary container, then the interstitial space between the bladder and the tank shell should be fitted with either a vacuum maintenance system or leak detection.

11.4.3.7 Inspection

Visual inspection of seams is one of the most common methods of monitoring the seam quality. Other tests can be performed including spark testing of the liner.

11.4.4 Vapor Bladder Tanks

11.4.4.1 Applications

For gasoline distribution, it is common to have vapor recovery systems based upon activated carbon, refrigeration and thermal oxidation processes. When tank trucks, tank cars, or marine vessels are loaded, the displacement vapor often cannot be returned to storage because the storage tanks are usually of floating-roof construction. This vapor must be processed or treated in accordance with the Clean Air Act or other local requirements.

Vapor bladder tanks are used to store vapors instead of liquids. They have been used to collect vapors from tank evaporation and for various vapor recovery operations including marine and tank truck loading operations. These systems provide vapor surge capacitance which can be useful to

- Expand the capacity of an existing vapor recovery system by reducing the peak flows into the system by time-averaging the flows

- Allow for vapor recovery maintenance or downtime without stopping operations.

- Reduce the potential size of new vapor recovery systems by averaging the flow rates.

11.4.4.2 Basic design elements

In vapor bladder tanks, an existing tank shell may be used to support the bladder from the tank shell. However, a self-supported roof is required above the bladder to allow it to expand above the tank shell and to provide maximum volume for the vapors. When existing tanks are converted to bladder storage, the existing tank roof is removed and a self-supporting steel or geodesic dome roof is installed. The bladders are stabilized by attaching weights and cables to the center of the bladder to lower its center of gravity in all operating positions. Bladders operate at typically about 0.5 to 0.75 in WC internal pressure. The expected life of vapor bladder tanks is from 7 to 10 years.

There are two different styles of vapor bladders:

- *Hemispherical.* This style is shown in Fig. 11.4.4.

- *Truncated cone.* This is shown in Fig. 11.4.5.

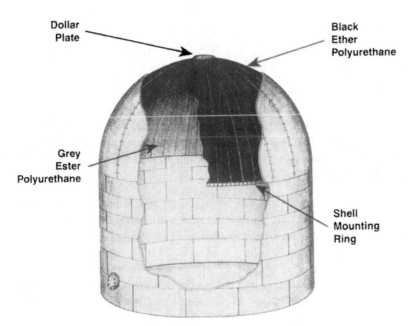

Figure 11.4.4 Hemispherical vapor bladder tank. (*Courtesy of Mesa Rubber Co.*)

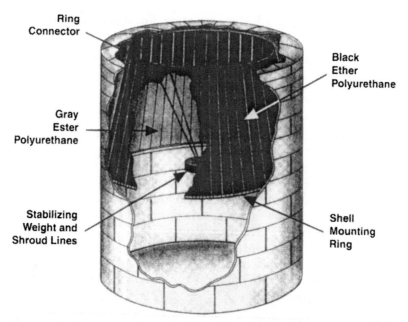

Ring Connector

Gray Ester Polyurethane

Stabilizing Weight and Shroud Lines

Black Ether Polyurethane

Shell Mounting Ring

Figure 11.4.5 Ring-type truncated cone vapor bladder tank. (*Courtesy of Mesa Rubber Co.*)

The truncated cone design has the following advantages over a spherical bladder design:

- It provides about 30 to 40 percent of the tank shell volume as opposed to about 80 to 90 percent for a truncated cone.

- It costs more due to the complexity of the seaming operations.

- It is more unstable than the truncated cone in two ways. First, when filled, it can slide around, abrading the fabric against the roof. Second, as the vapor volume is reduced, it can suddenly collapse to new positions, causing a significant shock load to the fabric and reduced life.

The spherical design is an older technology which has largely given way to the truncated ring. The truncated cone design uses a stable, free-floating fabric piston which is stiffened by a ring slightly smaller than the internal diameter of the tank. The bladder is supported by a ring fastened to the tank shell which either is perforated for accepting bolts and clamping rings or is suspended by a clamping ring. The ring stiffener is required to prevent rubbing of the fabric on the roof and instability problems with the natural spherical shape that the membrane would otherwise assume.

Fabric is more efficiently used in the truncated cone design due to the straight seams. The fabric is attached to the shell by installing a bolting ring or clamping device to the shell. The top of the fabric piston is allowed to touch the roof because the stresses on the fabric can be dissipated uniformly over the area of the roof, increasing the life of the fabric.

Since the membrane can rub against the shell and the roof, it is recommended that all interior tank surfaces be free of sharp edges, weld splatter, or other discontinuities that could abrade or tear the fabric. Generally, riveted tanks should not be used for bladder applications unless precautions are taken. If used, riveted tanks should be lined with a spray-on polyurethane coating at least 50 mils thick.

11.4.4.3 Fabrics

In the past, nylon or polyester coated with nitrile rubber having vulcanized seams was used. Now, the fabrics used are typically polyester urethane calendared coatings over a nylon scrim on the vapor side and polyether urethane on the air side. These layers are chemically bonded to each other. The polyether urethane has better air-side hydrolytic and ozone resistance while the polyester urethane has better fuel and aromatics resistance. These materials can handle oxygenated fuel vapors. Alcohols have a deleterious effect on the membranes. MTBE and ETBE present no special problems. Other materials are available which should be chosen on the basis of specific service conditions.

The fabric is considered a thermoplastic and uses heat-fused seams. The seams are stronger than the fabric. Seams are inspected and tested by use of a vacuum box and soap solution.

Polysulfide sealant is often used to provide the seal between the support ring and the fabric. Surface preparation prior to caulking with polysulfides is critical. The steel should be prepared by wire-brushing loose rust and scale. The fabric side must be etched with methyl ethyl ketone (MEK) to provide an adequate bonding surface.

11.4.4.4 Safety considerations

The space between the vapor bladder and the tank will eventually show some concentration of hydrocarbon vapor due to fabric permeability. Although numerous permeability tests can be conducted based upon permeation rates of various substances, the real test of acceptability is driven by actual conditions. The space between the bladder and the tank should be monitored for hydrocarbon content for two reasons:

- It may be required to determine what contribution to the total plant emissions this tank bladder system represents and if any control measures are required.

- The development of a leak or tear in the fabric will be detectable by monitoring this space.

If this space is less than 10 percent of the lower explosive limit (LEL), then this is usually due to permeation of the bladder vapors. The actual monitored vapor levels in this space will tend to fluctuate due to slight changes in permeability with temperature and ventilation that occur in this space. If there is a steady rise in LEL and it exceeds 10 percent, then there is probably a leak in the membrane.

The operation of the bladder should be such that it is always rich in vapors. However, development of flammable concentrations of vapors inside the bladder is a possibility. If the bladder is nearly empty and is filled by several trucks which contained diesel which has a very low vapor pressure, then it is possible to develop this condition. Systems which saturate the vapor stream entering the bladder eliminate this possibility and are recommended. All metallic components used in the bladder system such as the cables, weights, support rings, or clips should be electrically bonded and grounded to the tank, which should be earth-grounded. The recommendations of API 2003 should be complied with. Drains ports just above the membrane attachment to the shell must be provided to allow atmospheric condensation to drain away.

Bladder tanks may be placed within the diked storage areas of facilities. The fire codes and other codes do not prohibit this. However, because these tanks are never filled with liquids, they are subject to buoyant forces should the secondary containment area fill. When existing tanks are converted to bladder tanks, the load conditions including buoyancy and wind loading should be evaluated and documented. Anchoring is one acceptable way to meet all load conditions, if the evaluation shows that the tank shell is subject to uplift due to these forces.

References

Section 11.1

1. American Iron and Steel Institute and Steel Plate Fabricators Association, *Steel Plate Engineering Data*, vol. 1, *Steel Tanks for Liquid Storage*, rev. ed., Washington, 1992.
2. American Petroleum Institute, Draft Copy of Stainless Appendix, Task Group Copy Agenda Item, 650-334, Chairman, Don Duren, March 28, 1995, Tulsa, Okla.

Section 11.2

1. ASME B96.1, *Welded Aluminum Alloy Storage Tanks,* American Society of Mechanical Engineers, New York.
2. *Alcoa Structural Handbook.*
3. Jawad ad Farr, "Structural Analysis and Design of Process Equipment."
4. Moody, "Analysis and Design of Plastic Storage Tanks," *Transactions of the ASME,* May 1969, p. 400.
5. Aluminum Association, "Aluminum in Storage," Washington.
6. Aluminum Association, "Aluminum Is Compatible with Plastic and Resins," Washington.
7. Aluminum Association, "Water and Aluminum," Washington.
8. Reynolds Metal Company, *Structural Aluminum Design,* 1962.
9. Metal Handbook, 9th ed., vol. 2, *Properties and Selection: Nonferrous Alloys and Pure Metals,* American Society for Metals, 1979.
10. Hatch (ed.), *Aluminum Properties and Physical Metallurgy,* American Society for Metals, 1984.
11. H. H. Uhlig (ed.), *The Corrosion Handbook,* Wiley, New York, 1948.
12. Aluminum Association, *Aluminum Construction Manual,* 1956.
13. American Society of Metals, *Metals Handbook, Desk Edition,* 1985; LaQue and Copson, *Corrosion Resistance of Metals and Alloys,* 2d ed., American Chemical Society Monograph Series, Reinhold Publishing Corporation, New York, 1963.
14. H. H. Uhlig, *Corrosion and Corrosion Control, An Introduction to Corrosion Science and Engineering,* 2d ed., Wiley, New York, 1963.

Section 11.3

1. API 12 P, *Specification for Fiberglass Reinforced Plastic Tanks,* 1st ed., September 1, 1986.
2. ANSI/ASCE 7-88, *Minimum Design Loads for Building and Other Structures,* November 27, 1990.
3. ASTM C 581, *Test for Chemical Resistance of Thermosetting Resins Used in Glass Fiber Reinforced Structures,* ASTM, Philadelphia.
4. ASTM C 582, *Specification for Contact Molded Reinforced, Thermosetting Plastic (RTP) Laminates for Corrosion Resistant Equipment,* ASTM, Philadelphia.
5. ASTM D 229, *Method of Testing Rigid Sheet and Plate Materials Used for Electrical Insulation,* ASTM, Philadelphia.
6. ASTM D 2990, *Test for Tensile, Compressive, and Flexural Creep and Creep-Rupture of Plastics,* ASTM, Philadelphia.
7. ASTM D 2150, *Specification for Woven Roving Glass Fabric for Polyester Glass Laminates,* ASTM, Philadelphia.
8. ASTM D3583-90, *Standard Test Method for Indentation Harness of Rigid Plastics by Means of a Barcol Impressor,* ASTM, Philadelphia.
9. ASTM D 2584, *Test for Ignition Loss of Cured Reinforced Resins,* ASTM, Philadelphia.
10. ASTM 4097, *Standard Specification for Contact-Molded Glass Fiber Reinforced Thermoset Resin Chemical Resistant Tanks,* ASTM, Philadelphia.
11. ASTM D740, *Tests for Flexural Properties of Unreinforced and Reinforced Plastics and Electrical Insulating Materials,* ASTM, Philadelphia.
12. ASTM D3299, *Standard Specification for Filament Wound Glass Fiber Reinforced Thermoset Resin Chemical Resistant Tanks,* ASTM, Philadelphia.
13. ASTM D 638, *Test for Tensile Properties of Plastics,* ASTM, Philadelphia.
14. ASTM D648, *Test for Deflection Temperature of Plastics under Flexural Load;* Royce A. Currieo, "How to Choose Process Equipment Made from Reinforced Plastic," *Chemical Engineering,* August 1993.
15. API Standard 2000, *Venting Atmospheric and Low Pressure Storage Tanks: Nonrefrigerated and Refrigerated,* 4th ed., September 1992.

Tank Inspection, Repairs, and Reconstruction

The emphasis in this chapter is on tanks that have been already placed into service. This is the area which the relatively new API Standard 653 addresses and upon which this information is based. The discussion of inspection applies primarily to existing tanks, but there are some aspects of shop and field inspection for new tankage.

12.1 Background

The trend in the domain of the public and legislative bodies is to view aboveground storage tanks (ASTs) as environmentally hazardous equipment. Several recent AST accidents have contributed to this viewpoint, the most notable being a sudden and catastrophic fuel oil spill that occurred in Floreffe, Pennsylvania, releasing over a million gallons into the Monongahela River, a drinking water supply for various municipalities. These incidents have helped contribute to the current AST regulatory atmosphere that exists on local, state, and federal levels. Any leak or spill that causes contamination of subsurface waters or navigable waters will most likely result in severe financial and legal penalties, including the potential for new and stricter regulations affecting all industrial tankage. It is interesting to note that the stringent new regulations and rules are generally corrective rather than prescriptive or preventive measures. *Corrective* in this context means that rather than the prevention of AST leaks and spills being addressed through design and inspection, the issues are addressed by requirements for secondary containment and regulatory requirements that go into force only after a spill or leak has occurred.

Incidents such as those described above often serve to illustrate areas that the current body of industrial standards do not properly address. To respond to this, the American Petroleum Institute (API) issued several new recommended practices and standards.

12.2 Industrial Standards

To get a perspective on the role of API 653 among all the numerous API codes and standards governing ASTs, it is helpful to understand the intended scope of these documents. In planning to design and construct new tankage, the engineer has available an ample set of industrial standards that will provide a state-of-the-art tank design. Among these standards are API 650, API 620, AWWA D-100, UL-142, etc. All these standards are geared to provide an agreement on design and fabrication between the supplier of the tank and the purchaser. These standards are based upon accumulated industrial experience and will provide a tank that is reasonably free from the chance of failure when placed into service. However, the standards were not intended to eliminate maintenance and inspection procedures that can mitigate against failure from long-term service effects such as corrosion, foundation settlement, changes of service, or improper structural modifications made to the tank. Once the tank is placed into service, individual operating companies and facilities were left on their own to determine the level of inspection and assessment required to ensure against failures such as leaks, spills, and brittle fracture. The catastrophic failure of the fuel oil storage tank in 1988 in Pennsylvania showed a need for standardized industrial guidelines directed at preventing failures of tanks already placed into service. API 653 was thus created to fill this void in available standards. Although API 653 relies heavily on the principles of API 650, it extends to many areas that are not addressed by API 650.

The three recent documents issued by API to ensure long-term suitability for service and to reduce environmentally related problems caused by tank failure are

- API Standard 653, *Tank Inspection, Repair, Alteration and Reconstruction*
- API Recommended Practice RP 651, *Cathodic Protection*
- API Recommended Practice RP 653, *Interior Linings*

It should be understood that API 650 pertains to new construction only. The API 653 document applies to tanks that have been placed into service.

Because API Standard 653 is considered an inspection document, we focus on it. However, it is really a comprehensive prescriptive, as opposed to a corrective, approach to the problem of AST leaks and spills and should be considered a first line of defense against the various tank failure modes.

API publications commonly used in conjunction with API 653 include

API RP 570, *Piping Inspection*

API RP 651, *Cathodic Protection for Aboveground Petroleum Storage Tanks*

API RP 652, *Lining of Aboveground Petroleum Storage Tanks*

API Standard 650, *Welded Steel Tanks for Oil Storage*

API Standard 620, *Design and Construction of Large, Welded Low-Pressure Storage Tanks*

API Standard 2000, *Venting Atmospheric and Low-Pressure Storage Tanks*

API RP 2003, *Protection against Ignitions Arising out of Static, Lightning, and Strong Currents*

API Publication 2015, *Cleaning Petroleum Storage Tanks*

API Publication 2207, *Preparing Tank Bottoms for Hot Work*

API Publication 2217, *Guidelines for Continued Space Work in the Petroleum Industry*

12.3   Intent of API Standard 653

API 653 outlines a program of minimum requirements for maintaining the integrity of welded and riveted, nonrefrigerated, atmospheric storage tanks. It addresses tank inspection, repairs, alterations, and reconstruction. Most existing ASTs including the foundation, bottom, shell, structure, roof, appurtenances, and nozzles fall within its scope.

Although numerous topics are covered in detail, the standard cannot possibly cover all the conditions encountered in the field. Frequently, the original standard of construction and even the materials of construction for a tank under consideration are unknown. Therefore, API 653 guides us by stating that the principles of API Standard 650 for new construction should be followed when in doubt.

Note that in spite of the many procedures, considerations, and guidelines outlined in API 653, none should be considered superfluous if the goal is to prevent AST failures. API 653 cannot provide the AST owner with cookbook answers to all problems encountered and is therefore considered the *minimum required* to ensure AST integrity.

Implementing API 653 accomplishes what cannot be done in any other way. It is the best current technology that is reasonable and cost-effective to ensure that

- Storage tanks do not leak while they are in service.
- They do not catastrophically fail due to structural reasons or by brittle fracture.
- The efforts and costs are minimized, yet effective.

12.4   How Does API 653 Prevent Tank Failures?

API 653 provides three mechanisms to prevent potential storage tank failures:

12.4.1   Assessment

Standard 653 places great emphasis on engineering experience and evaluation of an AST's suitability for service. It also requires that this evaluation be conducted by "organizations that maintain or have access to engineering and inspection personnel technically trained and experienced in tank design, fabrication, repairs, construction, and inspection." Suitability for services does not necessarily involve a physical change to the tank. One example would be a change of service temperature. These are typical cases where an evaluation should be performed:

- Storage of fluids that are incompatible with the storage tank materials of construction, which could lead to pitting, unpredictable rates of corrosion, stress corrosion cracking, etc.
- A change in the stored product density
- Distortion of the shell, roof, or bottom
- An observable change or movement in shell distortions
- Very high fluid transfer rates into or out of the tank
- High, low, or variances in service temperatures
- Local thin areas in the shell
- The presence of cracks
- Brittle fracture considerations
- Foundation problems

12.4.2   Inspection

The philosophy of the standard is to gather data and to perform a thorough initial inspection in order to establish a baseline for each tank against which future inspections may be used to determine the rate of corrosion or changes that might affect its suitability for service. Key to the philosophy of the standard is the observation of changes and rates of change that affect the physical tank. From these data, the experienced tank engineer may be able to judge the suitability for continued service or the need for repairs. See Sec. 12.10 for more details on inspection.

12.4.3 Repair, alteration, and reconstruction guidelines

Standard 653 provides guidelines for many of the common repairs and alterations that are done to tanks, including

- Alteration of existing nozzles
- Patch plates
- Bulges and out-of-round conditions
- Bottom repairs
- Replacement of the bottom
- Roof repairs
- Floating-roof seal repairs
- Hot taps
- Repair of defective welds

12.5 Responsibility and Compliance

The owner or operator has ultimate responsibility for complying with, or not complying with, the provisions of API Standard 653. Certain tasks can be assigned to others by the tank owner or operator, but the limits of responsibility for these tasks must be clearly defined prior to commencing work. In some cases, the tank owner or operator may have requirements more stringent than those described in API 653. In other cases, the owner/operator may not endorse certain practices described in API 653.

Since compliance with a standard such as API 653 requires an expenditure of time, effort, and money, a likely question is, Is compliance really mandatory? In some states the answer is a clear-cut yes because compliance with API 653 is legislated. However, for most facilities, the answer is not so clear. A standard in itself is rarely mandatory but is frequently made so by reference to *industrial standards* and *good engineering practice* in the words of local, state, or federal authorities. Remember that since API 653 gives minimum requirements, it may be virtually required merely because nothing better exists. An example of this is EPA's Spill Prevention Control and Countermeasures (SPPC) regulations. This regulation, which is applicable to facilities near navigable waterways, requires that tanks be inspected at regularly scheduled intervals and that the inspections be documented. EPA's SPCC program does not mandate the use of API 653. However, because API 653 is the only recognized industrial inspection standard for ASTs, it would seem that compliance with the standard is, by default, required, unless the owner or operator is already doing all the things outlined in API 653. OSHA, in Process Safety Management regulation 1910.119, specifically mentions that employers must maintain "written ongoing integrity procedures" and that they shall follow "generally accepted good engineering practices." It also states that employers shall "document each inspection." It would appear that compliance with API 653 would satisfy all these requirements.

API 653 clearly places the burden of determining long-range *suitability for service* on the owner or operator. In many ways the responsibility for compliance to a standard such as API 653 is like an insurance policy. It increases the operating costs associated with ASTs, but in the long run these costs can be more than recouped. Many facilities have paid the cost of site remediation resulting from AST leaks and spills and the heavy fines levied by the EPA costing much more than the preventive costs to implement and maintain a program such as described in API 653. Many of the notable AST catastrophes would not have occurred if they had been on a fitness-for-service program as outlined in API 653. Not only would the individual owners not have suffered financial responsibility and interruptions to their operations, but also the current costs of compliance with many new AST regulations would be much lower than they are now. To ensure that long-term suitability for service is achieved, API 653 clearly specifies the degree of quality required by qualifying the inspection personnel and by documentation of findings at the time of inspection.

12.6   How Long Will It Take to Implement the API 653 Program?

Because of the amount of work associated with compliance for a large facility with many tanks, the issue of how long compliance should take is left undefined in the standard. Since the standard requires that internal inspections be performed at intervals not exceeding 10 years (greater periods are an exception), it would seem that a facility should have scheduled all its ASTs for the first comprehensive, internal inspection not more than 10 years from now. Many companies are proactive in their plans to comply within a 3- to 5-year time frame. In any case, the initial efforts should be started now. During the period of getting into compliance, any major tank failure would not only be devastating to the owner company's public image but also raises jurisdictional concerns about why that tank was not in compliance with an existing industry standard. In other words, the sooner you comply, the better your chances for reducing possible leaks and failures, public image problems, environmental civil and possibly even criminal penalties and liabilities.

12.7   API 653 and Costs

Although it is not immediately obvious, API 653 can have an impact on new tank designs in achieving the lowest capital and operating costs desired. Since the interval between internal inspections is governed by such factors as the use of a liner, amount of corrosion allowance, cathodic protection, and leak detection, these factors should be taken into account when a new tank is being designed. The cost of implementing API 653 can be impacted by the up-front planning and thinking that go into setting up such a program. These costs are expected to fall into three broad categories:

12.7.1   Costs for internal inspections

These costs are very significant and can run in the millions of dollars per year for a large integrated oil company. Most of the costs can be attributed to the cost of preparing the tank for an internal inspection and interrupting its in-service operations. API 653 has various methods of increasing the time interval between internal inspections that can cause the interval to range from a few years to as many as 20 years.

12.7.2   Costs based on inspection findings

In assessing suitability for service, there is an opportunity to save costs. For example, either an existing tank that has many violations of current standards can be brought up to date without regard to costs, or an engineering evaluation can be done which determines just what has to be brought up to date to make the tank fit for service. The cost difference between these two approaches can be significant.

12.7.3   Costs associated with recordkeeping

Set up and maintain a recordkeeping system. One of the goals should be to develop and standardize a recordkeeping system that optimally would include standardized software to minimize overall costs to the owner or operator.

12.8   In-House versus Contract Inspection

When API 653 implementation is considered, a common question turns on whether a contractor should be used.

A large facility such as an oil refinery usually can afford to staff full-time API 653 certified inspectors. The advantage of running an in-house inspection program is that it can be controlled, it provides uniform results, and it has the potential to be the lowest-cost system for maintaining tank integrity. However, there is the potential for conflict of interest. If operations is pressured not to take tanks out of service, then the owner's tank inspectors will be pressured not to take tanks out of service as well. API 653 attempts to resolve this by making a statement in the standard that "such inspectors shall have the necessary authority and organization freedom to perform their duties."

Doing the work, on the other hand, by contracting it has the advantage of contractors who specialize in this work, who have the expertise and equipment to perform efficiently. The disadvantage to contracting the work is that the inspection agencies usually perform the recommended repairs as well, generating an inherent conflict of interest. The contracted inspection agencies also tend to be more conservative in their recommendations as a result of product liability issues.

Many companies have developed a hybrid of these options, contracting for the data collection efforts by contractors who have the appropriate inspection equipment and performing the engineering and design assessments on their own.

12.9   Thoroughness of Inspection

The scope of inspection is subject to interpretation, and so the degree of inspection is an issue that frequently arises. For example, are two spot readings for thickness per floor plate adequate, or should the entire surface be scanned 100 percent? Since a single pit can escape all but the most comprehensive, time-consuming, and costly examinations, the element of risk is introduced. The inspection program should consider the specific situation, such as whether the tank has a leak detection system and the potential liability caused by a leak, to determine exactly how to inspect the tank. Some organizations have classified their tank inspections as *full* or *cursory* and specified the conditions for which these types of inspections will apply in their programs.

12.10   Getting Started

Getting going on a program such as this might be a tough problem, but it is certainly doable. To help start, here are some of the items that should be considered up front:

- Make a decision or policy to comply.
- Establish a budget for compliance.
- Schedule the first external and internal inspections.
- Determine in-house versus contracted inspection.
- Establish what types of situations should involve a tank engineer.
- Establish a recordkeeping procedure.
- Establish a procedure for the operator's monthly inspection.
- Gather data including tank ages, last inspections, problems, construction data, drawings, etc.
- Establish a procedure for the periodic external inspections.
- Establish a procedure for the internal inspections.
- Establish corrosion rates for use in determining inspection intervals, or determine them based on tanks in similar services and conditions.
- Be sure to consider safe entry of tankage for internal inspections.

References

1. American Petroleum Institute, API Standard 653, *Tank Inspection, Repair, Alteration, and Reconstruction,* 1st ed., January 1991, and Supplement 1, January 1992.
2. American Petroleum Institute, API RP 651, *Cathodic Protection of Aboveground Petroleum Storage Tanks,* 1991.
3. American Petroleum Institute, API RP 652, *Lining of Aboveground Petroleum Storage Tank Bottoms,* 1991.
4. American Petroleum Institute, API RP 575, *Inspection of Atmospheric and Low Pressure Storage Tanks,* CRELetter Ballot 36/1991 Edition, January 21, 1992.
5. P. Myers, "Best Practices for Aboveground Storage Tank Inspection," *The Reporter* (Pacific Energy Association), Spring 1994, pp. 15–17.

Settlement

13.1 Introduction

Fortunately, tanks are relatively flexible structures and can tolerate a surprisingly large amount of settlement without showing signs of distress. The percentage of reported incidents involving settlement problems confirms this. However, the flexibility is not without limits. There are numerous examples of tank failures that have resulted in inoperative floating roofs, shell and roof buckling damage, leaks, and, at worst, a complete loss of tank contents.

The complete elimination of settling is not possible. Foundation design, soil conditions, tank geometry and loading, as well as drainage all have a significant effect on settlement. Large petroleum tanks are generally constructed on compacted soil foundations or granular material. Smaller tanks in the chemical industry are often constructed on concrete slabs.

This chapter applies to settlement of large tanks over 50 ft in diameter constructed on compacted soil or ringwall foundations. The reason for this is that most of these tanks are constructed on compressible foundations where the thickness, elasticity, and compressibility of the foundation and subsoil layers can vary enough to produce nonplanar distortions with the application of uniform loading. However, the basic principles apply to all tanks, especially uniform settling and planar tilt.

The foundation loading pattern under static conditions is such that when filled, tanks will uniformly load the foundation beneath the tank as the result of hydrostatic pressure in a disk pattern. The edge of the tank under the shell, however, has increased loading due to the weight of the shell and the part of the roof supported by it. It also has other local loading effects such as twisting of the plates under the shell due to shell rotation. The edge of the foundation is also unlike the interior areas because it terminates near the outside of the tank and is often constructed of materials other than those of the interior (ringwall, etc.). This also allows for different soil properties such as moisture and compaction under the tank and outside its footprint, which promotes nonuniform settlement.

For these reasons a significant portion of the foundation and settlement problems occur at the outside edge of the foundation. In fact, the only assessment that can be made with regard to settling without entering the interior of the tank is the elevation readings made at the base of the tank. Since most tanks are not taken out of service for periods of years, it is advantageous that one key indicator of settlement (elevation monitoring at tank base) is also the area involving the majority of problems. This concept should not lull one into thinking that there are no settlement problems if external indications of settling are not present. There have been cases where failures have occurred in the tank bottom interiors due to interior settling that did not show any undue settling as monitored by the elevations around the base of the tank.

The threat of settlement failure poses serious consequences to the surrounding property and safety of life. Until the mid-1950s, tanks were limited to about 200,000-bbl capacity. Since then, the capacity has increased such that 800,000- and 1,000,000-bbl storage tanks are not uncommon. Given the size, it is important that reasonable criteria be available to guide the owner/operator when settling has occurred.

However, the exact mechanisms associated with tank bottom failures due to settlement are complex and seem to defy a rational, all-encompassing design guideline that would prove useful and reliable under all conditions.

13.2   Settlement and Tank Failure Mechanics

Failure due to settling can be defined by the occurrence of these effects:

- Roof binding on floating-roof tanks
- Damage or early wearout of floating-roof seals
- Shell buckling in fixed- or floating-roof tanks
- Roof buckling in fixed-roof tanks
- Loss of support of roof support columns in fixed-roof tanks
- Cracking of welds
- Loss of acceptable appearance

- Overstress of connected piping

- Accelerated corrosion due to drainage pattern changes outside the tank

- Inoperative or less effective drainage on the interior of the tank, especially where cone-up or cone-down or single-slope bottoms are used

- Increased susceptibility to seismic damage as a result of distorted, overstressed, or deformed bottoms

- Leaks in bottom or shells resulting from settling

The most serious failure mode results in leakage or loss of contents. The presence of even a small crack in the tank bottom can pose a serious threat to the integrity of the tank. Several notable settlement failures that have occurred involved the following sequence:

1. Development of an initial leak, caused by a crack in the tank bottom.

2. Washout of foundation support immediately near the initial leak location. This causes the crack to grow due to the lack of support, and the leakage increases.

3. The leak flow increases, and the support under the tank is undermined to the point where the bottom plates separate from themselves or the shell where the foundation has washed away.

Prior to several incidents,[1] leakage was seen emanating at the chime, but the contents could not be pumped out before a major failure occurred.

13.3   Different Kinds of Settlement

The kinds and modes of settling in tank bottoms are numerous. However, the numerous settlement configurations can be more easily dealt with by categorizing them as follows:

Uniform settlement

Planar tilt

Differential shell settlement

Global dishing

Local interior settling

Edge settlement

Other special cases

13.3.1   Uniform settling

In this type of settling, the soil conditions are relatively uniform, and it is soft or compressible. A storage tank under these conditions will slowly, but uniformly, sink downward, as shown in Fig. 13.1. In and of itself, there is no significant problem with indefinite uniform settling. However, two important side effects result from this kind of settling:

Water ingress.   Water ingress occurs when a depression or water trap is formed around the periphery of the tank where it meets the soil. When it rains or floods, moisture accumulates under the tank bottom near the shell or chime region and acts to corrode the bottom. Also any moisture under the tank may condense, may be unable to escape to the atmosphere, and may be trapped under the shell, causing corrosion.

Piping.   Piping connected to the tank will eventually become overstressed by the movement of the tank relative to the piping and its supports.

It is possible to assess the degree of uniform settlement by simply monitoring elevations at the base of the tank.

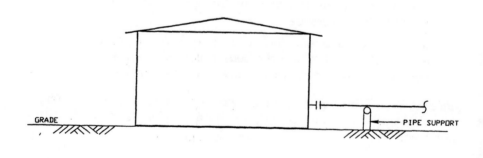

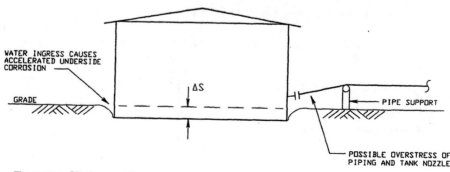

Figure 13.1   Uniform settlement.

13.3.2 Planar tilt

In this mode the tank tips as a rigid structure. (See Fig. 13.2.) This mode can be, and usually is, combined with other modes of settling. Often planar tilt accompanies uniform settlement. As well as the concerns addressed for uniform settling there are several additional phenomena that occur as the tilt becomes severe:

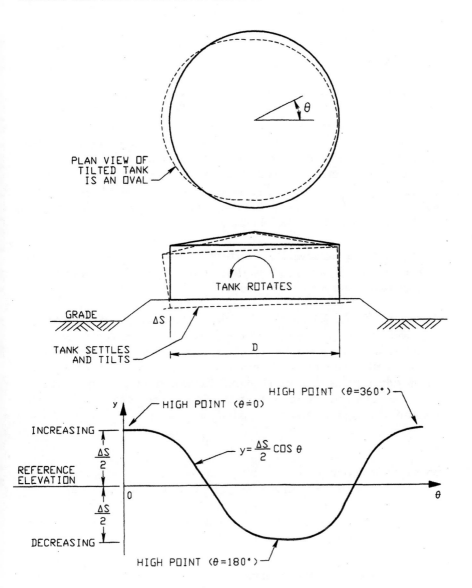

$$\Delta S = \text{MAXIMUM EDGE-TO-EDGE SETTLEMENT}$$

Figure 13.2 Planar tilt settlement.

Appearance. The human eye is sensitive to vertical lines. With a relatively small angle of tilt, the tank begins to look strange. This can cause the public, employees, or anyone else to begin to question the safety of the tank and the operating and maintenance practices of the owner/operator. Planar tilt limited to $D/50$ is a reasonable plumbness specification that provides an acceptable tank appearance.

Note that API 650 limits new tank construction plumbness to $\frac{1}{200}$. API 653 limits the total plumbness to $\frac{1}{100}$ for reconstructed tanks with a limit of 5 in maximum. These are both well within the visually noticeable limits of $\frac{1}{50}$.

Hydrostatic increase. The tilt of the tank results in an increase in hydrostatic head as shown in Fig. 13.2. The increase in hydrostatic head may be estimated approximately by $D \, \Delta S/2$, where D is the tank diameter and ΔS is the high-to-low difference in tank bottom elevation. The effect is to increase the shell hoop stress slightly.

In practice, increased hydrostatic pressure due to tilt is rarely a consideration or a problem. Planar tilt can be assessed from an external tank inspection conducted by taking elevation readings at several locations around the base of the tank.

Storage capacity reduced. Since the design liquid level is often just beneath the roof or an overflow, the maximum liquid level and capacity may have to be reduced to account for the planar tilt.

Ovalizing. If a tank tilts, the plan view will be an ellipse, as shown in Fig. 13.2. Since floating-roof tanks have specific clearances and out-of-round tolerance for their rim seals to work properly, the possibility of planar tilt's causing a seal problem exists. However, from a practical standpoint, the amount of planar tilt would have to be extreme in order for ovalizing due to planar tilt to become a problem. The amount of ovalizing can be estimated by

$$\Delta S = 2\sqrt{TR} \qquad (13.1)$$

where ΔS = maximum acceptable settlement, ft,
T = radial tolerance on floating-roof tank shell, ft, and
R = tank radius, ft. In practice, this factor is not important because the amount of planar tilt would have to be substantial to create a problem in this area.

13.3.3   Differential shell settlement

One of the most obvious and least difficult methods of assessing the existence of settlement problems is to take elevation readings around the circumference of the tank shell on the projection of the bottom beyond the shell.

Differential settlement results in combination with or independently of uniform settlement and planar tilt. The basic idea is that the bottom of the tank is no longer a planar structure. Figure 13.3 shows the result of differ-

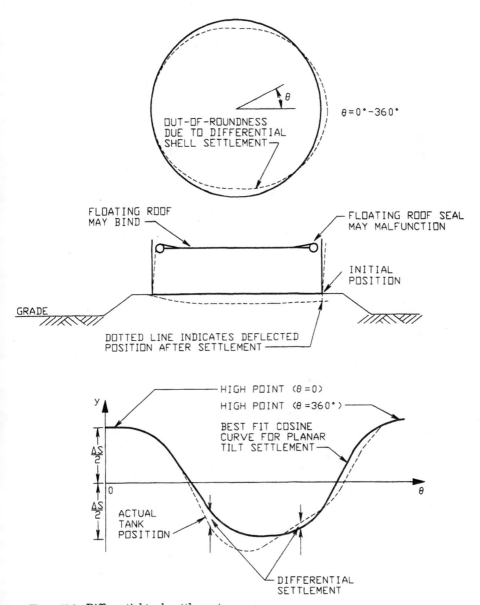

Figure 13.3 Differential tank settlement.

ential shell settlement. The readings that are taken around the circumference can be plotted as shown in the figure. If the bottom of the tank is planar, then a cosine curve may be fitted through the measured points.

Differential shell settlement, as might be expected, is of a more serious nature than uniform or planar tilt settlement because deflection of the structure on a local scale is involved which produces high local stresses. Differential edge settlement results in two main problems:

Ovalizing due to differential shell settlement. As shown in Figs. 13.3 and 13.4, the differential settlement that occurs in the tank bottom near the shell produces out-of-roundness in the top of tanks which are not restricted in movement as, e.g., in a fixed-roof tank. One of the most noticeable and serious problems with differential edge settlement in the bottoms of floating-roof tanks is the operation of the floating roof. Because floating-roof seals have specific tolerance limits between the edge of the roof and the tank shell, ovalizing can interfere with the operation or even destroy the seal itself.

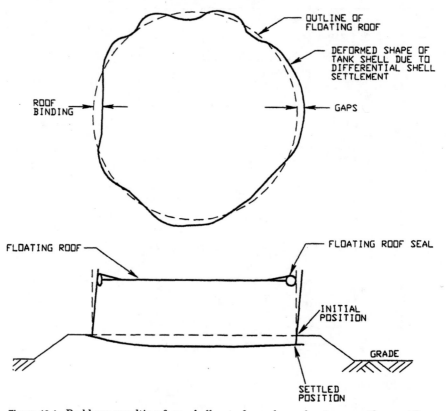

Figure 13.4 Problems resulting from shell out-of-roundness due to nonuniform settlement.

If the bending stiffness of the tank is much less than the extensional stiffness (thin-wall structure), then the theory of extensionless deformations may be used to compute the relationship between differential settlement and radial deformation at the top of the tank. A relatively simple differential equation that expresses the radial displacement at the top of the tank to the differential settlement at the base of the shell has been developed.[2]

$$r = \frac{DH}{2} \frac{d^2s}{dx^2} \tag{13.2}$$

where r = radial displacement at height H due to settlement, ft
   x = circumferential shell coordinate
   D = tank diameter, ft
   H = tank height, ft
   s = vertical settlement displacement, ft

This equation applies to uniformly thick cylinders with free ends. It is therefore applicable to open-top floating-roof tanks. Although these tanks are constructed of varying-thickness shell courses and the wind girder stiffens the top, experimental research shows that there is reasonably good agreement between the theory of extensionless deformations and the actual radial deformations that occur as a result of settlement in typical floating-roof tanks. For fixed-roof tanks, the top is not free to move radially, and therefore the settlement increases to a point at which the stresses in the shell near the top become sufficiently high to cause buckling which produces dents in the shell.

It has been found that since discrete elevation readings are taken around the circumference of the tanks, a finite difference form[3] of Eq. (13.2) is more practical:

$$r_i = \frac{HN^2}{\pi D} \Delta S_i \tag{13.3}$$

where r_i = radial shell displacement at station i
   N = number of stations or readings
   H = shell height at which radial displacements are calculated
   D = tank diameter
   ΔS_i = settlement at station $i = u_i - \frac{1}{2}(u_{i+1} + u_{i-1})$
   U = difference between best-fit cosine curve (planar tilt) and actual settlement S_i

Use of these equations is illustrated in the examples.

Shell stresses due to differential shell settlement. Nonplanar differential settlement may result in high shell stresses being generated.

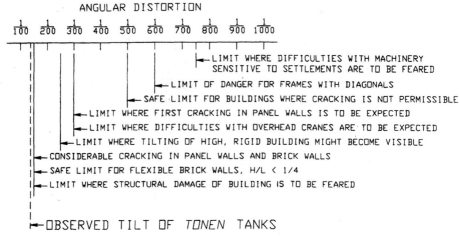

ANGULAR DISTORTION

$\frac{1}{100}$ $\frac{1}{200}$ $\frac{1}{300}$ $\frac{1}{400}$ $\frac{1}{500}$ $\frac{1}{600}$ $\frac{1}{700}$ $\frac{1}{800}$ $\frac{1}{900}$ $\frac{1}{1000}$

←LIMIT WHERE DIFFICULTIES WITH MACHINERY SENSITIVE TO SETTLEMENTS ARE TO BE FEARED

←LIMIT OF DANGER FOR FRAMES WITH DIAGONALS

←SAFE LIMIT FOR BUILDINGS WHERE CRACKING IS NOT PERMISSIBLE

←LIMIT WHERE FIRST CRACKING IN PANEL WALLS IS TO BE EXPECTED

←LIMIT WHERE DIFFICULTIES WITH OVERHEAD CRANES ARE TO BE EXPECTED

←LIMIT WHERE TILTING OF HIGH, RIGID BUILDING MIGHT BECOME VISIBLE

←CONSIDERABLE CRACKING IN PANEL WALLS AND BRICK WALLS

←SAFE LIMIT FOR FLEXIBLE BRICK WALLS, H/L < 1/4

←LIMIT WHERE STRUCTURAL DAMAGE OF BUILDING IS TO BE FEARED

←OBSERVED TILT OF *TONEN* TANKS

Figure 13.5 Limiting angular distortion. (*Adapted from Berrum, 1963.*)

These high stresses are generated near the top of the tank and may result in buckling of the upper shell courses.

An early approach to determining allowed differential settlement was to arbitrarily limit the differential settlement to a constant representing the ratio of the settlement to the span between consecutive settlement measurements. Figure 13.5 shows how various structures, particularly buildings, are damaged when the slope represented by the deflection-to-span ratio exceeds some value.

One commonly used limit[4] is

$$\Delta S = \frac{l}{450} \tag{13.4}$$

where l = length between settlement readings, ft, and ΔS = allowable settlement. In practice, local slopes limited to approximately $\frac{1}{450}$ to $\frac{1}{350}$ have proved greatly conservative and often result in tanks being releveled when this is usually more harmful to the integrity of the tank than doing nothing or and when further settlement could be tolerated.

The tolerable differential settlement upon which the API 653 method is based assumes that the shell is unrolled into a flat shell plate. It assumes that the shell is equivalent to a large elastic beam with the depth equal to the height of the tank and the width equal to the thickness of the shell, and it does not twist out of a planar configuration. This concept is illustrated in Fig. 13.6.

From the theory of pure bending in beams

$$\sigma = \frac{Ey}{R} \tag{13.5}$$

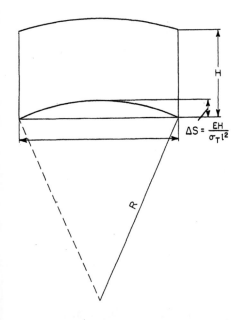

$$\Delta S = \frac{EH}{\sigma_T l^2}$$

Figure 13.6 Portion of a shell.

where E = modulus of elasticity, psi
y = distance of fiber from neutral axis, in
R = radius of curvature of bent beam, in

Substituting $H/2$ for y gives

$$\sigma = \frac{E}{2}\frac{H}{R} \tag{13.6}$$

From geometry, the relationship between the chord length and height inscribed in a circle and the radius is

$$l^2 = 2R\,\Delta S - \Delta S^2 \tag{13.7}$$

Neglecting the second term on the right, we have

$$\Delta S = \sigma\frac{l^2}{EH} \tag{13.8}$$

Alternatively, in terms of strain in the shell,

$$\Delta S = \varepsilon\frac{l^2}{H} \tag{13.9}$$

For 36-ksi steel, the axial strain at failure equals approximately 0.11. By writing Eq. (13.9) and using

$$\Delta S = 0.11\frac{l^2}{EH^2}\frac{\sigma_T}{E}\frac{E}{\sigma_T} = 58\frac{l^2}{E}\frac{\sigma_T}{H} \tag{13.10}$$

where σ_T = tensile stress at failure = 58,000 psi, Marr et al.[14] suggested that Eq. (13.10) be written in the form

$$\Delta S = K \frac{EH}{\sigma_T l^2} \qquad (13.11)$$

The coefficient K accounts for the nonlinearity of the stress-strain curve between yield and tensile failure and is simply a parameter to correlate with differing stress levels.

By substituting strain levels for tensile failure, yield, and strain hardening as shown above into Eq. (13.9), these values of K are found:

Condition	Strain level	Constant K
Yield	0.00125	0.6
Strain hardening	0.014	7
Tensile failure	0.11	58

Marr et al.[14] recommend a limiting value of $K = 7$. Substitution of σ_y for σ_T results in

$$\Delta S = 11 \frac{\sigma_y l^2}{EH} \qquad (13.12)$$

API 653 uses a factor of safety of 2.0 and uses Eq. (13.10) modified for this:

$$\Delta S = \frac{11}{2} \frac{\sigma_y l^2}{EH} \qquad (13.13)$$

API 653 suggests using Fig. 13.7 as the basis for establishing acceptable settlement. This essentially involves determining the out-of-plane differential settlement by computing the variation of a point from the average of two adjacent points. D'Orazio and Duncan[5] have shown that when the settlement readings are closely spaced, then the errors in measurement completely overshadow the actual tolerable settlement. Figure 13.8 shows how the errors cause the uncertainty of K to be unacceptably high for distances less than 50 to 60 ft. This is consistent with findings reported by D'Orazio and Duncan.[6]

Duncan and D'Orazio have also shown that differential settlement occurs by a *fold mechanism* which leads to sagging or saddle-shaped bottom profiles. This results from the fact that the shell is stiff in comparison to the load that is able to be imposed on it by the limited annular area which can act downward on the tank shell.

It is proposed, therefore, rather than using the distance l for determining the radius of curvature, that the maximum out-of-plane settlement be determined by setting the distance between adjacent

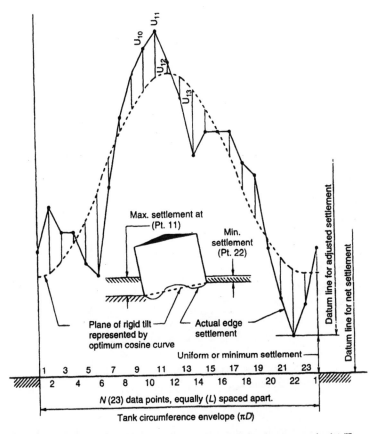

Figure 13.7 Graphical representation of tank shell settlement.

nodes of shell bending equivalent to the second harmonic of the Fourier series. This can be done by substituting $\pi D^2/8$ into Eq. (13.12):

$$\Delta S = 1.7 \, \frac{\sigma_y D^2}{EH} \tag{13.14}$$

For a height of 40 ft and a 36,000 psi yield strength, the formula is

$$\Delta S = 52.8 \times 10^{-6} D^2 \tag{13.15}$$

The advantage of this formula is that it is independent of the number of readings taken and is not overobscured with data error uncer-

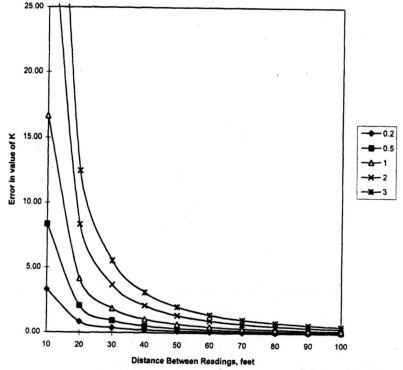

Figure 13.8   Errors in settlement readings.

tainty. A tank that is 100 ft in diameter would have an acceptable ΔS of 0.52 ft, or 6.24 in. Based on experience, this is reasonable. Although there are relatively large allowable settlements for large tanks (say, 40 in for a 250-ft-diameter tank), there is no experience to show that these kinds of tolerances are not acceptable. Rather, it is usually other factors such as radial tolerances caused by settlement which cause seal problems or roof binding problems that would cause the tank owner/operator to modify the tank so that it would not be subject to these problems.

13.3.4   Global dishing

In global dishing, the entire bottom settles relative to the shell. This may occur alone or in combination with other forms of settlement. Note that there is no one general form of global settling. However, the majority of tank bottoms do tend to form a dished shape, as shown in Fig. 13.9. This type of global tank bottom settling configuration is covered because it generalizes the problem of global settling. However, it should be realized that there are several other common global settling

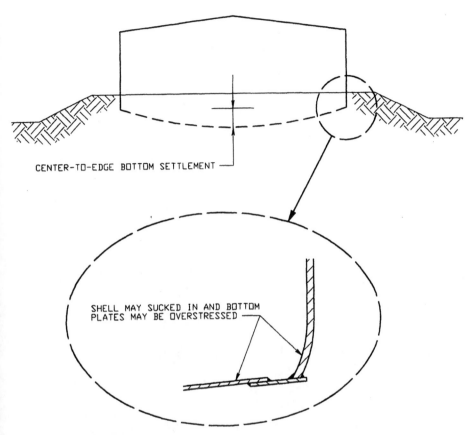

SHELL MAY SUCKED IN AND BOTTOM
PLATES MAY BE OVERSTRESSED

CENTER-TO-EDGE BOTTOM SETTLEMENT

Figure 13.9 Dish settling.

patterns. Investigators of the various modes of general global settling have recommended criteria for each type, as shown in Fig. 13.10.[7]

The problems associated with general global settling are

- High stresses generated in the bottom plates and fillet welds
- Tensile stresses near the shell-to-bottom welds that may cause shell buckling
- Changes in calibrated tank volumes (strapping charts and gauges)
- Changes in the drainage of the tank bottom profile and puddling when attempting to empty tank

If the tank bottom can be idealized as a circular plate subject to uniform loading, it can be analyzed by large deflection theory (this is applicable when the deflection is significantly larger than the plate thickness). Indeed, this is the approach that has been used in the literature to attempt to establish acceptable settlement deflection lim-

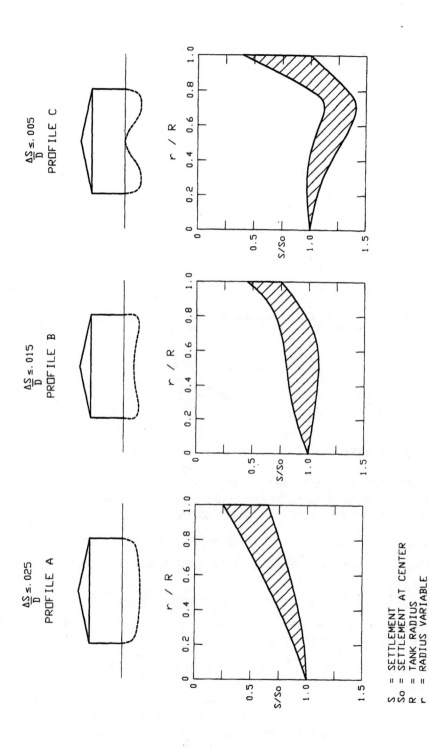

S = SETTLEMENT
S_o = SETTLEMENT AT CENTER
R = TANK RADIUS
r = RADIUS VARIABLE

Figure 13.10 Normalized settlement of tank bottom.

its. The membrane stress for a simply supported, fixed edge which is not free to move radially is given by

$$\sigma = 0.905E\left(\frac{s}{r}\right)^2 \tag{13.16}$$

where s = maximum settlement deflection, ft, and r = radius, ft. Since many tanks have an initial slope or are by design sloped (cone-up or cone-down), it is possible to include a term for this. Rewriting Eq. (13.16) to take into account the magnitude of the absolute elevation difference in the bottom s_0 and including a factor of safety, we have

$$s = \left(s_0^2 + 0.28\sigma_t\,\frac{D^2}{FS \times E}\right)^{1/2} \tag{13.17}$$

where s = tolerable settlement, ft

s_0 = initial rise or depression from center of tank to perimeter, in

σ_t = membrane tensile stress developed in bottom due to dishing at failure

FS = factor of safety against failure

E = Young's modulus, 29,000,000 psi

D = tank diameter, ft

The *factor of safety* is defined as the ratio of the specified minimum tensile stress to the actual stress. From Eq. (13.17) it will be noticed that the settlement is proportional to the square root of the inverse of the factor of safety. In other words, a factor of safety of 4 corresponds to a settlement value of one-half what it would be at tensile failure. Setting the factor of safety to 2 corresponds to a settlement of 0.7 of the value it would be at failure. Selection of the factor of safety may be made based upon the initiation of yielding. This means that the allowable settlement would be set equal to the stress level that would result in general yielding. Since the ratio of yield to tensile strengths varies from about 0.5 to 0.7, a factor of safety based on yielding would be between 2 and 3.

If 40-ksi steel is selected and a factor of safety of 4 is used, a simplified criterion for allowable settlement may be made.

$$s = \frac{D}{88} \tag{13.18}$$

In the literature, values for global dishing range from $D/50$ to $D/100$ depending on the foundation type, safety factor, or empirical data. The value stated in the 1st edition of API 653 is $D/64$. This corresponds to a factor of safety of about 1.6, and for global dishing these values appear to be reasonable. A 100-ft-diameter tank, using the provisions of Appendix B of API 653, would have a total dish settlement of $B = 0.37R$, where B is in inches and R is in feet, or 18.5 in. However, for values of R less than 3 to 5 ft, these limitations are not really applicable to local settling, as explained later.

The theory presented above is based upon large deflection theory of circular flat plates with edges that are not free to move radially. However, when the difference in settlement between the center and the periphery of the tank is large, there are often indications that the bottom membrane does move inward radially, as the shell will be pulled in, as shown in Fig. 13.9. From theoretical considerations the difference in membrane stresses generated between a circular plate simply supported with a fixed edge and an edge that is free to move radially is approximately a factor of 3.[8] This means that the stresses will be one-third as high for bottom plates that are free to slide as for those that are not. When the tank is loaded with liquid, the bottom plates are probably held in place more securely, and it may not be a valid assumption to use the free-edge condition in spite of the fact that this has often been observed as a pull-in of the shell because this has not occurred consistently.

For other modes of global settling it has been suggested[9] that different allowable settlements be provided for different configurations. The bottom slope configuration falls into three definite patterns, as shown in Fig. 13.10.

13.3.5 Local interior settling

Local settling that occurs in the interior of tanks often takes the form of depressions shown in Fig. 13.11. This type of settling poses problems similar to those listed in the section on global dishing, above. The proposed methods of assigning a tolerance to this type of settlement are again based upon the theory of large deflection. Some of the methods include a relaxation when the settling occurs near the tank wall to take into account the freedom of the plate near the shell to slide radially inward as the depression increases.

It is important to note that the fabrication process itself leads to buckles and bulges in the bottom plates. When the tank is filled with liquid, these tend to level out, but often they reappear when the liquid is removed. Most of the models proposed in the literature to develop settlement criteria to address this kind of settlement do not take into account the initial waviness of the bottom. These models which are based on uniformly loaded circular plates have been overly stringent where bulges with small radii are under consideration because the models do not take into account the initial curvature.

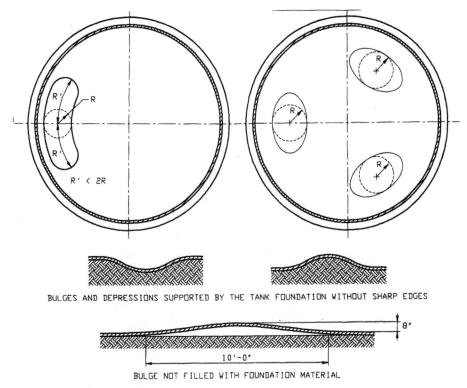

BULGES AND DEPRESSIONS SUPPORTED BY THE TANK FOUNDATION WITHOUT SHARP EDGES

BULGE NOT FILLED WITH FOUNDATION MATERIAL

Figure 13.11 Bottom settlement.

This type of settling is inevitable in compacted earth foundations because the composition and thickness of the soil layers under the tank vary. These deformations are usually formed gradually without sharp changes in slope so that the bottom plates are adequately supported in most cases. The risk of failure due to this type of settlement is minimal unless there are serious problems with the welding integrity.

When there are large voids under the tank bottom, the bottom plates may lift off the soil completely, as shown in Figs. 13.11 and 13.12. Although this is not usually a problem, if the amount of void is excessive, this can lead to localized rippling effects (covered later). A discussion of tank releveling below covers the problems associated with filling these void spaces with grout.

13.4 Sloped Bottoms

The previous sections on settling in general apply to a flat-bottom tank. However, in practice many tanks have intentional slopes built into the bottom that fall into three categories:

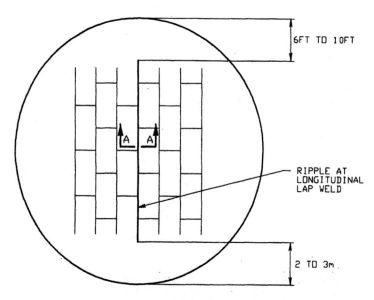

TYPICAL LOCATION OF A RIPPLE IN TANK BOTTOM

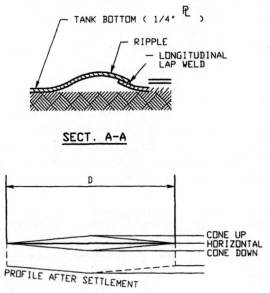

Figure 13.12 Tank bottom ripples.

1. Single-slope
2. Cone-up
3. Cone-down

Since the design slope of these bottoms averages about 1 in per 10 ft, they can still be considered to be flat bottoms, and the previous sections still apply.

However, there is one special situation that arises because the bottom is sloped. *Cone-up* bottoms subject to general dish settlement can tolerate more total settlement than flat-bottom or cone-down or single-slope bottoms. As settling occurs, the bottom goes into compression as it goes from cone-up to flat.

As the bottom settles below the flat surface that coincides with the shell base, the compressive stresses that were generated are relieved until its condition is cone-down approximately equal to the magnitude of the original cone-up condition. See Fig. 13.12.

However, if the degree of initial cone-up slope is significant, the settling relatively uniform, and the bottom constructed with lap-welded joints, then a phenomenon known as *rippling* can occur. This often occurs during the initial hydrostatic test on newly constructed tanks. Because the bottom is in compression, the bottom membrane buckles. Because of the linear layout of the bottom plates and the use of fillet welds, large moments are introduced into the lap joints, and a crease or a fold that traverses large fractions of the diameter can form, as shown in Fig. 13.12. The ripples are typically unidirectional and occur in the long direction of the bottom plates. The crease may be very severe (a radius of curvature of approximately 1 ft is not uncommon). The ripple indicates that yield stresses have been exceeded. The ripple can act as a stiffening beam and cause increased differential settlement and deformation to occur adjacent to the ripple, leading to bottom failure.

In this simplistic view, the allowable settlement of the cone-up condition should be more than twice what a flat or otherwise sloped bottom tank should be. Of course, from a practical view, it is desirable to maintain the cone-up configuration, as this is usually selected not solely for the increased allowable settlement but also for operating and drainage reasons.

A design criterion may be established that limits rippling and provides a basis for allowable settling for cone-up tanks. Since the ripple tends to form symmetrically about the welded lap joint, it suggests that the lap-welded plates in compression are affected by the eccentricity caused by the lap. Bending stresses in lap-welded joints are at least 3 times the tensile stresses. Therefore, if one-third of the yield is used as a criterion for maximum stresses, this can provide a basis for providing a maximum slope such that one-third of 40-ksi yield material is the failure limit. Using a factor of safety of 4 and inserting these parameters into Eq. (13.17), we have a maximum slope of $\frac{1}{175}$, or approximately $\frac{3}{4}$ in/ft. If the maximum cone-up slope is limited to about 1 in per 10 ft, then ripples should not occur as a result of settlement. Using this criterion allows for the cone-up configuration to pass through the flat configuration without rippling.

13.5   Edge Settlement

Edge settlement occurs in the bottom plates near the shell, as shown in Fig. 13.13. It is frequently difficult, if not impossible, to determine this condition from the exterior of the tank. However, from the inside

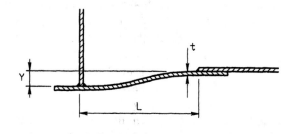

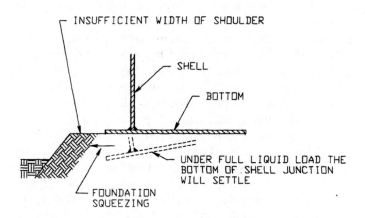

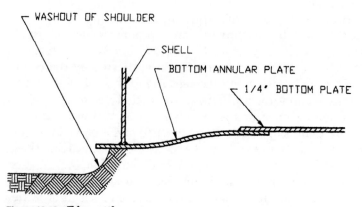

Figure 13.13 Edge settlement.

of the tank, this is one of the most prominent and obvious forms of settling. Because it is obvious deformation to the naked eye, it begs the question of what to do with the tank before putting it back into service when encountered on a routine internal inspection.

Edge settlement is observed frequently in northern latitudes where the roof snow loads are significant. Due to the extra live load acting to compress the shell downward, there is a knife cutting effect that may cause the bottom near the shell to exceed its soil-bearing capacity with resultant edge settlement. The combination of high snow loads and high moisture content during the spring aggravates edge settlement.

Edge settlement occurs most often on tanks that have been built on grade with compressible soils or materials. If the soil has not been compacted sufficiently or becomes soft when wet, the probability of edge settlement is increased. The main reason that edge settlement occurs is the increased loading on the foundation at the periphery due to the weight of the steel and inadequate soil-bearing strength. A condition that aggravates edge settlement is lack of sufficient extension of the foundation outside the tank radius to prevent lateral squeezing of the foundation (see Fig. 13.13).

Edge settling can occur locally where there are soft spots around the edge of the foundation. However, edge settling usually involves a rather substantial portion of the circumference of the tank. Edge settlement is rarely seen for tanks that are constructed on ringwall foundations of reinforced concrete, although it is not uncommon where the tank is built on a crushed-stone ringwall foundation.

The fillet welding process between the annular plate and shell and the bottom plates induces shrinkage and residual stresses that tend to warp the bottom plates or annular plates near the shell into an upward direction, leading to bulges near the shell. Although this is not strictly edge settlement, it may contribute to it by providing an initial slope in the annular plate and may cause the bottom of the tank under the shell to apply greater downward pressure on the soil than would otherwise be present. The initial slope may be counted as edge settlement when indeed it was simply the initial slope caused by the welding process. Proper welding procedures, careful selection of welding sequence for all welds in the bottom annular plate, and careful fit-up should minimize this problem.

13.5.1 Current settlement criteria

To date, there has been no really appropriate method developed to provide a reasonable basis for the tolerable amount of edge settlement. There are, however, numerous tanks in service with edge settlement with magnitudes of 6 to 18 in over a span of 1 to 2 ft and functioning without leaks or failures. API conducted an informal survey of members on the API task group for settlement and determined that there have been few, if any, cases of failure due to edge settlement. However, there was one major failure, but in this case the tank had been releveled which introduced an inflection into the profile of the bottom plates.

Note that this mode of settling is unlike the other modes of settling. API 653 as well as other proposals for defining edge settlement criteria is based upon a model that is similar to the dishing models described above. There is enough difference, however, in the settling mechanism that these approaches may lead to criteria that are excessively conservative. Any model that uses an allowable-stress basis for limiting settlement is maybe conservative because this type of settlement involves substantial yielding of the bottom plates, as is apparent from the large deflections over short spans. A strain-limiting approach may, therefore, be more appropriate.

13.5.2 Strain-limited criteria

Experience shows that there have been very few tank bottom failures attributable to edge settlement in spite of the notable local curvature and yielding that occur. Edge settlements on the order of 6 in within a few feet of the shell are not uncommon. The criterion for edge settlement in API 653 is

$$B = 0.37R$$

where B = allowable edge settlement, in, and R = distance between shell and start of edge settlement, ft. There are two apparent problems with this formula. The first is that as the value of R becomes small (i.e., less than 1 to 2 ft), the value of tolerable edge settlement also becomes small. In fact, small values of edge settlement less than approximately 2 in have never been a problem. The fact is that bottom plate tolerances and warpage caused by welding can produce apparent "edge settlement" when in fact it is not settlement at all. The second problem is that this formula is known to be highly overconservative. Other countries have standards for edge settlement which seem satisfactory in the area of edge settlement and are far less conservative. For example, the criterion for edge settlement in EEMUA Publication 159 (see References) is

$$B = 2R$$

This is over 5 times the API allowable value.

Since edge settlement usually occurs over a small distance (typically on several feet at most), any approach to setting a criterion for it based upon allowable stresses is going to be overly conservative, since yielding has already occurred when edge settlement is detected. A strain-limited approach is presented here as a criterion and one that has often been used in the United States.

Consider Fig. 13.14. It was one of the first approaches used to establish settlement limits quite successfully in the industry after the publication of API 653 limits was known to be in error. The approach is to model the tank bottom which has undergone edge settlement as a guided cantilever beam. Most ordinary tank bottom steels have a yield point at a strain of about 0.0012 and a strain at which hardening occurs of 0.014. Failure occurs at a strain of approximately 0.11. If a value halfway between the yield point and the initiation of hardening is chosen, the strain is 0.007. This gives a factor of safety of 15 against failure and of 2 against the initiation of strain hardening. ASME allows a strain of 0.05 before stress relieving is required.

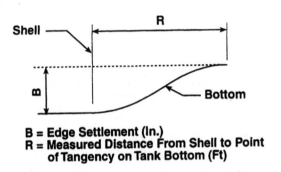

B = Edge Settlement (In.)
R = Measured Distance From Shell to Point of Tangency on Tank Bottom (Ft)

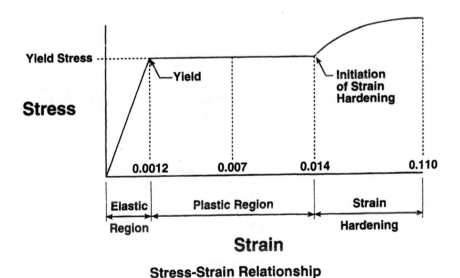

Stress-Strain Relationship

Figure 13.14 Strain-limited edge settlement limit. (*Courtesy AEC Engineering, Minneapolis.*)

Edge Settlement Analysis

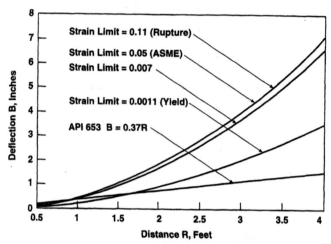

Figure 13.15 Comparison of strain-limited approaches to bottom settlement. (*Courtesy AEC Engineering, Minneapolis.*)

Under the assumptions of a ¼-in bottom, a yield stress of 33 ksi, and a modulus of elasticity of 29 million psi, a formula based on the model shown in the figure is

$$B = 0.41R^2$$

This equation is compared to other values of strain as well as the API formula in Fig. 13.15. Note that in the figure the strain-limited criteria are not activated unless the amount of settlement exceeds at least 2 in.

In 1995 and 1996, API performed some axisymmetric finite element modeling of the tank bottom that showed that for a constant strain of 1.2 percent the stress and the amount of edge settlement are functions of both the width of the settled area R as well as the diameter D. These curves can be used in lieu of the formulas in API 653 for allowable edge settlement. However, a lower cutoff of 2 in should be included so that nuisance alarms regarding edge settlement do not occur. This lower bound has been added to the plot shown in Fig. 13.16. On this plot is also shown the current limit represented by $B = 0.37R$.

13.6 Designing for Settlement

Depending on the degree and nature of settlement expected, which may be determined from similar installations in the area or from soil surveys and investigations, there are several means of coping with expected settlement, in order of increasing effectiveness:

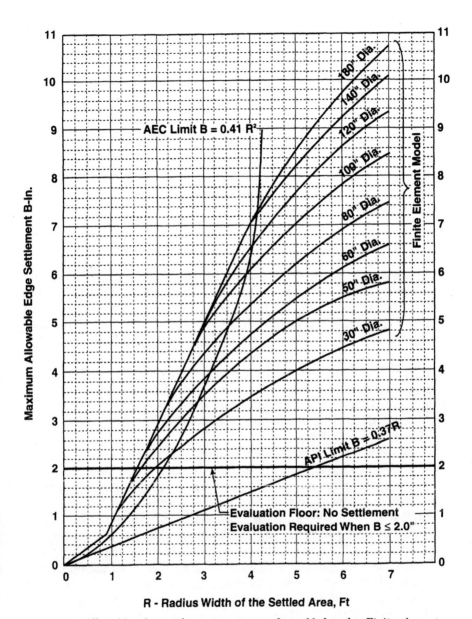

Figure 13.16 Allowable edge settlement nomogram for welded tanks. Finite element, asymmetric model, 1.2 percent strain, bottom plate thickness = 0.25 in.

- Standard lap-welded bottom
- Annular plates with lap-welded bottom
- Butt-welded bottoms

Although each succeeding construction method is more effective, it is also more costly. Unless used for other reasons than high settlement, the butt-welded tank bottom is generally ruled out on a cost/benefit basis.

Additional measures can more effectively be included such as deeper levels of soil compaction, crushed-stone ringwalls, reinforced-concrete ringwalls, or slabs on ringwall foundations.

The use of annular plates reduces edge settlement. The use of concrete ringwalls virtually eliminates edge settlement and greatly reduces differential settlement.

Since the standard lap-welded tank bottom is economically the bottom choice in most cases, there is a strong tendency to use this design for locations even where significant settlement is expected.

13.7 Releveling Tanks

Releveling of tanks is a fairly common procedure followed when the tank owner/operator feels that there is excessive settlement that might lead to problems, such as buckling of shell plates, leakage in the bottom plates, excessive out-of-roundness, and high stresses. Releveling can also correct some of these problems. When floating-roof tank bases have experienced differential settlement, the roofs can bind and seals may not be effective or may be damaged. Releveling can often cause the tank to return to a round shape. Tanks that have buckled due to settlement or tanks that have been constructed with initial out-of-roundness are usually not susceptible to much improvement by releveling.

Unless a tank is a floating-roof type and binding has resulted from the ovalizing due to settlement, it is usually better not to relevel the tank. Releveling introduces numerous uneven spots into the bottom, particularly when grout or sand pumping is used. Because the flow cannot be controlled, there will be an actual increase in the number of low points and high contact points. Releveling has resulted in an inflection in the tank bottom profile curve, in some cases developing very high bottom plate stresses. These were believed to have resulted in catastrophic failure of the bottom after a crack leaked and washed out supporting soil and increased in severity until complete failure of the tank occurred.

When releveling is contemplated, it is best to evaluate the advantages and disadvantages carefully under the guidance of a professional engineer experienced in tank design.

13.8 Methods of Releveling

Companies specialize not only in tank releveling but also in the various methods discussed below. It is important to deal only with reputable contractors who have carefully planned a shell releveling procedure as well as to confirm the adequacy by design and experience.

All the releveling procedures have some factors in common:

- Careful consideration should be given to an alternative to releveling the tank, because not only does the releveling process introduce construction hazards and safety problems but also the bottom is subjected to traumatic forces which can lead to cracks and potential failure after the tank is put back into service. Alternatives are to conduct more analysis and a nondestructive examination in order to avoid the releveling, modification of piping to accommodate settlement, more frequent inspections, and other ways as well.

- For floating-roof tanks, an evaluation should be made to determine if the roof should be supported from the shell to avoid excessive stresses and the possibility of cracks occurring from differential movement during the releveling process. Figure 13.17 shows an example of supporting a floating roof from the shell.

- When large tanks are lifted or jacked, the bottom flexes, inducing membrane stresses into the bottom and radially compressive forces in the shell. Calculations should be made to determine if a perimeter lift is acceptable or if additional supports need to be installed on the interior under the bottom.

- When tank jacking methods are used, it is possible to jack tanks up to approximately 10 ft high. This allows bottom inspection, cleaning, removal of contaminated soil where leakage has occurred, rebuilding of the foundation if necessary, or coating of the bottom from the underside.

- Consideration must be given to fixed-roof supports so that roof buckling and damage do not result.

- Consideration must be taken of the amount of differential jacking so that shell buckling or weld damage in the corner welds or in the bottom plates does not occur.

- In all the tank releveling procedures, large forces are involved and any mishaps could cause injuries or unanticipated costs. Any work of this nature should be carefully reviewed from a safety, environmental, and good-practice stance. The owner/operator should also convince herself or himself that those performing the work have direct experience in the methods to be used.

- Almost without exception, a large-diameter tank which has been releveled should be hydrostatically tested. There might be a few cases in small tanks where the shell stresses are so low that testing is not necessary or where the amount of jacking was extremely small that might allow for an exemption.

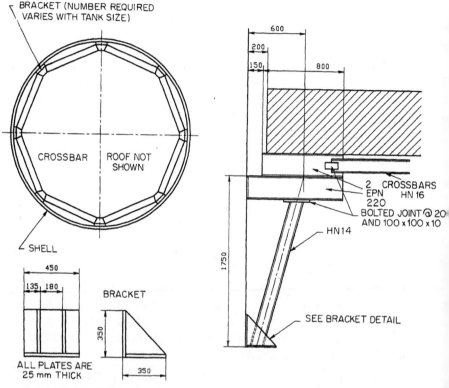

Figure 13.17 Floating-roof support.

- Any piping that is connected to the tank should be assessed as to whether the releveling work requires its disconnection or whether it will produce excessive stresses, causing equipment damage. Underground piping should be exposed adjacent to the tank, if connected to the tank, so that it can be monitored as above.

- When a tank is elevated or jacked up, it is common practice not to support the bottom plates. Although substantial in-plane stresses develop and the bottom sags visibly, there are no particularly adverse effects. However, a hydrostatic test must be conducted to ensure that new cracks have not developed. In addition, all welds should be reinspected after the shell-jacking process.

- A registered professional engineer should review and approve the methods and calculations made to determine the engineering feasibility and the safety of the jacking and/or lift.

13.8.1 Hydraulic tank lifting

Tanks are such light structures per unit of plan area that they may be lifted by pumping water under and around the tanks so that their inherent buoyancy lifts them off the ground. Since a tank will typically weigh only 15 to 30 lb/ft^2, this requires a submergence of about 6 to 12 in of water to cause the tank to float. Cribbing may be used to support the tank; or local shimming, grouting, or supporting may be done before the tank is set back in place. Most often the tank would be set aside and the foundation completely reworked.

Another common method is to wrap a skirt around the base of the tank and to apply air pressure by using standard blowers. The force of the air in the skirt and under the bottom will lift the tank, and it can be moved about for foundation repairs.

13.8.2 Air pillow lifting

This is a relatively new, promising method that uses cushions that are pumped up with controlled air pressure. The cushions or air pillows are spread around the circumference and optimized for weight and air pressure. When pressure is applied, the tank lifts. This method is comparable to shell jacking but does not require welding on the shell, which is a significant advantage.

13.8.3 Shell jacking

Shell jacking is a common method in which lugs are welded to the shell near the base, as shown in Fig. 13.18. Typical spacing is about 15 ft. Once the lugs are in place and a suitable jacking pad is set up, jacking proceeds around the tank circumference in small increments. The small increment prevents damage to the shell and to the bottom by warping the bottom excessively out of plane. Shims are installed as the jacks are moved around, and the tank can be raised to any desired elevation. The bottoms of these tanks will, of course, sag, and sufficient analysis is needed to determine what kinds of forces and stresses are acting on the bottom and on the shell near the bottom.

Typical releveling tolerances that are specified average about $\frac{1}{4}$ in of level for any measured point on the tank perimeter at the base.

The contractor should furnish, design, and install the lugs as well as carefully remove them and any weld arc strikes or remaining slag which should be ground out. The contractor should also recommend and propose the point loading under each shimmed area so that foundation damage or settling does not occur during the progress of the work. The recommended spacing of shims is approximately 3 ft.

The contractor should also propose how the sand or grout should be applied, if at all, to various low points under the bottom. If a large enough area is available under the tank (the tank has been jacked up high enough), a flowable grout or sand layer can be an effective planar foundation for the tank. As explained above, miscellaneous injec-

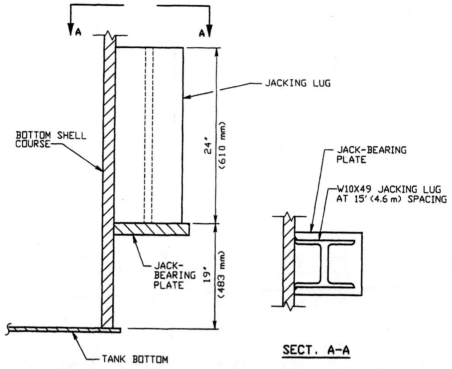

Figure 13.18 Jacking lugs used on large tanks.

tion of grout through holes cut into the bottom plates used to level low spots under the bottom is usually ineffective or makes the situation worse. If the work is meant to correct out-of-roundness, frequent monitoring of the radial tolerances as well as the effect of releveling on these tolerances should be required. At least eight equally spaced points at the top of the shell should be used for monitoring. Elevations and radial measurements should be recorded before and after the work. Detailed procedures and methods should be required of the contractor and all these factors examined in light of the specific circumstances.

13.8.4 Under-the-shell method

The under-the-shell method uses jacking under the bottom of the shell. In this method, small pits must be excavated under the shell of the tank in order to place a jack. Figure 13.19 shows a typical arrangement. The principal objection to this method is that pits must be excavated beneath the tank shell. In soil foundations, this may cause a loss of compaction on the order of 40 to 50 percent. Another

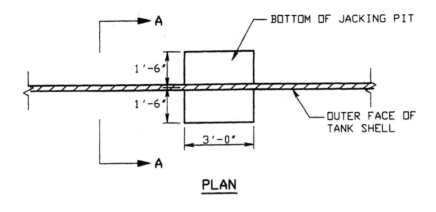

PLAN

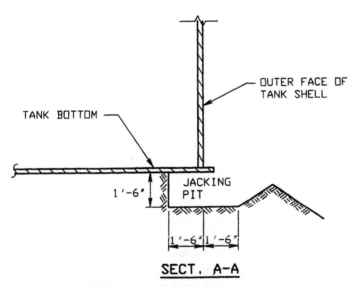

SECT. A-A

Figure 13.19   Jacking pit dimensions.

problem is that the spacing for shims and for jack points must be greater than that for the shell jacking method and therefore will provide higher soil stresses while the work is in progress.

The same procedures, specifications, precautions, and testing as covered under shell jacking should be observed.

13.8.5   Tank leveling by pressure grouting

Tank leveling by pressure grouting or sand pumping is used to force the low spots or settled areas upward. For this operation a nozzle is attached through the bottom. This method can be used to raise small areas or large areas in tank bottoms that are low. The contractor forces sand or grout under pressure into these areas, to stabilize the bottom plates. Where the involved areas are small and numerous, this method is usually ineffective because the mixture will flow through the areas of least resistance and lift the plates off even farther in these areas. It also causes the tank to rest on points (rather than more uniformly, if nothing were done). However, in some cases grout can be effectively used.

Pressure grouting has been effectively used to level areas under fixed-roof supports.

A tank owner/operator considering this method should examine a step-by-step proposal from the contract to ensure that good practices are followed and that all safety and environmental regulations are considered. When the existing bottom is cut for injecting grout, precautions should be taken for the possible existence of flammable liquids or toxic substances that could have been stored or leaked through in the past.

13.9 Conclusions

Settlement is a complex topic, and there is no universal consensus about how much deformation is acceptable. Note that there are numerous modes of settlement, and each involves somewhat differing mechanics as well as effects on the tank structure. Table 13.1 shows the various modes of settlement, the effects of the settlement based on the type of tank, and the basic theories that have been used to set allowable settlement criteria.

While the edge settlement criteria of API 653 appear at this time to be based on an incorrect model and are therefore overly conservative, API 653 seems to be appropriate for all the other modes of settlement. However, in the area of differential edge settlement, the determination of differential edge settlement based on $S_i - \frac{1}{2}(S_{i+1} + S_{i-1})$ appears to lead to inordinately large errors when the number of data points is excessive. By using the criteria presented in this work, the radius-of-curvature errors introduced by a high number of readings N can be reduced to a realistic level. The same error is introduced in the radial deflection equation, and more work needs to be done in this area.

Practical examples are provided at the end of this chapter to guide the user as to reasonable and current practices for handling tank settlement problems.

Example 13.1: Best-Fit Cosine Curve and Radial Shell Displacements In part 1 we examine the theoretical basis for determining a best-fit cosine curve to a settled tank bottom. In part 2 we take a practical example to show how to apply the theory to real cases of settlement.

Type of settlement	Application	Effect	Basis for settlement criteria	Recommended criteria	Symbols
Uniform	EFRT* IFRT FRT	Piping may become over-stressed.† Water may pool at base, causing corrosion.†	No basis for setting a repeatable design limit.	Determine acceptability on a case-by-case basis. Limit piping stresses. Consider corrosion.	
Planar tilt	EFRT IFRT FRT	■ Increased hydrostatic pressure. ■ Aesthetics. ■ Liquid may overflow at tank design capacity. ■ Radial tolerances affected.	Same as uniform. Also, increased hydrostatic head, aesthetics, and ovalizing need to be considered, although in many cases they will not be important.	■ Reduce liquid level so overflow does not occur. ■ Limit $\Delta S = D/50$ for aesthetics. ■ Limit stresses $\sigma_D + \Delta\sigma \leq \sigma_T$ ■ Limit piping stresses. ■ Consider corrosion.	ΔS = allowable settlement D = tank diameter σ_D = allowable design shell stress $\Delta\sigma$ = increase in stress caused by tilt σ_T = allowable hydrostatic test stress
Differential shell settlement	EFRT IFRT	■ Ovalizing causes seal problems. ■ Induced stresses at top. ■ Shell buckling.	Uses lesser of: ■ Limits for maximum radial tolerance. ■ Limits based on shell stresses.	■ Based on radial tolerance ΔR $$\Delta S = \frac{l^2}{HD}\,\Delta R$$ Use 8 points.	ΔS = allowable settlement, ft l = distance between measurement points σ = stress in shell E = modulus of elasticity H = tank height
	FRT	■ Roof damage. ■ Shell buckling.	■ Limit shell stresses.	Limit stress in shell. $$\Delta S = \frac{11}{2}\,\frac{\sigma y l^2}{EH}$$ (API 653) Alternatively $$\Delta S = 52.8 \times 10^{-6}\,D^2$$	

*EFRT: external floating-roof tank; IFRT: internal floating-roof tank; FRT: fixed-roof tank.
†Same comment applies to all succeeding rows where the elevation of the shell-to-bottom weld changes.

TABLE 13.1 Allowable Tank Settlement Criteria (*Continued*)

Type of settlement	Application	Effect	Basis for settlement criteria	Recommended criteria	Symbols
Global dishing	EFRT IFRT FRT Cone-down or flat or single-slope	■ Rupture of bottom.	■ Membrane stress theory.	Use API 653 $$\Delta S = \frac{D}{65} + \Delta S_0$$	ΔS_0 = initial cone-down measured from center to edge
	Cone Up	■ Ripple.	■ Set limit on slope to limit maximum compressive stresses and subsequent bottom buckling.	Max. slope = $3/4$ in per 10 ft $$\Delta S = \frac{D}{65} + 2\,\Delta S_0$$	ΔS_0 = initial cone-up measured from center to edge
Local settling	EFRT IFRT FRT	■ Rupture of bottom.	■ Maximum stress theory. However, for very small D, theory is inapplicable due to significance of thickening of lapped plates and local bending.	Use API 653 as follows: $$\Delta S = \frac{D}{65} \quad \text{for } D > 8 \text{ ft}$$ For $D < 8$ ft, Max. ΔS = 2.0 in	
Edge settling	EFRT IFRT FRT	■ Rupture of bottom or shell-to-bottom weld.	■ Limit strain to one-half of strain-hardening limit.	Use $B = 0.41\,R^2$ or Fig. 13.16	

*EFRT: external floating-roof tank; IFRT: internal floating-roof tank; FRT: fixed-roof tank.
†Same comment applies to all succeeding rows where the elevation of the shell-to-bottom weld changes.

554

part 1 Settlement data taken around the periphery of a tank base can be represented by a settlement function $S = f(x)$, where x is the coordinate measured around the circumference of the tank. A best-fit plane can be fitted to the data by using Fourier analysis.

A periodic function can be represented by the Fourier series

$$f(x) = a_0 + \sum_n (a_n \cos n\theta + b_n \sin n\theta) \tag{13.19}$$

where the Fourier coefficients are

$$a_0 = \frac{l}{2\pi} \int_{-\pi}^{\pi} f(x)\, dx$$

$$a_n = \frac{l}{\pi} \int_{-\pi}^{\pi} f(x) \cos n\theta\, dx \tag{13.20}$$

$$b_n = \frac{l}{\pi} \int_{-\pi}^{\pi} f(x) \cos n\theta\, dx$$

For $f(x)$ with period 2π, the rectangular rule of calculus may be used to evaluate the Fourier coefficients. First, divide the interval of the domain of the function, $f(x)$, $-\pi \le x \le +\pi$ into N equal parts with $\Delta x = 2\pi/N$ and the points at x_0 ($= -\pi$), $x_1, x_2, \ldots, x_N$ ($= +\pi$). The Fourier coefficients then become

$$a_0 = \frac{l}{q} [f(x_1) + \cdots + f(x_N)]$$

$$a_n = \frac{2}{N} [f(x_1) \cos n\theta_1 + \cdots + f(x_N) \cos n\theta_N] \tag{13.21}$$

$$b_n = \frac{2}{N} [f(x_1) \sin n\theta_1 + \cdots + f(x_N) \sin n\theta_N]$$

We are interested only in the first harmonic of the series which represents a plane of tilt for the tank bottom. All other harmonics are neglected. For the first harmonic, the terms above reduce to

$$f(x) = a_0 + a_1 \cos \theta + b_1 \sin \theta \tag{13.22}$$

where

$$a_0 = \frac{1}{N} \sum_{i=1}^{2} S_i$$

$$a_1 = \frac{2}{N} \sum_{i=1}^{N} S_i \cos \theta_i$$

$$b_1 = \frac{2}{N} \sum_{i=1}^{N} S_i \sin \theta_i$$

and where $S_i = f(x_i)$. This is the basis for the example below in determining the best-fit cosine curve to the data points of measured settlement.

The equation of the tilt plane can be simplified by letting

$$a_1 = r \sin \alpha \tag{13.23}$$

and

$$a_2 = r \cos \alpha$$

where

$$r = \sqrt{a^2 + b^2} \quad \text{and} \quad \alpha = \arctan \frac{a}{b}$$

Then Eq. (13.22) can be rewritten as

$$f(x) = a_0 + r \sin (\theta + \alpha) \tag{13.24}$$

The term a_0 represents the uniform settlement component, and the second term is the tilt plane component. The phase angle α represents the orientation of the plane with respect to the origin or the first data point in the series. While this form may be used, it is more practical to use the equivalent form, Eq. (13.26), below.

Since the angular coordinate x represents the distance around the tank bottom, we must define it according to

$$\theta = \frac{2x}{D} \tag{13.25}$$

where D is the tank diameter and θ is the angle of the measured settlement point. Equation (13.22) may then be rewritten as

$$S = a_0 + a_1 \cos \frac{2x}{D} + b_1 \sin \frac{2x}{D} \tag{13.26}$$

part 2 Table 13.2 is a sample hand computation for the determination of the best-fit cosine curve to settlement readings. Figure 13.20 is a plot of the best-fit cosine curve as well as the radial displacement caused by the differential shell settlement.

The settlement data can also be used to determine the radial deflections at the top of the tank which may cause floating-roof binding and serve as a limit to acceptable settlement. Eq. (13.2) in finite difference form is used to calculate the radial deflections.

$$r_i = \frac{HN^2}{\pi D^2} [u_i - \frac{1}{2}(u_{i+1} + u_{i-1})] \tag{13.27}$$

Note in the example that the settlement readings were in centimeters but the units of both diameter and height were feet. The units of radial displacement will be the units selected for settlement u_i. The problems with the radial movement calculations are the same as with the settlement calculations in that the greater the number of data points N that there are, the greater the significance of the errors. It is recommended that for tanks up to 220 ft in diameter only 8 points of settlement taken at 45° be used and that above this diameter 16 points be used to minimize the spurious error problem. An alternative recommendation is to use Eq. (13.15).

Example 13.2 Figure 13.21 shows a tank with various pockets of settlement. Determine which, if any, of the bulges or depressions should be releveled in accordance with API 653, Appendix B.

solution The first thing to realize is that the bulge and the depression are treated in the same way. Another important fact is that settlement readings will usually be reported according to standard surveyor's units, which are feet and decimal feet.

TABLE 13.2 Determining Best-Fit Cosine Curve and Radial Displacement due to Shell Settlement

Column no.	1	2	3	4	5	6	7	8	9	10	11
Description	Settlement readings	Relative settlement	Angle of point from origin	Cosine of angle	Sine of angle			Planar tilt best-fit cosine curve	Differential settlement	Finite difference	Radial shell displacement
Variable	S_i	s_i	θ_i	$\cos\theta_i$	$\sin\theta_i$	$S_i\cos\theta_i$	$S_i\sin\theta_i$	$a_1\cos\theta_i + b_1\sin\theta_i$	u_i	$u_i-(u_{i+1}+u_{i-1})/2$	Note 2
				$\cos(3)$	$\sin(3)$	$(1)\times(4)$	$(1)\times(5)$	$a_1\times(4)+b_1\times(5)$	$(2)-(8)$		$8.99(10)$
									4.09	See note 1	
0	1460	41.19	0.00	1.00	0.00	1460.00	0.00	19.64	21.55	10.51	94.4633
1	1466	47.19	0.39	0.92	0.38	1354.41	561.01	29.18	18.01	5.28	47.465
2	1457	38.19	0.79	0.71	0.71	1030.25	1030.25	34.29	3.90	-2.11	-18.971
3	1447	28.19	1.18	0.38	0.92	553.74	1336.85	34.17	-5.98	-2.10	-18.892
4	1436	17.19	1.57	0.00	1.00	0.00	1436.00	28.85	-11.66	-4.70	-42.228
5	1430	11.19	1.96	-0.38	0.92	-547.24	1321.15	19.14	-7.95	-0.96	-8.606
6	1423	4.19	2.36	-0.71	0.71	-1006.21	1006.21	6.52	-2.33	1.00	9.027
7	1413	-5.81	2.75	-0.92	0.38	-1305.44	540.73	-7.10	1.29	4.04	36.331
8	1396	-22.81	3.14	-1.00	0.00	-1396.00	0.00	-19.64	-3.18	-5.51	-49.504
9	1393	-25.81	3.53	-0.92	-0.38	-1286.96	-533.08	-29.18	3.37	-4.28	-38.473
10	1403	-15.81	3.93	-0.71	-0.71	-992.07	-992.07	-34.29	18.47	8.61	77.419
11	1401	-17.81	4.32	-0.38	-0.92	-536.14	-1294.36	-34.17	16.36	10.10	90.828
12	1384	-34.81	4.71	0.00	-1.00	0.00	-1384.00	-28.85	-5.96	-0.80	-7.228
13	1373	-45.81	5.11	0.38	-0.92	525.42	-1268.49	-19.14	-26.67	-12.04	-108.289
14	1389	-29.81	5.50	0.71	-0.71	982.17	-982.17	-6.52	-23.30	-12.00	-107.938
15	1430	11.19	5.89	0.92	-0.38	1321.15	-547.24	7.10	4.09	4.96	44.596
									21.55	See note 1	
Column sum	22,701		47.12	0	0	157.08	230.82				

$$*a_0 = \Sigma S_i = \frac{22{,}701}{16} = 1418.81$$

$$a_1 = 2\,\frac{\Sigma S_i \cos\theta_i}{N} = 2(157.10)/16 = 19.64 \qquad b_1 = 2\,\frac{\Sigma S_i \sin\theta_i}{N} = 2(230.81)/16 = 28.85$$

$$\Delta S = 8.99\,\Delta S_i = 8.99 \times \text{column 10}$$

Tank diameter = 150

Shell height = 52

$$N = 16$$

Note 1: The value 4.09 is the last value from the data point series and is brought up in order to be able to make the computation required. Also, the first calculated value in the series is brought to the bottom of the column so that the calculation may be made. So, for example, in column 10, the value 10.51 = 21.55 − ½(18.01 + 4.09).

Note 2: $r_i = \dfrac{HN^2}{\pi^2 D}\,\Delta S = 8.99\,\Delta S_i = 8.99 \times \text{column 10}$

TABLE 13.2 Determining Best-Fit Cosine Curve and Radial Displacement due to Shell Settlement (Continued)

Column no.	Definition of column entry	Formula
Column 1	Actual settlement readings which may be in any units are listed in the first column. The subtotal is recorded for determination of the relative settlement in column 2.	S_i $a_0 = \dfrac{\Sigma S_i}{N}$
Column 2	Relative settlement is obtained by subtracting from the settlement readings the average of the sum of the settlement readings of column 1 (22,701).	$s_i = S_i - a_0$
Column 3	The angle in radians for each data point as measured from the origin or initial data point.	θ_i
Column 4	The cosine of the angle of each point which is the cosine of column 3.	$\cos \theta_i$
Column 5	The sine of the angle of each point which is the sine of column 3.	$\sin \theta_i$
Column 6	The relative settlement (column 2) times the cosine of the data point (column 4). Used to determine a_1.	$S_i \cos \theta_i$ $a_1 = 2 \dfrac{\Sigma S_i \cos \theta_i}{N}$
Column 7	The relative settlement (column 2) times the sine of the data point (column 5). Used to determine b_1.	$s_i \sin \theta_i$ $b_1 = 2 \dfrac{\Sigma S_i \sin \theta_i}{N}$
Column 8	The best-fit cosine curve or planar tilt.	$a_1 \cos \theta_i + b_1 \sin \theta_i$
Column 9	The difference between the actual settlement and the best-fit cosine curve u_i.	$s_i - u_i$
Column 10	This is the finite difference coefficient used to obtain radial displacement. Note that the last value in the column must be brought up ahead of the first value and the first value carried to the bottom. This simply makes the data points circular. In other words, the point after the last point is the first point in the series.	$u_i - \frac{1}{2}(u_{i+1} + u_{i-1})$
Column 11	This is the radial shell displacement. Here the units of H and D must be consistent.	$\dfrac{HN^2}{\pi D}\,[u_i - \frac{1}{2}(u_{i+1} + u_{i-1})]$

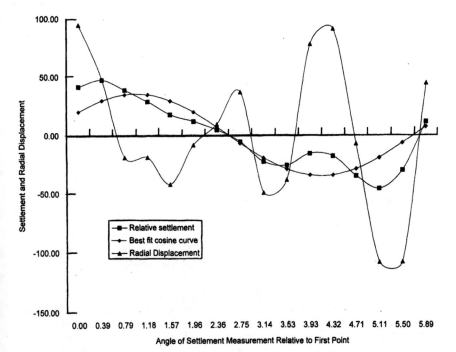

Figure 13.20

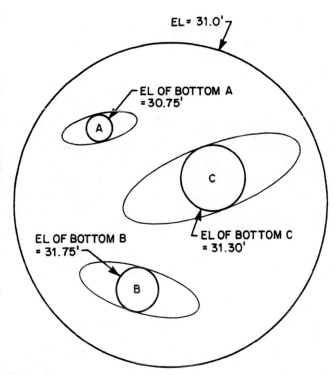

Figure 13.21

	Actual settlement	Allowable depression $B = 0.37R$	Acceptability
A	$(30.75-31) \cdot 12 = 3.0$ in depression	$B = 0.37(8/2) = 1.48$ in	Not acceptable
B	$(31.75-31) \cdot 12 = 9.0$ in bulge	$B = 0.73(12/2) = 2.22$ in	Not acceptable
C	$(31.3-31) \cdot 12 = 3.6$ in bulge	$B = 0.37(20/2) = 3.7$ in	OK

practical tip Unless the bulge or depression exceeds 2 in, most likely the distortion is not generating membrane stresses. Therefore, in practice, none of the depressions or bulges in this case should be modified or tampered with. However, on the examination for API 653 inspectors, the answers should be made in accordance with the table above.

Example 13.3 A tank 115 ft in diameter and 40 ft high is to be checked for settlement. How many readings are required around the perimeter?

solution API 653 requires that there be at least 8 equally spaced settlement measurement points with a maximum spacing between the points of 30 ft. In API 653, paragraph 10.5, measured settlement during hydrostatic testing, a method is given for determining the number of required points N. Presumably, this method also applies to settlement points required in Appendix B for internal or external inspections. The method states that $N = D/10$, where D is the tank diameter in feet.

In the solution of this example, we demonstrate the API method and a simpler method which is recommended for inspector examination questions, since it is simpler and easier to use and remember.

API 653 method of solution $N = D/10 = 115/10 = 11.5$. According to API 653, this must be rounded up to 12. Next check to make sure that the distance between points is greater than 30 ft. This is accomplished by determining $\pi D/12 = 3.14(115)/12 = 30.09$ ft. Since this is greater than 30 ft, the number N must be increased to 13. Next we check the spacing: $3.14(115)/13 = 27.7$, so this is OK. The minimum number of points is 13, which is greater than 8, so this is OK.

simpler method By setting $N = \pi D/30$, the points are automatically spaced no more than 30 ft apart. Since the number N will rarely be an integer, it is rounded up to the next higher integer, and this represents the minimum number of settlement points as long as N is greater than or equal to 8.

Using the previous example, $N = 3.14(115)/30 = 12.03$, which rounds up to 13. Since this is greater than 8, this meets the criterion of API 653. This method is much more reasonable than the method presented in API 653 and makes more sense.

Example 13.4 There are different types of settlement which can often be predicted in advance by using soil tests. This settlement mode does not introduce stresses into the tank structure, but piping and attachments must be given consideration to prevent problems caused by such settlement. What kind of settlement is this?

solution Although both uniform settlement and planar tilt are good answers, the most exact answer is uniform settlement. Planar tilt introduces a slightly higher liquid level and therefore shell stress into the tank, so uniform settlement is the best answer. This can be verified by referring to API 653 paragraph B.2.

Example 13.5 A tank 127 ft in diameter and 40 ft high has settlement readings along 14 equally spaced points (28.5 ft apart). It is desired to determine the maximum differential settlement in accordance with API 653.

solution Since no information was given about the tank material, the yield stress Y should be assumed to be 30,000 psi in accordance with API 653, paragraph 2.3.3.1. The value E is the modulus of elasticity, and it can always be assumed to be 29×10^6. The basic equation used to solve the problem is gotten by substitution of the values into the API 653, Appendix B equation for acceptable differential settlement s:

$$s = \frac{11L^2Y}{2EH} = 11(28.5)^2 30,000/2 \times 29 \times 10^6 \times 40 = 0.116 \text{ ft} = 1.38 \text{ in}$$

Note that the tolerable settlement is sensitive to the distance between settlement points. Although this solution would be appropriate for an inspector examination question, the proper approach in this case would be to make other evaluations. Since releveling is rarely, if ever, warranted for settlement less than 2 in, an alternative solution should be sought. Using the alternative solution proposed by Eq. (13.15), we have

$$s = 52.8 \times 10^{-6} \times D^2 = 0.85 \text{ ft} = 10.2 \text{ in}$$

Example 13.6 A 137-ft-diameter tank has edge settlement which has a very characteristic dropoff of the bottom plate near the shell. It extends 40 or 50 ft around the circumference. Determine the acceptability of edge settlement, using the API method, the method employed by the EEMUA (European Equipment Manufacturer's Association), and two different strain-limited methods. The amount of settlement observed is 6 in vertically at the shell (B), and the radius over which it occurs is 4 ft.

solution API 653 method:

$$B = 0.37R = 0.37(4) = 1.48 \text{ in}$$

EEMUA method:

$$B = 2R = 2(4) = 8 \text{ in}$$

AEC engineering method:

$$B = 0.41R^2 = 6.6 \text{ in}$$

Using the method of Fig. 13.15:

$$B = 7 \text{ in}$$

Note that the variation of solutions is a factor of over 5. Which one is right? As we have discussed, settlement of 2 in or less is not really worthy of consideration since it may be introduced at the time of construction and does not represent the result of mechanical forces of settlement but rather fabrication tolerance and foundation tolerances. The other three answers are somewhat in agreement. Due to the simple method of the AEC, which is generally conservative in comparison with the strain-limited approach of the curves, this method is recommended for determining edge settlement acceptance. In addition, this method had stood the test of time.

References

1. James S. Clarke, *Recent Tank Bottom and Foundation Problems*, Esso Research and Engineering Co., Florham Park, N.J., 1971; Phil Smith, Texaco.
2. Z. Malik, J. Morton, and C. Ruiz, "Ovalization of Cylindrical Tanks as a Result of Foundation Settlement," *Journal of Strain Analysis*, 12(4):339–348 (1977).
3. F. A. Koezwara, "Simple Method Calculates Tank Shell Quotation," *Hydrocarbon Processing*, August 1980, shows how to apply this formula to a real tank settling problem.
4. E. E. DeBeer, "Foundation Problems of Petroleum Tanks," *Annales de L'Institut Belge du Petrole*, 6:25–40 (1969).
5. Timothy B. D'Orazio and James M. Duncan, "Criteria for Settlement of Tanks," *Journal of Geotechnical Engineering*, 1983.
6. Timothy B. D'Orazio and James M. Duncan, "Distortion of Steel Tanks due to Settlement of Their Walls," *Journal of Geotechnical Engineering*, 115(6):875, June 1989.
7. T. D'Orazio and J. Duncan, "Differential Settlements in Steel Tanks," *Journal of Geotechnical Engineering*, 113(9), Dec. 4, 1986.
8. S. Timeshenko, *Theory of Plates and Shells*, table 82, 2d ed., McGraw-Hill, New York, 1959.
9. Timothy B. D'Orazio and James M. Duncan, "Differential Settlements in Steel Tanks," *Journal of Geotechnical Engineering*, 113(9), September 1987.
10. American Petroleum Institute, API Standard 653, Appendix B, Washington, 1991.
11. J. Buzek, Suitability for Service of Existing API 650 and API 12C Tanks, AEC Engineering, Minneapolis, Minn., 1991.
12. J. M. Duncan, and T. B. D'Orazio, "Stability of Steel Oil Storage Tanks," *Journal of Geotechnical Engineering*, 110(9), September 1984.
13. EEMUA (Engineering Equipment and Materials Users Association) Document no. 159 (draft), Tank Inspect, London, 1992.
14. W. A. Marr, J. A. Ramos, and T. W. Lambe, "Criteria for Settlement of Tanks," *Journal of Geotechnical Engineering*, 108(GT8), August 1982.
15. Publication 159, *Recommendations for In Service Periodic Inspection of Aboveground Vertical Cylindrical Steel Storage Tanks*, Engineering Equipment and Materials Users Association, London, 1993.
16. R. A. Sullivan and J. F. Nowicki, "Differential Settlements of Cylindrical Oil Tanks," *Proceedings*, Conference on Settlement of Structures, British Geotechnical Society, Cambridge, England, 1974, pp. 402–424.

14

Groundwater Protection

SECTION 14.1 LEAK DETECTION BOTTOMS AND RELEASE PREVENTION BARRIERS

14.1. Background

The tank population is increasing in average age because fewer tanks are being built. At the same time, the most vulnerable component on a tank for a leak-type failure is the tank bottom. The tank bottom is subject to both internal and bottomside corrosion.

Countless hours of effort have been applied to reduce the risk of unchecked contamination from leaking tank bottoms. Much of this effort has gone into corrosion prevention, maintenance standards, inspection methods, and leak detection and monitoring. Frequently, when a tank is shut down and an internal inspection performed, it is found that the remaining life, if any, in the tank bottom is shorter than the intended continuous operating interval planned. This immediately poses the questions of not only how to repair or replace the tank bottom but also how to monitor for leakage so that the maximum life of the tank may be enjoyed. Additionally, an increasing number of states now have some form of regulation that requires undertank leak detection or monitoring.

The focus of this section is on existing tank bottom replacement, but the principle applies to a relevant degree to new tanks as well.

14.1.2 Some Definitions

A number of terms should be defined before any serious discussion can ensue relating to tank bottom, liners, and leak detection.

14.1.2.1 Leak detection

Leak detection has been used in two contexts. The most common and widely used definition of leak detection is a system that incorporates a liner or release prevention barrier to redirect leaks to a perimeter system where they may be visually observed. A second context has been in the arena of systems that depend on instrumentation rather than visual inspection to determine if a leak exists, including such technologies as acoustic emissions, precision volumetric and mass methods, tracers, and ground-penetrating radar.

The American Petroleum Institute has recognized the standard definition by creating API 650, Appendix I, which defines a leak detection system which must direct leaks to the perimeter of the tank where they shall be capable of detection by visual examination.

14.1.2.2 Leak containment

Leak containment has been referred to as *subgrade protection* in Appendix I of API 650. Leak containment is the prevention of groundwater contamination by ensuring that a leak is trapped by the liner or by containers or pans until the tank can be brought down. It can be assumed that any tank with a release prevention barrier includes leak containment as long as the leak is not allowed to spill out onto the ground. Therefore, double-bottom tanks and tanks with liners that can contain a small leak for a period of time are considered leak containment systems as well.

14.1.2.3 Secondary containment

Secondary containment refers to a containment structure usually constructed of dikes or impervious walls to contain the tank contents in the event it is drained out. Most of the regulations such as NFPA 30 or SPCC require that the dikes be sized to contain the largest AST volume plus some freeboard for rainwater. Secondary containment is not covered in this section. In some cases the secondary containment may be constructed by providing an outer tank that can withhold the entire inner tank's contents, should a rupture develop. These types of tanks are usually rare for larger tanks and are considered for applications with extremely high hazards such as liquefied petroleum and hydrocarbon gases. See Fig. 14.1.1. Double containment tanks are commonly used underground and for smaller chemical tanks.

14.1.2.4 Double bottom

Double bottom refers to two steel bottoms constructed by slotting the tank shell above the lower bottom and installing a new bottom above

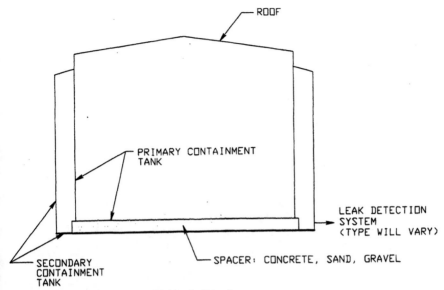

Figure 14.1.1 Double-bottom double-shell tank.

it. A triple or quadruple bottom is simply the addition of additional slots and new bottoms. Figure 14.1.2 shows a double-bottom and a triple-bottom tank. When a triple bottom is installed, it is because the second bottom has used up its remaining life.

14.1.2.5 Liner or release prevention barrier

Liner and *release prevention barrier* are terms used interchangeably to describe the system that causes a leak in the tank bottom to redirect horizontally to a telltale system. The liner may be made of any number of materials including steel, concrete, elastomeric materials, and clay. Figure 14.1.2 shows the placement of the liner or RPB in the bottom.

14.1.3 Performance Criteria for Leak Detection and Leak Containment

API 650, Appendix I, applies the following criteria to visual perimeter leak detection systems:

- Leaks through the bottom must be directed to the perimeter where they are visually detectable. If a leak does occur, it must not spill on the ground.

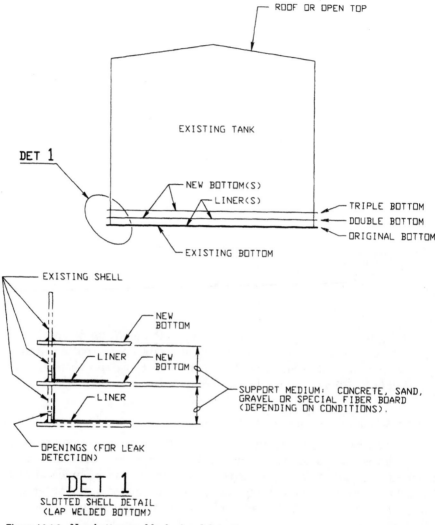

Figure 14.1.2 New bottom and leak visual detection.

- Electronic sensors and detectors may be used, but they must be in addition to the requirements for leak detection at the perimeter.
- Materials used for leak detection must be compatible with the range of products, and the stockside temperature ranges and material in contact with the subgrade must be suitable for below-grade service.
- The permeability of the liner shall be less than 10^{-7} cm/s.

14.1.4   Undertank Materials, Liners, and Cushions

14.1.4.1   Spacer materials for double bottoms

For double-bottom designs, a common question is, Should concrete or sand be used? Although concrete is more costly on a unit volume cost basis, it has a number of advantages and overall tank bottom construction costs may be lower. Since concrete has little or no zero-void space, a leak will quickly show up in the leak detection telltale system. Sand, however, has a void space of approximately 40 percent. This means that for a 250-ft-diameter tank with a 6-in sand cushion, at least 73,000 gal must leak out to saturate the void space before the leak flows freely to the telltale system. Sand as a filler material between the two bottoms may be considered a hazardous material if it is contaminated with product leaks. The removal of the sand will probably be more difficult.

Concrete is advantageous from an installation viewpoint because it can be easily and permanently controlled to provide the drainage slope required. It is easy to create an accurate surface with concrete. Sand, however, is not easy to slope or control because it is too easily disturbed by people and equipment pathways. When the bottom plates are laid down, the concrete acts as a rigid foundation and aids in the process of laying out plates and forming them, when required by impact or by cutting and welding operations. Sand is hard to keep out of the weld joints and can contaminate welds. Dragging plates across the sand has caused the plate to dive down through the sand and cut the liner on occasion. Plate dragging also has a high potential to cut cathodic protection wiring buried in the sand. The elastomeric liner under the sand can be more easily damaged during construction than if it is placed under concrete.

Although the concrete system cannot accept a cathodic protection system, the concrete itself is considered to be a factor inhibiting corrosion since it is alkaline. Experience shows that unless the steel bottom is allowed to sit in standing water or has saline runoff from the tank shell in marine environments, it lasts at least 20 to 30 years, if not more.

However, concrete has drawbacks, too. For relatively large tanks subject to soil settlement, cracking of the concrete can cause failure of the elastomeric liner placed under the concrete. One problem with concrete spaces subject to large settlement is the low pockets introduced by nonuniform settlement, where pockets of standing water accelerate corrosion. Another drawback is that acoustic emissions testing companies claim that a concrete pad makes finding leaks more difficult for them than a sand pad or filler does.

14.1.4.2 Underbottom materials

Concrete. Concrete has most of the advantages that are listed above for double-bottom fill materials such as reduced corrosion, quicker bottom plate layout and installation, slope and flatness control, the ability to install leak detection grooves, and low void space.

Sometimes, concrete is indispensable as a liner, thus fulfilling the need for a leak detection barrier when an elastomeric liner will not suffice. This is the case for hot tanks. Since the liner should be designed for the stockside temperature, which may be well above any ordinary elastomeric liner design limits, the concrete, if reinforced, may be considered a liner.

Sand or soil. Sand or soil as an undertank material has the advantage of reduced material costs. However, the washed sand or soil should be selected with minimal amounts of minerals and salts that could accelerate corrosion rates. Sand or soil under a tank can accommodate large amounts of local settlement without suffering any adverse effects.

14.1.4.3 Liners and RPBs

It is important to remember that the purpose of the RPB is not to function as a tank. It is not required to contain any significant hydrostatic head (the driving force for leakage), nor is it required to contain a leak for any long period of time. Presumably, when the leak detection system is activated, then the tank is shut down, inspected, and repaired. The same standards that apply to bladder liners should not be applied to RPBs for this very reason. However, very poorly installed or designed RPBs can have undesirable consequences. In one instance a leak developed in a newly installed double-bottom tank with a high-density polyethylene (HDPE) caulked-in-place liner. Leakage was found under the RPB emanating from the foundation. No indication of a leak was found in the telltale system that monitored the fluid space above the RPB. This caused a significant concern that the leak was flowing through a breach in the caulking at the perimeter of the seam and not filling the leak containment system sufficiently to activate the telltale system. In this case, sophisticated analytical methods were required to determine whether the tank was leaking.

Clay. There are a lot of concerns about the usage of clay. First, it must meet the permeability standards required by API 650, Appendix I. Clay can crack when dry and lose its properties as a liner. The cracks caused by shrinkage in the clay would allow large quantities of groundwater to migrate up to the bottom of the liner of the tank, causing potential problems. It is recommended that a geotextile fabric be installed between the clay and the liner, because the effects of clay shrinkage in direct contact with the liner on the integrity of the liner are not known.

There is not a lot of experience with clay liners. Claymax bentonite liners, which have been used extensively for diking requirements, are subject to changing permeability when exposed to certain conditions of pH or chemistry. Some states do not allow clay to be used as a liner because of its known expansion and contraction problems which affect the liner integrity. Clay is a poor conductor when dry and, therefore, will have a variable effect on any cathodic protection systems that must pass current through the clay. Also, it is hard to visualize a good method of ensuring a leakproof joint between the ringwall and the clay when shrinkage contraction occurs.

High-density polyethylene. HDPE is a common material specified for undertank and double-bottom linings as well as for secondary containment, because it is the most chemically resistant material and has the lowest cost compared to other materials. There are a number of problems with HDPE liners that must be weighed against its benefits.

A membrane thickness of at least 40 mils is required in several states. General practice and practical considerations for field handling require a thickness of at least 60 to no more than 100 mils. Installation can be facilitated by the thicker material in warmer climates and by the thinner material in colder climates.

The design should minimize the lineal footage of seams and the stresses developed not only during installation but also during service. HDPE has a relatively high thermal expansion rate, and numerous problems develop if the contractor installing the material is not familiar with its properties. For example, installing and seaming the material warm but without adequate allowance for shrinkage will cause the membrane to become taut on cooling down and to tear at the seams. Installing it cold and allowing the sun to warm it up later in the day will lead to wrinkling. Ideally the liner is installed in the afternoon, and it is at is greatest elongation. The last closing seams should be done so that excess tension is not developed in the membrane.

The seams on membranes greater than 80 mils should be beveled and lightly sanded or ground to remove a thin layer of oxide that affects good adhesion of the thermal bonding process. Prior to production seam welding, the contractor should qualify the weld procedure by test welds. The test specimens should be cut and pulled to failure. There have been instances where the welding procedure has been properly performed but the pull test failed because of the delivery of faulty material where the resins used to manufacture the membrane were improper. The weld must be stronger than the membrane in the pull test.

A representative of the tank owner should visually inspect all liner seams and ensure that seam preparation has been performed in accordance with specifications. A vacuum test using at least 5 psi is recommended to ensure that there are no leaks in the seams. Some contractors use a spark test which is comparable in effectiveness to determine the integrity of the seams. Destructive test coupons should be removed from the seams throughout the tank at specified intervals and tested to ensure that there is no problem with the liner manufacturing processes and the installation processes.

Sealing the membrane to the tank dead shell is the main problem of using HDPE. Figure 14.1.3 shows a number of ways of sealing the liner to the old tank bottoms. There are many variations of these details. However, there are some fundamental differences as well which are covered here. Unless the membrane comes up the side of the dead shell, it is unlikely that any free liquid will be contained by the liner. The thermal expansion as well as the settling and movement of the tank during loading and unloading will cause the liner to break loose from the caulking and will allow fluid leaks. When tanks with this detail have been in service and examined in this area, it is not uncommon to find the liner has shrunk inward as much as several inches. When caulk is used with HDPE, it does not stick.

When a ringwall is used, then the details shown in Fig. 14.1.4 are effective in producing a good leaktight containment.

Other liner materials.   Polyurethane fabric on a scrim (fabric reinforcement) is frequently used to line the interstitial bottom space. The advantage of these materials is that the integrity of the fabric seams is easier to guarantee and forming the material up the side of the dead shell is much easier. Because this material is not stiff and can be formed, there may not be a need to seal it by using a baton strip against the dead shell, provided that leak detection ports are below the top of the fabric.

Other methods tried have been to lay a geotextile fabric and spray a polysufide material to form a continuous coating of the old bottom

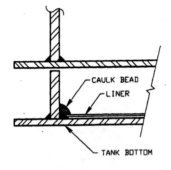

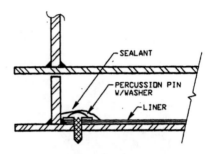

DETAIL A

- SUBJECT TO LEAKS
  @ CAULK BEAD AREA
- LOW COST

DETAIL B

- SUBJECT TO LEAKS
  @ PERCUSSION PIN
- LOW COST

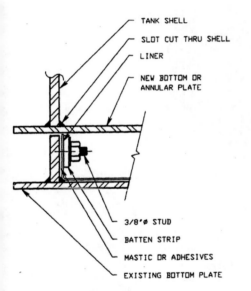

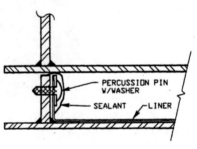

DETAIL D

- PERCUSSION PIN LIKELY
  TO ACCELERATE CORROSION
  AND ALLOW LEAKAGE UNDER
  LINER

DETAIL C

Figure 14.1.3 HDPE liner attachment details.

area and dead shell. There is not enough experience with this method to consider its merits or effects at this time.

Glass-reinforced plastic (GRP) liners have been used as a substitute for new steel bottoms. However, the quality and reliability of GRP liners are extremely sensitive to contractor knowhow and numerous additional variables that have caused many companies to move away from using these liners. Inspecting steel under the GRP liner becomes

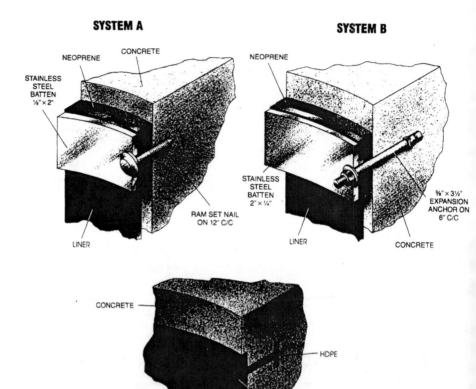

Figure 14.1.4   HDPE-to-ringwall attachment details. (*Courtesy of Gundle, Inc.*)

a significant problem, and removing GRP liners is also a significant task.

14.1.5   Miscellaneous Considerations

14.1.5.1   Grounding

When flammable liquids are stored, the tank should be grounded to prevent buildup of a charge on the tank and to reduce the possibility of arcing. NFPA 78 provides guidance for grounding tanks. When vertical metallic tanks over 20 ft in diameter rest directly on earth or concrete, they are considered grounded by the large contact surface between the bottom and earth. However, if a dielectric liner such as HDPE or other elastomeric liners are placed between the old and new bottoms or under the tank, then additional precautions need to be taken.

If the liner is placed inside the tank above the old bottom and if the new bottom is welded to the dead shell, then additional grounding is not necessary. However, if the new bottom is not welded to the dead shell or if a liner is placed under a tank bottom, then grounding should be achieved by bonding the tank through a minimum of two ground terminals at maximum 100-ft intervals along the tank circumference. Another way to ground the tank is to connect it to grounded piping; however, some piping is isolated from tanks when cathodic protection systems are present. Figure 14.1.5 shows one way of grounding a tank.

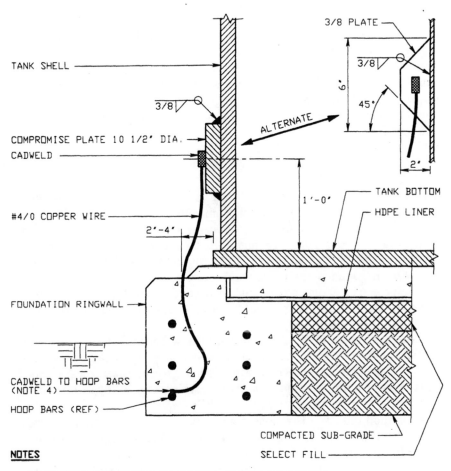

NOTES

1. FOUNDATION CONTRACTOR TO INSTALL COPPER WIRE PRIOR TO POURING RINGWALL.

2. TANK CONTRACTOR TO CADWELD COPPER WIRE TO SHELL.

3. TANK TO BE GROUNDED THROUGH A MINIMUM OF TWO LOCATIONS, AT A MAXIMUM OF 100-FT. INTERVALS ON TANK PERIMETER. (PER NFPA)

4. PER NFPA 780, 3-16.2 CONNECT (CADWELD) THE COPPER WIRE TO THE BOTTOM HOOP BAR. THIS BAR JUST BE AT LEAST 20 FEET LONG OF ELECTRICALLY CONNECTED REINFORCING BARS OF 1/2 INCH DIAMETER MINIMUM.

Figure 14.1.5 Dielectric RPG tank grounding method.

14.1.5.2 Bottom slope

Flat-bottom tanks can use a cone-up, cone-down, or single-slope configuration where the slope is 1 to 2 in per 10 ft of diameter. Most of the discussion revolves on the cone-up tanks because these are most common. The slope provides a means for leakage to easily find its way to the perimeter telltale system. Sometime radial grooves are used on either new foundations or concrete spacers. The leak detection system simply comprises the slots cut into the dead shell for viewing a leak.

When a cone-down bottom is used, care must be exercised in the design to ensure that the bottom or center sump does not remain in a pool of water, which causes accelerated corrosion. In fact, when concrete spacers have been used, in almost every case where the bottom was severely corroded, the bottom sump and plates were intermittently in water due to poor drainage or maintenance. In the cone-down designs, water that accumulates in the sump must be pumped out periodically.

14.1.5.3 Double bottoms

The design engineer should evaluate the condition of the *dead shell* that will exist between the old bottom and the new bottom. In most cases, the dead shell is in good condition with little corrosion except at the very bottom. However, should it be severely thinned or pitted, the following factors may apply. A weakened dead shell may not transfer the dead loads or seismic loads to the foundation. It may buckle or warp. It also might not withstand a buildup of hydrostatic pressure that could occur should there be a severe leak in the new bottom.

A common error in the installation of new bottoms is to attempt to install the bottom inside the tank; fillet-welding it to the interior surface of the shell, as shown in Fig. 14.1.6, is prohibited by API 653. This type of joint is subject to shell rotation and will fail either on first filling or after fatigue of the new fillet weld.

A common question that arises with the design is whether to caulk or weld the underside of the new bottom. Unfortunately, this issue is far from simple and involves a number of parameters. The newly issued API 650, Appendix I, states that welding or caulking a double bottom under the new bottom is required. However, the discussion in I.4.1 of Appendix I requires that an analysis and evaluation be performed if the new double bottom is not uniformly supported both inside and outside the shell.

The idea of the caulk or weld is to seal out moisture which may enter by a number of mechanisms. Rainfall may flow around the new chime and by capillary action migrate into the leak containment

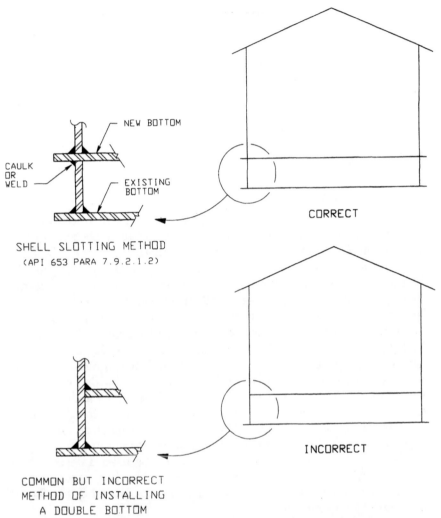

NEW BOTTOM

CAULK
OR
WELD

EXISTING
BOTTOM

SHELL SLOTTING METHOD
(API 653 PARA 7.9.2.1.2)

CORRECT

INCORRECT

COMMON BUT INCORRECT
METHOD OF INSTALLING
A DOUBLE BOTTOM

Figure 14.1.6 Correct and incorrect methods of retrofitting double-bottom tanks.

space. If the foundation is in flood areas, the flood level may rise
above the new tank floor, which will also flood the underside of the
new bottom. Slight thermal variation may cause a breathing of mois-
ture-laden air and cause a moisture pumping under the tank. Most
people agree that this space needs to be sealed off from the atmosphere.

   A single $\frac{3}{16}$-in fillet weld pass would be the most economical weld to
make. However, accessibility to the weld and control of the gap space
can often be a problem. To really do it right may require the welder to
use mirrors, slowing down the welding speed considerably. A properly

done weld should have a life expectancy approximately equal to that of the service life of the tank. Whether this is attained in practice is debatable.

To reduce initial capital expenditures, an alternative to seal welding is to seal the joint with caulking. Caulking does not have the life span of seal welding and is sensitive to surface preparation, flexure of the joint, sunlight, chemical environment effects, etc. However, it is probably cheaper to install on an initial-cost basis than to seal-welding the floor to dead-shell joint. Here are some comparisons between the two methods of sealing this space:

14.1.5.4 Triple bottoms

It has been reported that some tanks have up to four bottoms installed, above one another. Although it is certainly possible to add three or more bottoms, it should be realized that if the old bottoms ever have to be removed, the work will be more costly and difficult. As the number of bottoms increases, the likelihood of having to relocate tank appurtenances will increase and, of course, the usable volume of the tank is reduced. Although the structural effects are not really understood, however, the proof that they do not seem to be adverse is the large number of operating years of experience with three or more bottoms. One problem that has occurred is that the oldest bottoms continue to deteriorate, and since the void space was filled with sand, it washed out. This caused buckling of the dead shells and created a very difficult repair job. If it is determined that the second bottom is deteriorating and a new bottom is required, then consideration should be given to more effective corrosion prevention techniques.

It is recommended that triple bottoms not be used unless a thorough analysis is done, taking into account all loading conditions as well as assurances that the welds and materials are adequate for all anticipated load conditions.

14.1.6 Configurations of Tank Bottoms

14.1.6.1 Bottom configurations

There are numerous configurations for tank bottoms. The most basic classification is based on whether there is one steel bottom or two. The use of single or double steel bottoms is applicable to either new or existing tanks; however, the vast majority of double bottoms have resulted from a need to replace a single steel bottom that has reached the end of its useful life. All the systems described below are based upon the use of a release prevention barrier or liner to deflect leakage to a telltale system. The performance requirements for these bottoms are described in API Standard 650, Appendix I, *Undertank Leak Detection and Subgrade Protection.*

14.1.6.2 Single steel bottoms

General considerations. Several designs for single steel bottom tanks which have provisions for leak detection are shown in Fig. 14.1.7. In the steel single-bottom designs, the use of a concrete ringwall reduces settlement and movement and provides a good means of creating a pan or containment to collect leakage where it may be directed to telltale or visual leak detection systems.

Leak detection or containment may be installed on new tanks or existing tanks. Appendix I of API 650 applies to new installations since the standard is applicable to new tank construction only. API 653, *Tank Inspection, Repair, Alteration, and Reconstruction,* which applies only to tanks that have been in service, makes but a brief reference to leak detection or containment: "If a tank bottom is to be replaced, consideration should be given to installing a leak detection [telltale] system that will channel any leaks in the bottom to a location where it can be readily observed from the outside of the tank."* It was the intent of the API Committee to allow the criteria from Appendix I of API 650 to be used for either new or existing construction.

14.1.6.3 Double-bottom designs

A general practice is to install double-bottom tanks on existing tanks where the bottom is being replaced and to use a single bottom for new tanks. The details are shown in Fig. 14.1.8. This probably has the greatest economic payout if all things are considered. However, there are cases where double bottoms are installed on new installations. Sometimes this is the result of local regulatory compliance or an opinion that the double steel bottom offers the best, longest-lasting means for achieving effective leak detection.

14.1.7 Retrofitting Tank Bottoms

Tank bottoms are usually replaced with ¼-in-thick A36 steel plate. Annular plates or lap-welded bottoms are used. There are two fundamental methods of replacement. In the first method, the old bottom is cut loose, removed entirely, and replaced in kind. In the second, much more common method, a new double bottom placed 4 to 6 in above the old bottom is installed into the tank by the "shell slotting method."

---

*API 653, *Bottom Leak Detection,* paragraph 2.4.5, American Petroleum Institute.

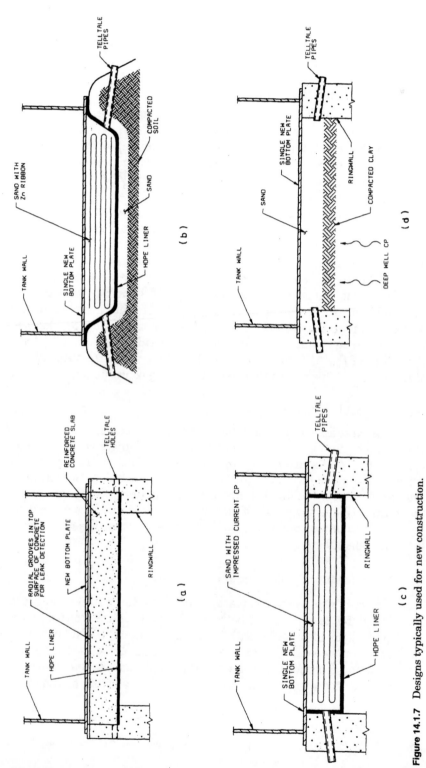

Figure 14.1.7 Designs typically used for new construction.

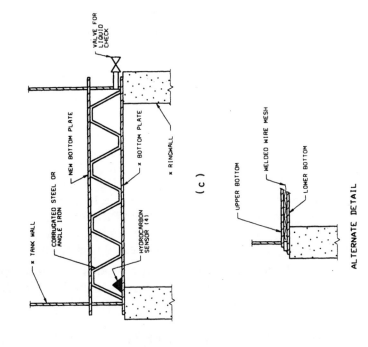

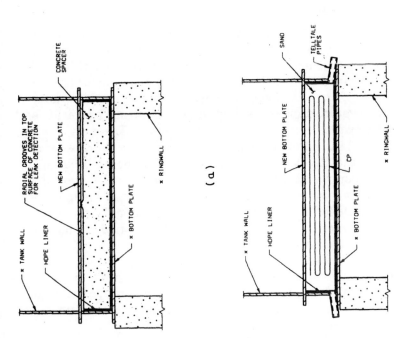

Figure 14.1.8 Designs typically used for retrofit.

579

The use of a double bottom has a number of advantages:

- The old bottom when used with a liner can function as a leak detection system.

- The new spacer which is located between the old and new bottoms provides a known substrate upon which corrosion prevention methods may be controlled and monitored.

- The interstitial space between the bottoms can function as a small containment space to hold a certain amount of leakage. For example, several tanks have developed critical leaks which have double bottoms in ethanol service. Due to the critical nature of the operation the tank was kept in service and the leakage returned to the tank. There was no possibility of groundwater contamination since the old bottom had been lined and tested to contain leaks.

- The new bottom provides a chance to easily reprofile the bottom slope for drainage, especially if concrete is used as the spacer material between the bottoms.

SECTION 14.2   CURRENT LEAK DETECTION TECHNOLOGIES FOR ABOVEGROUND STORAGE TANKS

14.2.1   Introduction

14.2.1.1   Historical background

Before the discovery of groundwater contamination at a number of sites that received national attention, tank owners and operators made reasonable but mostly ineffective attempts to prevent leaks in both new and existing tanks. There was no good way to determine the extent of leakage or the technology to detect leaks. When contamination of groundwater supplies resulted in public outrage and calls for reform, the growth of regulations increased rapidly. The most sweeping change affecting tanks occurred as a result of the Resources Conservation and Recovery Act (RCRA) which imposed broad new controls on underground tanks. The uniform federal rules applied to underground tanks brought together a wide array of disciplines and expertise in order for effective technologies to be used for the underground leak detection tank situation. RCRA applied a phase-in approach that required ever-increasing improvement in the detection and prevention of leaks.

Although the aboveground tank situation was recognized as potentially damaging to the environment in the same way as underground tanks, federal rule making did not capture any specific requirements for these tanks because there was no sweeping federal act upon which to tack on rules applicable to aboveground tanks. In addition, it was not, and still is not, clear as to the extent of environmental damage resulting from leaks from aboveground tanks. However, in recent years there has been an increasing call to nationally legislate rules for aboveground tanks.

While much of the groundwork for leak detection has been established as a result of the federal rules for underground tanks, much of it is inapplicable to aboveground tanks. An aboveground tank can be so much larger than an underground tank that the sensitivity levels required for effective leak detection are orders of magnitude greater.

Since the technology of underground tanks did not really apply to aboveground tanks on a practical basis, there was a leak detection technology vacuum during the late 1970s and early 1980s. At the same time, the need for leak detection on aboveground tanks was becoming acute, as the average age of the tank population was increasing and the potential for leaks was increasing in addition to an increasing number of litigations involving aboveground tank leaks. The petroleum industry applied the first successful leak detection technology to existing tanks in need of new bottoms, using a technology called a *double-bottom* tank design. In this design, the corroded old bottom was filled with sand or gravel, and a new bottom was welded 3 or 4 in above the old bottom. Extensive experience and use of this design led to variants that include leak detection liners placed over the old bottom which function to prevent galvanic corrosion and to ensure the containment of leaks and the use of concrete as the spacer material which has various construction benefits. In addition, liners have been used under new tanks. Any system using a liner of any kind is generically referred to as a tank with a release prevention barrier (RPB).

The use of RPBs including the double-bottom tank leak detection system has proved to be entirely adequate and reliable. In fact, the American Petroleum Institute saw enough use and justification of the double-bottom leak detection system to formulate performance criteria for leak detection systems in general, which are outlined in API Standard 650, Appendix I.

However, the double-bottom solution is inadequate to address all the leak detection problems for all situations. While it is considered to be essentially 100 percent effective in detecting leaks before they can do damage to the environment, installation of a double bottom requires that the tank be removed from service, cleaned, and hydrostatically tested after the bottom is installed. This is a very costly process and one that should not be undertaken unless a new bottom is really needed. Without installing a double bottom, the industry is left with no feasible options for detecting leaks for operating tanks.

A fundamental tenet of tank leak detection is to detect it as close to the source as possible. Unfortunately, one of the favored methods that state and local regulators require for the prevention of leaks is one that violates this tenet and is one of the least desirable methods from an engineering viewpoint—the use of monitoring wells. Monitoring wells are effective for leak detection only after massive quantities of leakage have occurred. In addition, they cannot discriminate leaks from any particular source, meaning that unrelated leaks may be indicated.

All the methods described in this chapter are compliant with the principle of detection close to the source. Methods dependent on large plumes for leak detection should be considered only as a means of last resort and are therefore not covered in this chapter. Such methods include monitoring wells, ground-penetrating radar, and slant-drilled wells under tanks.

14.2.1.2 The basic problem of leak detection in ASTs

One of the reasons that the innovative creative powers of U.S. technology were not focused on leak detection was, in part, due to a belief that the leak would be diverted by the ground and end up at the periphery of the tank shell where it would "come out on your shoes." This theory has been thoroughly discredited with experience and testing. However, to accomplish just that, the double-bottom tank was developed. Short of the double bottom, many companies tried the inventory control approach to tank leak detection. In the inventory method the net inflow and outflow from a tank are monitored, and the net difference is compared against the inventory change in the tank. But the problems proved to be significant, as indicated in the following quote.

> To illustrate how difficult it is to detect small leaks in large tanks, consider that today's tank gaging accuracy is sometimes as good as one part in two thousand. Thus, if we are exceedingly careful, from a typical 100,000 barrel tank, we might detect a 50 barrel loss. So in a 6-hour test, we might detect a leak as small as 850 gallons/hour; in a 24-hour test maybe 90 gallons/hour. To detect the "0.2 gallon/hr" leak proposed today in some regulations would take an out-of-service test period of over a year! Our traditional inventory control methods are perfectly suitable for their intended purpose, stock loss control. But leak detection by today's inventory control methods is simply technologically unfeasible.[1]

Because the volume of leakage measured over any reasonable test period (usually a couple of days) is so much smaller than the tank volume, the problem of leak detection becomes essentially one of statistics in which the leak or the signal must be separated from the unrelated events or noise that occurs during the measuring phase of the test. Figure 14.2.1 illustrates the concept more clearly.

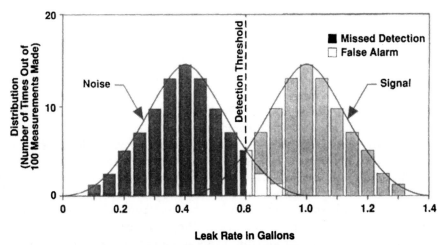

Figure 14.2.1 Leak-testing signal, noise, and threshold levels.

Because there are two real conditions that exist (the tank is leaking or it is not) and two possible measurements (either true or false) there are four possible outcomes that can result from a leak detection test, as shown in Fig. 14.2.2. The system performance is measured in terms of probabilities as follows:

$$P_d = \text{probability of detection}$$
$$P_f = \text{probability of false alarm}$$

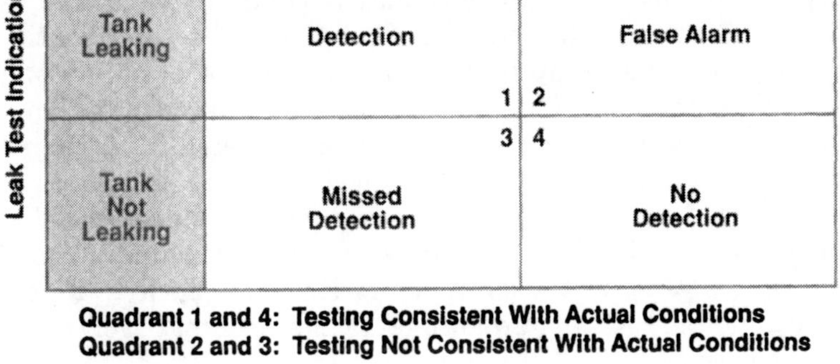

Actual Conditions

	Tank Leaking	Tank Not Leaking
Tank Leaking	Detection　　　　　1	2　　　　False Alarm
Tank Not Leaking	3　　Missed Detection	4　　No Detection

(Leak Test Indication)

Quadrant 1 and 4: Testing Consistent With Actual Conditions
Quadrant 2 and 3: Testing Not Consistent With Actual Conditions

Figure 14.2.2 Four possible outcomes from leak testing.

This leads to the identity

$$P_f + P_a = 1.0$$

The EPA has required that the probability of detection for underground tanks be at least 95 percent, resulting in a probability of false alarm of less than 5 percent. For aboveground tanks this level of performance using the same leakage rates is much harder to achieve owing to the large tank volume and small leakage rates.

The fundamental problem of leak detection is therefore to maximize the signal-to-noise ratio and pick an appropriate threshold signal for which a leak is indicated. Each of the test methods used has its own peculiar "noise" components. For example, in acoustic emissions the noise may result from background noise from product flows in other piping systems, structural changes in the tank occurring as a result of thermal expansion producing acoustic emissions, droplets of condensate splashing onto the liquid surface, etc. In volumetric tests, the main source of noise is the thermal fluctuations, evaporation, and structural changes resulting in tank volume changes. To separate the signal from the noise, a threshold value is chosen such that the smallest possible leak is detected with an acceptably low probability of false alarm. This concept is illustrated more clearly in Fig. 14.2.2.

14.2.1.3 Consequence of being wrong

As seen from the matrix of possible leak detection outcomes, there are two possible errors. Probably the more significant error to the tank owner would be the possibility of false alarm. A test indicating a leak would result in the shutdown of the tank, cleaning, and internal inspection. This can cost up to several hundred thousand dollars for just the tank cleaning costs alone in addition to the cost of lost production or capacity due to the tank's being out of service. This result would make the tank owner less likely to rely on leak detection in the future as a deciding factor in repairing suspected tanks. The other possibility is that a leak that exists is missed. Under these conditions the owner would not take the tank out of service, and it would continue to operate most likely up to 10 years before the next internal inspection. While the immediate costs to the owner are low, the long-term liabilities could be significant. One company recently concluded that remediation costs resulting from tank leaks were greater than $100,000 at least 90 percent of the time and greater than $1 million at least 5 percent of the time. This does not even take into account the ill will that is generated when this information is made available to the public and the costs of burdensome regulations that evolve as a result of these leaks.

14.2.2 Basic Leak Detection Methods

Five different basic technologies are covered including the use of RPBs, volumetric, acoustic emission, soil vapor monitoring, and enhanced inventory. A summary of these technologies can be found in Table 14.2.1.

14.2.2.1 Release prevention barrier system

This category of detection includes tanks with any liner including elastomers, double bottoms, or even properly designed concrete foundations which may function as a liner. The RPB concept is simply to lay an impermeable barrier between the tank bottom and the ground which directs leaks to the perimeter for visual indication. At the same time, leaks that do occur do not go into the environment. For this reason the use of an RPB is typically considered to have a probability of detection of 100 percent (whether or not this is strictly true). However, numerous possible sources of error can cause either false alarms or missed detection. As an example of a false alarm, some caulking compound was left in the double-bottom space of a heated tank, and, when the tank was started up a few weeks later, a leak was indicated by some oozing of hydrocarbon at the leak detection slots at the periphery of the tank. Careful investigation indicated that this was a false alarm. Sometimes the presence of water which enters above the liner and collects in the leak detection ports mimic false leaks.

Although there have been no studies or tests conducted which indicate probabilities of false detection or alarm as compared to other leak detection methods, the use of an RPB is considered the primary and most reliable method of leak detection. API has endorsed this technology and has collected wide industrial experience indicating satisfactory results. The use of an RPB is recommended for new tank construction by the American Petroleum Institute. Its installed cost on a new tank or a tank retrofitted with a double bottom is less than a few percent of the total capital costs.

A major advantage of this method of leak detection is that the tank operation is not affected by the test which is essentially continuous, needing only the watchful eye of an operator. It has also gained acceptance as the method of choice by both industry and regulator alike.

14.2.2.2 Volumetric and mass

Volumetric leak detection is applicable to fixed-roof tanks and to floating-roof tanks. However, the large inaccuracies introduced into these methods by the floating roof essentially make these methods impractical for tanks with floating roofs.

TABLE 14.2.1 Comparison of Leak Detection Technologies

	Release prevention barrier	Volumetric methods, level and mass	Acoustic emissions	Soil vapor monitoring	Enhanced inventory leak detection
Basic concept	Divert leak by using impermeable membrane to edge of tank or to wells for visual indication	Still tank and see if compensated level or pressure is dropping	Look for impulsive sound measurements that are characteristic of leaks	Inoculate a tank with tracer and attempt to detect its presence in soil beneath tank	Record all flows into and out of tank and compare net total flows to inventory change in tank for discrepancy
Effect on tank operations	Operation unaffected	Still tank Set liquid level to 3–5 ft Blind valves	Still tank Blind leaky valves	Operation unaffected Install probes under tank	Operation unaffected
Test time	Ongoing	24 to 48 h	4–8 h	Several hours to several weeks	Over some period such as a month
Instrumentation	None, visual detection	For level: level sensor and at least one vertical temperature array For mass: a differential pressure sensor and a temperature sensor for the DP cell and a supply of regulated air or gas	Array of accelerometers mounted externally or hydrophones mounted internally	A gas chromatograph or fiber-optic sensor; on active systems, a vacuum pump required	Flowmeters on all piping to and from tank. Tank gauging and level sensors

Weaknesses	Costly to retrofit	Requires tank shutdown Night measurements required Precision instrumentation required Inapplicable to floating-roof tanks unless they are removed Liquid level must be <10 ft	Hard for user to evaluate Few tests or documentation for effectiveness on ASTs	Costly for larger-diameter tanks Soil permeabilities under tanks unknown Water layers and pockets can interfere with diffusion	Accuracy Long test periods Tracking of flows and reconciliation of data Well-calibrated instruments Unknown accuracy dependent on site specifics
Advantages	No shutdown of tank—does not interfere with operations Highest probability of detection and lowest false alarms	Requires tank shutdown Relatively accurate	Requires tank shutdown Can pinpoint location of suspect leak	No shut down of tank—does not interfere with operations Most accurate other than RPB system	No shutdown of tank—does not interfere with operations Relatively low-cost—uses existing instrumentation
Relative contracted costs not including operating costs	Low if installed with new tank or with new double bottom Very high if retrofitted	Low	Medium	Medium on small-diameter tanks High on large-diameter tanks	Medium
API research results	No work done on this	Detectable level 1.9 gal/h for 24-h test in 117-ft-diameter tank filled to 10 ft 1.0 gal/h for 48-h test $P_d = 95\%$	Detected leak of 15–20 gal/h in 40-ft-diameter tank by 25-ft-deep water with a 2-mm test hole in bottom. Smaller test holes plugged with debri	Untested by API Vendors claim rates to 1 gal/day detectable	Untested by API Accuracy specific to site conditions

The basic idea of volumetric leak detection is that the temperature-compensated volume in a tank should slowly be reduced by leakage occurring through the bottom. By shutting all flow into and out of a tank, its volume and liquid level should remain constant if the thermal expansion of the tank and the liquid is accounted for. While in theory this sounds simple, there are many complications in practice. Even with the best instrumentation only 90 to 99 percent of the unwanted noise fluctuations can be removed. Therefore, longer test periods favor more accurate measurements. The three primary mechanisms affecting the volumetric techniques are

1. Product expansion and contraction

2. The expansion and contraction of the tank shell (both thermal and pressure-induced)

3. Measurement error associated with the instrumentation

Other sources of noise result from thermal gradients. There are thermal gradients in both the vertical and the horizontal directions. The tank itself is subject to volume changes caused by growth of the shell in the radial direction as a result of thermal changes and internal pressure caused by the liquid in the tank. Strong winds can deflect the tank shell, causing a variation in liquid level at constant stock volume. The diurnal volumetric changes caused by the thermal changes in the ambient air temperature and by solar radiation all impact the attempts to accurately measure leakage.

To date there are essentially two basic forms of volumetric leak detection. The temperature-level method depends on measuring the drop in level that results from a leak, after compensation for the thermal expansion of the liquid. The mass method measures the pressure head caused by the liquid level. These schemes are shown in Figs. 14.2.3 and 14.2.4.

The basic concept of the precision temperature-level approach to leak detection is to measure the liquid level accurately, compensate for thermal expansion or contraction, and look for a drop in the temperature-compensated level that results from a leak. The liquid temperature is determined by using a vertical array of temperature sensors to compensate for vertical thermal gradients.

In mass-measuring methods of leak detection, the pressure acting near the bottom of the tank is measured. The pressure corresponds to the mass above the measuring point and should be independent of liquid-level changes caused by thermal expansion. This was thought to be a significant advantage over the temperature-level method. However, testing has shown that both systems are comparable in accuracy.

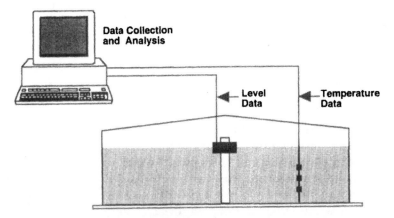

System monitors level of product. Wires connect the float to a computer. Temperature sensors monitor the horizontal and vertical extent of the product. Data from the temperature sensors are also transmitted electronically to the computer.

Figure 14.2.3 Volumetric level and temperature measurement system.

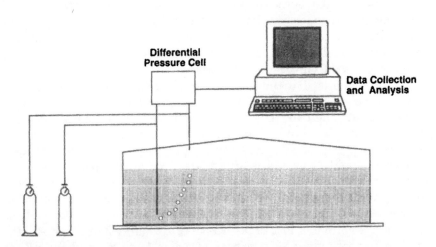

A "bubbler" system forces air or gas into a tube whose outlet is at the bottom of the tank and also into a second (or reference) tube whose outlet is in the vapor space. The differential pressure cell measures the amount of pressure necessary to force air through the tubes, and these readings are input to a computer for analysis.

Figure 14.2.4 Volumetric mass measurement system.

The mass-measuring systems should not be subject to the thermal variations that the level systems are. However, the transducers used to measure pressure are significantly affected by the sensor's temperature. The installed sensitivity of the differential pressure cell sensor to thermal ambient temperature changes is 3 to 5 times greater than the uncompensated volume changes measured by the level sensor.

Mass-measuring systems also do not compensate for any thermally induced product changes below the lowest pressure-sensing port. While this problem may be eliminated by design, it has been found in practice that this remains the limiting factor for accuracy of pressure measurements.

By keeping the liquid level low and making measurements at night when the horizontal thermal gradients are low, the level-and-temperature method is approximately equal to the mass-measuring method. The primary disadvantage of this method is the 24- to 48-h out-of-service stilling period required to conduct the test.

In spite of the problems, API testing[2] indicated that leaks as low as 1.9 gal/h could be detected in a 117-ft-diameter tank with a probability of detection of 95 percent conducted using a 24-h test. The detectable leak rate was reduced to 1 gal/h for a 48-h test.

14.2.2.3 Acoustic emissions

As applied to aboveground tanks, acoustic emission leak detection technology listens for the characteristic noises created by a leak from the bottom of a tank. The passive acoustic system operates essentially by detection and location of noise signals consistent with the types of signals emitted from tank bottom leaks.

The problem is that the intensity of the leaking noise signal is so low compared to that of other ambient noises that it is almost drowned out. The leak signal is barely detectible and decays rapidly with distance. In addition, multiple reflection paths confuse attempts to locate the leak signal. However, the development of sophisticated algorithms and signal processing have allowed this technique to be considered feasible. Information such as duration, propagation mode, and spectral characteristics can be used to reject noise contamination of the signals. In controlled tests[3] sponsored by the API, the smallest hole in the bottom of test tanks was approximately 2 mm in diameter with a corresponding leakage rate of about 15 to 20 gal/h in a 60-ft diameter tank. It is thought that the detectible leaks can be much smaller than this; however, this has not yet been verified by the API testing program.

Because this technology is sophisticated and not understood well, numerous problems arise in its successful implementation and application. It can have a significant probability of false alarms. Although several companies offer leak detection based upon passive acoustic emissions, very little technical information has been published about the performance characteristics of these systems or the nature of the signals produced or the probability of false detection. Figure 14.2.5 shows the typical setup for an acoustic leak detection test.

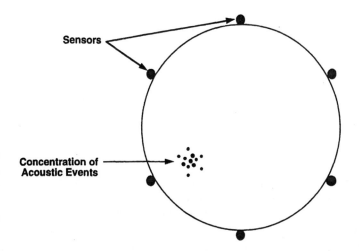

Sensors

Concentration of Acoustic Events

Impulsive acoustic events that exceed a certain threshold are plotted on a map of the tank floor. A concentration of these events indicates not only the existence of a leak but also its location.

Figure 14.2.5   Acoustic emissions testing.

When a tank bottom leaks, three distinct types of noise are created. The first is the *persistent leak signal* produced by turbulent flow through the hole in the tank bottom. This signal accompanies all tank floor leaks. The second is the *impulsive leak signal* produced by air bubbles collapsing in the backfill under the hole. The responses of sensors to these noises are similar, whether the sensors are placed in the liquid itself or are located on the exterior wall of the tank.

The impulsive leak signal is greater in magnitude than the persistent signal and is therefore easier to detect. However, the impulsive signal occurs only when air is entrained into the leak flow field beneath the tank. This means that the soil beneath the tank must be relatively well drained. If there is a condition of saturation of liquid beneath the leaking tank floor, then this signal does not exist. It is not known what the probability of existence for the impulsive leak signal is for tanks that have existing leaks.

The persistent leak signal is dependent on the soil conditions under the tank and on the flow rate of the leak. It is created by the flowing turbulence of the liquid through the hole in the bottom plate, and its characteristics are altered by the conditions beneath the tank. Multipath reflections of signals inside vertical cylindrical tanks can be stronger than the direct path signal itself, thus masking the signal in noise. The ability to discriminate the persistent leak signal because of multipath signal noise may be a limiting factor in its practical application in leak detection.

Noise generated in typical processing plants such as traffic, leaking valves, control valve flows, and piping noise can mask the acoustic leak signal. However, most of the noise is confined to a frequency below 10 kHz. The persistent leak signal may be detectable above the ambient noise levels if the signals are considered above 10 kHz. The impulsive signal is about 10 to 20 times larger than the average background noise. Most noise can be avoided by careful selection of measurement period, sensor location, and data collection and signal processing.

For using the impulsive leak signals, multiple sensors spaced at various locations allow computation of the time it takes for the impulse to traverse the distance from the leak to the sensor. From this computation the location of the leak signal can be estimated. Because of multiple reflections, various signal processing methods and algorithms are used to filter out or correct for this.

Some potential false signals are from

- Impulsive signals appearing to be from the floor but generated elsewhere such as in roof drains, pivoted float arms, and roof supports
- Impulsive signals generated by condensate dripping onto the product surface
- Impulsive signals generated by the movement of the floating roof
- High winds
- Thermal excitation of the tank shell

Interpretation of signal data affects the potential for false alarms:

- The data collection and analysis have a significant impact on the potential for false acoustic leak events.
- Identifying the propagation mode for signals received is critical to locating the source of the signal.

One company recently examined 345 tanks using the acoustic emissions leak detection method. Of the tanks 21 were indicated to be leaking, of which 19 were given an internal inspection. Of the 19 potential leakers inspected, 16 had leaks. This gives a probability of detection of .84 and of false alarm of .16.

14.2.2.4 Soil vapor monitoring

Soil vapor monitoring techniques depend on sampling the vapors from beneath a tank floor and testing them for the presence of the stock liquid. Although these methods have the potential to detect leaks in aboveground tanks, results may be dependent on the permeability of the soil. In tests conducted by API on tanks with oil sand foundations, the low permeability of this backfill material prevented the target vapors from reaching the sensors. However, longer tests or closer sensor spacing may yield improved results.

Another factor that affects the use of soil vapor monitoring is the presence of a water heel in the tank bottom. With water instead of stock leaking from the bottom, the detection of stock components would be more difficult, if not impossible. This can be a significant problem because many petroleum tanks have several inches of water in the bottom as a result of entrainment into the tank with the incoming flow or atmospheric condensation.

The following basic features apply to these systems:

- The number of sampling points required is a function of the tank diameter, the decay time of the target vapor, the permeability, and the diffusivity of the backfill. Some backfills provide poor conditions for diffusion of the target vapors.

- Obtaining of vapor samples has been based upon both diffusion of the vapors into the sample pipes and application of a vacuum to withdraw vapor into the sample pipes. If the system depends on diffusion, it is called a *passive system*. If the system uses vacuum to assist the withdrawal of undertank vapors, it is called an *active system*.

- If a water layer is present in the tank bottom and the target vapor cannot adequately penetrate it, then the concentration of the target vapors beneath the tank will be reduced to an impractical low level.

- The target vapor must be unique to the background environment of existing compounds beneath the tank floor. A survey to assess the uniqueness of the target vapor should be conducted prior to leak-testing the tank.

One of the most successful variations of the soil vapor method has been with the use of chemical marker compounds. A marker compound is injected into the stock, and soil vapor monitoring is used to detect the presence of this marker. Although chlorinated fluorocarbons such as Halon 1202 or 2402 have been used, the markers of choice are now perfluorocarbons because they do not affect the ozone depletion in the atmosphere, are nontoxic and nonflammable, and have little effect on the stored liquid. Other uses for these compounds are typically as electrical parts-cleaning solvents. Markers are injected into the tank at rate of 1 to 10 ppm. The typical vapor pressure of markers is 40 to 300 psi, making it extremely volatile. Since these compounds do not occur in nature and are not in typical stocks, the detection of the marker is an indication of a leak. Manufacturers of these systems say that they can detect 1 gal/day in any size tank. Tests may take up to 2 weeks to perform. To increase the probability of detection and reduce the probability of false alarms, the suppliers of this technique run a leak simulation test using a marker compound different from the leak test marker so that it cannot contaminate the soil with the selected marker. The equivalent of 5 to 10 gal of a simulated leak is injected, and the response is recorded.

Some issues associated with this technique concern the addition of a foreign substance to the tank products. In most cases there should be no detrimental effect. Studies are being conducted on the presence of these markers in jet fuels and the effect on jet engines. However, many tanks storing jet fuels have been tested using these techniques.

The cost of these systems goes up approximately as the area of the tank bottom or the square of the diameter because the sensor coverage is constant (i.e., one sensor may have to cover a fixed area of about 500 ft^2). The advantage of the marker system is that it is highly sensitive and depends on unique markers which do not exist underneath tanks. Another very important advantage is that the test can be conducted while the tank is in service, unlike the volumetric or acoustic emissions method.

14.2.2.5 Enhanced inventory methods

Figure 14.2.6 shows the schematic arrangement of the enhanced inventory method. The concept behind this method of leak detection is very simple. All flows into or out of the tank are added, and the net difference is compared to the volumetric change in the tank. Without a leak

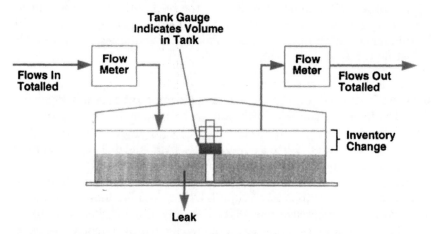

Leak detection by inventory reconciliation is an accounting process. The total flow in less the total flow out should equal the inventory change. If it does not, then a leak is indicated.

Figure 14.2.6 Enhanced inventory leak detection.

the inventory change should match the net difference between total receipts and total disbursements. Of course, all the factors that create noise in the volumetric methods apply to this system. In addition, there is the instrument error introduced by all the inflow and outflow metering. With enough readings statistical analysis can show that the random errors tend to cancel while the systematic errors do not.

Very few actual test data have been collected, and careful reports of this method have not been documented for aboveground storage tanks. More work is needed to establish the testing requirements as well as the associated probabilities of detection and false alarm.

This method is important for the industry to develop because most tanks already have all the instrumentation needed to gather the appropriate data. By development of guidelines based upon the type and accuracy of existing information, a methodology could be established to determine what level of leak detection is possible given the local specific instrumentation and tank size as well as operations.

14.2.3 Future of Leak Detection

The tank with a release prevention barrier has the best and only leak detection system required. However, it is not practical or cost-effective to shut down all tanks without them and to require that an RPB be used. However, for tanks being retrofitted with new tank bottoms or for new tanks, consideration should be given to installing a release prevention system. Fortunately, the premium for a release prevention barrier is small in comparison to the cost of either a new tank or a tank retrofitted with a double bottom.

Because most chemical and petroleum companies have adopted the inspection requirements of API Standard 653, which requires internal inspections at intervals of less than 10 years for the most part, severely corroded and leaking tanks will ultimately be repaired. However, in the interim, other better measures are needed to determine the presence of a leak. This is the primary use of the methods described in this chapter.

The selection of methodology can be extremely challenging, however. All the variables affecting each technology must be considered. For example, if a test is to be conducted in a site with high degrees of process noise generated (such as from wind) in the same frequency range as acoustic emissions, it might be prudent to rule out acoustic emissions leak detection. If the tank diameters are very large or the hydrogeology of the soil and foundation conditions mitigate against uniform soil permeability under the tank, then soil vapor monitoring might be ruled out. Each of the technologies must be considered in connection with the site-specific factors in order to establish the best.

method(s) to use. Indeed, API has recommended considering multiple tests using different, independent technologies to confirm leakage before the tank is taken out of service. Although not enough data collection or testing has been done to establish the effectiveness of this approach, it would seem to be a relatively cheap insurance that intuitively boosts the probability of detection and reduces the probability of false alarm where the stakes are high.

Because these technologies are all emerging and little is really known about their effectiveness for aboveground tanks, it is important that legislation regarding these technologies be postponed until each technology advances and is thoroughly understood. It would be a mistake to take the work for underground tanks and apply it to aboveground tanks without adjusting it for the differences. More institutional work such as that being undertaken by API and ANSI needs to be done regarding probabilities of detection and of false alarm as well establishing the acceptance levels for these criteria. This can only be accomplished by proper test facilities and the concerted efforts of individuals, the vendor community, the tank operators and owners, and any other stakeholders.

References

Section 14.1

1. B. Colle, J. Lierly, J. Meyers, and R. Rials, "Aboveground Storage Tanks: A Cost Effective, Life Cycle Approach for Protection of Soil and Groundwater," *Proceedings,* vol. 2, *Pipelines, Terminals and Storage,* '95 Conference, Houston, TX, January 31–February 2, 1995.
2. NFPA 78, *Lightning Protection Code,* 1989 edition, National Fire Protection Association, Quincy, MA.
3. R. Rangle and J. Lindsay, "System Provides Secondary Containment for Aboveground Storage Tanks," *Oil and Gas Journal,* October 29, 1990.
4. David H. Park, "Cathodic Protection of Aboveground Tank Bottoms with Secondary Containment Liners," *Material Properties,* April 1991.
5. Timothy C. Dyring, "Liner for Aboveground Storage Tanks: Secondary Containment, Leak Detection," *Geotechnical Fabrics Report,* September/October 1968.

Section 14.2

1. Jim Seebold, "Small Leaks, Large Tanks," presented at the NPRA Maintenance Conference in San Antonio, TX, May 24, 1990.
2. S. D. Curran and P. E. Myers, "Leak Detection for Aboveground Storage Tanks," presented at the University of Wisconsin, Madison, Department of Engineering Professional Development, 1994.
3. American Petroleum Institute, "A Guide to Leak Detection for ASTs," draft, October 1994.
4. American Petroleum Institute, "An Engineering Evaluation of Acoustic Methods of Leak Detection in Aboveground Storage Tanks," January 18, 1993.
5. American Petroleum Institute, "An Evaluation of a Methodology for the Detection of Leaks in Aboveground Storage Tanks," December 15, 1993.

6. American Petroleum Institute, "Chapter 16, Mass Measurement, Section 2, "Standard Practice for Mass Measurement of Liquid Hydrocarbons in Vertical Cylindrical Storage Tanks by Hydrostatic Tank Gauging," draft, Washington, January 27, 1994.

7. P. Eckert, N. Fierro, and J. Maresca, "The Acoustic Noise Environment Associated with Aboveground Storage Tanks," Vista Research, Palo Alto, Calif.

8. N. Fierro, P. Eckert, and J. Maresca, "Experimental Evaluation of Volumetric Methods of Leak Detection in Aboveground Storage Tanks," Vista Research, Palo Alto, Calif.

9. H. A. Raiche, "An Update—Tank Testing Utilizing Acoustic Emission," Ashland Petroleum Company, presented to the API Operating Practice Committee, fall session, October 1994.

15

Miscellaneous Topics

SECTION 15.1   SAFE TANK ENTRY

15.1.1   Need for a Tank Entry Standard

15.1.1.1   Overview

More and more federal and environmental regulations require that personnel enter the interiors of tanks so that the tanks can be inspected. However, the risk of accidental deaths resulting from entering confined spaces such as storage tanks is substantial and has increased. As a result, federal rules governing such activities, such as OSHA Confined Space Entry Rules,* have proliferated.

These and other considerations caused the National Fire Protection Association (NFPA) in late 1992 to initiate a new nationally enforceable standard that would ensure the fire safety aspects of tank inspections. NFPA decided to create this standard primarily because existing standards that cover tank entry and cleaning were in the form of advisory documents and were unenforceable.

However, the creation of a new nationally enforceable document developed for the general industry could create serious problems for the petroleum industry by causing the requirements for a general service standard to apply to the unique problems of the petroleum industry tanks. It would also create duplication of standards-writing efforts and a confusing situation that might even prevent petroleum industry tank owners or operators from inspecting their own tanks.

At the time that the NFPA standard was proposed, the American Petroleum Institute (API) already had a standard that covered tank entry and cleaning, API Publication 2015, *Safe Entry and Cleaning of*

---

*29 CFR 1910.146.

Petroleum Storage Tanks, 4th edition. However, this document was not enforceable on a national basis and not up-to-date enough to address all the current rules regarding safe entry into tanks.

API decided to revise API Standard 2015 to preclude NFPA from creating a competing standard which might be damaging to the petroleum industry.

15.1.1.2 Scope

The fifth edition of API Standard 2015 is a comprehensive review of all factors that typically affect aboveground storage tank entry and cleaning, from decommissioning to recommissioning. The 5th edition applies to atmospheric aboveground storage tanks (up to 15 psig). It covers safe practices for preparation, emptying, isolation, ventilation, testing, cleaning, entry, hot work, permitting, and recommissioning associated with internal tank work. It applies to all chemical, petroleum, petrochemical, and related tanks. It does not apply to underground tanks, pressure vessels, cryogenic and refrigerated tanks, or process equipment and tanks.

The standard is intended to be consistent with and, if necessary, more stringent than federal regulations such as OSHA's Confined Space Entry Rules. In the event that conflicts arise, the standard yields to the regulations.

15.1.1.3 Format

The standard is divided into three principal parts. Section 1 comprises the first segment. It covers the scope and purpose of the document. It also includes a comprehensive reference list of applicable federal rules, industry standards, and definitions. Table 15.1.1 lists some of these references.

The second part is written in mandatory language. Its provisions are such that local and state agencies can easily enforce the document by incorporating it by reference. This section is written very generally and should be applicable to all tanks regardless of the industry in which they are used.

The appendices to the standard comprise the third part. They satisfy the specific needs of the petroleum industry and provide useful information of an advisory nature applicable to any industry. This section deals with some unique hazards associated with the petroleum tanks as well as some common methods of dealing with them.

15.1.2 Basic Requirements of API 2015

An overview of the areas covered by the standard is presented in Fig. 15.1.1. The format of the document roughly follows the typical sequence of events that would be implemented as they occurred in the field.

TABLE 15.1.1 Reference Publications

	ACGIH
	Threshold Limit Values for Chemical Substances and Physical Agents and Biological Exposure Indices and Documentation (published annually)

	ANSI
Z49.1	*Safety in Welding and Cutting*
Z88.2	*American National Standard for Respiratory Protection*

	API
2003	*Protection against Ignitions Arising out of Static, Lightning and Stray Currents*
2009	*Safe Welding and Cutting Practices in Refineries, Gas Plants and Petrochemical Plants*
2026	*Safe Descent onto Roofs of Tanks in Petroleum Service*
2027	*Ignition Hazards Involved in Abrasive Blasting of Atmospheric Storage Tanks in Hydrocarbon Service*
2202	*Dismantling and Disposing of Steel from Aboveground Leaded Gasoline Storage Tanks*
2207	*Preparing Tank Bottoms for Hot Work*
2219	*Safe Operation of Vacuum Trucks in Petroleum Service*
2220	*Improving Owner and Contractor Safety Performance*

	NFPA
70	*National Electrical Code*
77	*Static Electricity*
325M	*Fire Hazards Properties of Flammable Liquids, Gases and Volatile Solids*

	OSHA
1910.38	*Employee Emergency Plans and Fire Prevention Plans*
1910.132-6	*Subpart I, Personal Protective Equipment (General Requirements)*
1910.134	*Respiratory Protection*
1910.146	*Permit-Required Confined Spaces*
1910.147	*The Control of Hazardous Energy (Lockout / Tagout)*
1910.251	*Subpart Q, Welding, Cutting and Brazing*
1910.1000	*Subpart Z, Toxic and Hazardous Substances (PELs)*
1910.1028	*Benzene*
1010.1025	*Lead*
1926.62	*Lead Exposure*

	NIOSH
	Manual of Analytical Methods

15.1.2.1 Administrative controls

The first and most important aspect of tank entry is the up-front, pre-planning and scoping studies. During this phase, administrative controls are put into place such as written procedures; permitting systems; development of checklists; prejob discussions with all involved

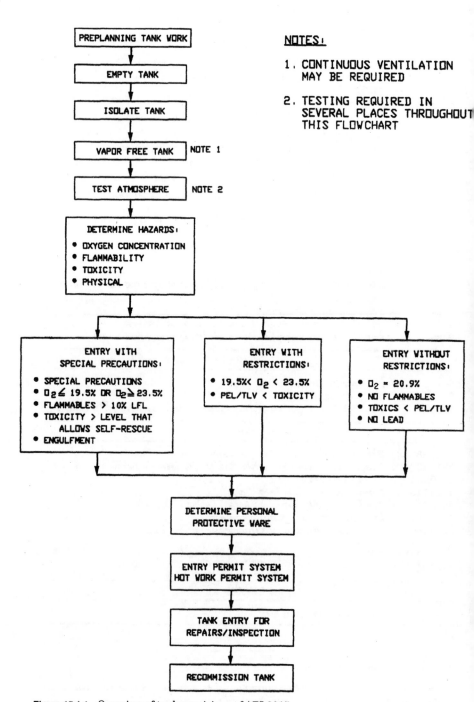

NOTES:

1. CONTINUOUS VENTILATION MAY BE REQUIRED

2. TESTING REQUIRED IN SEVERAL PLACES THROUGHOUT THIS FLOWCHART

PREPLANNING TANK WORK

EMPTY TANK

ISOLATE TANK

VAPOR FREE TANK   NOTE 1

TEST ATMOSPHERE   NOTE 2

DETERMINE HAZARDS:
- OXYGEN CONCENTRATION
- FLAMMABILITY
- TOXICITY
- PHYSICAL

ENTRY WITH SPECIAL PRECAUTIONS:
- SPECIAL PRECAUTIONS
- $O_2 \leq 19.5\%$ OR $O_2 \geq 23.5\%$
- FLAMMABLES > 10% LFL
- TOXICITY > LEVEL THAT ALLOWS SELF-RESCUE
- ENGULFMENT

ENTRY WITH RESTRICTIONS:
- $19.5\% < O_2 < 23.5\%$
- PEL/TLV < TOXICITY

ENTRY WITHOUT RESTRICTIONS:
- $O_2 = 20.9\%$
- NO FLAMMABLES
- TOXICS < PEL/TLV
- NO LEAD

DETERMINE PERSONAL PROTECTIVE WARE

ENTRY PERMIT SYSTEM HOT WORK PERMIT SYSTEM

TANK ENTRY FOR REPAIRS/INSPECTION

RECOMMISSION TANK

Figure 15.1.1 Overview of tank provisions of API 2015.

employees and contractors; qualification standards for various key personnel; training, hazardous substance testing, equipment methods, and frequency requirement; the use personal protective equipment and training; and hazards assessment.

The standard provides useful guidance and requirements such as references from federal regulations that require specific training, how to work with contractors, and the nature of the qualifications for various job functions. In most incident investigations, it is these critical administrative controls that seem to be missing or poorly planned, which would have prevented the mishaps.

15.1.2.2 Classification of hazards

Because of the numerous combinations of hazard conditions encountered in tank entry work, the standard categorizes them so that the task of planning the work can be simplified and dealt with prior to beginning the actual work.

There are four recognized hazards associated with entry of petroleum tanks:

1. Fires and explosions

2. Oxygen deficiency or enrichment

3. Toxic substances

4. Physical hazards

Table 15.1.2 is a listing of some hazards that are often unanticipated and that support the concept of up-front research of hazards as covered in the administrative controls prior to the undertaking of tank cleaning work. Probably the hazards most difficult to assess have to do with the nature of the vapors and fumes in the tank. Since these cannot be assessed without the use of instrumentation, the standard relies heavily on atmospheric testing.

To classify hazards, the standard provides three degrees of hazardousness, as shown in Fig. 15.1.1:

1. Entry with special precautions

2. Entry with restriction

3. Entry without restrictions

Entry with special precautions. This is the most hazardous condition found inside storage tanks. It corresponds to OSHA's Permit Required Confined Space Entry Rules, and OSHA's provisions should be implemented completely in this case. This condition of hazard classification applies if any one of the following conditions exist:

TABLE 15.1.2   Various Tank Entry Hazards

This brief background on the hazards of entering and cleaning tanks is by no means a complete listing, nor does it recommend, endorse, or provide any guidance. It is simply the noting of the disguised form which some of the hazards may take in an attempt to point out the importance of careful research of all potential hazards prior to attempting tank entry and/or cleaning operations.

Fires and Explosions

The potential for the presence of flammable substances inside "cleaned" tanks is high. There have been many accidents resulting from the existence of flammable liquids or vapors that were thought not to be present. Here are some examples:

- Soil under tank bottoms may be contaminated with hydrocarbon. If the bottom plates have any cracks, pinholes, or leaks, then flammable substances can migrate back through the bottom plates and into the tank after cleaning. Work on bottom plates warrants the use of special procedures.
- Floating-roof pontoons are the source of trapped flammable liquid and vapors. These must be carefully tested to ensure that all vapor and all liquid have been removed. At least one fire has been initiated by removal of fillet-welded patch plates that had liquid trapped between the plates. Aluminum pontoons may contain trapped liquids that enter through pinholes caused by corrosion or weld seam cracks. Several explosions have resulted when welders worked on these pontoons, assuming that no flammable liquid was present. The pontoons should be gas-tested before any hot work commences inside the tank. Other internal piping, fittings, support columns, and internal appurtenances are a high-risk source of flammable vapors and liquids.
- Floating-roof seals may contain substantial amounts of absorbed flammable liquids.
- Although relatively uncommon, oxygen-enriched atmospheres resulting from leaking welding cylinders and equipment are potentially dangerous because the oxygen concentration lowers the flammable range of the vapors.
- Since condensates form and solidify to the underside of roofs and on rafters, they are a common source of tank fires.
- Pyrophoric substances should be dealt with using specialized procedures developed with personnel experienced in this area.
- Any form of high-velocity ejection of material into the tank interior such as fogging with water, steam, sandblasting, and water spraying has the potential to generate static electricity. Particularly vulnerable are operations involving cleaning when the vapor space passes through the flammable regime.

Toxic Substances

Toxicity and symptoms for the materials to be cleaned from tanks should be thoroughly understood. Commonly found toxic substances are

- Hydrogen sulfide
- Lead hazards from paint removal, leaded gasoline, tanks that have lead compound deposits on the tank walls
- Dusts resulting from grinding, paint dust, insulation particles including asbestos
- The stored chemicals and sludges
- Welding fumes and the compounds generated as a result of the high temperatures and the residual deposits decomposing

TABLE 15.1.2   Various Tank Entry Hazards (*Continued*)

Physical Hazards
▪ Tank internal components such as mixer props, steam heating coils, sumps, and piping are hazards that may cause trips and falls. These types of physical hazards are often amplified because of poor lighting.
▪ Heat stress is a high-potential hazard during hot weather because of the enclosed space and the need for protective clothing and respiratory gear.
▪ Psychological effects can be important. A welder required to wear air-supplied respiratory equipment may be out of his or her comfort zone. Under stress, workers may make mistakes and use poor judgment.
▪ Missing or improper bracing for heavy objects that can fall on people such as floating roofs, temporary scaffolding, swing lines, tank shells prior to having wind girders installed, or any heavy object used during construction.

- Oxygen concentration outside the limits of 19.5 to 23.5 percent

- Flammables concentration greater than 10 percent lower flammable limit (LFL)

- Toxicity greater than a level that would allow for self-rescue

- An engulfment or physical entrapment hazard

Entry with restrictions.   This intermediate hazardous condition applies if either of the following conditions is true:

- The oxygen concentration is between 19.5 and 23.5 percent.

- The toxic substances concentrations are above the PEL/TLV* but not so high as to inhibit self-rescue efforts.

Entry without restrictions.   This condition allows tank entry without personal protective equipment or other restrictions. A permit must still be issued. The conditions allowing this classification are the simultaneous presences of

- Oxygen concentration at 20.9 percent

- Flammable vapors at 0 percent LFL

- Toxic substances at or below PEL/TLV limits prescribed by the more stringent of employer, local, state, or federal rules

It should be understood that the worst hazardous condition applies (entry with special precautions) to all tank entries unless it is proved

---

*PEL is OSHA's designated *permissible exposure limit*; see 29 CFR 1910.1000. TLV is the American Conference of Governmental Industrial Hygienists' *threshold limit value*.

otherwise by atmospheric testing. Any tank that has been cleaned and reclosed and rendered inactive for any period of time must be treated as though it required entry with special precautions, until it is retested.

15.1.2.3 Atmospheric testing

To determine what level of hazard exists and when it exists in the confined space within a tank, the standard relies upon atmospheric testing to help assess the hazardous condition described above. To ensure that correct data are gathered on the hazardous nature of the atmosphere, the standard requires the following:

- Intrinsically safe equipment must be used (good for class I, division I, group D).
- Each instrument must be calibrated and its calibration documented.
- The user must understand the limitations and behavior of all the instruments.
- The user must be qualified and trained in the proper use of the equipment.
- The tests must be run in the following sequence: oxygen, flammable, and toxic.
- Redundant testing must be done when ventilation equipment is stopped.
- Redundant monitoring of the atmosphere inside the tank must be done.

The appendices give the peculiarities of these types of instruments as well as factors affecting accuracy. A qualified person determines what types of atmospheric testing are done and how often they are performed.

15.1.2.4 Preparing the tank for entry

In addition to hazard assessment and testing requirements, the standard specifies conditions for performing the actual work itself.

Emptying the tank. The standard gives guidance for emptying various kinds of tanks including fixed-roof, external floating-roof, and internal floating-roof tanks.

Isolation. All lines must be blinded off or separated from the tank so that there is no possibility of leakage into the tank. Cathodic protection systems should be turned off, electrical systems locked and tagged out, and all cleaning equipment grounded to the tank. Specifications for electrically bonding the tank are given.

Control of ignition sources. Specifications for electrical equipment used in the work are given in addition to grounding and bonding requirements. The standard requires that if an electrical storm is in progress or imminent, all work be stopped.

Vapor freeing. Vapor freeing of a tank is the process of removing atmospheric breathing hazards. Most often this is accomplished by ventilation equipment using air as a diluent. However, the nature of the tank contents and local regulations may restrict this application. Figure 15.1.2 gives some guidance on tank ventilation or vapor freeing.

Once air-moving equipment is shut down, toxic or flammable vapors can quickly reestablish themselves in tank interiors. This can

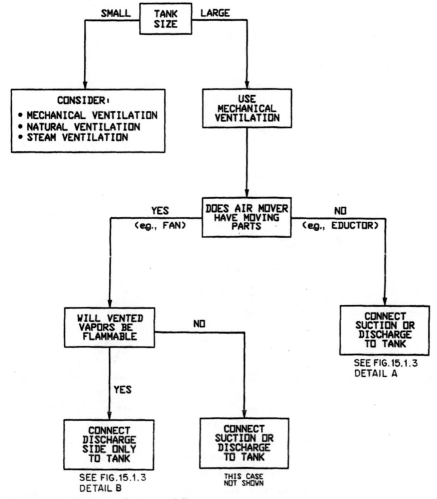

Figure 15.1.2   Tank ventilation guidelines.

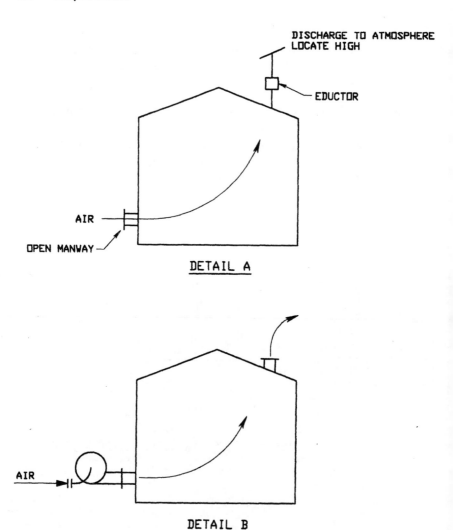

Figure 15.1.3 Typical ventilation arrangements.

result from the existence of sludge, pipe columns that retain liquids, cracks in bottom plates that allow liquids or fumes from under the tank to enter, leaking pontoons, and as many other possibilities. For this reason, the standard cautions apply: Ventilation equipment should be left on during the entire time that work is being performed if there is any doubt as to the ingress of hazardous fumes.

Whenever a tank is vapor-freed and sealed up for any period of time, it must be assumed that its atmosphere has become hazardous again unless testing shows it to be otherwise.

Personal protective equipment.   Once hazards have been assessed and classified based upon testing, the work may begin. In case of entry with special precautions or entry with restrictions described above, workers will use various levels of personal protective equipment:

- Impervious clothing
- Reflective clothing (for visibility)
- Flame-resistant coveralls
- Clothing materials that do not produce static electricity ignition sources
- Hard hats, gloves, boots, goggles, etc.
- Respirators

The employer must provide this equipment. The employee shall be trained in the use of respirator equipment. The employer must maintain a respiratory protection program in accordance with 29 CFR 1910.134. Every respirator must be fit-tested to the individual using it. The individual must also be trained in its use. Detailed requirements for breathing and breathing air supplies are given in the standard.

Permitting.   The standard outlines a system to permit entry by anyone into a tank. It essentially follows the rules given in OSHA 29 CFR 1910.146 for confined-space entry. It also gives the requirements for hot work permits. It cautions that other kinds of work such as liner installation or spray painting may require other forms of control.

Recommissioning.   Precautions for safely refilling a tank are given.

15.1.3   Miscellaneous Guidance

The appendices cover commonly occurring hazards in petroleum tanks. Some of these are

- Hydrogen sulfide
- Lead hazards
- Petroleum fumes
- Welding fumes
- Lead-based paints
- Pyrophoric substances
- Toxic dusts
- Physical hazards

Guidance is given concerning vapor freeing of tanks, control of ignition sources, hazards specific to tank type (external floating-roof tanks, cone-roof tanks, etc.), the space above floating-roof tanks but within the shell, problems in dealing with leaded gasoline tanks, sludge removal, and so forth.

One of the most directly usable appendices is the planning checklist. It covers all aspects of the work from decommissioning to recommissioning.

15.1.4 Conclusion

Entry into aboveground storage tanks will become more common as the inspection intervals become more frequent, as a result of increasing frequencies of tank internal inspections. As a result, the probability of injuries and deaths has increased. However, the new edition of API 2015 can be used to minimize these risks. While the use of API 2015 is not mandatory, it is written so that it may very easily become so if local regulatory agencies incorporate it by reference. While this is expected to be the case, it would be worthwhile for anyone involved in this kind of work to read, comprehend, and consider all aspects of this standard simply to reduce the exposure to risk. The bottom line is that using the standard will go a long way toward reducing injuries, accidents, incidents, and deaths associated with this kind of work.

SECTION 15.2   TANK DEGASSING, CLEANING, AND SLUDGE REDUCTION PRINCIPLES

15.2.1 Overview of Tank Bottoms and Sludges

15.2.1.1 Role of tank cleaning today

Because more tank owners are concerned with leaks and contamination, more and more internal tank inspections are being done. The use of API 653, which is largely voluntary but is also being regulated as a minimum requirement in many states, causes the requirement for an increased number of internal inspections. To conduct the inspection, the tank must be desludged and cleaned. However, this is not the only reason that dealing with sludges has become a necessity. The concerns of health and safety play a key role in how tank cleaning is done. For example, API and ANSI have standards which prohibit tank entry above 10 percent of the lower flammable limit. However, a number of companies prohibit entry altogether if there is any level of flammables in the tank. This has driven the cleaning industry toward automated nonentry tank-cleaning methods.

15.2.1.2 Why sludges form

Fluids entering a tank become quiescent, and the liquid velocities become very low in spite of the liquid entering and leaving the tank. In other words, the tank acts as a separation chamber for finely suspended particles. If the particles or droplets have a lower specific gravity, they will tend to phase-separate at the top of the liquid surface. If they are more dense than the bulk liquid, they tend to drop out.

Some operations such as dewatering depend on this fact to make a separation between the liquids and the solids. For example, an oil-water separator skims the oil and light floc that forms at the surface of the tank, and sediment is collected from the bottom. However, special provisions are often made to collect the sludges or heavy phases that form at the bottom. Sometimes, a tank will have a conical bottom that allows the solids or sludges to migrate to a low-point outlet where they can be pumped out. Other designs use a rake that sweeps the bottom of the tank toward a center withdrawal system. This design is often used in municipal water treatment plants where sludges are removed from the liquor at fairly high rates. In fact, in these operations, alum, ferric chloride, sulfate, or lime is used to form insoluble precipitates that will drop out of the sewerage. Air and agitation permit flocculation of disbursed suspensions to occur which can be skimmed from the surface.

However, in most cases sludges are a nuisance or even a serious problem. The petroleum industry is besieged by a variety of sludge problems. Producing tanks and, particularly, crude oil tanks tend to generate the most sludge. The reason is that these materials are derived from oil wells. These wells carry high-velocity gases and liquids from underground, and sand and fine particulate matter are brought up with it. The particulate matter separates in the tank and forms sludges. Since water is also present in these types of tanks, the water, oil and particulate form emulsions with a considerable solids content.

Sludges form in tank bottoms due to biological slime growth in either hydrocarbon or water tanks. Sometimes chemical reactions that form precipitates will generate sludges. Interestingly, the tank itself often contributes to the components of sludge. Rust in open-top tanks erodes from the shell in the form of a fine powder. Sometimes the tank wall acts as a nucleation site for the precipitation of iron sulfides or other compounds in oil storage tanks which are continually scraped off the walls by the movement of the floating roof. This material all ends up as sludge.

The composition of sludges widely varies even within a specific category of storage tank. For example, crude oil sludges can have the following rudimentary compositions:

Percentage of solids	1–60
Percentage of water	5–75
Percentage of hydrocarbon	5–90

15.2.2 Problems Caused by Sludge

15.2.2.1 Physical problems

Sludge has the following effects on tanks:

- It promotes pitting corrosion.
- It may contaminate the products when it approaches the tank outlet nozzle.
- It reduces the tank's capacity.
- It is "purchased" with tank receipts.
- It is extremely costly to handle and dispose of.
- It introduces inaccuracies in tank volume measurement.
- There is increased predesalter exchanger fouling in crude oil producing tanks.
- It may interfere with the operation of the floating roof.
- Drainage and outflow problems can occur.
- Sludges prevent inspection or repair of tanks internally.

Sludge formation in some cases has built up to 10 ft or more in the bottoms of large crude oil tanks. In one case, a roof being landed capsized because the uneven mountain of sludge on the tank bottom caused the roof to tip as the liquid level dropped.

It is doubtful that there are any beneficial effects of sludge buildup in the bottoms of tanks except that it has the potential to plug leaks. However, the pitting that results in the leaks may, in large part, be due to the sludge.

15.2.2.2 Environmental and regulatory problems

As a result of the Resources Conservation and Recovery Act (RCRA), sludges from tank bottoms must often be treated as hazardous wastes. Since the disposal of hazardous wastes has a high cost per pound (up to $1000 per ton), most tank owners recognize the importance of reducing the volume and mass of sludge as much as possible. For example, sludge removed from crude oil, gasoline, or other petroleum tanks almost always qualifies as a RCRA hazardous waste due to benzene levels typically found in the sludges. Therefore, a primary consideration is how to separate the oil and water for recovery in existing plant systems and minimize the volume of RCRA solid waste. The water can often be treated in the refinery's water treatment plant, the oil can be recycled back into the refining processes, and the greatly reduced volume of the remaining sludge is drummed and shipped out as hazardous wastes. Sometimes finishing steps to dry the solid waste result in lower costs for handling the RCRA waste.

Also, many companies are taking a proactive approach in reducing potential future liability by prohibiting tank entry of any kind when there are chemicals which are known to be highly hazardous or carcinogenic. This is generating the need for more sophisticated techniques for cleaning tanks without tank entry.

15.2.3 Source Reduction and Mitigation

The kind of thinking process that should be applied to sludge reduction and minimization is shown in Table 15.2.1. Note that *source reduction* is a term that may mean not the elimination of the components that create tank bottom sludges, but merely the passing on of

TABLE 15.2.1   Tank Sludge Management Considerations

Basic program components	Steps to implement	Attendant considerations
Determine bottoms characteristics	Sampling previous experience Determine what others in similar circumstances are doing	Obtaining representative samples Is sludge RCRA hazardous waste? Are sludge components reusable/recyclable?
Estimate sludge volume	Dipstick Gauging Experiencing/time	Impact of volume on costs, cleaning methods, recoverable components
Develop desludging plan	Regulatory requirements? Residuals meet standards? Required to protect human health and the environment?	Safety considerations Possible sludge volume reduction Cost-effectiveness Recoverable components for recycle/reuse Economics
Long-term waste minimization	Studies of methods	Design or operating changes that can reduce sludge volume—mixers, flow rates, separation equipment, etc.
Residuals' disposal	Disposal sites available and requirement	Alternatives include treatment of residual

these components to the next operation. For example, a producing operation that yields high levels of solids that would normally form sludges can be kept in suspension by chemical means or agitation. However, this passes the material to the next operator, in this case the refiner. Although it is effective for the producer, the solids content now becomes the problem of the refiner. The refiner will often introduces techniques such as mixing to keep the solids in suspension. Needless to say, at some point in the overall operation this material needs to be dealt with.

Refineries that have coking units are able to use the tank bottom sludge as coker feed to eliminate or substantially reduce the amount of actual hazardous waste that must be shipped off site. Table 15.2.2 is a summary of various methods often used to reduce and/or mitigate sludge. Some of the more common and important methods are discussed below.

15.2.3.1 Mixing

Although mixers may be used to homogenize and blend stocks to attain product uniformity, their other primary function is sludge reduction and minimization. For large tanks, side-entry mixers are one of the most effective ways to reduce sludge. Some refiners experiencing 1 to 2 ft of sludge generation have found that by adding side-entry mixers the sludge level has been reduced to a negligible amount. Side-entry mixers either come in a fixed configuration or are equipped with an actuator that allows the shaft to swing from side to side. This causes the complete scouring and sweeping of the tank bottom, resulting in good cleaning action. If a fixed position is used, then the sludge tends to build up in dead zones or on one side of the tank.

The disadvantages of mixers are as follows:

- It is rotating machinery which requires maintenance. Failure to properly install mixers and seals results in leakage, since the tank cannot be drawn down easily.

- Leaks at the shaft seal often persist for long periods of time and can result in environmental problems.

- Improper use, failure to adjust mixer position, or improper sizing will result in sludge buildup.

- The owner or operator must depend on the judgment and experience of the vendor because mixing is more an art than a science.

Note that when mixers are used where there are water bottoms, as in a crude oil tank, it is advisable to remove the water prior to mixing. If water is left in the bottom while the tank is being agitated, the

TABLE 15.2.2   Sludge Source Reduction and Mitigation Measures

Method		Applicability	Characteristics
Chemical additives		Introduced at tank or upstream of tank	May be costly Changes in physical properties of particulates, colloids, or suspensions
Tank agitation		Good for particles in suspension such as producing or crude oil tanks	Keeps solids in suspension Passes solids to next operation May use prop mixers, air sparging, or eductors to agitate
Conical bottoms	Steep angle	For easily separated material	Limited to small-diameter tanks
	Shallow angle	Large tanks	Greatly reduces cleaning efforts and costs
Operating methods		Some operations may have the opportunity to affect sludge formation by operations. For example, producing operations. High producing rates yield high silt and sand flows. Cutting rates reduces this. Some variables to consider are residence time (tank volume and throughput), temperature, and recirculation	While higher producing rates are usually wanted, high sludge formation rates are not. This needs to be optimized or negotiated.
Tank design		Specialized tanks	Placement of inlet and outlet piping, design velocities, height-to-diameter, and other tank design variables will affect the formation of sludges, but the effect is usually small except in specific applications.
Auxiliary operations		Equipment designed for separation of solids can be used including hydroclones, filters, and centrifugation	The appropriateness of these need to be considered in the entire context of the operation.

water will be disbursed into the stock and form emulsions or create other problems.

15.3.3.2 Eductors

Eductors or jet mixers are devices which use the principle of a venturi to multiply the flow rate of a high-pressure jet as pressure energy is converted to velocity energy. They are more effective for smaller-diameter tanks. Figure 15.2.1 shows how a typical tank might be agitated with eductors. The advantages of eductors over mixers for tank agitation are that:

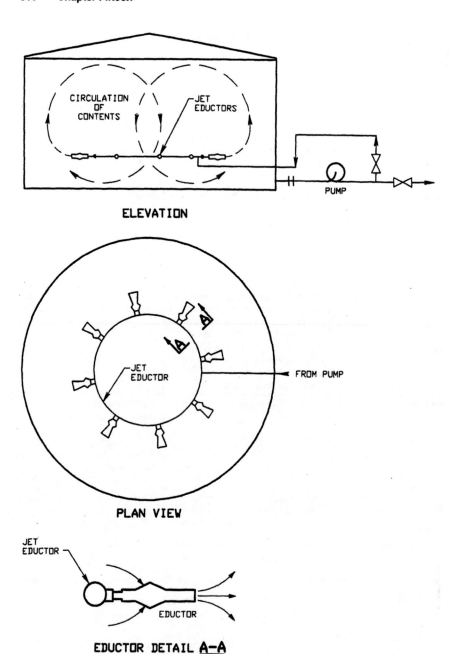

Figure 15.2.1 Eductors used for tank agitation.

- There are no moving parts inside the tank.
- There is no rotating shaft passing through the tank wall, so there is less chance of leakage.

The drawback of eductors is that they are much less efficient in terms of gallons pumped per horsepower than mixers. Eductors are often hooked up to a pump which recirculates the stock.

15.2.3.3 Air agitation

Sometimes a sparging ring, which is pipe bent into a radius with holes in it, is placed near the bottom with air pumped through the sparger for the purpose of agitating the tank. Usually, the power consumption is very high per gallon of pumped fluid. In other words, it is an inefficient process. However, it may be used in processes that require the product to be contacted by oxygen for process or quality improvement reasons. It is applicable only to very small tanks.

15.2.3.4 Cone-bottom tanks

Step angle bottoms. For small-diameter tanks, a cone bottom often works well to provide a means of withdrawing solids buildup. The criteria are that the ratio of solids to liquids must be relatively high, the solids must settle fairly rapidly, and the solids must not stick to the sloped surfaces. The disadvantages of this system are the:

- Need for periodic flushing of bottom buildup, requiring either operator attention or rather carefully instrumented automated systems

- Possible plugging due to consolidation of sludge or solids buildup

Shallow angle bottoms. Large tanks most often have a slope of about 1 or 2 in in 10 ft built into them in either a cone-up or cone-down configuration. While this has no effect on the formation of sludges, it does greatly help reduce the effort required to clean tanks. It has been reported that savings up to 20 percent of contracted services costs are possible when a sloped bottom is available as opposed to one which does not have good drainage. When sludge-cutting operations are performed by using jets of water or diesel to clean tanks, the bottom slope helps to direct the runoff into a sump or low point where it can be pumped out. Squeegee operations are also facilitated by the slope.

15.2.3.5 Design volume

An overlooked or unplanned but often effective way of dealing with sludge is to design to handle the sludge in the tank. The volume of sludge expected to be accumulated over the life of the tank can be designed for, but consideration should be given to keeping the outflow lines clear and treatment of corrosion and other sludge-related problems.

15.2.3.6 Floating suction lines

When sludge and bottom contamination or loss of product quality is a possibility, the use of floating suction has often been effective. For example, on diesel storage tanks the water and turbidity are often much higher at the bottom than at the upper levels of the tank. If fact, many operators depend on the tank's ability to improve the product quality by causing a phase separation. Withdrawing diesel from just under the floating roof by using a floating suction line has proved effective for transferring product that is clear and bright, meeting required specifications.

15.2.3.7 Chemical methods

Introduction of chemicals such as emulsion breakers, biocides, scale and corrosion inhibitors, coagulants, flocculants, antifoam agents, surfactants, dispersants, and paraffin control agents helps in some cases to reduce problems resulting from particulate or liquid suspension or to improve separation of product and water bottoms. However, introduction of chemicals on a once-through basis usually tends to be cost-prohibitive. The tank owner is also dealing with "black art" and does not have a way to really judge the effectiveness and results of these chemical treatment methods. Introduction of chemicals that are foreign to plant operations also requires a complete investigation of the effects of these chemicals and their by-products on the corrosion resistance and integrity of any equipment with which they come in contact.

15.2.4 Vapor Freeing and Degassing

15.2.4.1 Introduction

Tank degassing and *vapor freeing* are terms used to describe the reduction in concentration of chemical fumes that exist inside a tank to a level acceptable for personnel entry. Vapor freeing is the discharge of the tank vapors to atmosphere to reduce hazards and to provide acceptably low levels of tank fumes, so that entry into the tank personnel is within required limits, such as percent of LEL or permissible exposure limits for toxic fumes.

The term *degassing* usually implies legal requirements as a few states require "degassing," which means that the fumes must not be discharged to atmosphere without treatment or recovery. In addition, if floating-roof tanks are landed under the degassing rules, the vapor space that is created when the liquid drops below the landed roof must be processed through degassing equipment. In addition, for tank-entry reasons, vapor freeing is required by some local authorities to reduce air pollution.

Vapor freeing is done by the following methods:

- Mechanical ventilation
- Natural ventilation
- Displacement by water or inert gases

Degassing may be accomplished in a number of ways including

- Air or steam ventilation
- Thermal oxidation of the fumes
- Recovery of vapors by refrigeration
- Removal of fumes by carbon adsorption

15.2.4.2 Vapor-freeing operations

In vapor-freeing operations, the tank is ventilated to the atmosphere. Most often mechanical ventilation is used where air is forced into the tank, becomes vapor laden, and is expelled from the tank. Air movers used for tank ventilation with the potential for flammable vapors are driven by air or stream and are either the eductor type or an axial fan, like the ones shown in Fig. 15.2.2. For small tanks, natural ventilation may be used when there is both a manway near the bottom and one on the top and when the contents are sufficiently volatile to rapidly dissipate from the tank interior. Vapor freeing by displacement is not usually practical for tank entry because the water or inert gas must then be removed and may become a hazard.

API Standard 2015 covers specific requirements for vapor-freeing the following tank configurations:

- Fixed-roof tanks
- External floating-roof tanks
- Internal floating-roof tanks

Steam ventilation should generally be avoided. Steam nozzles produce significant potential for the generation of charged vapors which can be an ignition source. Also, because steam condenses it can cause the pressure in the tank to drop enough to collapse it, if there is insufficient opening to allow for makeup air. Steam ventilation may be considered for very small tanks storing materials in which the cleaning process is significantly facilitated by the condensing steam. These would typically be small chemical tanks.

Natural ventilation is generally not applicable to tanks in flammable service because rich fumes may be free to flow into a remote ignition source and flash back to the tank. It may be applicable to a limit-

Figure 15.2.2   Tank air movers for flammable atmospheres. (*Courtesy of Tuthill Corp.*)

ed class of small chemical tanks. Inert-gas displacement is rarely used.

15.2.4.3 Degassing operations

The vast majority of degassing on petroleum tanks is done by the thermal oxidation process because it is the simplest and lowest-cost method, considering the relatively large volume of vapor to be treated as well as other factors. However, some facilities may use refrigeration because they do not want a potential source of ignition inside the facility. Other reasons are that the noise level of the refrigeration unit may be less than that of a thermal unit, and where residential proximity might result in noise complaints, the quietest method is selected. Activated carbon is rarely used for cleaning large tanks because of the high cost and the large volumes of activated carbon required.

However, it may be used to treat the breathing and displacement vapors from smaller tanks as well as for degassing operations. Water displacement is often used when the product in a tank is changed and it is not desired to bring in a degassing contractor. Displacement by inert gases is rarely done.

The thermal oxidation and the activated carbon methods have a risk of causing an explosion or fire in degassing tanks with flammables.

Thermal oxidation

Flame oxidation method. This process uses a flame to oxidize the hydrocarbon fumes in a tank as shown in Fig. 15.2.3. This process is generally a contracted service provided by tank degassing contractors. Since vacuum trucks are usually operating at the same time as degassing units, the vapors from the vacuum truck can be discharged back to the tank space or into the degassing unit. A portable trailer unit that contains the oxidizing furnace, blowers, piping and controls, and a portable propane fuel supply is brought to the site and connect-

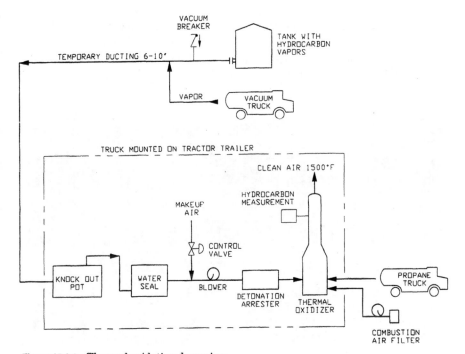

Figure 15.2.3 Thermal oxidation degassing.

ed to the tank. A flexible line made of plastic pipe or flexible ducting is connected from a tank manway to the thermal oxidizer. It is important for other manways on the tank to be open, to allow airflow into the tank, or else the blower that draws tank fumes into the oxidizer could create a vacuum and collapse the tank. Some tank degassing contractors use a vacuum breaker which allows air to flow into the degassing unit suction line, should the pressure drop too low. The tank fumes pass through a flame-arresting device so that a flame cannot propagate from the oxidizer through the ducting and into the tank, then through the blower and into the oxidizing chamber. Sufficient volume (residence time) must be provided to allow sufficient oxidation of the fumes to meet required regulations. Some regulations require at least 0.5 s of residence time. In addition, the temperature must be high enough, which is usually 1400 to 1500°F. Degassing tanks over 150 ft in diameter can take anywhere from 8 to 24 h or more. The typical flow rate for currently available degassing services is from 1000 to 3000 scfm. The process should not be stopped until the vapor space of the tank drops well below 10 percent of the LEL. When this occurs, the tank may be ventilated by forcing air through it unless local regulations prohibit this. There are numerous safety precautions that must be taken to avoid accidents.

Internal combustion engine method. Instead of using a flame to oxidize the hydrocarbon fumes, an internal combustion engine can be used as shown in Fig. 15.2.4. One supplier of this service uses two 460-in³ V-8 Ford engines which operate at about 150 hp each on a continuous basis. The vapors pass through a knockout pot to drop out liquids and water. A sys-

Figure 15.2.4 Truck-mounted engine's oxidation unit. (*Courtesy of WEECO.*)

tem of mixing valves is used to obtain the proper air/fuel ratio by introducing makeup air when the tank vapors are rich. A filter removes particulates and dust. Then the clean, ready-to-burn air-vapor mixture is passed through an isolating flame arrestor and into one or more internal combustion engines. Just prior to entering the engine, propane may be introduced to maintain the proper air-fuel ratio when the tank vapors are below the LEL. The engines must convert the power used to some form of energy. Most often, the engines are connected to a blower which has no function other than to consume energy. The blowers that draw air from the tanks are powered by the engines to conserve energy. The amount of vapors processed is proportional to the energy consumed by the engines. Typical vapor processing rates are less than those of the flame oxidation units, as they can only process about 400 cfm per engine. On larger tanks at least two or more engines are used.

Refrigeration.   Just as in thermal oxidation, the vapors are withdrawn from the tank. The refrigeration lowers the vapor stream sufficiently to drop out a significant amount of volatile organic compounds by condensation, and the remaining air and vapors are recirculated to the tank. The refrigeration system recovers the liquid in the vapor, which does not amount to very much, however. Usually, several stages of cooling are required. The water must first be removed by cooling the vapor stream to about 35°F so that additional stages of cooling do not freeze. Once the water is removed, the temperature may be further dropped to about −40 to −100°F. The cooling is supplied either by special low-temperature refrigeration units or by liquid nitrogen.

Carbon adsorption.   In this method, the vapor stream is fed into an activated-carbon unit. While activated carbon does an excellent job of reducing the levels of hydrocarbon vapors, it is consumed in the process. Typically, portable canisters are valved in parallel and, as they are used, switched to fresh sources. Once spent, the carbon must either be regenerated or disposed of. In either case the cost is too high for practical applications.

Balancing methods

Water balancing.   In this method, either water or a relatively nonvolatile stock such as diesel is used to displace the vapors that would otherwise form when a floating-roof tank is landed. This method is not useful for emptying tanks for personnel entry for inspection or repairs of the tank.

Inert gas balancing.   The specific gravity of a gas is proportional to its molecular weight. Most light hydrocarbons filling the vapor space of tanks, vessels, and barges have a specific gravity that is less than that of carbon dioxide (molecular weight 44). This means that carbon dioxide may be used to displace the hydrocarbon fumes from enclosed spaces. While nitrogen is frequently used to "inert" vapor spaces for flammability control, it probably would not function as efficiently as carbon dioxide to displace hydrocarbon vapors. In addition, carbon dioxide may be disposed to the atmosphere without special permits or stringent requirements. A few companies are using this method to displace hydrocarbon vapors from barges and are looking at this concept for tanks. By using approximately 1 to 2 times the volume of the vapor space in displacement gas, removal of over 90 percent of the hydrocarbon fumes is possible. However, if there is any residual liquid or sludge, then there is a possibility of the continuing evolution of hydrocarbon vapors. This method has not been proved for practical use on aboveground storage tanks. It may have good potential for tanks with very specific conditions and requirements.

15.2.5 Tank Cleaning

Although all tanks have similar problems associated with cleaning them, the discussion of tank cleaning is limited to chemical and petroleum tanks. Without doubt, these pose the most difficult cleaning problems.

15.2.5.1 Cleaning chemical tanks

Chemical tanks have several characteristics that affect the tank cleaning operations. First, many chemicals are highly hazardous or toxic. These tanks are also much smaller than the typical petroleum tank. The small dimensions make cleaning the tank through manways without entry feasible. This is important to reduce potential exposure to the hazardous chemicals which are usually present in chemical tanks.

Many chemical tanks can be cleaned by noninvasive methods because the operations people are well aware of the solvents and the chemistry of the stored products. However, because fairly good tank-cleaning methodology has been established for cleaning tanks, more and more chemical companies are contracting out these services to tank-cleaning companies.

The sludges are often unique with unusual properties. They may be extremely resinous and tightly adhering, as in the case of styrene tanks. In this case jets of halogenated hydrocarbons are the best solvents. In this case because of the toxicity and flammability problems, the cleaning may be conducted under an inert atmosphere with no personnel entry. Sometimes the residue to be cleaned requires a high temperature to melt it out. This is sometime the case in the plastics business. When entry is necessary, supplied-air breathing apparatus as well as the proper protective wear and suits will often be required.

15.2.5.2 Cleaning petroleum tanks

This section focuses on cleaning petroleum and, particularly, crude oil tanks because these represent the most difficult and costly cleaning jobs. The principles apply to all tanks, however. Chemical tanks can usually be cleaned by flushing or washing with the appropriate solvents or neutralizers. Sometimes high-pressure water washing is adequate. The more serious problem is, What should be done with the rinse streams or wash water streams when these operations are performed?

In general, the goals of tank cleaning are as follows:

- Preclean as much as possible prior to contracting for tank cleaning services by use of solvents, agitators, or other means.

- Minimize use of personnel entering the tank for physical removal of sludge by instead using washes, solvents, and other methods.

- Reclaim and recycle as much of the usable components of the sludge as possible.

- Minimize hazardous waste.

- Maximize safe operations and minimize health and safety risks as well as pollution.

- Consider the impact of gaseous emissions resulting from the cleaning operations.

- Optimize the overall costs, which often means not using the lowest-cost tank cleaner because other costs such as the overall shutdown time or the risks of an accident or explosion may far outweigh the cost savings on low-bid contractor-performed work.

Although various steps of tank cleaning appear to be discrete, the process of cleaning tanks should be considered a complete operation in itself. Although most tank-cleaning operations involve similar steps and principles, the actual processes can be quite different because of the many options once the sludge is removed from the tank. Once removed, the sludge must be processed. The greatest savings in cost associated with tank cleaning may often depend on the recycling of sludge components and reduction of the volume of hazardous waste. It is here where study of the costs per benefit should be conducted in connection with the overall operating facility and the effects of cleaning on it.

Tank cleaning can be performed in many different ways, but we classify them by simple principles as follows:

- Noninvasive procedures
- Tank cleaning and sludge removal
- Sludge treatment processing

Noninvasive cleaning procedures. *Noninvasive* in this context means removal of sludges and sediments by circulation and agitation of the tank contents prior to opening the manways or by opening manways and cleaning the tank through the manways without entering the tank. If a noninvasive procedure can be used to clean a tank, this is often the best choice because tank entry is not required, the risks and hazards are reduced, and the costs may be the lowest. In these methods the sludge is removed by pumping solvents (sometimes hot), diluents, or special chemicals that have the ability to penetrate or break up the sludge. Turning on the agitator or installing temporary agitation may allow the sediment to be fluidized where it can be pumped out and treated. When the remaining sludge can no longer be reduced by this method, the next step is to operate the cleaning process from the manways. Nozzles directing streams of cutter will be directed through the manways to cut the remaining sludge.

The methods are generally applicable to sediments which are easily fluidized. Unfortunately, although this method is preferred, it is usually insufficient by itself, especially in large-diameter tanks with high quantities of sludges.

Sludge removal methods

Preliminary steps that reduce physical sludge removal steps. Prior to physical sludge removal, it is best to attempt to remove as much of the sludge as possible by resuspending it and pumping it out. This is particularly true if the sludge level is above manways. A solvent such as light gas oil, clean light crude, or diesel might be added to the tank and then heated and pumped into the tank. The mixers would be set to agitate and hopefully fluidize the majority of the sludge into a pumpable slurry. These precleaning steps not only reduce worker exposure to toxic compounds such as H_2S or benzene but also reduce the costs. This step is often performed by operating company personnel, using their knowledge of the sludge, suitable cutters, diluents, or solvents and the best place to route the effluent or the best way to treat it. By the time the tank-cleaning contractor arrives on site, the sludge level should be as low as possible when a noninvasive approach is used.

Cutter oil techniques

Through-the-manway operations.   This discussion applies general-ly to the point at which a tank-cleaning service company would be called in. Since tank entry is prohibited when the flammable vapor level is above 10 percent of the lower flammable limit, as required by national codes and standards, the cutting operation must be conduct-ed from shell nozzles or roof nozzles. Ventilation usually occurs dur-ing the entire operation if the local area allows this; otherwise, spe-cial degassing procedures must be used.

Dispersant surface active additives and paraffin solvents may be used to assist in the penetration of the sludge. If steam coils are available, they can be used to bring up the temperature; or external heating can be applied in the recirculation stream. At this time all electrical connec-tions and piping should have been isolated from the tank. Agitators can-not be run because the prop may be only partially submerged, it may pose a hazard to workers entering the tank, and it may be a source of ignition. If the normal operation is to blend recovered oils with incoming crudes, the recovered slop oil can be blended at a low rate such as 5 to 7 percent without affecting the operation downstream.

Conventional in-the-tank sludge-moving operations.   Once the LEL is brought below 10 percent, it is usually most efficient from an eco-nomical and operating viewpoint to get workers into the tank. This is particularly true for larger petroleum tanks so that cutter may be applied where needed. A first step is to consider removal of the flu-idized sludge. This may be withdrawn through a water draw, a sump, or portable pumps. If a suction sump can be established in the sludge itself, this can be used as a starting point for the cutting operations. Workers either will use fixed (but portable) nozzle jets that rotate or will manually direct hose streams of solvents at the sludge. When it is manually done, the nozzle stream can be directed at the sludge from beneath the liquid cutter surface to minimize the agitation of the sludge and prevent rapid evolution of flammable fumes. For crude oil tanks, hot diesel is most effective and may well use less total quantity than cold diesel. For gasoline tanks the diesel should not be heated because the evolution of flammable vapors can increase beyond the control and ability of the tank-cleaning personnel.

When sludges are built up higher than the level of the manways, then several options are possible:

- Hot taps can be added near the manway as a means of pumping cutter in and out to clear the area of the manway.

- The liquefaction operation can be attempted from roof manways.

- A door sheet can be cut into the shell, and heavy equipment used to mine the solids. This is particularly applicable if the sludge is a dry cake in large volumes. Use of a water cutting operation which has become routine for companies specializing in hydroblasting has worked well and eliminates the ignition source. The water cutting is applicable to any thickness tank shell, and its average cost is about $3000 for a door sheet.

Robotics.   An innovative technique that works well under specific conditions is the use of mechanical robotic devices to fluidize the sludge. A robotic device is used to carry the cutter oil hoses and nozzles to the desired location and jet the solvent locally to cut sludge. When one area is cleaned, the robot moves to the next. Since local application of a pressure jet of cutter is much more effective than directing jets from a manway, this method has significant advantages in the cleaning operations and uses much less cutter than conventional operations. The technician operates the robotic device by remote video monitoring. Robotic devices are now equipped with low-light video cameras, lighting systems, and lens washers that enable the operator to view the status of the cleaning operation. It also is possible to perform the desludging operation under an inert atmosphere. The robots have safety devices including nonsparking components, grounding cables, and ground fault interrupters. They can also be provided with H_2S, O_2, and LEL detectors.

However, the robotic devices may have problems negotiating a tank with steam coils and numerous columns or roof leg supports and could become entangled.

Hot water chemical technique.   This method minimizes waste through source reduction and has many highly valuable characteristics. However, the application is limited to specific sludges, and it has other limiting requirements, discussed below. It should be considered in the early planning phases of tank cleaning.

When this technique is used, an evaluation should be conducted to ensure that the water generated can be treated by the existing refinery facilities. If this is possible, then this method becomes much more feasible. This method was developed to reduce the amount of hydrocarbons used in the tank-cleaning process. It has been tried in Japan and the United States. To establish and maintain the required temperature may be difficult or even unfeasible. Overall time varies from 10 days at best to an average time of 3 or 4 weeks. Heating the tank to the required temperature is a major consideration in this process.

This program begins with a feasibility study using a sample of the sludge. If the sludge has low solids content, this process may be effective in that it is achieves a high degree of separation of the hydrocarbon from the solids. As a result, if the solids contain metals and contaminants that would cause the sludge to be classified as a hazardous waste, the hazardous component, usually being the solid component, will be substantially reduced. The sample is also used to determine the effectiveness of the chemicals used which consist of wetting agents, emulsion breakers, and dispersants. A laboratory simulation is usually conducted as well.

If the simulation is successful, proposals may be made by vendors who specialize in this work. If feasible, planning is done for the full-scale operation based on introducing water charged with the chemicals and cutter oil in proportions approximately equal to the volume of sludge. The water introduces the chemicals into the sludge surface and provides a means of heating and mixing the tank. The oil is used as a solvent for the recovered sludge hydrocarbons. Once introduced into the tank, the water is heated to approximately 150°F. The sludge begins to break down into its components, as the penetrants act on the sludge surface. The oil components float to the top where they are miscible with the oil layer there. Emulsion-breaking chemicals act to separate oil-water emulsions. The process can routinely achieve over 90 percent removal of the oil and hydrocarbon from the sludge.

Heat may be generated either by live steam injected into the tank or by recirculation through external heat exchangers. Often temporary operations are required to heat the tank, including pumps, heat exchanges, and piping. If mixers are on the tank, they are activated. Typically, the recirculation and mixing occur over a period of several days.

The oil layer may be pumped back into the crude unit or into other oil recover operations in the refinery. The water is usually sent to the API separator. The remaining solids can be flushed out with a fire hose and vacuum truck.

Physical sludge removal and handling. The regulatory status of tank bottoms should be thoroughly investigated prior to starting tank cleaning operations. RCRA rules for hazardous waste are the most obvious requirements. However, the manner in which components of the sludge are recycled may be regulated. State and local authorities may impose other rules.

There are several possibilities for removing sludge from the tank. The oldest and most obvious methods are to physically enter the tank and move the sludge by shovel or other means. Sometimes the quantities of sludge are staggering. For example, 10,000 to 20,000 bbl of sludge is not uncommon in crude oil tanks. This is larger than the volume of many marketing tanks. In some cases, tank door sheets have been cut into the shell and front-end loaders used to move the sludge from the tank to the outside where further processing occurs. In general, cutting a door sheet into a tank and operating heavy equipment on the tank floor would not be recommended because of the high costs, the need to rehydrostatically test the tank, and the potential damage to the tank floor.

Past practices often were to dispose of the sludges by placing them on the ground as landfill material. This practice is now almost wholly obsolete because of the associated liabilities for the employees performing the desludging operation and because of long-term risks. Not only must there be concern for the hazardousness of the sludge, but also placement of this material on the ground poses the possibility of requiring ground cleanup later. It has been estimated that there is potential liability associated with past dumping of petroleum sludges ranging from $100 per ton to well over $300 per ton.

The initial analysis of the sludge will indicate whether the residual solids will be RCRA hazardous waste, whether they are treatable by such processes as neutralization, or any other specific problems that must be dealt with.

Probably the most important variable influencing the processes to be used for tank cleaning is whether the waste residuals are hazardous RCRA wastes. Even though various states implement their own solid waste programs, they mostly are equivalent to the federal rules.

In petroleum operations the cleaning sludge and waste will almost always be hazardous and primarily because the benzene levels are sufficient to trigger the applicability of RCRA. This also applies to chemical tanks.

Many chemical plants have the ability to reuse the sludge or treat it so that they often end with no RCRA wastes to dispose of. In the pulp and paper business, it is common to neutralize the waste and bring the end residuals outside the scope of RCRA.

Analysis for metals should be conducted on sludges to determine if the sludge can be amended with soils for agricultural use.

Common ways of handling sludge. Once sludge has been removed from the tank by fluidizing and pumping it out, there are various methods to handle it. If the refinery has a coker unit, then once an initial separation is made between the solids and the liquids, usually by centrifugation, the solids can be concentrated while pumpability is maintained and this material is pumped to the coker unit. The advantage is that there is no limit or specification on the quality or composition of the sludge as long as it is pumpable.

The process of separating the sludge components is usually done by the tank-cleaning vendors for large crude oil tanks because of the difficulty of the operation and the specialized equipment required. Although there are many methods for treatment of sludge when it is removed from the tank, the most common for petroleum tanks are covered below. These tank-cleaning services use generally the following methods to reduce the solids content and to reclaim the recovered oil.

Centrifugation is a commonly used method of treating the sludge once outside the tank. An initial volume of cutter is supplied from existing tanks or portable tanks supplied by the service company. It is heated and recirculated into the tank. The stream leaving the tank is put through centrifuges which assist in the solids-liquid separation. Part of the stream is recirculated, and part goes to recovery. Figure 15.2.5 shows a typical sludge treatment setup in a refinery using centrifugation or filtration. Centrifugation yields a cake which is dry to the touch and has the consistency of dried mud but which still contains approximately 45 to 55 percent moisture. When the waste is hazardous, a drying operation is often used to bring the moisture level to below 10 percent because of disposal costs of hazardous wastes which can be up to $1000 per ton.

Another method of dealing with the solids is to use a filtration operation to remove them. This technique may produce wastes with a greater percentage of solids content which is useful when the solids must be treated as hazardous wastes.

The clean liquor decanted from these operations consists of water, hydrocarbon, and sometimes an emulsion. A simple phase separation is conducted with the hydrocarbon returned to the recovery oil

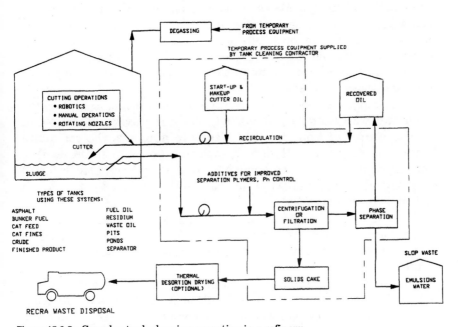

Figure 15.2.5 Complex tank-cleaning operation in a refinery.

processes, the water sent to the water treatment system, and the emulsion broken or recirculated to the tank where it eventually dissipates.

Centrifugation has a greater throughput with lower cost per barrel, but it yields a slightly higher moisture content. Filtration has the disadvantage that diatomaceous earth is added to the process, resulting in increased solids handling and disposal costs.

Special sludge problems

Naturally occurring radioactive materials (NORM). NORM does not normally pose significant risk to employees. However, if there is concern, baseline surveys should be conducted to determine the ambient levels as well as the levels of radioactivity in the sludge. NORM is typically present only in producing tanks and to a lesser degree in crude oil tanks.

Leaded tank sludges. Prior to 1980 it was common to directly bury leaded tank sludges near the tank or to spread them around near the tank to let the lead compounds oxidize to a less hazardous level. This practice has resulted in later remediation and has proved to be an expensive option. In many cases the remediation has not yet begun.

In the 1980s leaded tank sludges were usually landfarmed or disposed of at hazardous waste landfill sites. If landfilled, the sludge might have been stabilized with portland cement and/or fly ash to eliminate the free liquids and reduce the mobility of the lead.

For tanks which have been in leaded service, most refineries now incinerate the sludges at commercial waste incinerator facilities, as this is one of the few options that meet the current rules for hazardous waste management. Ash from these incinerators may have to be stabilized to reduce the migration of lead and metal compounds prior to disposal at a landfill.

If the sludge has enough recoverable hydrocarbons, common practice is to dewater and deoil tank sludges, using gravity separation, filtration, centrifugation, or other means to minimize the remaining hazardous solids and sludge. However, when a gasoline tank is cleaned, there may not be enough sludge or hydrocarbon to justify anything other than removing the wash water and sludge to a hydrocarbon recovery operation.

The first step in any program to manage tank bottoms is to characterize the waste itself. This not only provides an avenue to determine how to treat it but also characterizes whether the waste is indeed hazardous.

If a tank has been in leaded gasoline or related service in the past, it must be assumed to contain lead unless an analysis is conducted to show that it is free of lead. Employee blood tests for lead are one way to ensure that there is no exposure as the work progresses. The work is often done using supplied-air breathing apparatus as well as level A protective suits. Although the OSHA rules on lead-related operation would provide a means of monitoring with a lesser degree of protective equipment, some companies have found that it is less costly and simpler to assume the work will take place with high lead levels and operate accordingly, rather than comply with the monitoring and paperwork required otherwise.

15.2.6 Safe Vapor-Freeing, Degassing, and Cleaning Operations

15.2.6.1 General

The careful application of safe practices is essential for not only the tank operator but also the service contractor and anyone associated with the tank-cleaning operation. Tank-cleaning operations pose at least these hazards:

- Asphyxiation by the tank atmosphere (oxygen deficiency, inert or nonoxygen gases, oxygen enrichment, etc.)
- Fire and explosions in tanks handling flammables
- Poisoning through contact with the toxic sludge or atmosphere in the tank
- Mechanical hazards such as tripping or falling
- Any other hazards associated with a confined space

Preventing accidents requires the careful planning of the tank-cleaning job by both tank operator and service contractor. The principles of safe tank entry and cleaning described in API Standard 2015 should be reviewed and implemented. Since vacuum truck operations are usually involved, a careful review of the safety requirements listed in API Publication 2219, *Safe Operation of Vacuum Trucks in Petroleum Service,* is in order.

Note that petroleum and chemical tanks that have been in service with low-flash-point liquids will contain vapors that are too rich to burn. However, as the vapor-freeing or degassing processes progress, the tank vapor space will pass through the explosive range. Extraordinary precautions are in order to prevent major accidents.

For tank owners with numerous tanks, various policies can be implemented to reduce the chances of incidents. Some of the following have been used:

- No tank personnel may enter when the percentage of LEL is greater than some value X. Although 10 percent is most common, some companies have chosen 0 percent and others have gone as high as 20 percent. A good practice for anyone working in a tank that has the potential for flammable atmospheres is to continuously monitor the LEL as local LELs can rise to dangerous levels in spite of the average LEL measured at the discharge of the tank ventilation point.

- Proper use must be made of personal protective equipment and respiratory apparatus.
- No entry is permitted if the sludge or bottoms have a flash point below some value X.
- Continuous monitoring of internal atmosphere is necessary rather than once per shift or once per day.
- Air-driven lights or explosionproof lights should be used for interior visibility. Air-driven lights use an air motor to generate power for the lights. However, the amount of light given off is much less than that in other forms of lighting.
- Any equipment inside the tank should be bonded and grounded, and its resistance should be measured.
- Reducing agents should be used on tanks that may contain pyrophoric irons including potassium permanganate solutions.
- The vapor space should be made inert.

These are only a few ideas to consider in the development of a program for safe entry. Not all degrees of risk can be removed, as with most other operations and actions in life, and the risks must be weighted against the resources and methods available. The specific circumstances will determine the actual optimal method of conducting tank-cleaning operations.

For example, in one cleaning operation the choice was to take a period of 2 or 3 months to clean a crude oil tank, allowing no tank entry, or to reduce this to 2 or 3 weeks by allowing short periods of entry to move a diesel. The no-tank-entry option is not necessarily safer because there is a longer period in which abnormal operations on the tank are being conducted.

In all the vapor-freeing or degassing operations, there is significant potential to collapse the shell or roof of either floating- or fixed-roof tanks because of the blowers used. When air is drawn from a tank, there is the potential to increase the vacuum in the tank to collapse it. This must be guarded against by ensuring that manways are opened or that vacuum breakers are used. Most often these incidents occur just as the operation is being started. Since the low manways should not be opened to the atmosphere, allowing fumes to escape, the blower is started, but there is insufficient open area at the manway to allow air inflow. This causes the tank pressure to drop. A carefully planned procedure or the use of equipment to prevent vacuum will prevent this.

15.2.6.2 Vapor-freeing safety

As with any of and all these operations, proper bonding and grounding with a thorough check of electrical conductivity is the best assurance against ignition sources arising from static electricity. Another important hazard in vapor freeing occurs because at the start of the operation the expelled vapors are rich. Almost always, the vapors are more dense than air and will travel as invisible water streams to low points. Numerous tank fires and explosions have been started because the flammable vapor was ignited by a passing automobile or a flame far from the tank. Once the vapor stream is ignited, it can flash back to the tank, causing a fire or explosion. By discharging tank vapors with adequate velocity near the top of the tank, there is sufficient dilution to eliminate this hazard. Also, mechanical ventilation is preferred because there is more mixing of the rich vapors that are expelled to the atmosphere and the operation takes less time, which reduces the time-induced hazard.

The equipment used in mechanical ventilation must have the electrical classification for the operation and be driven by air or steam. The moving parts of the blower should be designed to minimize sparking. By placing the blower so that only fresh air is entering it, the possibility of a spark's arising inside the fan, causing the vapor to ignite, is reduced. Eductors which have no moving parts are an alternative to blowers.

15.2.6.3 Degassing safety

Another critical and important aspect of hazards has to do with the vapor space toxicity and flammability. Only the fire and explosion hazard is covered here.

The basic hazard of petroleum tank precleaning operations lies in getting the flammability level in the vapor space down to acceptable figures (10 percent or less) and keeping it down. Many petroleum tanks when first opened are above the upper flammable limit (UFL). This means that if the very common practice of degassing a tank by air ventilation is performed, there will be some period of time in which the entire tank volume will be explosive. Should an ignition source occur while the tank is between the LFL and the UFL, the tank will explode. However, the probability of occurrence of a spark can be reduced to acceptable levels by proper grounding and bonding and use and selection of equipment, precautions, and procedures.

Since the tank is a closed steel structure, outside ignition sources will have no effect on the interior vapor space, and for this reason there have been few accidents due to degassing at this critical time due to an ignition source. However, flames could travel from the exiting flammable ventilation fumes under the right conditions, if proper precautions are not observed.

Although in a small minority, some companies prohibit ventilation, using air to reduce the flammable vapor concentrations. However, this makes the job much more difficult and poses new risks introduced by using inert atmospheres or other degassing methods. At this time the majority of tank degassing is done by ventilating the tank with outside air. However, environmental regulations are beginning to require that other methods such as thermal oxidation be used.

Since many tank-cleaning operations are conducted through the roof manways, the hazards of descending onto a floating roof should be studied, and the considerations described in API Publication 2026 should be reviewed. It is important that contractors performing this work be thoroughly informed of the hazards and safe procedures as well. These are outlined in API RP 2220.

The possibility of accidents occurring from the degassing unit itself is significant. The equipment of the degassing contractor should be thoroughly inspected to understand how it operates; if it uses proper flame or detonation arrestors and whether they are properly installed; if all bonding and grounding procedures are followed; and if the personnel are experienced and trained in the hazards of operating these units.

15.2.6.4 Activated carbon hazards

On small chemical, hydrocarbon, or waste water tanks, activated-carbon canisters are often used to degass or to treat the breathing fumes so that the tank is in compliance with environmental rules. Activated carbon is derived from coal, pyrotized hardwood, or coconut shells. The carbon is broadly classified as to whether it has been chemically impregnated to remove specific compounds. For example, removal of H_2S from hydrocarbon vapor streams requires the impregnation of the carbon with a compound such as potassium permanganate. The treatment of such streams may use a single bed which has been chemically impregnated or multiple beds in which some are pure activated carbon and others have been treated.

As the hydrocarbons adsorb onto the surface of the carbon, a certain amount of heat is generated, which is called the *heat of adsorption,* and it is roughly equivalent to the heat of condensation of the vapor. The amount of heat generated depends on numerous variables such as the concentration of hydrocarbon, the type of hydrocarbon, the flow rate through the bed, and the activity of the carbon itself.

Hydrocarbons above 1000 ppm can cause the beds to heat up to dangerously high temperatures. If the hydrocarbon vapor stream contains acetone, methyl ethyl ketone (MEK), or cyclohezonon, the oxidation of these compounds is accelerated in these beds and heat rises even more. Chemically impregnated carbons increase the oxidation rate and thus temperature in the carbon bed. Sulfur compounds such as H_2S can oxidize with air to form sulfur, using the carbon as a catalyst. The heat liberated from this reaction can also generate dangerously high temperatures in the activated-carbon bed. Vapor streams containing more than 100 to 200 ppm H_2S have been known to overheat canisters.

The heat generated in the bed by these processes is usually carried away by the sensible heat of the flowing vapor stream. However, if the flow is reduced or stopped, then heat continues to be generated and the temperature may rise, becoming an ignition source. If the vapor in the connecting ductwork is within the explosive range, then a flashback to the tank may occur, causing an explosion in the vapor space of the tank. This has occasionally happened in the past.

Most carbon manufacturers recommend installing flame arrestors immediately upstream of the carbon if this situation is possible. However, flame arrestors may not operate as intended due to the possibility of overdriven detonation within the ducting. Careful application of detonation arrestors is required for this approach to be effective.

The following is advised:

- A systematic review of each process that uses carbon should be conducted to evaluate the potential hazards and to determine the potential for flashbacks.

- The bed temperature should be monitored, although local hot spots may not be detected.

- Install detonation arrestors, but ensure that the installation is such that it will function as intended.

SECTION 15.3   SECONDARY CONTAINMENT

15.3.1   Legal Requirements

It is important to define what secondary containment means in a general way to distinguish it from other concepts related to the term *containment. Secondary containment* is the capturing of the entire contents of the largest tank in the containment area in the event of a spill in order to allow sufficient time for cleanup of leaks and spills without endangering human health or the environment. For petroleum products, the most well-known and recognized requirement covering secondary containment is NFPA 30, *Flammable and Combustible Liquids Code.* This code requires that class I through class IIIA liquids (liquids with flash points below 200°F, which means most petroleum liquids other than heavy fuel oils) be contained in the event of a spill or rupture and that the volume be large enough to hold the contents of the largest tank. It further specifies that the containment area shall be constructed of earth, steel, concrete, or solid masonry designed to be liquidtight.

The Uniform Fire Code requires that a tank or tanks in a group be diked when the accidental discharge of liquid could endanger adjacent tanks, adjoining property, or waterways. The NFPA and UFC requirements are essentially duplicated in OSHA's Flammable and Combustible Liquid Code (29 CFR 1910.106). However, the implementation of NFPA requirements is usually enforced and interpreted on a local basis. The fire chief may alter or waive these requirements when in her or his judgment a tank does not constitute such a hazard.

On the federal level, the driver for using secondary containment is the Clean Water Act in which the spills prevention controls and countermeasures (SPCC) plans are required to be implemented with the intent of spill containment and prevention. The SPCC plans cover facilities that handle oil or petroleum products. The Oil Pollution Act of 1990 (OPA 90) expanded the financial responsibility of tank owners as well as requirements to mitigate and respond to a worst-case discharge of oil or other hazardous substances. Although secondary containment is required, a liner consisting of other than native soil is not specified. State laws, however, may require the use of an "impervious" secondary containment liner.

When toxic or hazardous substances that are regulated by the Resource Conservation and Recovery Act (RCRA, 40 CFR paragraphs 260–282) are stored, secondary containment is required on a federal level. Secondary containment is required for hazardous waste treatment, storage, and disposal (TSD) facilities.

Several states have rules mandating various requirements for secondary containment. Examples are Alaska, New Jersey, New York, and Florida.

15.3.2 General Considerations

Secondary containment usually consists of some combination of dikes, liners, ponds, impoundments, curbs, outer tanks, walls, or other equipment capable of containing the stored liquids. These are some important considerations for evaluating existing facilities:

- Regulatory requirements
- Required minimum containment volume
- Permeability of the containment system and liner, if used
- Substance(s) being stored and its (their) properties
- Site conditions, environmental sensitivity, leak detection and monitoring systems
- Spill cleanup

Design of secondary containment areas must take into account existing and related facilities. Often, the treatment for the secondary containment stormwater runoff or spills is common to other areas including pump and equipment slab areas, tank car and truck loading areas, or process areas. A process diagram that indicates flows and design criteria as well as the treatment capabilities should be developed to document and understand the limiting cases and flows. The capabilities of the treatment processes including phase separation of oil and water, oxidation, biofiltration, and dehalogenation should be considered in estimating the worst-case spills and how they will be handled as well as in identifying bottlenecks and modifications that should be implemented.

15.3.3 Secondary Containment Capacity

The required capacity of secondary containment is determined by the authorities having jurisdiction. As a minimum, industry codes require that the secondary containment area be large enough to contain the contents of the largest tank within the containment area. Most state regulations require that in addition it be sized to contain some level of precipitation, usually about 6 in. The SPCC program requires that the secondary containment area contain the volume of the largest tank and an additional 10 percent to contain precipitation. The safest possibility is to size it for at least the contents of the largest tank volume plus 10 percent.

When two or more tanks are permanently manifolded together so that they are hydraulically normally connected such that tank levels move together, then the sizing of the secondary containment area should be based on the connected tank volume.

15.3.4 Environmental Protection Considerations

To reduce the risks of environmental pollution and groundwater contamination substantially by requiring secondary containment is not unreasonable. To require no leakage is not possible. Risks should be balanced with a cost/benefit study on a case-by-case basis.

Secondary containment areas can range from simply mounding up earth into earthen dikes to an elaborate concrete slab designed to be watertight and coated with epoxy finishes. In many cases an earthen dike system is all that is necessary. Here is why. The total amount of contamination that can escape into the ground below a spill in a secondary containment area depends on

- The size of the spill
- The time for which the spill remains uncleaned
- The permeability of the containment area liner material
- The number and size of cracks or breaches in the liner
- The hydraulic depth

While the exact relationship is complex, the basic idea is that decreasing any one of the variables (time, permeability, or depth of fluid) in the containment area can reduce ground contamination.

Most, if not all, companies have a policy to immediately clean up and correct a spill that occurs. In addition, most local regulations require that the spill be stopped and cleaned up as soon as practical. This is the primary purpose of secondary containment. It provides sufficient time for the operator to clean up a spill without risking contamination of the ground outside the secondary containment envelope. Cleaning up a spill immediately after it has occurred within a secondary containment area is the best way to prevent pollution.

The remaining variable is the permeability. The concept of permeability seems simple, but it must be related to several other concepts to understand how foreign substances can escape through the secondary containment envelope into the environment. There are two modes of transport through a barrier membrane. The most commonly referred to mode in reference to secondary containment is the liquid transport mode (hydraulic conductivity). The other mode is the vapor transport mode, which occurs by molecular diffusion or permeation. Soils and clays transport most of the foreign material by hydraulic conductivity whereas geotextile membranes can transport vapors but are essentially impervious to liquid permeability. It is important to understand the context in which permeability is used, especially when it is legal.

For secondary containment the hydraulic conductivity or the ability of the liner to resist the transport of fluid through it is the important factor to be considered. State regulations tend to be vague about the required permeability. Often *impervious* or *sufficiently impervious* is the only guideline specified by the regulations. A review of the state standards indicated most regulations require from 10^{-6} to 10^{-7} cm/s as a minimum liner permeability. The weak link in membranes lies in the seams and the possibility of tears or punctures caused during installation or by traffic and settling after installation. API Reference 6 has an excellent discussion of both vapor and liquid permeability and methods of testing it.

For practical purposes, simple rules of thumb may be sufficient to select the liner. Generally, materials such as sand, aggregate, or other open-grained soils have a relatively high permeability regardless of the liquid spilled. In these cases a major spill would seep rapidly into the aquifer below. However, claylike soils or other tight soils can prevent liquid penetration if the spill is cleaned up quickly. Whether an earthen containment area is adequate depends on the permeability of the earthen containment soils, the liquids stored, and other things. A key factor is the most hazardous material to be stored and its properties.

15.3.5   Spilled-Liquid Hazardous and Physical Properties

The toxicity, mobility, and persistence of the liquids stored play a major role in determining the need for and the type of liner necessary in a secondary containment area. Materials that can move quickly through soils and are highly toxic merit special considerations in the liner selection. Low-viscosity, highly toxic, water-soluble components and those that do not degrade easily are more likely to need a leakproof liner than are substances with high viscosity and low toxicity that are not very water-soluble. Examples of the former are liquids such as pure benzene, methanol, and some halogenated hydrocarbons. Examples of the latter might be fuels, heating oils, crude oil, or other hydrocarbons.

In reviewing the secondary containment liner options, a what-if analysis might be performed on a hypothetical spill of the liquid, its consequences, and remediation costs and efforts. For example, a spill of methanol poses a substantial cleanup effort. First, it is low-viscosity, toxic, and water-soluble, meaning that it can contaminate large quantities of groundwater quickly. Because of this, remediation efforts would be spread over large areas and would be very costly. In addition, methanol is toxic to bacteria in concentrations greater than a few hundred parts per million and would prevent the mobilization of other organic methods of remediating the soil or destroy bacteria that would degrade other compounds in a spill.

Other regulated chemicals pose unique handling problems if spilled, as they initiate other hazardous waste management requirements. Examples are benzene and toluene in pure form.

15.3.6 Basic Design Considerations

Permeability is generally not specified in rules regarding liners; rather, the design is specified, although some states may have regulations. When permeability requirements are specified, a common value is 10^{-7} cm/s. This kind of permeability can only be obtained by using clays or elastomeric or membrane fabrics. Concrete can also meet this specification as its permeability is about 10^{-10} cm/s for water.

Permeability is not necessarily the most important element to prevent material from entering the environment. The design of joints, inspection of the containment envelope, and operations are important variables. The most impermeable membrane, if punctured or put together with poor joints, is ineffective.

15.3.7 Secondary Containment Types

Secondary containment for aboveground tanks may be

- A diked area external to the tank which may or may not have an additional lining inside the secondary containment area.
- A vault.
- A double-walled tank.
- A tank that includes an integral diking system.
- A double-bottom tank that is monitored by both visual and chemical sensors. This may or may not qualify for "secondary containment" depending on the interpretation of the authority having jurisdiction.

The most common form of secondary containment is the use of dikes and/or berms, as shown in Fig. 15.3.1. Either masonry or reinforced-concrete diked walls or earthen berms with or without liners are used. Note that there are suppliers who specialize in prefabricated secondary containment panels which can be bolted together in the field to form the dike walls of a secondary containment area. Most of these systems have a means of attaching a synthetic flexible membrane, if used to form the impermeable bottom layer of the secondary containment area.

A vault is a concrete containment tank which houses the primary containment. It is often used for aboveground and belowground applications.

Figure 15.3.2 shows a double-wall tank which has a volume between the double walls equal to the volume of the primary containment. Double-wall tanks must have a volume at least equal to the inner tank volume.

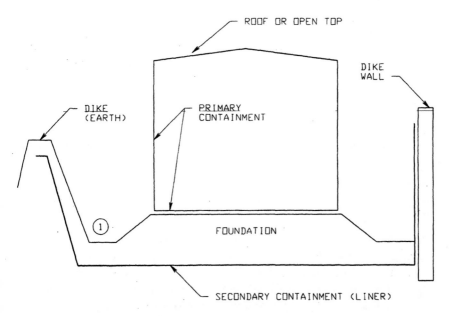

Figure 15.3.1 Secondary containment for tanks.

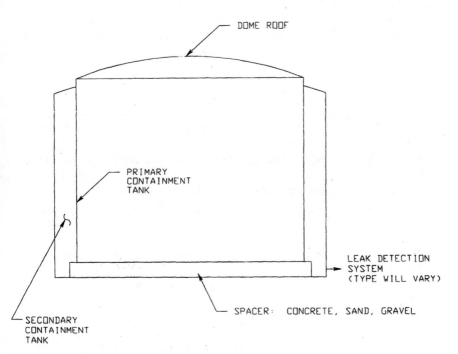

Figure 15.3.2 Double-wall secondary containment tank.

Underwriters' Laboratories in Standard UL-142 recognizes smaller tanks that are constructed with a double wall or with integral dikes or pans to fulfill the secondary containment function.

15.3.8 Materials for Liners

Choosing an appropriate liner is a difficult task dependent on many variables. Table 15.3.1 lists some considerations that impact the overall design of liner systems. Table 15.3.2 lists the characteristics of various liner systems.

15.3.8.1 Native soil

The primary factor to be considered when native soil is used for the secondary containment area and berms is its permeability characteristics. Table 15.3.3 shows typical values of permeability. Soil permeability tests on native soil can verify the quality of the soil for the purpose of secondary containment. Other options are to import alternative fill materials such as natural clays or bentonite. The generally accepted value for minimum clay permeability is 10^{-7} cm/s.

Selection of a suitable clay for use as a liner material is not easy. Clays are subject to many effects which may defeat the intended purpose including dying, cracking, deterioration, changes in permeability with time or with exposure to various substances, and inadequate thickness. The selection of clay should be made in consultation with an experienced soils engineer and chemist based upon tests.

TABLE 15.3.1 Liner Design and Selection Criteria

- The liner should be compatible with all the liquids stored and should have a life span consistent with the plant or facility design basis.

- The liner must accommodate rainfall without impairing the ability to monitor for and collect spills.

- The liner should accommodate equipment passage and roadways.

- The liner should have sufficient durability to accept normal traffic, fire exposure, weather exposure, maintenance and operations activities, and minor spill cleanup without major damage.

- The liner system should be compatible with existing standards such as those published by NFPA and API.

- The liner should be liquidtight up to the design maximum secondary containment level.

- The liner should have minimal impact on existing tanks and operations.

- The liner system should accommodate the use of cathodic protection in existing facilities, and its impact on the effectiveness of cathodic protection should be evaluated.

TABLE 15.3.2   Characteristics of Secondary Containment Liners

Surface material	Advantages	Disadvantages	Relative cost
Natural permeable soils	Readily available at the site, requiring only compaction Useful as a cover over impermeable surfaces such as synthetic liners, clays, or bentonites to protect these surfaces from exposure to ozone, sunlight, and vehicular traffic	Provides minimal environmental protection, useful by itself only in applications where a highly permeable surface can be tolerated	Low
Natural clays	An inexpensive material if readily available Commonly used material; standard testing procedures available High degree of impermeability can be attained: permeability will vary with thickness and degree of compaction	Subject to drying and cracking; must be protected with soil cover Subject to leaching of components when exposed to water or other solutions Subject to ion exchange when exposed to water-containing acids, alkalies, or dissolved salts Overcompaction can result in poor retention properties	
Soil cement	Good durability Resistant to aging and weathering	In-place soil usually used; permeability varies with type of soil May not be applicable in following soil situations: ■ Highly organic soils ■ Clays ■ Saline soils ■ Clean, well-graded gravel or crushed rock ■ Subject to degradation due to frost heaving of subgrade	Moderate
Bentonite	Low permeability Treated bentonites do not deteriorate with age	Requires protective soil cover Untreated bentonites may deteriorate when exposed to contaminant	Moderate

TABLE 15.3.2 Characteristics of Secondary Containment Liners (*Continued*)

Surface material	Advantages	Disadvantages	Relative cost
Asphalt	Widely available Provides good waterproofing membrane, but does not provide a high degree of impermeability	Has poor resistance to hydrocarbons Application on frost-susceptible soils is questionable High-temperature and long-term weather exposures may affect material properties	Moderate
Concrete	Easily applied to both rock and stable soil with good adherence Provides some degree of impermeability that can be enhanced with liners or coatings Good strength and durability Easily repaired	If poorly designed, concrete is susceptible to cracking when exposed to freeze/thaw cycles, or if differential soil settlement occurs	High
Synthetic polymeric membrane liners Synthetic polymeric membrane liners	High resistance to bacterial deterioration Specific polymers provide high degree of resistance to certain chemicals	Requires subgrade preparation and sterilization applied to earth to reduce the risk of puncture Must be protected from structural damage, particularly due to vehicular traffic May require protection from exposure to sunlight (ultraviolet light) and ozone Some polymers may be attacked by hydrocarbons and solvents Good oil resistance and good low-temperature properties do not normally go hand in hand	Moderate to high

SOURCE: *Technology for the Storage of Hazardous Liquids,* New York State Department of Environmental Conservation.

Natural clays are often available locally and may satisfy the permeability requirements. . The material must be placed in layers sufficiently thick to form a monolithic envelope and to reduce drying and cracking. Areas subject to high water tables are generally not suitable for clay linings. A climate which alternates from very dry to very wet also makes the use of clay unlikely. A soil cover is usually required to reduce drying and cracking damage as well as mechanical damage to

TABLE 15.3.3 Water Permeability of Various Soils
Used for Secondary Containment

Soil type	Water permeability coefficient, cm/s
Clean gravel	1×10^{-2}–1.0
Clean sand	1.0–1×10^{-2}
Silt	1×10^{-1}–1×10^{-4}
Stratified clay	1×10^{-3}–1×10^{-6}
Homogeneous clay	1×10^{-4}–1×10^{-9}
Impervious soils	1×10^{-6}–1×10^{-9}

the clay liner. Clay must be hydrated and kept moist if cathodic protection is to be maintained through it, as dry clay becomes a nonconductor. Clay liners are subject to porous areas because of the difficulty of ensuring a uniform material. Careful planning and testing is required to ensure that the clay envelope has no weak links in it. Clay is subject to cracking but can easily be repaired.

Bentonite is a commonly mined variety of clay material that either is applied directly as a liner or is blended into existing soil to enhance permeability. The main feature of bentonite is that it expands and swells when exposed to moisture, which can offer some self-sealing attributes. The blended soil-bentonite mixture must be compacted to form the liner. Bentonite may also be purchased as a liner sandwiched into a geotextile material, which allows moisture to absorb into the bentonite when placed on the ground. These bentonite liners are about ⅛ to ¼ in thick but expand when moistened. The seams are made by overlapping the sheets of bentonite at least 6 in. Native soil cover should be placed over the liner to protect it and to keep it moist.

15.3.8.2 Soil cement

Soil cement is a blended mixture of portland cement, selected soils, and water which results in a concretelike soil. If a high ratio of cement is used, shrinkage cracking will tend to develop. The soil cements are more permeable than bentonite or clays but less permeable than the native soils in general. Soil cement is subject to frost damage and should be avoided in areas with high groundwater levels as well.

15.3.8.3 Asphalt

When the secondary containment liner is subject to frequent damage, asphalt may be appropriate. Its chemical suitability must be examined because it will deteriorate, soften, and dissolve when exposed to various hydrocarbons and chemicals. Sealants can be selected to improve its resistance to degradation.

15.3.8.4 Concrete

Smaller facilities and those that are handling very hazardous chemicals or where aesthetics and durability are important considerations drive the use of concrete secondary containment. Concrete provides a high degree of flexibility and control of layout that make it ideal in many cases. Even though its initial costs may appear high, its many advantages will often result in the lowest overall long-term life-cycle costs.

Favorable properties of concrete include high compressive strength, fire resistance, chemical and weather resistance, and an efficient space diking system which reduces space requirements (compared to earthen berms). Structural piping and equipment support requirements can often be reduced by incorporating these requirements within the secondary containment structure. Another major advantage of concrete is that spills and leaks can be cleaned up without destroying the secondary containment lining system, unlike in earthen systems lined with clay. In earthen systems spill-contaminated material has to be packaged and treated or shipped to a treatment facility whereas little, if any, material is generated from a concrete containment area.

Although concrete has a low liquid permeability, proper design and testing are critical to ensure a sealed containment envelope by preventing cracking, allowing for thermal expansion, and sealing the cold joints and any penetrations. The biggest sources of leaks in secondary containment areas are from cracks, expansion, and cold joints.

Concrete secondary containment structures are usually either pour-in-place or prestressed panels. Standards covering the various methods are listed in the References. Pour-in-place systems are more appropriate for smaller areas where the layout is complex. Prestressed systems have been used on large, simple layouts effectively.

The integrity of the concrete containment depends on reducing the number and sizes of cracks and breaches through the concrete envelope. Concrete cracks for several reasons:

- It shrinks while it cures, resulting in tensile stresses which crack it. Shrinkage can be minimized by controlling water content and by using shrinkage control additives. In addition, by adding reinforcement beyond that normally required for strength alone the shrinkage cracking will be minimized.

- Thermal expansion and contraction cause cracking. Appropriate use of steel reinforcement and thermal expansion joints can reduce heat-related cracking. However, the number of expansion joints should be minimized since these have proved to be a major source of leaks in concrete containment areas.

- Frost penetration is another reason for cracking and spauling. The surface should be smooth and have good drainage to prevent pools of standing water. The concrete should not be overworked during placement, because the aggregate sinks and the reduced surface durability allows moisture to penetrate.
- Differential settlement can cause cracking. A compacted subbase and properly designed foundations reduce this potential.

Reinforcement and joint design are important to maintain a liquidtight structure. Design to include sufficient reinforcement to reduce cracking and to seal expansion and cold joints eliminate these problems. If specific engineering design standards are not available within a facility, a good general-purpose standard that has been applied to secondary containment structures is the American Concrete Institute standard ACI 350R-89, *Environmental Engineering Concrete Structures*. This standard contains rules for the amount of reinforcing required, the cement-aggregate ratio, maximum water content, as well as testing criteria.

Uncoated concrete is suitable for substances such as various hydrocarbons, including gasoline and petroleum products, acetone, alcohols, ethers, halogenated hydrocarbons, and numerous solvents. Some salts, strong acids, and alkalies may attack concrete. A liner placed over the concrete surface may be used to protect the concrete.

Concrete liners and required thickness are site-specific and depend on many factors such as chemical properties, exposure (intermittent or immersion), and ability to form pinhole-free surfaces. Often used lining materials are epoxies, urethanes, and fiberglass-reinforced plastic (FRP) type of linings. Table 15.3.4 lists various coating systems that can be applied to the surfaces of concrete.

15.3.8.5 Synthetic flexible membranes

A variety of materials are available for lining secondary containment areas including polyvinyl chloride (PVC), high-density polyethylene (HDPE), polyester, butyl rubber, neoprene, and polyurethanes. Selection of an appropriate membrane starts with chemical compatibility. Compatibility for the membrane should be examined against each of the possible spill chemicals within the containment area. Durability, costs, installation considerations, and deterioration by bacteria or other elements should be considered during the selection process. Also important is the ability to test the seams for strength and freedom from pinholes or tears or seam separation after installation. API Publication 315 lists some of the important properties, liner materials, and how to evaluate them.

TABLE 15.3.4   Concrete Secondary Containment Liners

A. *Coatings* (nonfilled/nonreinforced)
   1. Thickness: 5–25 mm
   2. Installation type: Spray or roller-application
   3. Generic resin types: Epoxy, polyester, vinyl ester, urethane, poly-
                                          sulfide, novalac epoxy, phenolic, silicon
   4. Fillers and reinforcement: None
   5. Installed cost: $3–$8 per ft^2
   6. Comments: These systems are simply heavy industrial
                                          paint coatings, giving minimal chemical and
                                          physical protection.

B. *Coatings* (filled/reinforced)
   1. Thickness: 15–80 mm
   2. Installation type: Spray, roller, or trowel application
   3. Generic resin type: Epoxy, polyester, vinyl ester, urethane, poly-
                                          sulfide, novalac epoxy, phenolic, silicon
   4. Fillers and reinforcement: Quartz, silica, carbon, graphite, flake glass
   5. Installed cost: $3–$8 per ft^2
   6. Comments: Although these systems are thin-film coat-
                                          ings, the addition of fillers and reinforcement
                                          enhances the coatings' chemical resistance,
                                          permeability, and durability. However, they
                                          are still thin-film coatings that have physical
                                          limitations.

C. *Toppings* (aggregate-filled)
   1. Thickness: $\frac{1}{8}$–$\frac{3}{8}$ in
   2. Installation type: Trowel application
   3. Generic resin type: Epoxy, polyester, vinyl ester, urethane, poly-
                                          sulfide, novalac epoxy, phenolic, silicon
   4. Fillers   Quartz, silica, carbon, graphite, and flake
                                          glass
   5. Installed cost: $6–$12 per ft^2
   6. Comments   Because these systems are thicker, they are
                                          normally more wear-resistant and durable.
                                          They are normally less permeable and han-
                                          dle standing chemicals better than thin-film
                                          coatings. However, as with thin-film coat-
                                          ings, if the concrete substate cracks, the
                                          cracks will telegraph through the topping.

D. *Toppings* (mat-reinforced and
   aggregate-filled)
   1. Thickness: $\frac{1}{8}$–$\frac{3}{8}$ in
   2. Installation type: Trowel application
   3. Generic resin type: Epoxy, polyester, vinyl ester, urethane, poly-
                                          sulfide, and novalac epoxy
   4. Fillers: Quartz, silica, carbon, and graphite, $1\frac{1}{2}$- to
                                          10-oz polyester and nexus mat
   5. Installed cost: $8–$15 per ft^2
   6. Comments: These systems are more conductive to con-
                                          taining chemicals, because of their low per-
                                          meability and concrete crack bridging char-
                                          acteristics. These systems are applied in
                                          multiple layers.

TABLE 15.3.4   Concrete Secondary Containment Liners (*Continued*)

E. *Polymer concrete* (overlays or full thickness)	
1. Thickness:	1 in and over
2. Installation type:	Cast in place by screening or vibration (similar to portland cement concrete)
3. Generic resin type:	Epoxy, vinyl ester, novalac epoxy, furan, silicate, sulfur, latex
4. Fillers and aggregate:	Quartz, silica, carbon, and graphite
5. Reinforcement:	Carbon steel or alloy mesh and rebar, fiberglass rebar, and plastic mesh
6. Installed cost:	$10–$30 per ft²
7. Comments:	Easily installed by general contractors. Reinforced and placed similar to portland cement concrete. Lower overall shutdown period. Can be placed as an overlay over concrete or used full thickness. Very durable
F. *Precast polymer concrete liners and structures*	
1. Thickness	1 in and over
2. Installation type:	Precast liners are set in place similar to precast portland cement concrete.
3. Generic resin type:	Epoxy, vinyl ester, novalac epoxy, furan, silicate
4. Filler and aggregate:	Quartz, silica, carbon, and graphite
5. Reinforcement:	Carbon steel or alloy mesh and rebar, fiberglass rebar, and plastic mesh
6. Installed cost:	$15–$75 per ft²
7. Comments:	Normally fabricated as trenches, sumps, manholes, pump pad covers, beams, etc. Can be a liner inside concrete or an independent, structurally engineered shape. Lower overall installation cost. Lower overall plant down time. Highly durable
G. *Acid brick liners*	
1. Thickness:	1 in and over
2. Installation type:	Bricklayer's method over membrane. Tile setter's method over thin set.
3. Generic types of mortars:	Epoxy, vinyl ester, novalac epoxy, furan, phenolic, silicate
4. Filler and aggregate:	Quartz, silica, carbon, and graphite
5. Brick type:	Red shale, fireclay, carbon, and silicon carbide
6. Installed cost:	$15–$75 per ft²
7. Comments:	Highly durable. Long-term lining solution. Labor-intensive, needing quality installers. Historically the most well-known and long-lasting in the chemically-resistant liner family
H. *Thermal plastics* (preanchored)	
1. Thickness:	⅛–¼ in
2. Installation type:	Prefabricated in shop and set in place in field
3. Generic types:	Polyethylene, PVC, polypropelene, kynar, teflon, halar

TABLE 15.3.4   Concrete Secondary Containment Liners (*Continued*)

H. *Thermal plastics* (preanchored) (*Cont.*)	
4. Anchors:	Prewelded to back side of sheet or extruded from back side of sheet (Anchors used to attach sheet to concrete)
5. Installed cost:	$18–$75 per ft^2
6. Comments:	Very flexible. Lower field costs and shutdown time. Moderate temperature limitations and must have certified plastic fusion welders

Another important consideration is whether the liner will be buried beneath a cover layer of soil or placed above the containment area. When the liner is placed on the surface, degradation by ultraviolet sunlight and ozone become important factors affecting life. Wind can also generate waves in the plastic, tending to tear and crack it. Liners above grade are also subject to fire damage.

Liners for new construction are usually placed before foundations, slabs, and equipment including tanks. In retrofit situations, the liner must be placed around existing equipment and foundations. Attaching the liner to existing structures requires careful design and considerable effort.

15.3.8.6 Sprayed applications

One alternative to premanufactured flexible membranes is a spray-on application of urethane or polysulfide. These spray-on applications are also effective coating systems for concrete containment systems. When they are used to line an earthen secondary containment system, the subgrade must be compacted and sharp objects removed from the surface. A geotextile fabric is often placed prior to the spray-on application of the urethane to increase the durability of the lining. The urethane should be sprayed when the ambient temperature is greater than 40°F. Two coats with a final thickness of 50 to 60 mils are required. Control of thickness is important and must be checked as the field lining is being installed. Spray-on coatings can be subject to failure to cure properly. The qualification of the contractor applying these materials becomes critical.

Sprayed polysulfide applications begin by placement of a geotextile which has been impregnated with polysulfide at the factory. The field spraying provides an additional 40-mil thickness as a chemical barrier and for durability and resistance to degradation. The same subsurface preparation is required for polysulfide as for the spray-on urethane applications.

SECTION 15.4   FOUNDATIONS FOR HOT TANKS

15.4.1   General

We now focus on hot tank bottom foundation designs. Although the principles are applicable to any hot tank, the designs considered here have been tailored for tanks storing hot asphalt products in the temperature range of 200 to 600°F. This chapter not only addresses the foundation design concepts but also incorporates provisions for leak detection and leak containment.

Tanks in which a large temperature gradient or frequent heating and cooling cycles are encountered are not addressed. For these conditions, special consideration should be given to fatigue, thermal expansion, and creep.

15.4.2   Undertank Temperatures

In some temperature distribution studies, high temperatures were found to exist several feet below the bottom of the tank. The initial temperature profile will vary from site to site because of the presence of moisture, different soil thermal conductivities, and other factors. Once a tank is put into hot service, it may take substantial time for the ground temperatures to reach steady-state conditions, perhaps months to years. However, it is clear that eventually high temperatures will extend to several feet below the tank's foundation.

Field tests also confirm high undertank temperatures. One tank owner found temperatures of 160°F at 30 in below some of the tanks after a relatively short period of service. Because moisture may be present or the steady-state temperature condition has not yet been reached, this temperature should be even higher. In another instance, an asphalt tank in a refinery tank foundation resting on a wood-piled slab foundation (wood piles are not recommended for hot tank foundations) had its piles charred to a depth of several feet below the tank's concrete slab.

15.4.3   Undertank Insulation

To counter the effects of high undertank temperatures, some designers have suggested using undertank insulation. However, a study of the temperature distribution indicates that insulation is relatively ineffective in reducing steady-state temperatures. This is because the thermal gradient across the insulation has to be large for the insulation to be effective. Unless the insulation's thermal conductivity is much less than the soil's, this will not happen. Also, the soil's thermal conductivity can vary widely and may be even lower than those used in our temperature study. Therefore, adding insulation may increase the time required to reach a steady-state condition, but eventually the temperature distribution will be almost the same.

From a practical viewpoint, insulation can create other problems such as increased settlement, moisture entrapment, tank bottom corrosion, and maintenance difficulties. Therefore, insulation is generally not used under hot tanks.

Other methods that remove heat by natural convection such as ventilated foundations might work, but no doubt would be expensive and probably are unnecessary.

15.4.4 Environmental Considerations

Many regulatory agencies now require release prevention barriers for tanks. This requirement extends to tanks storing hot substances. Release prevention barriers typically consist of undertank liners. These agencies also often require leak detection.

The API 650 specified method of leak detection requires that any leakage through the tank bottom be redirected to the perimeter of the tank where it can be observed. Having an undertank liner therefore serves to redirect the flow for the leak detection while at the same time acting as a release prevention barrier or liner.

Some materials such as asphalt, which is typically stored in a temperature range of 350 to 500°F, or molten sulfur, which is stored above its melting point of 115°C, are solid at ambient temperature. Some operators believe that no liner or secondary containment measures should be required for these materials since they solidify and do not leach harmful substances into the soils, should there be a leak. Indeed, both asphalt and sulfur have been used to pave highways, and so it is not likely that there is any environmental harm done by leaks from tanks storing these materials. For tanks with these substances it is recommended that tank owners negotiate a position with respect to leak containment and liners on a case-by-case basis. Other substances that are stored hot may be liquid at ambient temperature or may be toxic or harmful if a leak does occur, so that consideration should be given to lining these tanks.

15.4.5 Undertank Liners

15.4.5.1 High-temperature effects on undertank liners

For ambient temperature tanks, plastic liners are used for leak detection and leak containment. However, as shown by temperature studies, high temperatures can exist several feet below a hot tank. At these high temperatures, polymer-based liners—including HDPE—will melt or stretch and tear apart due to the tank's weight or the shifting soil. Therefore, plastic liners should not be used for hot tanks *unless designed for stock-side temperature.*

Although most elastomeric liners are only good to approximately 250°F, Teflon can withstand temperatures of up to 450°F. PFA Teflon sheets joined by heat-produced seams could be used. It is available in 60 to 90 mils in 4-ft widths by 100 ft or more long. However, the cost is very high, and this has not been tried. This is the reason that a double bottom (metallic liner) or a concrete liner described below is used for temperatures exceeding 250°F.

Although clay liners can withstand temperatures over 200°F without tearing or melting, they are susceptible to drying out and cracking unless kept continuously moist. At the high undertank temperatures, moisture could be driven away and the clay liner may crack.

It is apparent that providing a cost-effective and flexible liner for hot tanks is difficult.

15.4.5.2 Choosing a liner for a hot tank

For hot tanks, concrete, steel, and clay liners have been used. The choice should be based on economics, maintenance concerns, and local regulations.

If a clay liner is to be used, it should be placed as close to the water table as is reasonable. This will help keep the clay moist so it does not crack. The clay liner should be laid inside the ringwall and covered with chloride-free, dry sand prior to tank construction (Fig. 15.4.1). If leak containment is required, the preferred method is a double steel bottom.

15.4.5.3 Concrete as a liner

Concrete may be a release prevention barrier or undertank liner if it meets certain requirements. American Concrete Institute publication ACI 350R-89, *Environmental Engineering Concrete Structures,* presents these requirements and as well as recommendations for structural design, materials, and construction of concrete tanks, reservoirs, and other structures commonly used in water containment and industrial waste applications. Although permeability is not addressed in this document, watertightness is. For release prevention, a watertight concrete liner should fulfill the goal of preventing an environmental release. However, the local regulator has the final say as to what actually constitutes an acceptable release prevention barrier. In order to be watertight, the concrete cracking must be controlled by

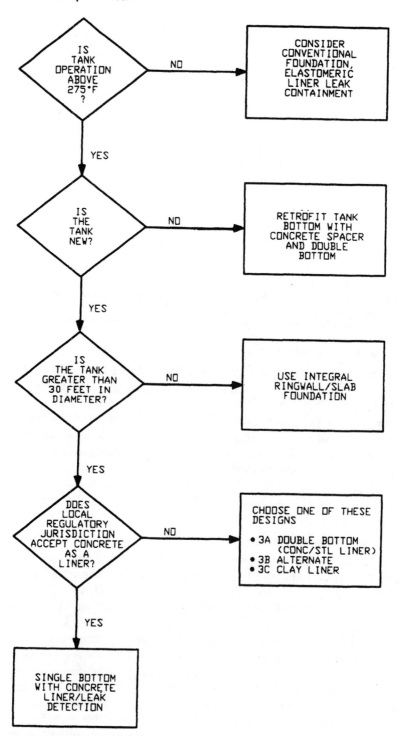

Figure 15.4.1 Flowchart of hot tank leak detection and leak containment foundations.

the use of temperature and shrinkage reinforcement. These specifications are given in ACI 350R-89.

15.4.6   Foundation Design

15.4.6.1   Designing with concrete at high temperatures

The compressive strength of concrete decreases as temperature is increased. This reduction in strength depends mostly upon temperature, moisture content, loading history, and type of aggregate used. As the concrete heats up, the aggregate and cement expand at different rates. This fact, coupled with the different stiffnesses for the aggregate and the cement, creates a complex interaction between the two.

For concretes with limestone or gravel aggregate, the strength reduction is very small up to about 600°F. However, concrete with other aggregates may see up to a 40 percent strength reduction at 600°F. At temperatures greater than 600°F, the cement starts to dehydrate, and its strength drops off more dramatically. Therefore, for temperatures higher than 600°F, special types of cement (such as alumina cement) should be considered.

Using alumina cement concrete for tank foundations with tank temperatures below 600°F is very costly and probably not necessary. Regular concrete with an appropriate strength reduction factor may be used for hot tank foundations less than 600°F. Tanks with temperatures in the range of 200 to 400°F should use 4000-psi concrete, while tanks in the range of 400 to 600°F should use 5000-psi concrete. In both cases, the foundation should be designed using a reduced strength of only 3000 psi for these concretes that effectively provides the required safety factor. Using a reduced compressive strength for design will be more economical than using alumina cement concrete and reducing the concrete degradation.

15.4.6.2   Concrete mix

The concrete to be used for hot tanks should be high-quality with a low water-to-cement ratio. The following design mixture is recommended: 0.4 water-to-concrete ratio, a minimum of 490 lb/yd^3 cement, a maximum of 5 percent entrained air, and no accelerators (especially accelerators with chlorides). Proper curing practice is essential and consists mainly of keeping the new concrete surface damp for at least the first 7 days. Locally available aggregate should be acceptable since the design already takes into account the reduced concrete strength at high temperatures.

15.4.7 Foundation Type

Figure 15.4.1 has been prepared to simplify the decision process required to select a hot tank foundation, taking into consideration the line, leak detection, and other important variables.

15.4.7.1 Single-bottom designs with concrete liners

For single-bottom designs, concrete slabs and/or ringwall foundations for hot tanks are recommended. It is recommended that a concrete slab be installed to cover the entire bottom of the tank. The purpose of the slab is twofold. It provides a release prevention barrier or liner under the tank. It also reduces the possibility of moisture under the tank bottom where accelerated corrosion could be a problem or cause uneven temperature variations which could create high local stresses on the shell-to-bottom welds and the bottom plates. Using reinforced concrete reduces the chances of differential settlement and failure. If the concrete is to also serve as a release prevention barrier, it is imperative that the concrete be properly reinforced.

Single-bottom designs with slabs under the tank are shown in Figs. 15.4.2, 15.4.3, and 15.4.4. The slab also affords the opportunity to install leak detection grooves that meet the requirements of API 650. See Fig. 15.4.5.

When the tank is under 20 ft in diameter, it will be usually more economical as well as effective to use the integral ringwall slab design shown in Fig. 15.4.4. Instead of a ringwall, a slab with thickened edges is used. The required reinforcing, leak detection, and thermal considerations are the same as those for larger tank foundations. For small tanks, this design accomplishes the same functions as the ringwall design at much lower cost.

Note that in Fig. 15.4.3 a leak will not be contained but will run out into the secondary containment area. However, this is probably not an important factor in the protection of the environment since the capacity of a double bottom does not really buy very much time after a leak is discovered.

The design of Fig. 15.4.3 includes an expansion joint to accommodate the thermal growth of the slab relative to the ringwall. The temperature range for this design is 200 to 600°F. Tanks with temperatures below 200°F should use the same concrete strength for design and construction.

The concrete foundation will act as a liner, creating a barrier to the tank and groundwater contamination. The foundation also includes leak detection grooves which will guide the leaking product toward the tank's periphery for easy detection. The concrete should be reinforced so that

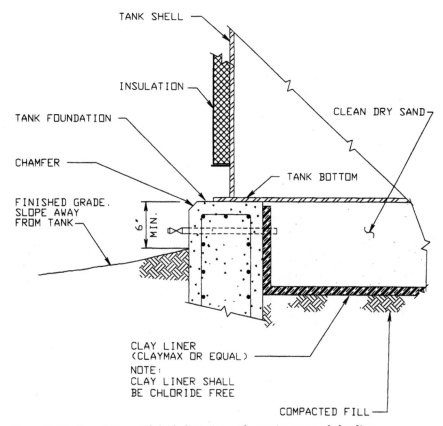

TANK SHELL

INSULATION

TANK FOUNDATION

CLEAN DRY SAND

CHAMFER

TANK BOTTOM

FINISHED GRADE.
SLOPE AWAY
FROM TANK

6" MIN.

CLAY LINER
(CLAYMAX OR EQUAL)
NOTE:
CLAY LINER SHALL
BE CHLORIDE FREE

COMPACTED FILL

Figure 15.4.2 Foundation with leak detection and containment and clay liner.

cracks cannot propagate and undermine the concrete's integrity as a liner, as described earlier. As with any other design, temperature steel should be included in the ringwall and concrete slab. However, because of thermal gradients, additional reinforcing steel should be placed in the circumferential (hoop) direction near the outside edge.

Figure 15.4.6 is an alternative to a slab under the tank. This design uses a curb to provide more leak containment. However, it is probably no more effective than the other designs while probably being more costly. Its use may be governed by local authorities in some locations.

15.4.7.2 Designs for tank bottom replacement or retrofitting

When one is upgrading or replacing the bottom of existing tanks for high-temperature service, Fig. 15.4.7 provides an economical as well as reliable and effective method of providing leak detection and a liner. A

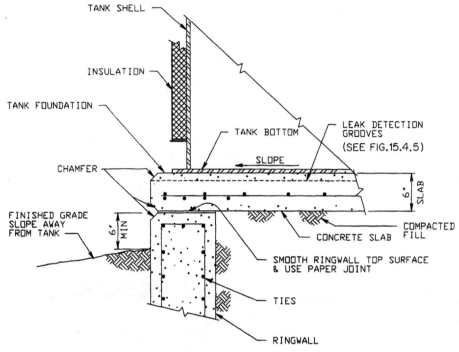

Figure 15.4.3 Foundation for large tanks (>20 ft in diameter).

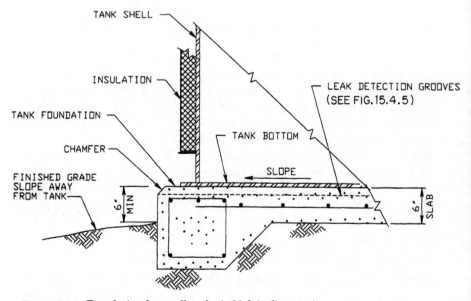

Figure 15.4.4 Foundation for small tanks (<20 ft in diameter).

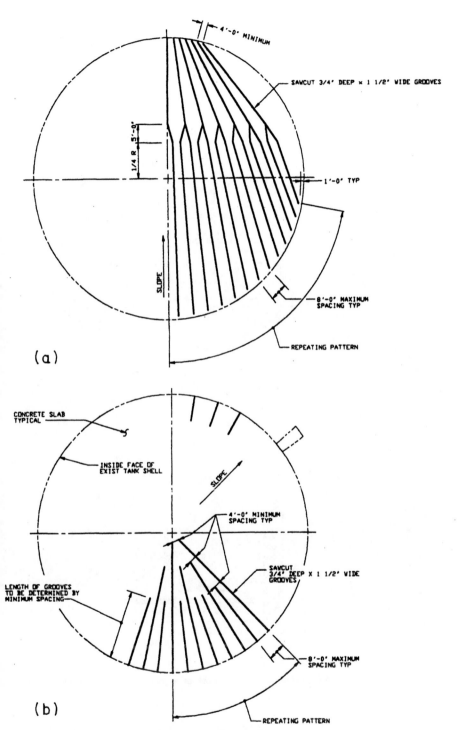

Figure 15.4.5 Leak detection foundation plan. (*a*) Single slope configuration; (*b*) cone-up configuration.

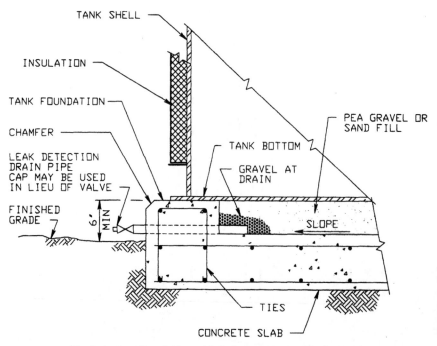

Figure 15.4.6 Single-bottom foundation with leak detection and leak containment.

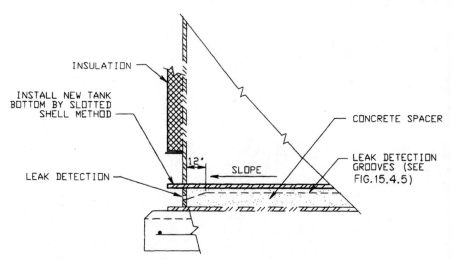

Figure 15.4.7 Retrofit existing tank with new bottom to include leak detection.

new concrete spacer at least 4 to 6 in thick is poured over the old tank bottom. The concrete liner should be reinforced according to ACI 350R-84 to provide watertightness and to prevent excessive cracking if the old bottom has corroded away or has significant leaks. Radial grooves are added for leak detection. For substances that may not be considered hazardous such as asphalt and sulfur, it would be adequate to reinforce the slab with welded wire mesh in lieu of rebars, as the cracking would not be an important factor in environmental protection.

15.4.7.3 Designs using double steel bottoms

Figure 15.4.8 can be used for new tanks or when the tank's bottom plate is replaced. Leak containment is accomplished by the second bottom and the coupling which may be plugged. Note that this system is placed over compacted fill soil. This design has leak containment in the form of a double bottom so that the tank bottom closest to the ground actually forms the liner or release prevention barrier.

15.4.7.4 Hot tank anchoring

In general, tanks should be designed with a low aspect ratio (height to diameter) such that anchoring is not required for the seismic load-

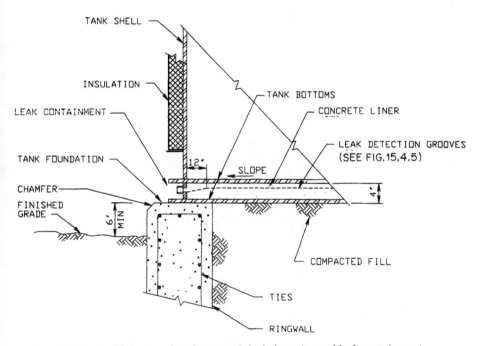

Figure 15.4.8 Double-bottom foundation with leak detection and leak containment.

ings specified by Appendix E of API 650. When it is not possible to keep the tank's aspect ratio low enough (approximately 0.4 to 0.5 in seismic zone 4), anchors may be required. The anchorage must be designed to accommodate the differential thermal expansion in the radial direction between the tank and the slab.

The detail of Fig. 15.4.9 should be used when a hot tank requires seismic anchorage. It allows for the different radial expansions that

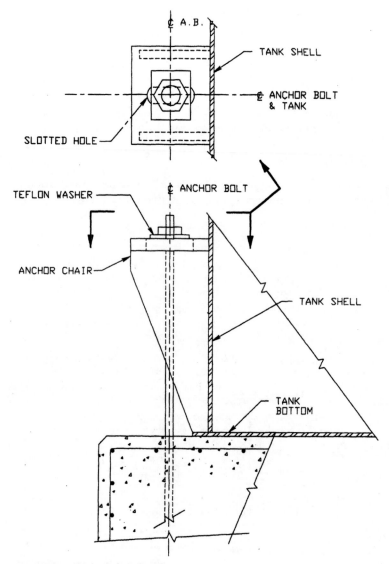

Figure 15.4.9 Tank anchor detail.

will occur between the tank and its foundation without generating significant bending stresses in the anchor bolts.

15.4.7.5 Hot tank sumps

Since withdrawing the maximum amount of product to facilitate tank cleaning, inspection, maintenance, and repair is particularly desirable for hot tanks where the contents can solidify or become hard to handle at ambient temperatures, tank owners often wish to install sumps in hot tank bottoms. However, in hot tanks, the indiscriminate use and design of tank bottom sumps or appurtenances have lead to failures due to the thermal expansion of the tank's bottom. Presently, for sumps or appurtenances to perform reliably without risk of failure, they must be designed on a case-by-case basis. One such design is shown in Fig. 15.4.10.

15.4.7.6 Hot tank corrosion

Tank corrosion in hot tanks can occur anywhere water is in contact with the tank' s bottom plate. Most of the time, the high undertank temperatures drive away existing moisture. However, in a location with frequent rains, a high water table, or an area subject to frequent flooding, water may be in contact with the tank bottom. Corrosion of hot tanks is generally limited only to the tank's periphery, since that is where water can come in contact with the tank's shell and bottom. Further toward the tank's center, moisture is driven off by the high soil temperatures.

The tank's edge may never become completely dry because of a phenomenon known as *moisture pumping*. It works like this: As the water under the tank is heated, it rises, pushing the water above it out of the way and drawing more water in to take its place. The tank's edge may never be completely dry because of moisture pumping, but moisture pumping can be minimized by placing a tank well above the water table. Also, a concrete pad or ringwall foundation should create an effective barrier which will minimize moisture pumping.

For the tanks in the temperature range being discussed, the water in contact with the bottom plate will probably turn to steam. Although steam is less corrosive than liquid water, its corrosive effects should not be discounted too heavily.

The best way to reduce undertank corrosion of hot tanks is to keep the tank's underside dry. Raising the tank 4 to 6 in above the adjacent grade—including future foundation settlement—should reduce moisture contact and therefore bottom-side corrosion.

In existing tanks, where the chime (the external part of the annular ring) sits in a puddle of water, severe corrosion can be expected.

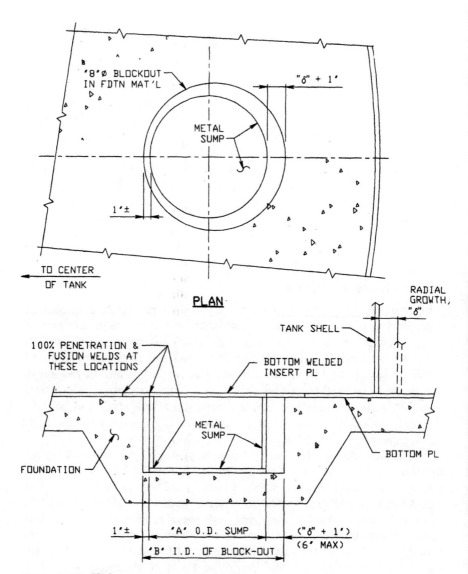

Figure 15.4.10 High-temperature sump detail.

With the combination of thermal stresses and corrosion, there is a potential for failure at this critical shell-to-bottom joint. For existing tanks, the tank perimeter should be excavated and drained as required to ensure that no standing water collects around the tank's base.

Cathodic protection under hot tanks is not recommended because the anode's life is greatly reduced at elevated temperature.

15.4.7.7 Design recommendations

- Plastic or clay liners should not be used unless required by law because they degrade when subjected to the high undertank temperatures.

- Concrete is the liner of choice because it can be designed to resist the high undertank temperatures.

- All liners (including plastic liners) should be designed for stock-side temperatures.

- A double steel bottom is the preferred method for leak detection and containment.

- Undertank insulation should not be used because it will have little effect on the steady-state temperatures.

- Tanks with elevated bottoms or forced convection cooling may not be an economically attractive alternative.

- Concrete slabs and/or ringwall foundations should be used.

- The concrete ringwall and slab should be designed with a reduced concrete strength to account for high-temperature effects.

- The tank foundation should be raised 4 to 6 in—including settlement—above the existing grade.

- Cathodic protection should not be used for hot tanks because of the limited anode life at high temperatures.

- The use of tank sumps or other bottom appurtenances is discouraged unless case-by-case analysis of the effects of thermal expansion is done.

- Additional circumferential reinforcing steel should be included in slab foundations.

References

Section 15.1

1. API Standard 2015, 5th ed., *Safe Entry and Cleaning of Petroleum Storage Tanks,* American Petroleum Institute, Washington, 1994.
2. API Publication 2026, *Safe Descent onto Floating Roofs of Tanks in Petroleum Service,* American Petroleum Institute, Washington, 1988.
3. API Recommended Practice 2220, *Improving Owner and Contractor Safety Performance,* American Petroleum Institute, Washington, 1991.
4. API 2013, *Cleaning Mobil Tanks in Flammable or Combustible Liquid Service,* American Petroleum Institute, Washington, 1991.
5. API Publication 2219, *Safe Operation of Vacuum Trucks in Petroleum Service,* American Petroleum Institute, Washington, 1986.

Section 15.2

1. NFPA 30, *Flammable and Combustible Liquids Code,* 1994.
2. *Uniform Fire Code,* 1991 edition, International Fire Code Institute.

3. American Concrete Institute standard ACI 350R-89, *Environmental Engineering Concrete Structures*, 1991.
4. American Concrete Institute Publication 318, *Building Code Requirements for Reinforced Concrete*, 1983.
5. ACI Publication 350, *Concrete Sanitary Engineering Structures*, 1983.
6. ACI Publication 301, *Specifications for Structural Concrete for Building*, 1984.
7. Post Tensioning Institute, *Design and Construction of Post Tensioned Slabs on Ground*, 1990.
8. Prestressed Concrete Institute, *Guide Specifications for Prestressed Precast Concrete for Buildings*, 1985.
9. API Publication 315, *Assessment of Tankfield Dike Lining Materials and Methods*, July 1993.
10. *Technology for the Storage of Hazardous Liquids*, New York State Department of Environmental Conservation, Division of Water, Bureau of Water Resources, Albany, N.Y., 1990.
11. *Toxic Substance Storage Tank Containment*, Ecology and Environment, Inc. Buffalo, NY, and Whitman, Requardt and Associates, Baltimore, MD, 1985.
12. *Technical Resource Document for the Storage and Treatment of Hazardous Waste in Tank Systems*, U.S. Environmental Protection Agency, Washington, 1990.
13. UL Standard 142, *Proposed Seventh Edition of the Standard for Steel Aboveground Tanks for Flammable and Combustible Liquids*, October 1992.

Section 15.3

1. M. S. Abrams, *Compressive Strength of Concrete at Temperatures to 1,600°F*, American Concrete Institute Special Publication SP-25, Temperature and Concrete.
2. Chicago Bridge and Iron Company, *Concrete Repair Procedure*, CBI Services contract 88017.
3. American Concrete Institute, *Environmental Engineering Concrete Structures*, ACI 350R-89.
4. Chicago Bridge Institute, *Project Specification—Foundation Construction*, CBI Services 88017.
5. Ed Sheppard, *Hot Asphalt Tank Foundation Designs*, Chevron Research and Technology Company, Nov. 17, 1991.

Section 15.4

1. OSHA 29 CFR Part 1910, Permit-Required Confined Spaces for General Industry; Final Rule.
2. API Standard 2015, *Safe Entry and Cleaning of Petroleum Storage Tanks, Planning and Managing Tank Entry from Decommissioning through Recommissioning*, 5th ed., May 1994.
3. API Recommended Practice 1141, *Guidelines for Confined Space Entry on Board Tank Ships in the Petroleum Industry*, 1st ed., March 1994.
4. Philip E. Myers, "OSHA Confined Space Reg Interpreted for Storage Tank Options," *Costings and Linings*, July 1994.

Tank Accessories

SECTION 16.1   LADDERS, PLATFORMS, STAIRS, AND ACCESSWAYS

16.1   General

Access to tanks is an important and somewhat complex topic that should not be glossed over in the design of new tanks or the assessment of existing tanks. Proper design minimizes injuries, helps in emergency response efforts, and even reduces the severity of calamitous events such as fires, explosions, and seismic activity. The information contained in this chapter covers the basic requirements and considerations for access to ASTs, including API and OSHA rules, common practices, various designs, and basic principles. Access to tank requires consideration of the following structural components:

- Stairs
- Platforms
- Handrails
- Ladders
- Toeboards
- Hoop guards, ladder safety cages, and drop bars
- Intermediate platforms and landings
- Gratings

It is important to understand that there are numerous details and those shown inevitably include preferences and prejudices that reflect only one of many ways of doing things. The minimum requirements that one should follow in these designs and layouts are the OSHA and API requirements.

Typically existing facilities need not be modified to comply, but new facilities or modifications to existing facilities should be designed in accordance with current applicable rules. However, a hazard risk assessment of existing facilities would indicate the need for upgrading. Most companies develop in-house standards and construction details so that proper clearances, details of construction, and legal requirements are in compliance with the authorities having jurisdiction and to prevent incidents.

16.1.2   Ladders and Stairways

16.1.2.1   Vertical ladders

These are used for internal floating-roof tanks and for external access to the tops of small tanks. Most internal floating roofs use a vertical ladder to provide access to the interior of the floating roof. Since internal platforms would interfere with the ladder, it is permitted to have ladders that exceed 30 ft. The vertical ladders are often made by welding rungs to the guide pole (typically 8-in pipe) and using a smaller 3- or 4-in pipe as the other rail.

Since internal aluminum floating roofs cannot accept the loading and forces of a rolling roof ladder, vertical ladders are also used. On steel roofs a rolling internal ladder is often used. This is particularly the case when an external floating roof tank was converted to an internal floating-roof tank because it was supplied that way.

The ladder should not be attached to the tank bottom because of the potential of bottom settlement. As with external ladders, the shell supports the ladder dead load plus live load.

16.1.2.2   Rolling roof ladders

The rolling roof ladder provides access to the external floating roofs. It is required to be supplied in an API 650 tank. It should be designed for a live load of 1000 lb at the midspan of the ladder in all positions. The top of the ladder is hinged to the gauge's platform, and the bottom rolls on tracts, as indicated in Fig. 16.1.1. The rolling roof stairs use two types of treads. The first type is closely spaced bars or pipes. In effect, this is an inclined ladder. The use of solid bars as opposed to pipe for rungs significantly reduces the potential for corrosion. The second type uses self-leveling treads which provide an actual tread to walk on instead of a rung. Although the self-leveling designs cost more, they are safer and much easier to use. The minimum width is 24 in, but better practice is to use 30 in.

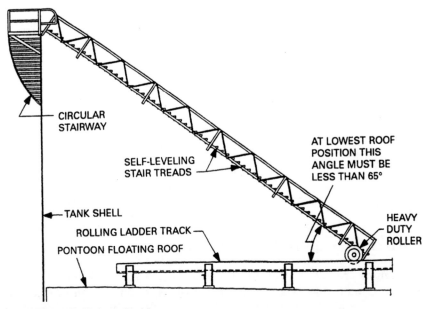

Figure 16.1.1 Rolling roof ladder.

The ladders should not be steeper than 65° when the roof is landed in the low position, and they must not bump into the pontoons, seal, or shell when the roof is in the high position. If the ladder approaches the top of the shell when the roof is in the highest position, then a handrail should be installed above the shell to prevent possible falls over the edge of the shell at the free end of the ladder.

16.1.2.3 Circular stairs

Ladders versus stairs. Either ladders or stairways are used to access tank tops. Stairways, as opposed to ladders, are preferred for open-top tanks because they allow the safe hauling of tools, instruments, or safety equipment to the top of the tank without undue measures. In additions, for tanks storing flammable liquids, the stairway often doubles as an access for portable fire hoses to fight rim seal fires. The spiral stairway is usually found on tanks over 30 ft high and larger than about 40 ft in diameter. Small tanks will often use ladders either for reasons of economy or because the tank is too small in diameter. As the tank diameter decreases, the width of the tread varies more, and the OSHA acceptable limits prevent the use of these on very small diameters. In addition, unless the tank was quite tall, the vertical height would be insufficient to accommodate the spiral stairway. Stairway layout and design follow the general rules specified by OSHA as well as local regulations with the exception of special con-

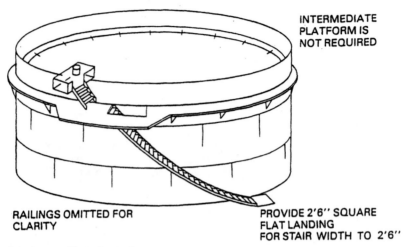

INTERMEDIATE
PLATFORM IS
NOT REQUIRED

RAILINGS OMITTED FOR
CLARITY

PROVIDE 2'6'' SQUARE
FLAT LANDING
FOR STAIR WIDTH TO 2'6''

Figure 16.1.2   Typical spiral stairway.

siderations created by the curved path. Figure 16.1.2 shows an example of a circular stairway. Note that stairways can ascend in either a clockwise or a counterclockwise direction. While it is preferable to have the direction standardized on one location, in order to miss appurtenances and optimize costs, the ascending direction may be changed. It is generally preferable for the stairway not to pass over mixers as it could interfere with maintenance operations.

Stairway angle.   Although stair angles as measured from the horizontal plane can range from 30° to 50°, the most common practice is to use 45° with an 8-in tread and an 8-in riser. In some jurisdictions 40° is the maximum stairway angle. Some companies have standardized on the use of a 40° or less angle to reduce the effort required to ascend the stairway.

Design loads.   OSHA spells out the required minimum load conditions for stairway design. State and local regulations may be more stringent. Stairs must be designed to carry 5 times the normal anticipated live load and not less than a 1000-lb concentrated moving load. Suggested design minimum for stair stringers is 100 psf or 1000-lb concentrated live load. Ladders should be designed for the anticipated load but not less than 200 lb. Railings should be designed to handle a 200-lb load at any point in any direction on the top rail.

Minimum stairway width and landing.   Stairways are required to be a minimum of 24 in wide. This includes clear space beyond any projections or appurtenances that might protrude into the stairway path. Better practice is to use a standard width of 30 in, and most companies have adopted this practice for reasons of practicality and safety. The minimum landing area at the base and the top of a tank is 24 in by 24 in but not less than the width of the stairway.

Intermediate landings. Intermediate landings are not required on circular stairways. However, some companies install them at specific intervals. When a stairway passes through intermediate wind girders, then it is usually a good idea to provide a landing at that point.

Overhead clearance. Overhead clearance must be at least 7 ft over all treads measured at the tip or leading edges, including the platform edge.

Method of attachment

General-purpose single-stringer stairways. Single-stringer stairways have one end of the tread welded directly to the tank shell. Spiral stairs must be supported completely from the shell. The single-stringer, as opposed to double-stringer, stairway is preferred because it is the economic choice. All components that are welded to the shell should be seal-welded to reduce potential corrosion and resulting rust streaking on the finished coating.

Double-stringer stairways for hot and insulated tanks. For insulated tanks the double-stringer stairway minimizes penetrations into the insulation. Sufficient clearance should be left between the stringer and the outside surface of the insulation. However, if the gap between the stinger and the outside surface of the insulation exceeds 8 in, then an inner handrail is required by OSHA.

Another application in which double-stringer stairways should be used is hot tanks or when cyclical thermal service is expected. Some stairs have failed by thermal fatigue cracking at the tread-to-shell weld, where frequent thermal excursions were experienced. This occurs because of the differential radial growth of the shell and the constant-temperature stairway.

When hot tanks are uninsulated, personnel must be protected from the hot surfaces, as shown in Fig. 16.1.3. To ensure that the treads are not above 140°F, a double-stringer stairway should be used. When treads are welded to the shell, the treads can exceed the maximum allowed surface temperature. Specific heat-transfer computations can be performed to determine how hot the treads will be near the tank shell, if necessary.

Bottom landings. The landing at the base of the stairway must be a square with one side equal to the width of the stairway, usually 30 in

AT OPERATING POINTS AROUND
BASES OF HOT UNINSULATED
EQUIPMENT AND ALONGSIDE
STAIRS, LADDERS,
OR WALKWAYS

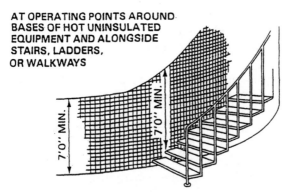

Figure 16.1.3 Hot-surface guard for hot tanks.

minimum. For tanks with a ringwall, the landing is usually placed at the same elevation as the ringwall, and the first stair step is located 6 to 12 in above the landing. Laying out the treads usually starts by working top down with the bottom step adjusted to fit the conditions. In addition, because of settlement of which a large portion occurs during the hydrostatic testing, it is wise to install the bottom landing pad sometime after the hydrostatic test.

16.1.3 Platforms and Walkways

16.1.3.1 General requirements

Platforms must be designed to support a moving load of 1000 lb. Many companies specify more complete loading conditions such as a deflection limit of $\frac{1}{2}$ in under load. The minimum width for a platform is 24 in. Common practice is to use 30 in instead. Headroom should be at least 7 ft. Flooring should be made of grating or nonslip material. It should be cautioned that checkered floor plate can be slippery when used around petroleum products. Most users handling thick, oily, or slippery substances use a grating or perforated plate with turned up edges that provide an adequate grip. Platform toeboards must be at least 3 in high and have a clearance of not more than $\frac{1}{4}$ in under them. The height of the platform handrail must be 42 in with a midrail approximately halfway up supported by railing posts not more than 96 in apart. The top handrail from an entering stairway must be bent to match the platform top handrail (it must not be offset).

When platforms are subject to unusually high loads such as snow loads, for insulated tanks or when the slope of the roof exceeds 2:12, the fixed roofs of tanks should not be used. Instead, independent, shell-supported platforms should be used.

16.1.3.2 Wind girders

Top wind girders can serve as walkways on large tanks. This is usually designed by letting the tank shell act as one handrail by setting the elevation of the wind girder at 42 in below the top of the tank. The purpose of the walkway is primarily for access by emergency response personnel responding to a rim seal fire and to support portable foam equipment and hoses during the exercise. The wind girder for these designs must be at least 24 in wide and have an outside handrail.

16.1.3.3 Gaging platforms

These are needed to access the gauge well as well as to provide access to the rolling roof ladder. Serrated steel grating is probably the most slip-resistant surface for conditions associated with petroleum products.

16.1.3.4 Interconnecting platforms

When several tanks are located in one area, it is often cost-effective to connect the tanks via a single system of walkways, as shown in Fig. 16.1.4. If the roof is used as the platform, then the slope must not exceed 7° (a 2:12 slope). If it does, then independent platforms are required. Insulated tanks must have independent platforms and walkways.

If a fixed-roof tank is to be used as a platform, it should meet the following criteria:

- The area to be used as a platform is adequately designed for the anticipated live load.

- The stored liquid is a noncorrosive, or regular roof thickness checks are made to ensure that there is little or no thinning occurring.

- The area serving as a walkway is clearly delineated as such by color-coding the area with painted lines and precautionary statements or by handrails.

When two or more tanks are interconnected, then secondary egress is provided if there is a potential for hazard or chemical exposure that can block the main egress.

16.1.3.5 Platform and walkway connections

It is important to ensure that platforms that connect tanks together allow for settlement and relative movement between the tanks. In areas of seismic activity, this point is particularly important because the swaying of the tanks can pull platforms off with the potential of tearing the shell. If sufficient free play cannot be designed for or the

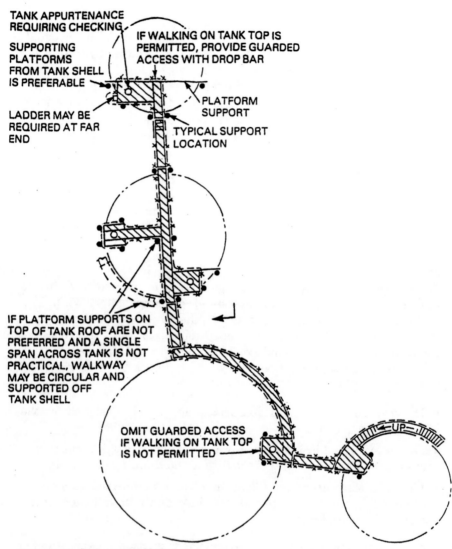

TANK APPURTENANCE
REQUIRING CHECKING

SUPPORTING
PLATFORMS
FROM TANK SHELL
IS PREFERABLE

IF WALKING ON TANK TOP IS
PERMITTED, PROVIDE GUARDED
ACCESS WITH DROP BAR

PLATFORM
SUPPORT

LADDER MAY BE
REQUIRED AT FAR
END

TYPICAL SUPPORT
LOCATION

IF PLATFORM SUPPORTS ON
TOP OF TANK ROOF ARE NOT
PREFERRED AND A SINGLE
SPAN ACROSS TANK IS NOT
PRACTICAL, WALKWAY
MAY BE CIRCULAR AND
SUPPORTED OFF
TANK SHELL

OMIT GUARDED ACCESS
IF WALKING ON TANK TOP
IS NOT PERMITTED

UP

Figure 16.1.4 Interconnected tanks.

criteria are not known, then the attachment to one tank should be
fixed and the other designed to break away. This consideration should
also be applied to frangible roofs which are designed to lift.

16.1.4 Railings

For small fixed-roof tanks over about 20 ft in diameter which have all
the appurtenances located in one area, the railing can be a section

extending about 8 ft on either side of the appurtenance. For smaller tanks it is most efficient to encircle the top of the tank completely. Notice that the top of the tank serves as the platform itself. The accessible areas should be coated with a nonslip surface.

16.1.5 Coatings

Typically, steel grating with proper surface preparation and coating is used. In coastal areas subject to marine corrosion, the use of galvanized gratings and components is effective. In chemical plants or for special conditions, aluminum or stainless steel components have been used. Treads are available with hot-dip-galvanized coatings. These do not need to be coated and provide a good service life.

SECTION 16.2 MISCELLANEOUS TANK APPURTENANCES

16.2.1 General

There are numerous devices and connections attached to tanks for piping, instrumentation, gauging access, and other appurtenances. When one is attempting to purchase new tanks, there are two useful tools for specifying appurtenances; one is a checklist of all appurtenances, and the other is a plan view indicating their locations. A generalized checklist of appurtenances is provided in Table 16.2.1. The other useful tool is a plan of the tank with appurtenances laid out on it. Often, it is possible to reduce the numbers of appurtenances by careful planning. For example, a roof manway or antirotation pipe may double as a gauging hatch. Layouts such as this can also be used to improve safety or emergency operations. For example, from the layout it is clear that should a fire develop at the mixer manway (a likely location), it would not cause the stairway to be blocked by flames. Layouts will also make the job of considering the effects of the prevailing winds which could carry toxic fumes to the tank gauger platform location.

16.2.2 Tank Openings

Tank openings are required to accommodate appurtenances such as manways, nozzles, and clean-outs as well as instrumentation and gauging openings. The requirements for these openings are thoroughly covered in the API standards.

Another type of opening accessory is the door sheet. Door sheets are usually cut into the tank shell when a double bottom or new roof is

TABLE 16.2.1   Typical Tank Appurtenances

Appurtenance	Comments
Manholes, 20- to 36-in diameter	API 650 figure 3-4A, detail A, B, or C *or* Shell nozzle (>20 in) per API 650, figure 3-5
Nozzles, threaded, ¾–2 in	API 650, figure 3-5
Nozzles, flanged, regular or low type, 1½- through 48-in; also nozzles with internal flanges or stubs	API 650, figure 3-5 150-lb flanges for 1½–24 in 125-lb flanges for 30, 36, 42, and 48 in Large flanges per ASME B 16.47
Bottom sump	API 650, figure 3-16 or purchaser-specified
Level indication	Gage board, float, guide wires, external piping, supports
Automatic gauging	Manufacturer Radar, float, bubbler, etc.
Swing piping including swivel joint and ballast pontoons, cables, and accessories	
Heat exchangers	Serpentine coils (field-fabricated) Bayonette or manway heaters (manufacturer) Heating pads externally wrapped (manufacturer) External (manufacturer)
Insulation	
Antirotation pipes	
PV valves	
Emergency vents	
Ladders, platforms, handrails	
Primary and secondary seals	Vapor fabrics, shoes, compression plates, shunts, wiper tips, etc.
Sampling	Many varieties from hatch to pumped systems external to tank

being constructed. For API 650, Appendix A, tanks a bolted, gasketed door sheet may be used. However, these have fallen into disfavor as the gaskets can leak and the labor required to bolt up door sheets can exceed the cost of using a welded door sheet.

Particular attention should be paid to the number and location of manways. These not only are used to perform maintenance but also play an important role in ventilating and cleaning a tank. A review of tank-cleaning practices should be undertaken when one is planning the number of manways that should be located. Since API does not specify the numbers of manways, Table 16.2.2 indicates a reasonable

**TABLE 16.2.2   Recommended
Manway Count versus Tank Diameter**

Tank diameter, ft	Number of manways
<20	1
20–100	2
100–150	3
>150	4 or more

approach to this problem. One manway should be at least 36 in to allow for a ladder for inspections.

Nozzle projections indicated in API 650 are minimum. This distance should be increased under these conditions:

- Insulation
- For bolt removal room

Swing pipes or swing lines are used to withdraw liquid from levels above the bottom for product purity reasons or for skimming the surface or for blending purposes. For example, diesel is withdrawn from the top because the clarity is greater and the impurities and turbidity are lower near the surface. For blending the swing pipe is often centered at about the midpoint of the liquid level, and the tank is recirculated. For fixed-roof tanks the swing line can be provided with buoyant pontoons to keep the suction line near the surface or can be operated by winches to maintain the end of the swing line at any predetermined level.

16.2.3   Mixers

Mixing technology is complex and hard to define. It is based more empirically on subjective evaluation than a set of physical rules based upon mathematics. Tank mixing is widely used in all industries for the following purposes:

- Blending of stocks or agitating them to promote chemical reactions
- Suspension such as to keep particles that form sludges in suspension to minimize sludge buildup
- Prevention of stratification and to maintain uniformity of bulk contents
- Liquid, solid, gas enhanced contacting for mass transfer or reaction
- To improve heat transfer

16.2.3.1 General fluid dynamic principles

A mixer is simply a large pump without the pump housing. There are two important processes that affect mixing applications with regard to the pumping action of the impeller. The first is pumped volume. This represents the average flow generated through a hypothetical aperture slightly larger than the impeller. It represents the flow that can be pumped by the impeller. The second process has to do with creating localized fluid turbulence, characterized by the liquid shearing rate.

Many of the characteristics that apply to pumps apply to mixers. For example, the power used by a mixer is the same as that of a pump, providing the variables are appropriately defined:

$$P \propto QH\rho$$

where P = impeller power input
Q = flow rate or pumping capacity
H = velocity head, equivalent of shear rate or turbulence
ρ = density of fluid

A mixer imparts a velocity field to the liquid contents of a tank. It is not just velocity that produces good mixing, it is the randomness of the velocity field that is actually responsible for maximum mixing. A measure of the randomness of the velocity field is the shear rate. It can be determined from the equation that at constant power the pumping rate and the shear rate vary inversely with each another.

About 80 percent of the blending applications are concerned primarily with pumping capacity or mass flow of the batch. These include blending and suspension applications. The remaining applications are very sensitive to the shear rate. These applications include mixing of fragile particles subject to degradation by too high shear rates, polymerization, pigment dispersions in paints, and gas-liquid contacting.

In the turbulent regime, the power consumption is proportional to N^3D^5, where N is the impeller speed and D is the impeller diameter. Pumping capacity is proportional to ND^3. The essential result is that for constant power consumption, a large impeller produces high flows and low shear rates and a small impeller produces low flows and high shear rates. The implies that higher-speed direct drive impellers are suitable where higher liquid shearing rates and slower gear-driven mixers are more suitable for high-flow applications.

16.2.3.2 Impellers

There are basically two types of impellers, radial and axial. Axial-flow impellers pump liquid primarily axially by screw action. These are more efficient than radial-flow impellers which direct flow radially outward. Radial-flow impellers are used primarily for gas-liquid mixing processes. Specialized impellers and various shapes are used for non-newtonian liquids such as pastes, glues, and other viscous materials.

16.2.3.3 Configurations

Mixers may be installed on the top, side, or bottom of tanks. The flow patterns vary significantly based upon the exact location and position of the mixer impeller, as well as its angle with respect to the centerline of the tank.

16.2.3.4 Portable mixers

Small portable mixers up to 3 hp are used on small tanks by clamping them to the rim or with a support bracket. The support structure should allow for a shaft angle of between 10° to 15° from the vertical and should allow its offset from the centerline to be adjusted for maximum performance.

16.2.3.5 Top-entry mixers

Top-entry mixers are used in vertical tanks and can be used with baffles, which eliminates vortices and increases overall circulation. Another way to reduce or eliminate the need for baffles is by installing the mixer off center and by inclining the mixer shaft at a slight angle. The impeller or impellers on a vertical mixer application may be placed at different elevations within the fluid, providing different efficiencies and mixing effectiveness. For a single impeller this is about midway between the top and the bottom. However, since the operating level varies, the ideal impeller position will rarely govern the location selected. Rather, operating-level considerations will govern the selected impeller locations.

It is usually preferable to use a top-entry mixer with a cantilevered impeller. This eliminates the need for a bearing at the bottom, which would be operating under adverse conditions. The cantilevered shaft must be heavy enough to resist lateral deflection, to increase bearing and seal life. In addition, a mixer located above a tank requires careful structural support. Because of these factors the top-entry configuration tends to have a higher initial cost than a side-entry mixer. The side-entry mixer tends to use more power for the same amount of mixing, however.

Baffles are used in vertical tanks with top-entry mixers to prevent swirling, vortexing, and air entrainment. They are required for low-viscosity mixing when top agitators are used on the tank centerline.

Typically four baffles, one-twelfth of the tank diameter in width, extending vertically along the shell, and located 90° apart are required. As the viscosity of the mixed liquid increases, then the viscous drag on the tank wall or internals increases, reducing the need for baffles as large or at all. If it is not certain that baffles will be required, it is convenient to include baffle clips in the tank so that they may be applied if needed after operations starts.

16.2.3.6 Side-entry mixers

The majority of side-entry mixers are used in large petroleum, petrochemical, and chemical tanks to control the accumulations of bottom sludge and water pockets, to maintain homogeneity and prevent stratification, to blend stocks, to aid in heat transfer while heating or cooling, and to maintain uniform temperatures. In general, the most demanding application for side-entry mixers is the blending of stocks.

Side-entry mixers are designed to be mounted in a manway or nozzle on the side of the tank. They are not used with baffles. Fixed-angle mixers are primarily used for blending, homogenizing, and maintaining uniformity of the stock including temperature. For tanks up to 50 ft in diameter they are mounted at an angle of 7° to the tank centerline for blending. For larger tanks an angle of 10° is used.

A common purpose of side-entry mixers is to scour the bottom to clean it or to prevent sludge bottoms from developing. An angle of 10° to 20° is effective. There are usually "dead" locations at the bottom for which the general circulation velocity is insufficient to keep the entire bottom sludge-free. However, mixer suppliers provide actuators that vary the angle of the mixers which eliminates the dead zones. This can be done manually or by automatic actuators. From a practical viewpoint, very few manually adjustable mixers ever get adjusted. Automatic actuators sweep the mixer from side to side in order to prevent buildup on the opposite-side cycle over a period of a few hours to a day. The sweep angle is usually about 30° on each side of the centerline. When very large tanks are involved, several mixers may be required.

In large tanks such as petroleum storage tanks, the side-entry mixer whose function is to blend or to reduce sludge buildup has become standard practice. The power levels required for these applications tend to be low, ranging from 1 to 5 kW/1000 m^3. Swivel-actuated mixers may require less power.

Side-entry mixers produce a spiral flow which sweeps across the bottom at relatively high velocities and becomes reduced at higher levels in the tank. Initially only the bottom is blended by the jet stream, but with time the lower-specific-gravity materials become blended as the circulation pattern rises. In applications, depending on the stock, there is a minimum required horsepower below which mixing will not be accomplished, no matter how long the mixer is operated. This minimum level of power required is about 1 kW/1000 m^3 for large product storage tanks.

The power requirements are a function of tank size, height/diameter ratio, viscosity, density, and the process mixing requirements. It is best to consult the mixer manufacturers to determine the required horsepower.

16.2.3.7 Other considerations for mixers

Noise levels should be specified for mixers at some distance away. A reasonable value is 85 dbA measured 1 m from the mixer.

Jet eductor nozzles are another way to mix a tank. By pumping liquid through the eductor, large volumes of liquid are pumped in the same way as a mixer pumps liquid. However, the power required per pumped volume is usually much greater for eductors than for mixers. While eductors have no moving parts, they receive their motive force from external pumps. Eductors have the potential to produce static electricity, especially if product containing water from the bottom is recirculated through them. The charges can accumulate on the surface and act as an ignition source for flammable stocks or when the liquid level is below the landed position of a floating-roof tank. They are generally not the optimal choice for tank mixing applications. Also not optimal are other mixing methods including air agitation and mixing tubes. Sometimes mixing is accomplished by recirculating stock through a swing line. When this is done, it is more effective to raise and lower the swing line while the stock is being recirculated.

An important consideration for using mixers in floating roofs is the issue of clearance. Many roofs and mixers have been damaged by landing the roof in the low-leg position. A cutout should be made in the floating roof in order to maintain at least 8 in from any part of the roof or the tank bottom.

References

1. J. H. Perry, *Chemical Engineer's Handbook*, 4th ed., McGraw-Hill, New York, 1963.
2. J. Y. Oldshue, *Fluid Mixing Technology, Chemical Engineering*, McGraw-Hill, New York, 1983.

Index